Manfred Reichenbächer · Jürgen Popp

Challenges in Molecular Structure Determination

With solutions at extras.springer.com

 Springer

Manfred Reichenbächer
Institute of Inorganic and Analytical
Chemistry
Friedrich-Schiller University Jena
Jena
Germany

Jürgen Popp
Institute of Physical Chemistry
Friedrich-Schiller University Jena
Jena
Germany
and
Institute of Photonic Technology
Jena
Germany

Solutions to the challenges presented in this book available at
http://extras.springer.com/2012/978-3-642-24389-9.
For further information, please consult the file README_978-3-642-24389-9.txt.

Revised and extended edition of the German textbook "Strukturanalytik organischer und anorganischer Verbindungen", Vieweg+Teubner 2007, 978-3-8351-0190-6

ISBN 978-3-642-24389-9   ISBN 978-3-642-24390-5 (eBook)
DOI 10.1007/978-3-642-24390-5
Springer Heidelberg Dordrecht London New York

Library of Congress Control Number: 2012933114

Printed on acid-free paper

Springer is part of Springer Science+Business Media (www.springer.com)

Challenges in Molecular Structure Determination

Preface

The present book is the first English version of the successful book *Strukturanalytik organischer und anorganischer Verbindungen* that was published in 2007 by Vieweg-Teubner Publishing. The present English edition is intended for a broader group of readers and users.

Sound knowledge and experimental experience of the application of modern spectroscopic methods to unravel the structure of organic and inorganic compounds are part nowadays of the scientific background of synthesis chemists, physical chemists, as well as of analysts, biochemists, and pharmacists among other disciplines. For this reason many very good books about spectroscopic methods can be found on the market. Nevertheless, these books focus mainly on the theoretical background of spectroscopy and are limited to the structural analysis of organic compounds. Only seldom can method-based exercises be found in these books.

The use of spectroscopic methods to analyze molecule structures and to infer relations between the structure and features of molecules requires basic as well as extensive applied knowledge about spectroscopy. On the one hand, the right combination of spectroscopic methods has to be tailored as a function of the specific question at hand for one method seldom permits one to solve the analytical problem encountered. On the other hand, the sound interpretation of the collected data is challenging. Unfortunately, these data cannot be, in most cases, interpreted using only the theoretical background of spectroscopy. Rather the experience necessary to solve basic as well as more complex structure analysis problems can only be acquired by intensive training. The main focus of this book is thus not to provide theoretical background but applied exercises presented as *challenges*. Consequently, the theoretical section is kept quite short and refers to the main relevant literature on the subject. The appropriate literature for each spectroscopic method can be found at the end of the respective chapter. There are many good books written by specialists in the field, therefore, we apologize to colleagues whose work we could not cite because of space limitations.

It may surprise the reader that equally long chapters have been devoted to each of the four spectroscopic methods, whereas in other books UV–VIS spectroscopy is hardly presented or even absent. In Sect. 3.5 the importance of this method is summarized and illustrated based on multiple examples coming from analytical practice. A prerequisite for the successful application of the spectroscopic methods in molecular structure determination is to have the necessary knowledge about the relations between chemical structure and spectrum. These relations will be presented in detail here, as the available books on the topic provide too general or no information on structure–spectrum relations. The reader may wonder why the four molecular spectroscopic methods are presented in approximately equally long sections, despite the fact that nuclear resonance and mass spectrometry play a more important role in the field of structure analysis than the two other methods. Indeed, their theoretical background is so large that we urge the reader to refer to the specific literature. The present book offers only the theoretical background necessary to be able to interpret the spectra.

Some considerations about methodological and didactical aspects have to be pointed out at this stage. This book begins with the chapter on mass spectrometry because it offers a good estimation of the general structure as well as to the number of double bonding equivalents which is necessary to obtain the correct structure of molecules. In the following chapters, vibration spectroscopy, electronic absorption spectroscopy, and NMR spectroscopy will be presented in this order. Some of the challenges related to each one of these methods will be presented based on spectral data of the preceding methods or the spectra will be complemented by synthetic or other types of information.

The theoretical materials will be illustrated in the whole book by challenges. Most of them originate from analytical practice. The solutions to these exercises are provided to the reader merely as an "indicator" to ascertain that the theory has been well understood.

Each of the most important exercises possesses a detailed solution/algorithm, which can then be used as a template to solve all the other exercises. Thus, this book constitutes a very good self-study reference. The reader will find the complete solutions on the website *extras.springer.com*, where the solutions are presented step-by-step as Power Point presentations.

The exercises consist of assignments to decode the molecular structure of inorganic compounds such as the geometry of small molecules or ions, classical complex bondings, the coordination of ligands, as well as the structure of organic molecules and biomolecules. Many assignments come from synthetic chemistry or analytical practice, such as for instance the analysis of HPLC/DAD products or products from the domain of environmental analytics. The reader is supposed to identify which method is the most appropriate to use to solve a particular analytical problem and the limitations of each method. Moreover, examples using exclusively UV–VIS spectroscopy are presented as this method is in some cases the only correct method to implement. Two additional examples will illustrate step-by-step how the whole range of methods from mass spectrometry to NMR correlation spectra gives the respective information; still open questions about the structure analysis of molecules remain. Finally, 22 sets of spectra are provided for the practice of molecular structure determination.

The issues surrounding structure analysis of inorganic molecules are clearly far less complicated than those for organic molecules. It is worth noting that the scope of the present book only includes *molecular* structure analysis using vibration and electron spectroscopy. It does not encompass the domains of nuclear resonance spectroscopy of soilds, ESR, and other methods.

Our book was mainly conceived as a starting tool in the domain of structure analysis of molecules. The difficulty level of the exercises was adapted in consequence of this goal and the theory necessary to solve the exercises was synthesized in an understandable form. Since modern experimental NMR techniques belong nowadays to the basic knowledge of many disciplines, COSY, HSQC, NOESY, and TOCSY spectra were included in the 22 exercises presented in Chap. 5. The solution for the first challenges is given in detail and should be the algorithm for solving the following challenges.

Because of the broad array of methods and theories, it was impossible to include all molecular spectroscopic methods such as ESR or chiroptical methods in this book. In our opinion, however, the methods presented here provide students with all the basics needed to determine the structure of low-molecular weight compounds and ease the understanding of more complex methods of analysis to tackle more complex issues.

Thus, this exercise book on structure analysis is devoted to chemistry students as well as to students of many other disciplines (e.g., pharmacists, biologists, geologists, etc.), who face similar issues.

We are grateful to Dr. Wolfgang Günther and Dr. Manfred Friedrich (Faculty of Chemical and Earth Sciences, Friedrich-Schiller University, Jena) for the measurement of NMR spectra. We would also like to thank PD Dr. Thomas Mayerhöfer (Institute of Photonic Technology, Jena) for his corrections of earlier versions of this book. Finally, we thank Dr. Pauly (Springer) for the excellent cooperation.

Jena, Germany   Manfred Reichenbächer
January 2012   Jürgen Popp

Challenges

PowerPoint presentations for solving the challenges are available at http://extras.springer.com

Please note that solutions are presented in the book for challenges marked by an asterisk.

(continued)

(*continued*)

Contents

List of Important Abbreviations

1°	Primary
2°	Secondary
3°	Tertiary
amu	Atomic mass unit
AO	Atomic orbital
b	Broad
CI	Chemical ionization
COSY	Correlation spectroscopy
CV	Combination vibration
DAD	Diode Array Detector
DBE	Double bond equivalent
DEPT	Distortionless enhancement by polarization transfer
dp	Depolarized Raman line
EE^+	Even number of electron ion
EI	Electron impact ionization
EN	Electronegativity
eV	Electron voltage
FID	Free induction decay
FIR	Far infrared
FM	Formula mass
gem	Geminal
HMBC	Heteronuclear multiple bond coherence
HMQC	Heteronuclear multiple quantum coherence
h-MS	High-resolution mass spectrometry
HSQC	Heteronuclear single quantum coherence
i. a.	Inactive
in	In-plane
IP	Ionization potential
IR	Infrared
LSR	Lanthanide shift reagents

M	Spin multiplicity
$M^{\cdot+}$	Molecular peak (ion)
m	Medium
MN	Mass number
MO	Molecular orbital
MS	Mass spectrum; mass spectroscopy
m/z	mass-to-charge ratio
NCI	Negative chemical ionization
NIR	Near infrared
NMR	Nuclear magnetic resonance
NOE	Nuclear Overhauser effect
NOESY	Nuclear Overhauser enhancement spectroscopy
$OE^{\cdot+}$	Odd number of electron ion
oop	Out-of-plane
OT	Overtone
P	Polarized Raman line
PCP	Polychlorinated biphenyls
R	Resolution
Ra	Raman
RDA	*Retro*-Diels-Alder
S	Total spin angular moment
sh	Shoulder
s	Electron spin
T	Triplet state
TAI	Trichloroacetyl isocyanate
TLC	Thin-layer chromatography
TMS	Tetramethylsilane
TOCSY	Total correlation spectroscopy
UV	Ultraviolet
vic	Vicinal
Vis	Visible
vs	Very strong
vw	Very weak
w	Weak

List of Important Symbols

Some important symbols and abbreviations are listed below. Further symbols and abbreviations the use of which is restricted to special sections are defined in those sections.

A	Absorbance (without unit)
B	Magnetic flux density (magnetic field strength)
c	Light velocity
E	Energy
f	Force constant
h; $\hbar$	Planck's constant; $\hbar = \frac{h}{2\pi}$
I	Intensity; nuclear spin quantum number
i	Charge-induced cleavage
J	Indirect nuclear spin coupling constant in Hz
$\vec{J}$	Nuclear spin
m	Atomic mass; magnetic quantum number
M	Relative atomic mass; multiplicity of a multiplet
n	Number of the atom; amount of substances
N	Number of the spins; number of atoms
r_H	McLafferty rearrangement
S	Singlet; incremental value
T_1	Spin–lattice relaxation time
T_2	Spin-spin relaxation time
v	Vibrational quantum number
α	Molar absorptivity
γ	Out-of-plane vibration (oop, δ_{oop}); magnetogyric ratio
δ	Deformation vibration; chemical shift
Γ	Representation of a class
λ	Wavelength
μ	Reduced mass
$\vec{\mu}$	Nuclear magnetic moment
v	Stretching vibration; wave number of the IR absorbance; frequency

$\tilde{v}, v$	Wave number in cm^{-1}
σ	Shielding constant
ρ	Polarization ratio
φ	Dihedral torsion angle

Superscript Indices
| * | Excited state |

Subscript Indices
a	Axial
as	Asymmetrical
b	Broad
e	Equatorial
eff	Effective
i	Running index
max	Maximum
rel	Relative
s	Symmetrical
0	Ground state
1	First excited State
$\perp$	Parallel
$\parallel$	Perpendicular

Chapter 1
Mass Spectrometry

1.1 Introduction

1.1.1 Principles of Mass Spectrometry

Mass spectrometry (MS) is an analytical technique for determining masses of free ions in high vacuum and for determining the elemental composition and chemical structures of molecules. In order to understand mass spectra knowledge is necessary on how they were created. Therefore, let us start with some remarks on the experimental techniques of various types of mass spectrometers.

A mass spectrometer consists of three modules:

- The *ion source* converts gas-phase molecules into charged molecules or molecular fragments.
- The *mass analyzer* separates the charged molecules or fragment ions according to their *mass-to-charge ratio* (*m/z*) by applying electromagnetic fields.
- The *detector* measures the value of an indicator quantity and thus, it provides data for calculating the abundances of each ion present.

The introduction of the sample into the ion source occurs directly or by coupling with a gas chromatograph and a HPLC (High-Performance Liquid Chromatography) device. Mass spectrometry produces various types of data. The most common data representation is the *mass spectrum*. It is the representation of the abundance of the ions versus the mass-to-charge ratio scaled on the base peak which is the peak with the highest abundance (intensity). Its relative intensity I_{rel} is set to 100%.

The *mass number* (**MN**) is the mass-to-charge ratio rounded to the nearest integer value, which is also called nominal mass. Thus, for example, the mass number MN is 104 for a unit charged ion with the exact mass 104.0876. The unit of measurement of the mass numbers is *amu* (atomic mass unit) which always differs from the chemical molecular mass which has the unit g mol^{-1}. In addition, the unit Thomsen (**Th**) is sometimes used.

M. Reichenbächer and J. Popp, *Challenges in Molecular Structure Determination*,
DOI 10.1007/978-3-642-24390-5_1, © Springer-Verlag Berlin Heidelberg 2012

1.1.2 Techniques for Ionization

1.1.2.1 Electron-Impact Ionization

Electron-impact ionization (EI) is the most important technique for ionization. In the EI process, the sample is vaporized in the mass spectrometer ion source, where it is impacted by a beam of electrons with sufficient energy to ionize the molecule. EI ionization requires that the molecules are vaporized before ionization. Therefore, EI is only appropriate for molecules that are volatile under the conditions of the ion source. The impact electrons are emitted by a filament and subsequently accelerated by the anode voltage. Most EI mass spectra are gathered using 70-eV electron energy.

The primary process consists of the ionization of the molecules M thereby generating a unit charged radical cation which provides the *molecular ion peak* $M^{\bullet+}$ in the mass spectrum according to Eq. 1.1

$$M + e^- \rightarrow M^{\bullet+} + 2e^-. \tag{1.1}$$

The generation of double positively charged ions is very unlikely and, in general, takes place only in case of aromatic or heteroaromatic compounds. Since ions are detected according to m/z, double positively charged ions of odd masses appear at half mass numbers. Thus, for example, a unit charged ion $m/z = 101$ amu provides a peak at the mass number 50.5 as a double charged ion.

Many peaks at half mass number indicate the presence of an aromatic or heteroaromatic compound.

Electron-impact ionization belongs to the so-called *hard ionization methods* because the energy of the impact electrons is several times higher than the energy of the chemical bond. Therefore, the high internal energy supplied to the molecule leads to fragmentation reactions which generate fragment ions that give rise to the *fragment ion peaks* in the mass spectrum. The fragmentation reactions dominate if the molecular ion is very unstable. In this case, the mass spectrum is characterized by intense fragment ions and a small molecular ion peak. Sometimes, the molecular ion peak does not occur in the spectrum at all because its intensity is nearly zero.

The EI mass spectrum is characterized by many and mostly very intense fragment ion peaks.

The ionization source contains a variety of charged and uncharged species, but only positively charged species (radical cations or cations) are focused into the inlet slit of the mass analyzer because of the negative acceleration voltage employed in the usual techniques. Uncharged species as well as molecules are removed by the high vacuum; therefore, such species do not provide signals in the mass spectrum.

Many chemical substance classes like alkanes, alcohols, diols, or aliphatic carbonic acids do not provide a molecular ion peak or provide peaks with very weak intensity because of the high impact energy. Thus, the mass spectrum is characterized by the pattern of the fragment ion peaks. In order to find the molecular ion peak one can reduce the impact energy. Reducing the impact energy reduces both the amount of fragmentation and the total number of ions formed but the molecular ion will show up before the production of ions ceases entirely.

A better possibility for the detection of the molecular ion peak is the application of *soft* ionization techniques. The chemical ionization techniques are appropriate methods for volatile molecules.

1.1.2.2 Chemical Ionization

In the Chemical Ionization (CI)-techniques ions are produced through chemical ion-molecule reactions via collisions in the ionization source. The ions are generated by a reagent gas through the application of a high energy value ($\approx$150 eV) during a primary reaction. Common reagent gases are methane, ammonia, and *iso*-butane. The primary process is shown in Eq. 1.2

$$CH_4 + e^- \rightarrow CH_4^{\bullet+} + 2e^-.(\text{Ionization}) \tag{1.2}$$

The collision of the ionized species with reagent gas molecules which are present in large excesses compared to the analyte provides protonated species, shown in Eq. 1.3

$$CH_4^{\bullet+} + CH_4 \rightarrow CH_5^+ + CH_3^{\bullet}.(\text{Protonation}) \tag{1.3}$$

The collision of the protonated reagent gas creates an ionization plasma consisting of positive and negative ions of the analyte. Some reactions of this type are shown in Eqs. 1.4, 1.5, and 1.6.

Protonation (favored for aromatic and heteroaromatic compounds)

$$M + XH^+ \rightarrow MH^+ + X \Rightarrow [\textbf{M} + \textbf{1}] \textbf{ peak} \tag{1.4}$$

Hydride abstraction (favored for aliphatic carbonic acids and alkanes)

$$M + XH^+ \rightarrow [M - H]^+ + X + H_2 \Rightarrow [\textbf{M} - \textbf{1}] \textbf{ peak} \tag{1.5}$$

Adduct formation

$$M + XH^+ \rightarrow MHX^+ \Rightarrow [\textbf{M} + \textbf{XH}] \textbf{ peak} \tag{1.6}$$

Protonation (Eq. 1.4) is favored in most cases. Accordingly, the mass spectrum does not show the molecular ion peak $M^{\bullet+}$ but the quasi-molecular ion peak $[M + 1]^+$ which is one mass unit higher than $M^{\bullet+}$. Because of the soft ionization

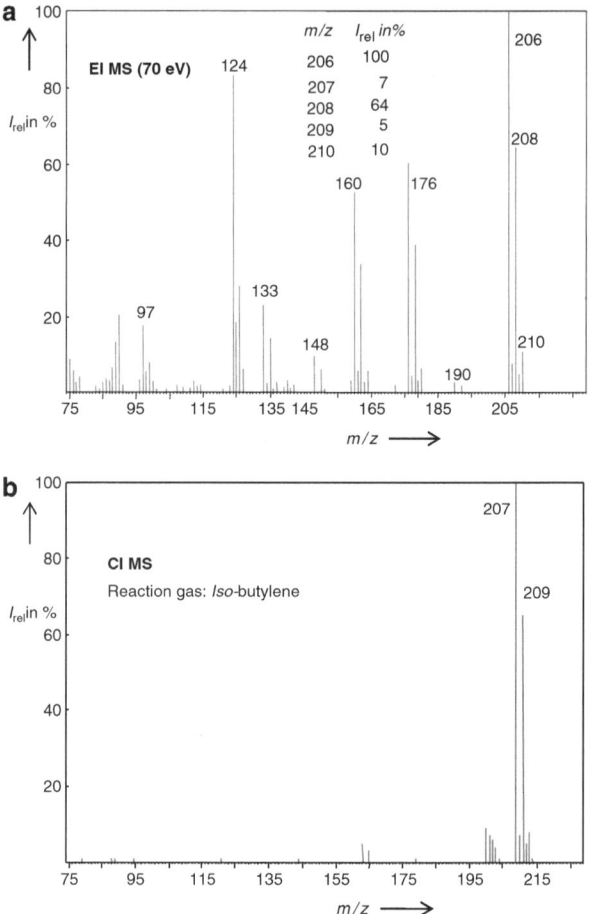

Fig. 1.1 EI and CI mass spectra of an initially unknown compound

the CI mass spectrum shows only a small abundance of fragment ions in comparison to the EI mass spectrum; compare both spectra in Fig. 1.1 obtained from an
initially unknown analyte.

*Bear in mind that the molecular ion peak can be determined and protected by the
CI techniques for volatile molecules which do not show the molecular ion peak in
the EI mass spectrum because of a high fragmentation cross section.*

Mass spectrometers equipped with a trap mass analyzer allow switching between
EI and CI techniques by a simple mouse click.

The frequency of the fragment ions depends on the difference between the
ionization potential (IP) of the reagent gas and the analyte in the case of CI
ionization. If the difference between IP(reagent gas) and IP(analyte) is greater
than 5 eV, then the amounts of fragment ions will increase appreciably. A reagent
gas with low IP increases the frequency of the molecular ions at the cost of fragment

ions. The order according to decreasing proton affinity is given in Eq. 1.7 for most of the used reagent gases

$$CH_5^+ > C_2H_5^+ > H_3O^+ > C_4H_9^+ > NH_4^+. \tag{1.7}$$

A variation is *negative chemical ionization* (**NCI**) whereby negatively charged species are transferred into the mass analyzer. In order to see a response by negative chemical ionization, the analyte must be capable of producing negative ions, for example, by electron-capture ionization. Because not all analytes can do this, the NCI-techniques provide a certain degree of selectivity that is not available with other techniques. Because of the high electronegativity of halogen atoms, NCI is a common choice for their analysis. This includes many groups of compounds which are present in environmental compartments, such as pesticides and PCBs. In addition, a higher sensitivity is accomplished for the quantitative determination of such analytes.

1.1.2.3 Further Soft Ionization Methods

Further soft ionization methods are available for heavily vaporizable molecules or macromolecules which cannot be transferred into the gas state without decomposition, such as fast atomic compartment (FAB), ionization by photons (Laser desorption/ionization, LDI), ionization of the dissolved sample by an electric field (electron spray ionization, ESI), and ionization of the solid sample by means of matrix-assisted desorption/ionization (MALDI). These methods are important in bioanalytical analysis, for example, for the determination of molecular weight and in sequence analysis, see the specialized literature.

1.1.2.4 Resolution, *R*

The resolution of a mass spectrometer indicates which mass difference (Δm) can be just separated at the mass m

$$R = \frac{m}{\Delta m}. \tag{1.8}$$

Two peaks are "separated" per definition if the valley between two peaks amounts to 10% of the peak with the least intensity in a sector field mass spectrometer and 50% in a quadrupole instrument. A larger value of R indicates a better separation of two peaks.

Challenge 1.1
Which resolution is necessary in order to separate ions with the following mass?

(a) $m/z = 200$ and 201 amu
(b) 168.1025 ($C_9H_{14}NO_2$) and 168.1012 amu ($C_7H_{12}N_4O$)

Solution to Challenge 1.1
The following results are obtained by using Eq. 1.8:

(a) $\Delta m = 1$
$$R = \frac{200}{1} = 200. \tag{1.9}$$

The resolution of ions according to the unit mass number must lie in the range from 500 to 1,000. This low resolution can be accomplished by all mass analyzers.

(b) $\Delta m = 0.0013$

$$R = \frac{168}{0.00013} \approx 130000. \tag{1.10}$$

A mass analyzer which has a resolution higher than 100,000 is necessary. Therefore, a high-resolution mass spectrometer (**h-MS**) must be applied.

1.1.3 Overview of the Most Important Mass Analyzers

There are several mass analyzers in combination with accordingly suitable detectors for the separation of ions.
Mass analyzers:

- High-frequency field of a quadrupole bar system (quadrupole mass spectrometer)

  Unit mass resolution
  Mass range: up to 4,000 amu
  Exact measurement of intensities

- Magnetic sector field (magnetic sector field mass spectrometer)

  Unit mass resolution
  Mass range: up to 1,500 amu

– Combination of a magnetic and an electric field in series (double focusing mass spectrometer).

Resolution: $>1{,}000{,}000$
Mass range: up to 4,000 amu

– Electric ion traps (ion trap mass spectrometer)

Unit mass resolution
Mass range: up to 40,000 amu
EI/CI change by mouse click
Appropriate for MS/MS techniques

– Time of flight instruments

Resolution: 10,000
Mass range: $>200{,}000$
Mostly in combination with MALDI (MALDI-TOF mass spectrometer)

Detectors:

– Conversion dynode with an SEM (secondary electron multiplier)
– Scintillation counter

Challenge 1.2
For solutions to Challenge 1.2, see *extras.springer.com*.

1. A compound does not show a signal in the EI mass spectrum in the range of the expected molecular peak. How can the molecular peak be made visible?
2. The molecular mass of a peptide in the mass range of 10 kDa is to be determined. Can the EI and the CI mass spectroscopy be used? If not, suggest an appropriate device combination.
3. Figure 1.2a, b, c show the EI, CI (reaction gas: CH_4), and CI (reaction gas: $CH(CH_3)_3$) mass spectra of phthalic acid di-*n*-hexyl ester (Structure 1.1) in arbitrary order. Assign the different ionization methods to the spectra and explain your decision.

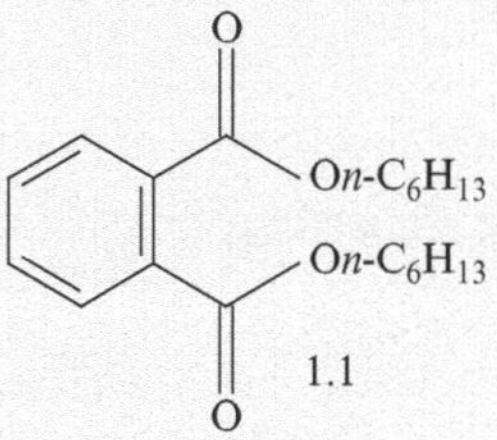

(continued)

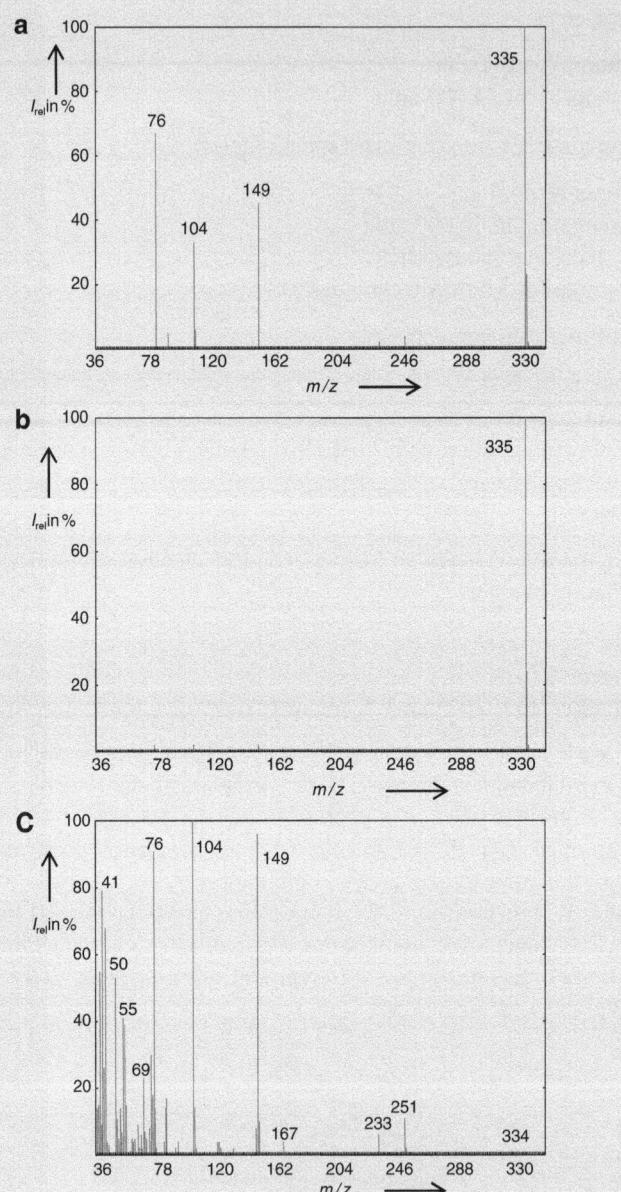

Fig. 1.2 (**a–c**) EI and CI mass spectra of phthalic acid di-*n*-hexyl ester (Structure 1.1)

4. What must the resolution of a mass analyzer be in order to separate the
   following ions: (a) $C_6H_5NO^+$ / $C_5H_5N_3^+$, (b) $C_7H_6O^+$ / $C_7H_5O^+$,
   (c) CH_3CO^+ / $C_3H_7^+$

1.2 Molecular Ion M˙⁺ (Molecular Peak) and Isotope Peaks

1.2.1 Definition

The molecular ion is derived from the neutral molecule by loss (or addition) of an electron. The molecular ion yields the *molecular peak*.

The molecular ion generated by means of EI ionization is always a radical cation, $M^{\bullet+}$.

For all organic compounds the molecular peak refers to the combination of the lightest isotopes which are also the most frequent forms of their natural compositions. These are the following isotopes: ^{1}H, ^{12}C, ^{14}N, ^{16}O, ^{32}Cl, ^{79}Br, and ^{28}Si. Note that ^{19}F, ^{31}P, and ^{127}I are monoisotopic which means that they have only a single stable isotope.

1.2.2 Intensity of the Molecular Peak

The correct assignment of the molecular peak is a requirement for a reasonable interpretation of a mass spectrum. Besides, it follows that the intensity of the molecular peak of the various chemical classes is very different (see Table 1.1). It may be the base peak for a highly stable molecular radical cation or it may not appear at all. This is the case if the electron impact generates a radical cation which has a very low activation barrier for decomposition reactions.

1.2.3 Protection of the Molecular Peak

The peak in the mass spectrum which belongs to the molecular ion must fulfill the following requirements:

– The molecular peak (with its isotope peaks) is the peak with the *highest mass number* which must contain all atoms of the molecule.

Table 1.1 Classification of organic chemical classes according to the intensity of the molecular peak under EI ionization. Note that the intensity decreases in each column from the top down

Strong	Mean	(Very) weak	Barely or not visible
Heteroaromatic compounds	Conjugate olefins	Short-chain alkanes	Tertiary alcohols
Aromatic hydrocarbons	Ar–Br	Long-chain alkanes	Tertiary bromides
Ar–F	Ar–CO–R	Unbranched alkanes	Diols
Ar–Cl	Ar–CH$_2$–R	Alk–OH	Hydroxycarboxylic acids
Ar–CN	Ar–CH$_2$–Cl	Alk–CO–OH	Dicarboxylic acids
Ar–NH$_2$		Secondary alcohols	Acetals; ketals

– *Mass differences* from the molecular peak to fragment peaks may not show values in the range **4–14** or **21–25**. These are so-called senseless mass differences because either four or more H atoms must be eliminated or two C–C bonds must be broken down simultaneously in order to obtain fragment ions in the range 4–14, but this is very improbable. Fragmentation reactions for ions that give rise to fragment peaks with the difference to the molecular peak in the range from 21 to 25 do not exist.

– The *isotope peaks* of the molecular peak must give rise to reasonable intensity relations appropriate to the natural frequency of the molecular peak.

– Being a radical cation the molecular peak must always be an ion with an *odd number of electrons*, so-called $OE^{•+}$ ions, see Eq. 1.11. Ions with an even number of electrons, EE^+ ions, can only be caused by the decay of fragment ions under loss of a neutral molecule according to Eq. 1.12:

$$\begin{array}{cc}
\xrightarrow{\ -N\ \text{(neutral molecule)}\ } OE^{•+} & (1.11)\\
M^{•+} \quad \\
\xrightarrow{\quad\quad\quad\quad} EE^+ & (1.12)\\
-R^{•}\ \text{(radical)}
\end{array}$$

1.2.4   Check on $OE^{•+}$ and EE^+ Ions

The general chemical formula $C_xH_yO_nHal_mN_zS_r$ is modified as follows:

– *Halogen* atoms are substituted by H.
– **O** and **S** are deleted without substitution.

Test criterion for the modified formula $C_xH_yN_z$ is:

– An $OE^{•+}$ ion is checked if Eq. 1.13 is valid:

$$x - 1/2y + 1/2z + 1 = 1 \cdot n. \qquad (1.13)$$

– An EE^+ ion is checked if Eq. 1.14 is valid:

$$x - 1/2y + 1/2z + 1 = 1/2 \cdot n. \qquad (1.14)$$

1.2.5   Isotope Peaks

Each mass analyzer separates the ions at least according to integer mass numbers so that the isotope mixture of the sample molecule is completely separated resulting in single peaks at the highest mass numbers in the spectrum. According to definition

Table 1.2 Combinations of isotopes of C_2H_5Cl

Combinations of isotopes	Mass number	Designation of the ions
$^{12}C_2{}^1H_5{}^{35}Cl$	64	$M^{\bullet+}$ (Molecular peak)
$^{12}C^{13}C^1H_5{}^{35}Cl$	65	$[M+1]^{\bullet+}$
$^{12}C_2{}^1H_4{}^2D^{35}Cl$	65	$[M+1]^{\bullet+}$
$^{12}C_2{}^1H_5{}^{37}Cl$	66	$[M+2]^{\bullet+}$
$^{12}C^{13}C^1H_4{}^2D^{35}Cl$	66	$[M+2]^{\bullet+}$
$^{12}C^{13}C^1H_5{}^{37}Cl$	67	$[M+3]^{\bullet+}$
$^{12}C_2H_4{}^2D^{37}Cl$	67	$[M+3]^{\bullet+}$
$^{12}C^{13}C^1H_4{}^2D^{37}Cl$	68	$[M+4]^{\bullet+}$

the molecular ion is the combination of the lightest isotopes for organic compounds and all combinations with heavier isotopes are isotope peaks which are assigned the symbols **M + 1, M + 2, M + 3**, and so on. All possible ions for the example C_2H_5Cl are listed in Table 1.2.

Because of the low natural frequency of the isotopes ^{13}C (1.1% related to ^{12}C) and ^2D (only 0.015% related to ^1H) combinations with more than one of these isotopes are very unlikely and, hence, only of very low intensity. Therefore, only peaks with the following mass numbers are visible for C_2H_5Cl in the mass spectrum: 64 $M^{\bullet+}$, 65 $[M+1]^{\bullet+}$, 66 $[M+2]^{\bullet+}$, 67 $[M+3]^{\bullet+}$, and 68 $[M+4]^{\bullet+}$, but this last peak shows only a very low intensity.

1.3   Determination of the Sum Formula as Well as the Number of Double Bond Equivalents (DBE)

The recognition of *heteroatoms* as well as the determination of the *sum formula* is the basis for the determination of the structure of an unknown compound. Mass spectroscopy is an important method to obtain this information.

There exist two methods for the determination of the sum formula, depending on which device is available.

1. Determination of the elementary composition of the molecular as well as fragment peaks by means of high-resolution mass spectroscopy (**h-MS**)

High-resolution mass spectroscopy using a double-focusing mass spectrometer enables the determination of the exact mass of ions, i.e., the determination of the mass in the millimass range.

Various elementary compositions of the same mass number differ in the millimass range, i.e., every individual elementary composition is unambiguously defined in the millimass range. The elementary composition is obtained by the comparison of the mass experimentally determined with that of the theoretical formula mass for a certain composition listed in corresponding tables.

Examples for calculated formula masses for ions with the atoms C, H, N, and O are listed in Table 6.5.

The exact determination of the mass is realized by *peak matching*, i.e., the unknown mass is compared with an exact mass of a reference substance. The peak matching method is of high diagnostic value for structural analysis because it can be used for the determination both of the molecular and fragment ions, only µg-amounts are necessary and the mass difference between the molecular and fragment peaks yields the corresponding elementary composition of the eliminated group. In this manner information about functional groups can be obtained, see Challenge 1.3.

Challenge 1.3

The structure of a synthesized steroid (Structure 1.2) is to be verified by instrumental analytical methods. Figure 1.3 shows a section of the mass spectrum in the upper range and the exact masses obtained by peak matching for some selected ions are listed in Table 1.3.

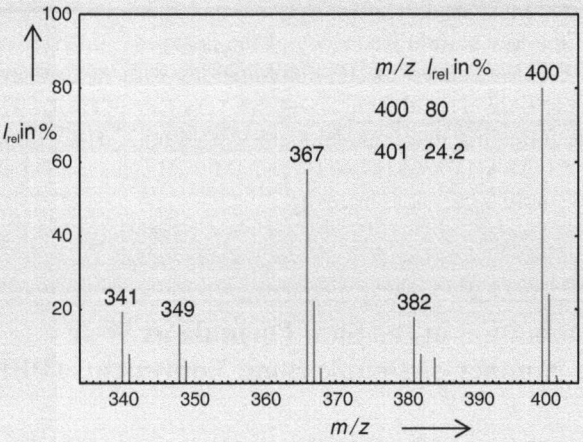

m/z	I_{rel} in %
400	80
401	24.2

Fig. 1.3 70-eV mass spectrum of a steroidal synthetic product with the assumed Structure 1.2 in the upper range

Table 1.3 Peak matching values for some selected ions of the mass spectrum of the steroids shown in Structure 1.2

Experimental mass in amu	Deviation from the theoretical mass in mamu	Sum formula	Information
400.33380	+0.3	$C_{27}H_{44}O_2$	Molecular peak $\Rightarrow$ **Sum formula**
382.32400	−0.5	$C_{27}H_{42}O$	Elimination of H_2O from $M^{\cdot+}$ $\Rightarrow$ **1 OH** is present
367.30230	−2.2	$C_{26}H_{39}O$	Elimination of CH_3 from $C_{27}H_{42}O$ $\Rightarrow$ **1 CH_3** group is present
349.28729	+2.2	$C_{26}H_{37}$	Elimination of H_2O from $C_{26}H_{39}O$ $\Rightarrow$ **1 further OH** is present

Check whether the following structural information can be confirmed:

Sum formula, two OH groups in the molecule, and CH_3 groups in positions which are favorable for the fragmentation.

Solution to Challenge 1.3
The elementary compositions of the molecular and fragment peaks obtained by h-MS are listed in Table 1.3. According to Table 1.3 the following results are obtained:

- The expected *sum formula* is confirmed.
- The presence of *two OH* groups is confirmed by the elimination of two H_2O molecules.
- The molecule contains at least *one CH$_3$* group in a position that is favorable for fragmentation.

2. Determination of the elementary composition of the molecular peak by means of a low resolving device

Low resolving mass spectrometers separate the ions only after integer mass unity; thus, the exact mass cannot be obtained. For this case the determination of the elementary composition is based on the natural ratio of isotopes of the elements, see Table 6.1.

This method, however, requires the possibility of the determination of correct intensities in the range of the molecular peak. Note that this procedure is limited to the determination of the elementary composition of the molecular peak, i.e., the sum formula of the analyte, because the intensity of isotope peaks of fragment ions are mostly superimposed by contributions from other fragment ions.

1.3.1   Calculation of the Number of Carbon Atoms, n_C

The number of carbon atoms in the molecule is determined by the relation of the intensity of the [M + 1] peak, I[M + 1] and the molecular peak, I[M] taking into account the natural frequency of the ^{13}C isotope (1.1%) according to Eq. 1.15

$$n_C = \frac{I[M+1]}{I[M] \cdot 0.011}.$$ (1.15)

Disturbances:

- High numbers of nitrogen atoms in the molecule. Note that three nitrogen atoms approximately correspond to the intensity of one carbon atom because of the natural frequency of the ^{15}N isotope.
- The presence of Si in the molecule. The amount of the intensity of the ^{29}Si isotope to the intensity of the [M + 1] peak must be computationally eliminated according to Eq. 1.18.

1.3.2 *Recognition of* **Sulfur** *and Calculation of the Number of* **Sulfur** *Atoms,* n_S

The isotope ^{34}S with an intensity of 4.4% relative to ^{32}S contributes to the [M + 2] peak. Therefore, a weak intensity of the [M + 2] peak indicates the presence of sulfur in the molecule. The number of sulfur atoms is calculated by Eq. 1.16

$$n_S = \frac{I[M+2]}{I[M] \cdot 0.044}.$$ (1.16)

Disturbances:

- The presence of chlorine and bromine. In this case, sulfur can be recognized if at a distance of 2 amu to the isotope peak with the combination of all ^{37}Cl and ^{81}Br atoms a further peak is visible with low intensity. This peak is also used for the calculation of n_S. This is, for example, the [M + 4] peak if one chlorine and one bromine atom, respectively are present in the molecule.

1.3.3 *Recognition of* **Silicon** *and Calculation of the Number of* **Silicon** *Atoms,* n_{Si}

The most frequent isotope is ^{28}Si. The isotope ^{29}Si ($I_{rel} = 5.1\%$) provides, in comparison with the number of carbon atoms, an [M + 1] peak with an unusually high intensity. Furthermore, ^{30}Si ($I_{rel} = 3.4\%$) results in an [M + 2] peak whose intensity cannot be reasonably assigned to sulfur.

The number of the silicon atoms in the molecule is calculated by Eq. 1.17

$$n_{\text{Si}} = \frac{I[M+2]}{I[M] \cdot 0.034}. \tag{1.17}$$

Disturbances are caused by sulfur, chlorine, and bromine.

With n_{Si} calculated by Eq. 1.17 the intensity of the [M + 1] peak caused by this isotope can be calculated according to Eq. 1.18

$$I[M+1] = n_{\text{Si}} \cdot I[M] \cdot 0.051. \tag{1.18}$$

This amount must be subtracted from the experimentally determined intensity of the [M + 1] peak in order to calculate the number of carbon atoms according to Eq. 1.15.

1.3.4 Recognition of Chlorine and Bromine and Determination of the Number of Chlorine and Bromine Atoms

The presence of Cl or Br in the molecule can easily be recognized by intensive isotope peaks at an interval of generally two mass units: [M + 2], [M + 4], and so on.

The number of Cl and Br atoms is determined by comparison of the experimental intensity ratio with that of the theoretically calculated and tabulated one, see Table 6.2, or graphically with the so-called halogen pattern.

1.3.5 Recognition of Fluorine and Iodine

Fluorine and iodine do not cause isotope peaks because both atoms are monoisotopes (^{19}F and ^{127}I).

Fluorine can be recognized by unusual mass differences of fragment peaks. Thus, the mass differences of $\Delta m = 19$ amu (elimination of F$^{\bullet}$) and $\Delta = 20$ amu (elimination of HF), are only possible if fluorine is present in the molecule.

Iodine can easily be recognized by the unusually high mass differences of $\Delta m = 127$ amu and $\Delta m = 128$ amu, because of the loss of I$^{\bullet}$ and HI, respectively.

1.3.6 Recognition of Nitrogen

In mass spectroscopy the *nitrogen rule* is valid:

The molecular peak is odd if the molecule contains an odd number of nitrogen atoms. The molecular peak is even if the molecule contains an even number or no nitrogen atoms.

Table 1.4 Natural isotope frequency of some 3d elements

Element	m/z in amu (I_{rel} in %)
Cr	50 (5.19), 52 (**100**), 53 (11.34), 54 (2.82)
Fe	54 (6.34), 56 (**100**), 57 (2.40), 58 (0.31)
Co	59 (**100**)
Ni	58 (**100**), 60 (38.23), 61 (1.66), 62 (5.26), 64 (1.33)
Cu	63 (**100**), 65 (44.57)
Zn	64 (**100**), 66 (57.41), 67 (8.44), 68 (38.68), 70 (1.24)

A molecular peak with an odd number provides an unambiguous result, i.e., the molecule contains one or three or five and so on nitrogen atoms. If the number of the molecular peak is even, the molecule contains no nitrogen or an even number of nitrogen atoms.

In this case, further criteria must be employed to decide whether the molecule contains nitrogen:

- An abnormally high number of fragment peaks with even mass numbers are observed.
- The presence of typical nitrogen-containing fragment peaks, for example, a base peak at $m/z = 30$ amu, which is caused by the ion $CH_2 = NH_2^+$. Further key fragment ions are listed in Table 6.4.
- The loss of nitrogen-containing radicals or molecules. For example, the mass difference $\Delta m = 46$ amu may be caused by the loss of NO_2 and $\Delta m = 27$ amu by the loss of HCN. Further so-called [M-X] peaks are presented in Table 6.3.
- Hints obtained by other spectroscopic methods, for example, characteristic vibrations of nitrogen-containing functional groups in infrared spectroscopy.

1.3.7 Recognition of Oxygen

Because the intensities of the isotopes ^{17}O and ^{18}O are very small the isotope peaks do not indicate the presence of O in the molecule.

The presence of oxygen can be obtained by the loss of oxygen-containing groups (see Table 6.3) and characteristic oxygen-containing fragment ions (Table 6.4) in the mass spectrum.

Note that infrared spectroscopy is the best spectroscopic method for the recognition of oxygen in the molecule.

1.3.8 Recognition of Phosphor

Because natural phosphor is monoisotopic (^{31}P) phosphor atoms do not cause any isotope peaks. Therefore, there are some characteristic isotope-free fragment peaks such as $m/z = 65$ amu ($PO_2H_2^+$), 97 amu ($PO_4H_2^+$), and 99 amu ($PO_4H_4^+$).

Such peaks observed in the mass spectrum suggest the presence of phosphor in the molecule. Note that NMR spectroscopy is the best method for the recognition of phosphor.

1.3.9 Recognition of Metal Ions

Most of the metals show isotope-rich isotopes whereby the lightest isotope is not the most frequent one. Some examples are given in Table 1.4.

1.3.10 Recognition of Hydrogen

The number of H-atoms cannot be determined by mass spectroscopy but it is obtained by the difference from the molecular peak and the sum of all other atoms. Note that the number of hydrogen atoms is best obtained by ^{1}H NMR spectroscopy.

1.3.11 Double Bond Equivalents

The number of double bond equivalents corresponds to how many *double bonds* or *double bond moieties* (threefold bond or rings) are present in the molecule.

Note that the numbers of rings in a molecule can only be obtained by the number of DBE.

Procedure for the determination of the numbers of DBE:

– Determination of the base hydrocarbon C_nH_x:

All **O** and **S** atoms are deleted without substitution.
All **halogen** atoms are substituted by H.
All **N** atoms are substituted by CH.
NO$_2$ and **SO$_2$R** are best deleted and are substituted by H. Thus, these structural groups do not provide an amount to DBE. But these groups give an amount of always one DBE if they are not eliminated.

– The hydrocarbon C_nH_x is the base for the calculation of the numbers of DBE according to Eq. 1.19

$$\text{DBE} = \frac{(2 \cdot n + 2) - x}{2}.$$ (1.19)

Table 1.5 Amounts provided
to the DBE by structural
moieties

Structural moiety	Amounts of DBE
Per C = X (X = C, N, O)	1
Per C≡X (X = C, N)	2
Per ring	1
Per benzene ring	4
NO_2, SO_2 (if it was not eliminated)	1

– Partition of the DBE

The numbers of DBE are listed in Table 1.5.

Challenge 1.4
Determine the sum formula for the compound in Structure 1.2 from the mass spectrum presented in Fig. 1.3 and calculate the DBE.

Solution to Challenge 1.4
Calculation of the numbers of carbon atoms according to Eq. 1.15 with
   I_{rel} [M] = 80% (m/z = 400 amu), I_{rel} [M + 1] = 24.2% (m/z = 401 amu)

$$n_C = \frac{24.2}{80 \cdot 0.011} = 27.5. \tag{1.20}$$

The number of carbon atoms n_c = 27 obtained by Eq. 1.20 results in a mass of 324 amu. Thus, the difference to the molecular peak is 76 amu which can only be assigned to two oxygen atoms (32 amu) for a nitrogen-free molecule. Thus, the residual mass of 44 amu must be attributed to 44 H atoms.

The sum formula is $C_{27}H_{44}O_2$ which is the same result as obtained by h-MS in Challenge 1.3.

For the calculation of the DBE the sum formula $C_{27}H_{44}O_2$ must be converted into the hydrocarbon C_nH_x by elimination of the two oxygen atoms which gives $C_{27}H_{44}$. The molecule contains six DBE calculated by Eq. 1.21

$$DBE = \frac{(27 \cdot 2 + 2) - 44}{2} = 6. \tag{1.21}$$

The DBE numbers are in accordance with the chemical structure: Two DBE for the two double bonds, the residual four DBE are assigned to the four rings.

Challenge 1.5
Figure 1.1a presents the 70-eV EI mass spectrum of an unknown compound. Determine the *sum formula* and calculate the *numbers of DBE*.

Solution to Challenge 1.5
- Assignment of the molecular peak:

  $m/z = 206$ amu.

- Protection of the molecular peak:
  All isotope peaks can only be reasonably assigned if the molecular peak is
  $m/z = 206$ amu.

- Assignment of the isotope peaks:

m/z in amu	206	207	208	209	210
I_{rel} in %	100	7.0	64	5	10
Assignment	M	M + 1	M + 2	M + 3	M + 4

- Calculation of the number of carbon atoms:

  The number of carbon atoms is $n_C = 6$ calculated by Eq. 1.15 with $I[M] = 100\%$ and $I[M + 1] = 7.0\%$.

- Search for halogen atoms:

  The ratio of the intensity $I[M] : I[M + 2] : I[M + 4] = 100 : 64 : 10$ detects unequivocally the presence of two Cl atoms in the molecule, see Table 6.2.

- Search for nitrogen atoms:

  Because the molecular peak is even, the molecule contains *no* or *two* nitrogen atoms. The hint of the presence of nitrogen is given by a large number of fragment peaks having an even mass number. The fragment ion series $m/z = \mathbf{160}/162/164$ amu is caused by the elimination of NO_2 ($m/z = 46$ amu) from the molecular ion. Note that the isotope pattern of this fragment series corresponds to the presence of two chlorine atoms. Thus, nitrogen should be present in the molecule and because of the even number of the molecular peak, a further nitrogen atom must be present. The fragmentation $m/z = \mathbf{160}/162/164$ amu into $m/z = \mathbf{133}/135/137$ amu ($\Delta m = 27$ amu) may be caused by the loss of HCN which is characteristic for arylamines and N-heterocyclic compounds.

  (continued)

– Listing of the detected atoms:

The addition of 6 C (72 amu) + 2 Cl (70 amu) + 2 N (28 amu) gives 170 amu. The difference to the molecular peak is 36 amu which can be assigned to two O and four H atoms. Thus, the loss of NO_2 proposed above is confirmed.

– Sum formula:

The sum formula is $C_6H_4N_2O_2Cl_2$.

– Check for $OE^{\bullet+}$:

The transformation of the sum formula into $C_xH_yN_z$ is verified by the elimination of the two oxygen atoms and the substitution of the two chlorine atoms by two hydrogen atoms resulting in $C_6H_6N_2$. The test value calculated by Eq. 1.13 is $6 - \frac{1}{2} \cdot 6 + \frac{1}{2} \cdot 2 + 1 = 5$ gives an integer number. Thus, the proposed sum formula is an $OE^{\bullet+}$ ion which means, $C_6H_4N_2O_2Cl_2$ can be used as the sum formula.

– Calculation of the numbers of the DBE:

The base hydrocarbon C_nH_x is obtained by substitution of two N by two CH atoms, two Cl by two H atoms and the elimination of the two O atoms which gives C_8H_8.
The number of DBE is 5 calculated by Eq. 1.19.

– Partition of the DBE:

As is shown later the unknown molecule is an aromatic compound for which four DBE are necessary. The residual one DBE must be assigned to the NO_2 group. There are neither further multiple bonds nor rings in the molecule. Note that the determination of the structure will be pursued step-by-step under inclusion of further spectroscopic methods.

Challenge 1.6
For solutions to Challenge 1.6, see *extras.springer.com*.

1. Determine the sum formula as well as the number of DBE of the unknown compounds which provide the 70-eV EI mass spectra shown in Fig. 1.4a, b, c, d.

(continued)

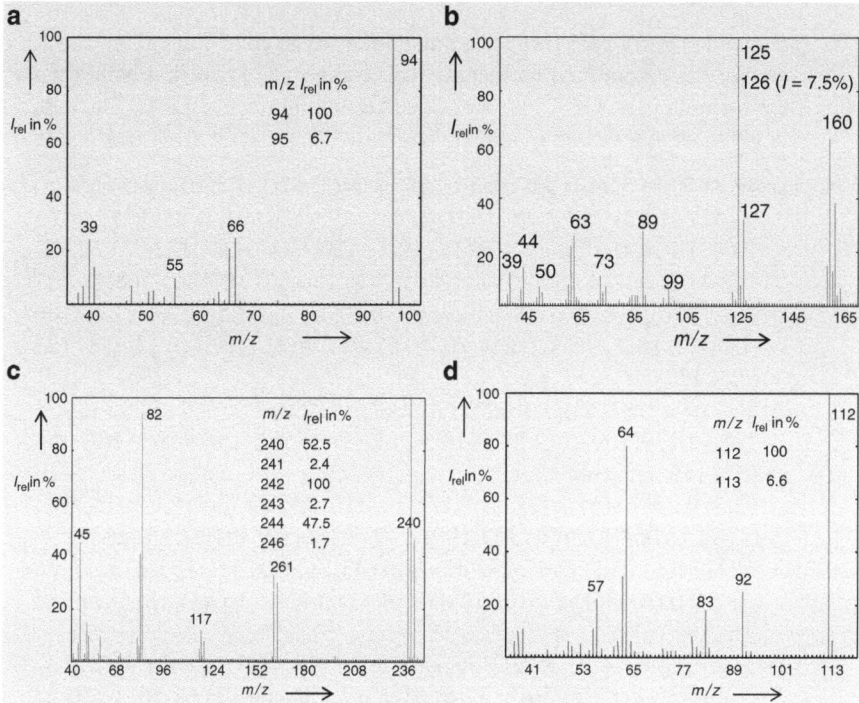

Fig. 1.4 (**a–d**) 70-eV EI mass spectra of unknown compounds

2. The elementary compositions are determined by means of peak matching for the following selected ions are:

   a.$C_{10}H_{15}O$ b.$C_{10}H_{14}O$ c.C_9H_7N d.$C_9H_{15}Cl_3$ e.$C_6H_5O_2N$

   Check which of these compounds could be the cause of the molecular peak.

3. The fragment ions $(CH_3)_2CH^+$ and CH_3CO^+ can be produced by the mass spectrometric fragmentation of 4-methylpentane-2-one (Structure 1.3). Both ions have the mass number 43 amu and show the same peak. How can it be determined experimentally which ion is given in the mass spectrum?

4. Determine the sum formula from the data obtained by the mass spectra in the upper area for the following ten compounds. Note that fragment peaks

   (continued)

are given only for peaks whose relative intensity is greater than 1%. Check whether the determined molecular peak can be valid, check whether the sum formula is an $OE^{\bullet+}$ ion, and calculate the number of DBE. The data are given as m/z in amu (I_{rel} in %).

(a) 49(3) 50(9) 51(13) 52(7) 53(6) 54(8) 61(3) 62(6) 63(10) 64(6) 65(26) 66(42) 67(12) 78(5) 79(2) 91(3) 92(24) 93(100) 94(7) 95(0.2)

(b) 103(3) 104(4) 105(30) 106(3) 119(1) 120(1) 121(30) 122(23) 123(2) 148(1) 149(100) 150(10) 151(1) 165(1) 166(80) 167(7.2) 168(1)

(c) 61(2) 62(6) 63(14) 64(3) 65(27) 66(2) 70(1) 74(2) 75(1) 76(1) 85(1) 86 (1) 87(1) 88(1) 89(8) 90(5) 91(95) 92(9) 118(2) 119(100) 120(88) 121 (7.8) 122(<1)

(d) 76(7) 77(9) 84(23) 85(6) 87(2) 110(2) 111(52) 112(5) 113(17) 114(2) 138(2) 139(100) 140(8) 141(33) 142(3) 155(1) 156(96) 157(8) 158 (31) 159(2.4) 160(<1)

(e) 62(3) 63(3) 68(5) 69(7) 70(3) 73(1) 74(10) 75(28) 76(3) 83(2) 92(1) 93 (2) 94(6) 95(67) 96(6) 123(100) 124(7) 140(70) 141(5.5) 142(<1)

(f) 37(1) 38(2) 39(15) 40(2) 41(45) 42(4) 43(2) 50(1) 51(1) 53(2) 55(5) 56 (2) 57(100) 58(4) 127(6) 128(2) 141(2) 142(1) 169(1) 183(<1) 184 (16) 185(0.7)

(g) 32(6) 33(9) 34(9) 35(25) 36(1) 37(1) 43(2) 44(2) 45(41) 46(21) 47 (100) 48(11) 49(6) 56(1) 57(8) 58(18) 59(34) 60(73) 61(46) 62(4) 63 (2) 64(3) 65(3) 66(2) 76(1) 94(88) 95(2) 96(8) 97(<1)

(h) 68(3) 69(1) 75(7) 76(5) 77(9) 90(32) 91(47) 92(5) 93(2) 105(2) 107 (15) 108(1) 120(7) 121(100) 122(9) 123(5) 137(7) 151(9) 152(59) 153 (5.6) 154(2.4)

(i) 63(1) 65(7) 66(0) 81(50) 82(32) 83(13) 91(15) 93(3) 97(1) 99(100) 100 (0.1) 109(35) 110(7) 111(14) 125(18) 126(6) 127(54) 128(1) 137(8) 138(9) 139(10) 153(4) 155(74) 156(4) 167(2) 181(2) 182(4.5) 183(0.3)

(j) 54(1.8) 56(26) 57(0.8) 81(5) 93(5) 94(5) 95(6) 119(3) 120(1) 121(42) 122(3) 123(<1) 127(1) 128(3) 129(3) 130(1) 184(8.8) 185(1.1) 186 (100) 187(5) 188(1)

1.4 Factors That Influence Ion Abundance

The intensity of the peaks is determined by the frequency of the positive charged species—radical cations and cations—generated in the ion source. The frequency depends on the stability of the ions (*enthalpy effect*) and steric factors (*entropy effect*). This law is the basis of various fragmentation mechanisms. These are important for structural analysis because they are attached to definite structural moieties/functional groups in the molecule. Furthermore, there are many unspecific fragmentations so that, even if with very low intensities, every mass number will show in the mass spectrum.

1.4.1   Stability of the Product ions (Enthalpy Effects)

1.4.1.1   Generation of Stable Ions

Favored fragmentation with a high-intensity cross section is caused by generation of resonance-stabilized ions, so-called *onium ions*. Such onium ions are immonium ions $RR'C = N^+R''$, acylium ions $R–C\equiv O^+$, benzyl ions $C_6H_5–CH_2^+$ and the valence isomer cyclic tropylium ion $C_7H_7^+$, allyl ions $RR'C = CH–CH_2^+$, cyclic ions $C_3H_3^+$ ($m/z = 39$ amu), $C_5H_5^+$ ($m/z = 65$ amu), $C_4H_4X^+$ with $X = Cl$, or Br ($m/z = 87/89$ amu and 131/133, respectively).

1.4.1.2   Elimination of the Largest Alkyl Group (Stevenson Rule)

If more alkyl groups are bound to a carbon atom, then the elimination of the largest alkyl radical is preferred.

Thus, the acylium ion $CH_3–C\equiv O^+$ ($m/z = 43$ amu) generated by the elimination of the larger pentyl radical according to the Stevenson rule is the base peak in the mass spectrum of 2-heptanone, see Eq. 1.22.b:

$$C_5H_{11}–C(=O)–CH_3^{\bullet+} \begin{array}{l} \xrightarrow{\,-CH_3^{\bullet}\,} C_5H_{11}–C\equiv O^+ (\textbf{2\%}) \qquad\qquad (1.22a) \\[2ex] \xrightarrow{\,-C_5H_{11}^{\bullet}\,} CH_3–C\equiv O^+ (\textbf{98\%}) \qquad\quad (1.22b) \end{array}$$

1.4.1.3   Stability of Neutral Products

Although the stability of the ion is much more important *low-energy radical products* or *neutral molecules* cause an additional influence resulting in abundant intensities in the mass spectrum.

Fragmentation reactions are favored by

- Stabilization of radicals by electronegative atoms such as oxygen with generation of alkoxy radicals $OR^{\bullet}$.
- Small stable molecules of high ionization energy are favored, such as CH_4, H_2O, C_2H_4, C_2H_2, HCN, CO, NO, CO_2, H_2S, HCl, $CH_2=C=O$ (ketene).

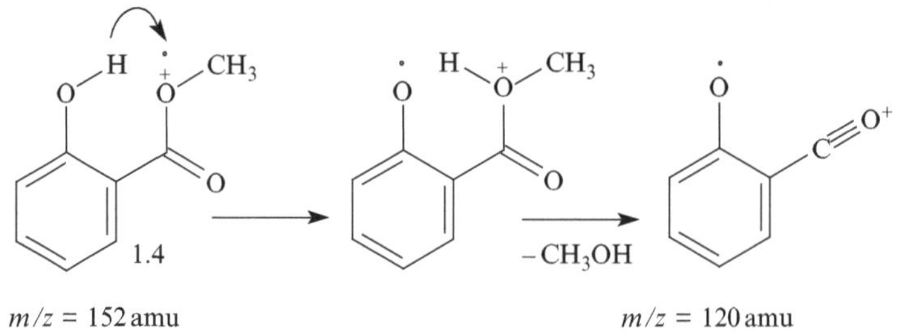

Fig. 1.5 *Ortho*-effect during the mass spectrometric fragmentation of 2-hydroxybenzoic acid methyl ester (Structure 1.4)

1.4.2   Steric Factors

1.4.2.1   Influence of the Structure of Transition State

Six and five rings in the transition state favor hydrogen rearrangement (see, for example, the McLafferty rearrangement) but not three or four rings.

The preferred ring size of the transition state is influenced by the covalent radius of the heteroatom at the hydrogen rearrangement. Thus, the 1,3-elimination of HCl and HBr passes via a five ring, the mass spectrometric elimination of H_2O from alcohols runs over a six-ring transition state (1,4-elimination).

1.4.2.2   Ortho-effect

Ortho-di-substituted aromatic compounds can eliminate neutral molecules by hydrogen rearrangement via a six-ring transition state. Thus, the *ortho*-isomer can easily be distinguished from the *meta*- and *para*-isomers. Figure 1.5 shows the elimination of CH_3OH from 2-hydroxybenzoic acid methyl ester via a six-ring transition state with generation of the intensive peak at $m/z = 120$ amu.

Challenge 1.7

For solutions that meet Challenge 1.7, see *extras.springer.com*.

1. Which intensive peaks (mostly the base peak!) are to be expected for the following compounds? Justify your decision.

a. CH_3–CH_2–CH_2–$NHCH_3$	b. $C_6H_5C(O)H$
c. C_6H_5–CH_2–CH_2–CH_3	d. CH_3–CH_2–CH_2–$C(O)CH_3$
e. CH_3–CH_2–$CH = CH_2$	f. HS–CH_2–CH_2–OH

2. Which low-energy neutral molecules are eliminated by the mass spectro-metric fragmentation of the following molecules?

(continued)

a. Aniline	b. $C_6H_5-CH_2-CH_2-CH_3$	c. $C_5H_5-C(O)H$
d. 2-Hydroxybenzoic acid	e. $C_6H_5-SO_2NH_2$	f. Nitrobenzene
g. N-Acetylaniline	h. γ-Butyrolactone	i. Ethyl phenylether

3. What are the specific characteristics of the mass spectra of the following chemical classes?

a. Aromatic compounds	b. N-heteroaromatic compounds
c. *n*-Alkanes	d. Branched alkanes
e. Aliphatic diols	f. Alkyl carbonic acids

4. What are the causes for the very different intensities of the peaks at $m/z = \mathbf{43}$ amu ($I_{rel} = 100\%$) and $m/z = \mathbf{169}$ amu ($I_{rel} = 2\%$) in the mass spectrum of dodecane-2-one $CH_3-(CH_2)_9-C(O)-CH_3$?

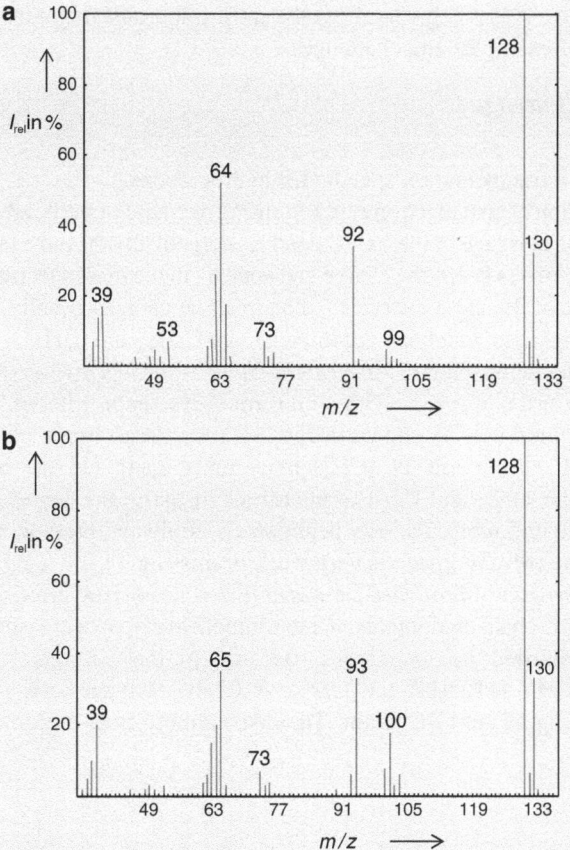

Fig. 1.6 (a–b) 70-eV EI mass spectra of two isomers of chlorphenol

(continued)

5. Figure 1.6a, b present the mass spectra of two chlorphenols. Assign the mass spectra **a** and **b** to the isomers *ortho-*, *meta-*, and *para*-chlorphenol and justify your decision. Can all three isomers be unambiguously distinguished?

6. Nitrotoluene shows an intensive peak at $m/z = 120$ amu in the EI mass spectrum. Which isomer is present? Formulate the fragmentation reaction for this peak.

1.5   Important Fragmentation Reactions

Besides many unspecific fragmentations there are a number of fragmentation reactions that follow certain laws caused by energetic and steric effects resulting in mostly intense peaks. These specific mass spectrometric fragmentations are the basis for the recognition of functional groups and structural moieties. In the following an overview is given about the most important fragmentation reactions on the basis of examples and challenges.

1.5.1   σ Cleavage

σ cleavage is a fragmentation specific for hydrocarbons.

The expulsion of an σ electron causes a strong decrease of the bond energy of the σ bond. Thus, the cleavage of the σ C–C bond is energetically the most favored process. One of the fragments keeps the charge and appears in the mass spectrum whereas the other is a radical. Because every C–C bond can be cleaved equally EE^+ ion series C_nH_{2n+1} are provided which are accompanied by C_nH_{2n} and C_nH_{2n-1} fragments.

The fragmentation pattern of *straight-chain saturated hydrocarbons* is characterized by clusters of peaks with a uniform curve shape without "unsteadiness" of the signal intensities as shown in Fig. 1.7a for the example of *n*-decane. The peaks of each cluster are 14 (CH_2) mass units apart and the most abundant fragments occur at C_3 and C_4. The intensities of the peaks decrease in a smooth curve down to the molecular ion peak which is always present, though in low intensities. The [M-CH_3] peak is very weak or missing.

The mass spectra of *branched saturated hydrocarbons* are principally similar to those of straight-chain compounds, but the smooth curve of decreasing intensities is broken by preferred fragmentations. Because of the different stability of the carbocations $CH_3^+ < RCH_2^+ < R_2CH^+ < R_3C^+$ the cleavage occurs at the position of the branching of the C–C chain. The discontinuity is of high diagnostic value

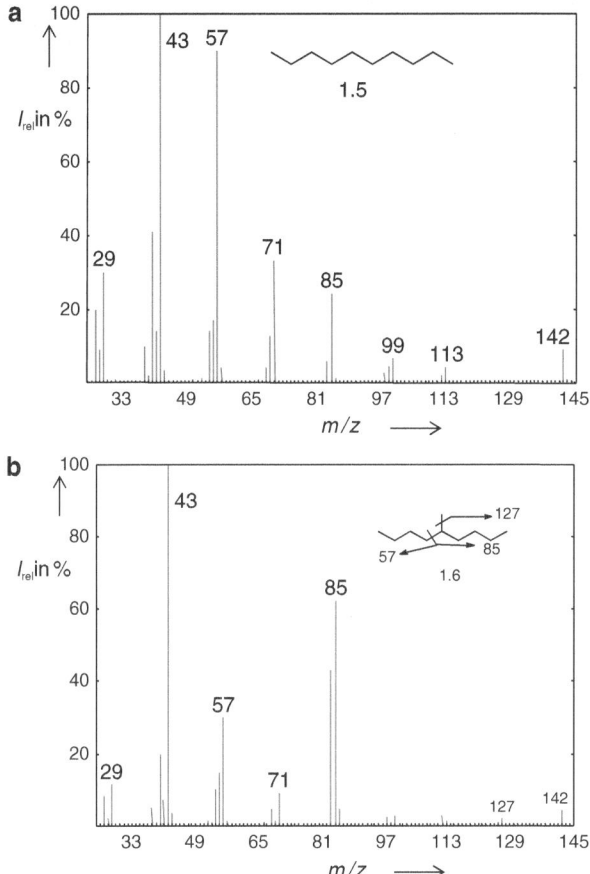

Fig. 1.7 (**a–b**) 70-eV EI mass spectra of *n*-decane (**a**) and the isomer 5-methyl-nonane (**b**)

because it indicates the position of the branch of the C–C chain as is shown for an *iso*-decane (Structure 1.7) in Fig. 1.7b. The high intensity peak at $m/z = 85$ amu and the abundant intensity of the peak $m/z = 57$ amu point towards the presence of a C_7 and a C_4 chain, respectively, and the appearance of the [M-CH$_3$] peak indicates that a CH_3 group is placed at the chain branching. Therefore, the structure 5-methyl-nonane is justified.

A secondary fragmentation of the primary obtained carbocations provides an elimination of olefins resulting in carbocations diminished by 28 amu units (C_nH_{2n} series), see Eq. 1.23:

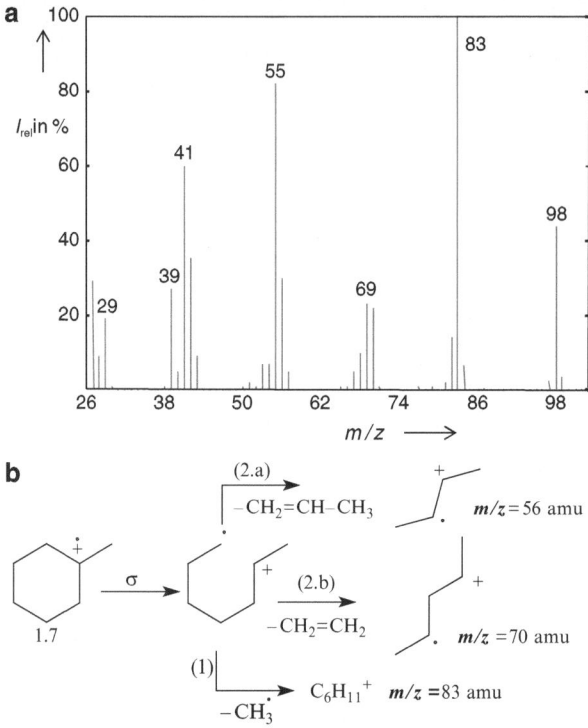

Fig. 1.8 70-eV EI mass spectrum and fragmentation routes of methyl-cyclohexane (Structure 1.7)

$$R\diagdown\!\!\!\diagup\!\!\!\diagdown\!\!\!\diagup\!\!\!\diagdown_{CH_2^+}\longrightarrow R\diagdown\!\!\!\diagup\!\!\!\diagdown_{CH_2^+} + CH_2\!\!=\!\!CH_2 \qquad (1.23)$$

The molecular ion peak of *saturated ring hydrocarbons* has a relatively abundant intensity as the peak at $m/z = 98$ amu in the mass spectrum of methyl-cyclohexane in Fig. 1.8. shows.

After cleavage of the ring two fragmentation routes can proceed:

Route 1: Generation of **EE$^+$** ions as a consequence of a simple σ cleavage with the ion series $m/z = 83, 69, 55, 41, 27$ amu.

Route 2: Generation of structurally important **OE$^{•+}$** ions due to the elimination of olefins resulting in the peaks $m/z = 56$ (route 2.a) and 70 amu (route 2.b).

Challenge 1.8

For solutions that meet Challenge 1.8, see *extras.springer.com*.

1. Figure 1.9a, b show mass spectra of two isomers with the sum formula
   C_8H_{18}. Deduce the structure for these two alkanes.

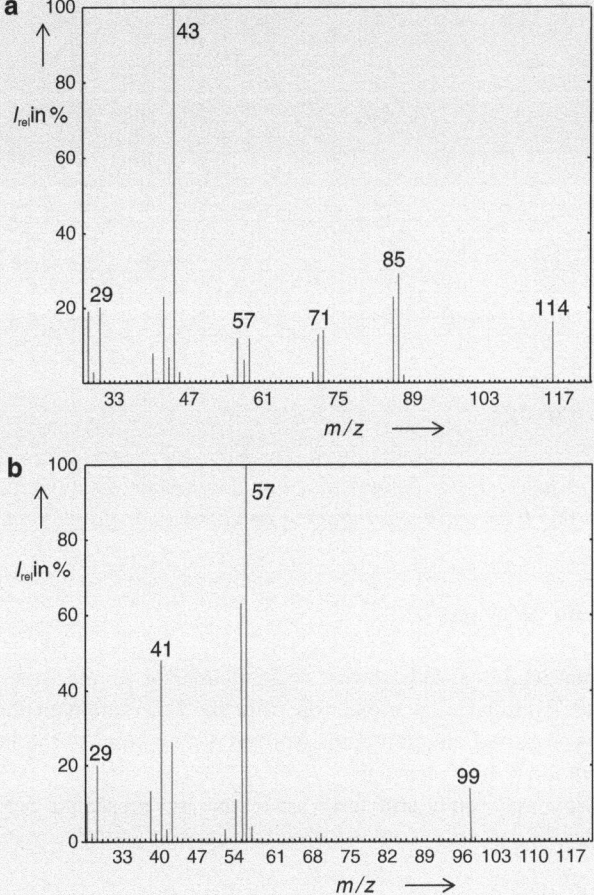

Fig. 1.9 (**a–b**) 70-eV EI mass spectra of two isomers of the sum formula C_8H_{18}

2. How is it possible to recognize the branching of C–C chains?
3. Figure 1.10 shows a mass spectrum of a branched alkane with the sum
   formula $C_{10}H_{22}$. Determine the position of branching.

(continued)

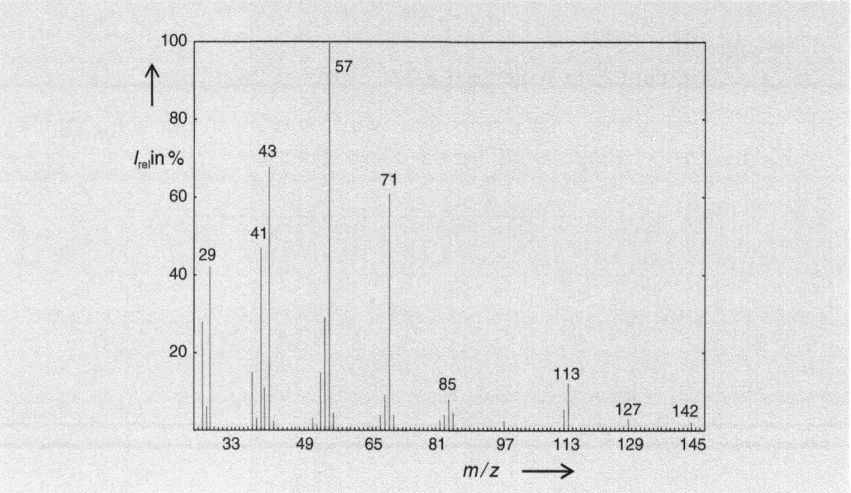

Fig. 1.10 70-eV EI mass spectrum of a branched alkane with the sum formula $C_{10}H_{22}$

1.5.2 α Cleavage

α bonds next to heteroatoms or next to π bonding are preferably cleaved, leaving the charge on this fragment containing the heteroatom or the π bonding.

1.5.2.1 Onium Cleavage

This fragmentation is caused by the strong tendency of electron pairing. The electron impact eliminates an *n* electron from the heteroatom or the π bonding. The remaining unpaired electron tends to form a new bond to the neighboring α atom, therefore the term α cleavage.

In the following, some structural moieties are presented for α cleavage fragmentation.

Saturated Heteroatom

The two possible routes are given in Eqs. 1.24 and 1.25:

$$CH_2 \longrightarrow CH_2 \longrightarrow \overset{+}{X}R \xrightarrow{\ \alpha\ } R \longrightarrow \overset{\cdot}{C}H_2 + CH_2 = X^+R \qquad (1.24)$$

$$\overset{\cdot}{C}H_2 \longrightarrow CH_2 \longrightarrow \overset{+}{X}R \xrightarrow{\ \alpha\ } CH_2 = CH_2 + \overset{+}{X}R \qquad (1.25)$$

The tendency for the generation of onium ions (route 1) diminishes with the decreasing electron-donating strength given in the series: $N > S$, O, unsaturated π system $> Cl > Br > H$.

Because of the high electron-donating strength (nucleophilicity) of nitrogen the onium ions produced by amines are extremely favorable and thus generate mostly the base peak as is shown in the mass spectrum of *n*-butylamine (Structure 1.8) in Fig. 1.11. The resonance-stabilized onium ion $CH_2=NH_2^+ \leftrightarrow {}^+CH_2-NH_2$ gives rise to the base peak at $m/z = \mathbf{30}$ amu.

Unsaturated Heteroatom

The general fragmentation reaction is given in Eq. 1.26

$$R-CR=X^{\bullet+} \xrightarrow{\alpha} R^{\bullet} + RC\equiv X^{+}. \tag{1.26}$$

The generation of the resonance-stabilized benzoylium ion at $m/z = 105$ amu gives rise to the base peak via α cleavage as can be formulated for acetophenone in Eq. 1.27

$$C_6H_5-C(O)-CH_3^{\bullet+} \rightarrow CH_3^{\bullet} + C_6H_5 - C\equiv O^{+}. \tag{1.27}$$

Bear in mind, in the case of alternative fragmentation routes, the elimination of the largest alkyl radical is the most facilitated fragmentation according to the Stevenson rule.

The α cleavage starts the fragmentation of alicyclic compounds but in contrast to the acyclic compounds, at first, an α-cleavaged molecular ion is generated which is rearranged into a resonance-stabilized ion via a six-ring transition state. This is the species used for following fragmentation reactions as shown in Fig. 1.12a, b for

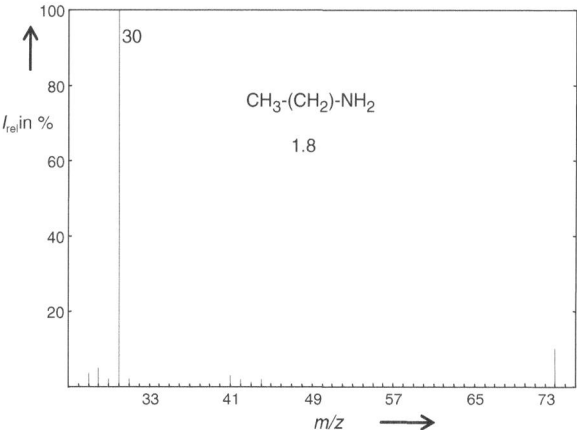

Fig. 1.11 70-eV EI mass spectrum of *n*-butylamine (Structure 1.8)

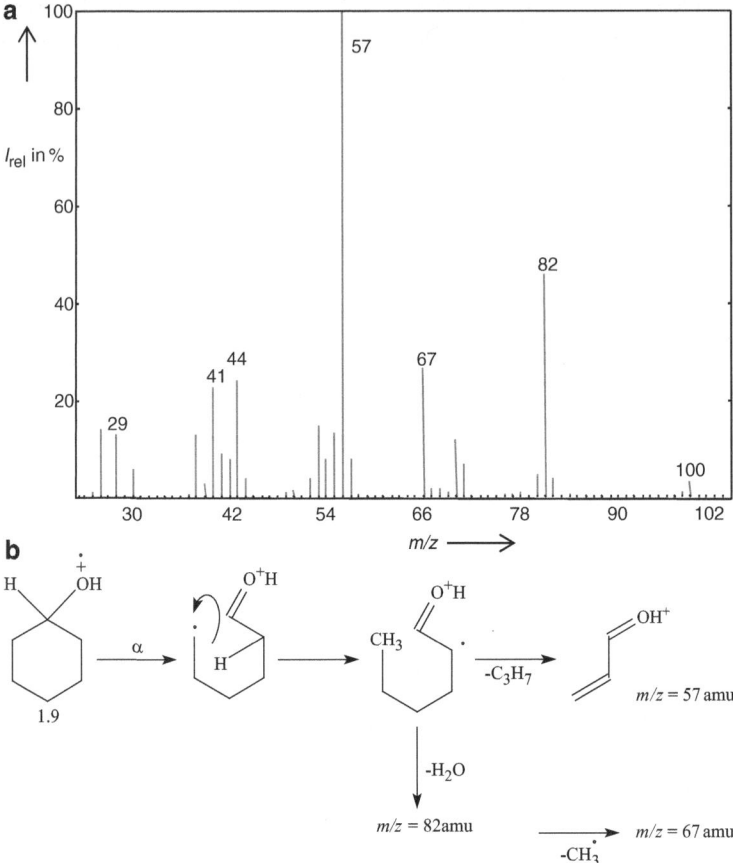

Fig. 1.12 70-eV EI mass spectrum and fragmentation reactions of cyclohexanol

cyclohexanol. After ring cleavage and subsequent rearrangement reactions the structurally important ions at $m/z = 57$ (base peak), 82, and 67 amu, are formed.

1.5.2.2 Allylic Cleavage

Whereas electron impact eliminates an n-electron of a heteroatom in the molecule a π-electron is eliminated in alkenes. Again the tendency towards electron pairing causes α-bonding cleavage with generation of a resonance-stabilized allyl ion as formulated in Eq. 1.28

$$CH_3 - CH_2 - CH - HC^{+\cdot} - R \rightarrow CH_3^{\cdot} + CH_2 = CH - HC^+R \leftrightarrow CH_2^+ - CH = CHR.$$

$$(1.28)$$

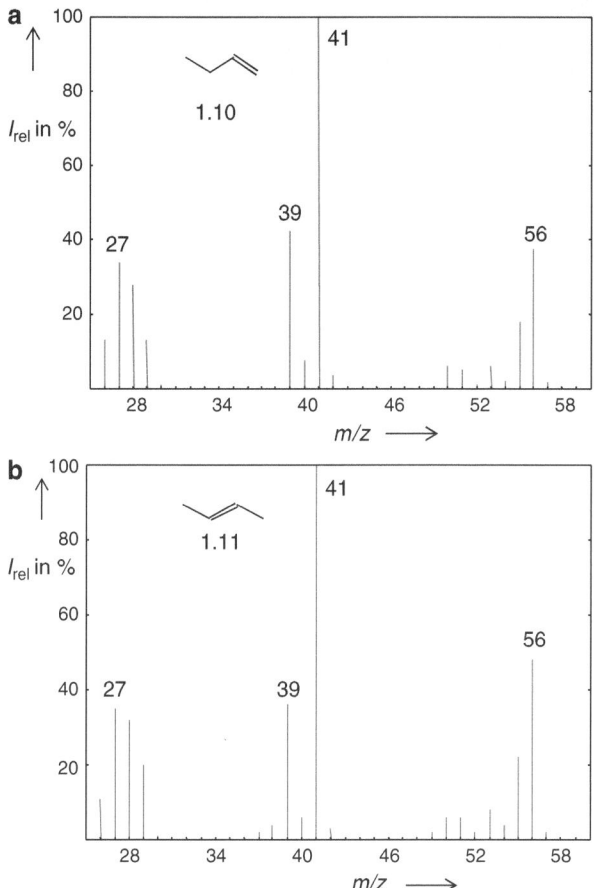

Fig. 1.13 (a–b) 70-eV EI mass spectra of butane-1 (Structure 1.10) (a) and butane-2 (Structure 1.11) (b)

Localization of the double bond in acyclic alkenes cannot be determined with certainty because of its facile migration in the primarily created radical cation. Thus, isomers mostly provide similar mass spectra; compare the spectra of butane-1 (Structure 1.10) and butene-2 (Structure 1.11) in Fig. 1.13a, b. Note that for example 2-methyl-propene-1 shows a similar mass spectrum. The base peak at $m/z = 41$ amu is the same in all three spectra. It is caused by the allylic ion $CH_2=CH–CH_2^+$. Furthermore, clusters of peaks at the interval of 14 units are present in the mass spectra of acyclic alkenes caused by the ions C_nH_{2n-1} and C_nH_{2n} which possess a higher intensity than those of the C_nH_{2n+1} ion series in acyclic alkanes.

1.5.2.3 Benzylic Cleavage

Besides heteroatoms also aromatic groups possess an activating influence for α cleavage. Thus, the benzylic ion $C_6H_5–CH_2^+$ or, more likely the valence isomeric tropylium ion $C_7H_7^+$, originating from cleavage at the bond β leads to the base peak at $m/z = 91$ amu as is shown in the mass spectrum of *n*-propylbenzene in Fig. 1.14. Branching at the α carbon atom provides peaks higher than 91 amu at the distance of 14 mass units.

A benzylic-like cleavage also dominates in the mass spectra of heteroaromatics as is shown in Fig. 1.15 for 2-*n*-propylthiophene.

The fragmentation reaction of the benzylic ion is controlled by loss of the neutral molecule $HC{\equiv}CH$. These peaks and peaks resulting from σ-fragmentations at $m/z = 77$, 65, 51, and 39 amu are key fragment peaks for the recognition of aromatics, see the fragmentation reactions of *n*-propylbenzene in Fig. 1.14. The peaks at $m/z = 51$ and 77 amu are typical for *mono*-substituted aromatics.

Sometimes the benzylic ion peak is accompanied by an intensive peak at $m/z = 92$ amu caused by a McLafferty rearrangement (see Sect. 1.5.3). This peak points towards the presence of a γ hydrogen atom in the side chain.

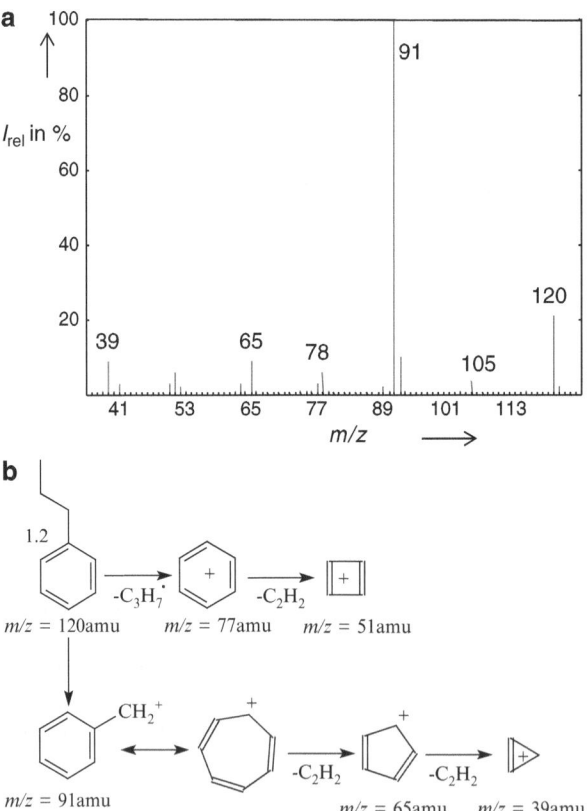

Fig. 1.14 70-eV EI mass spectrum and fragmentation reactions of *n*-propylbenzene (Structure 1.12)

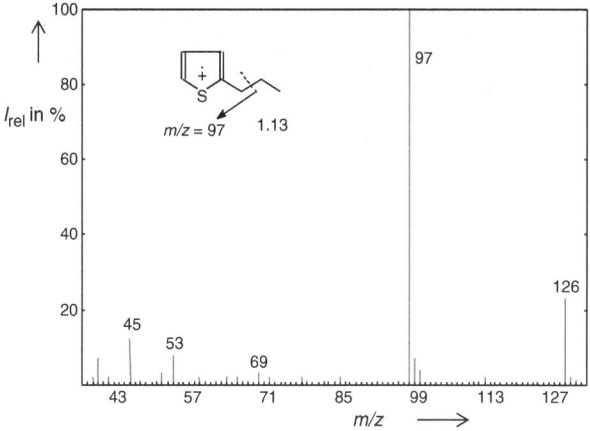

Fig. 1.15 Mass spectrum of 2-*n*-propylthiophene (Structure 1.13)

Challenge 1.9

For solutions that meet Challenge 1.9, see *extras.springer.com*.

1. In general, the loss of the highest-energy radical H• is very unlikely but the [M-1] peak is the base peak in the mass spectrum of hydrindene (Structure 1.14) which is shown in Fig. 1.16. Explain this fact and formulate the corresponding fragmentation reaction for this peak.

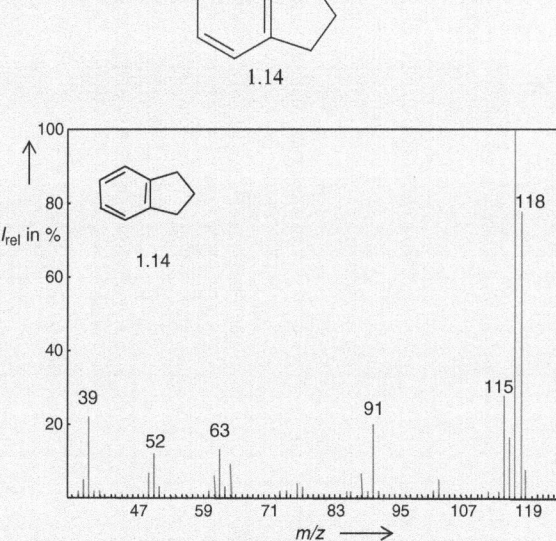

Fig. 1.16 70-eV EI mass spectrum of hydrindene (Structure 1.14)

(continued)

2. Figure 1.17 presents the mass spectrum of methyl *n*-butyl ketone CH_3–C (O)–$(CH_2)_3$–CH_3 (Structure 1.15). Formulate the fragmentation reactions for the following peaks and name these reactions: $m/z = 43, 57, 71,$ and 85 amu.

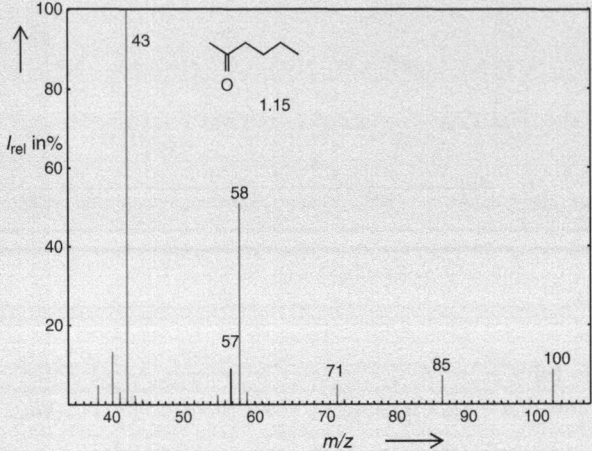

Fig. 1.17 70-eV EI mass spectrum of methyl *n*-butyl ketone (Structure 1.15)

3. Which of the four isomers with the sum formula $C_9H_{10}O_2$ (Structure 1.16–1.19) provides the mass spectrum presented in Fig. 1.18? Justify your answer.

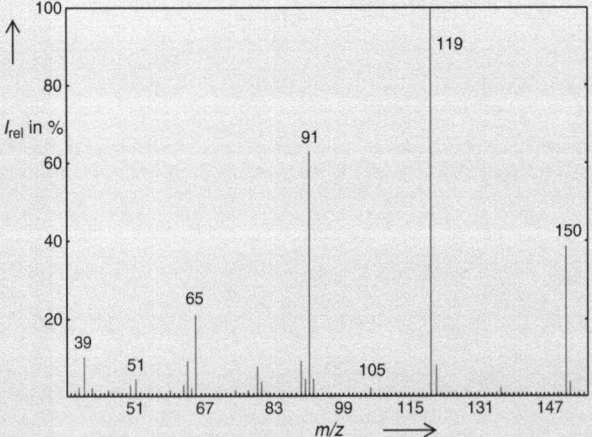

Fig. 1.18 Mass spectrum of an unknown aromatic compound with the sum formula $C_9H_{10}O_2$

(continued)

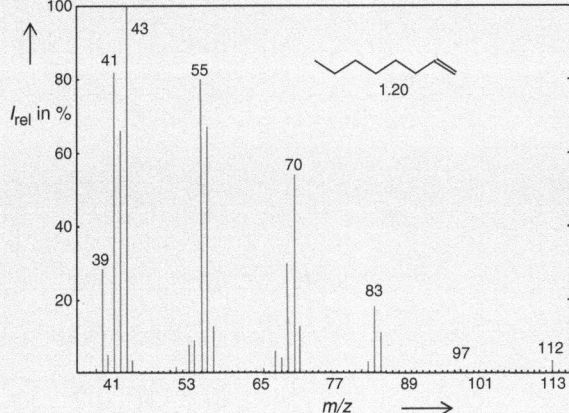

1.16   1.17   1.18   1.19

4. List the mass spectroscopic information for the structural evidence of octene-1 (Structure 1.20) whose mass spectrum is shown in Fig. 1.19.

Fig. 1.19 Mass spectrum of octene-1

5. The cyclic isomeric ketones (Structure 1.9 and 1.10) provide the mass spectra shown in Fig. 1.20a, b. List the relevant fragmentation reactions for these compounds and assign both mass spectra to the corresponding ketones. Formulate the fragmentation reactions for the ions at $m/z = 161$, 148, 145, 133, and 120 amu in spectrum A and $m/z = 161$ and 134 amu in spectrum B.

1.21   1.22

(continued)

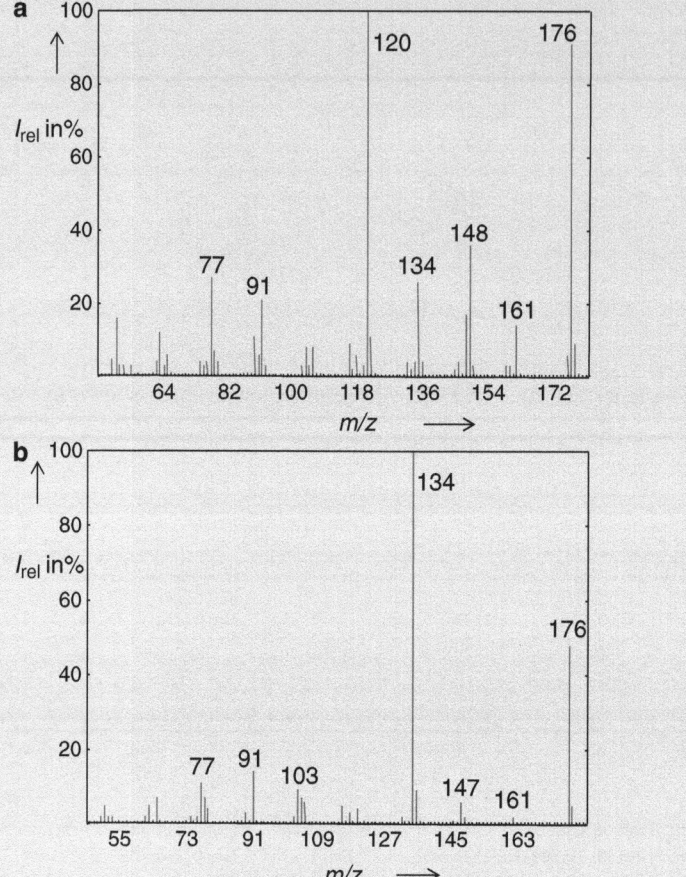

Fig. 1.20 (**a–b**) Mass spectra of 7-methoxy 2-tetralone (Structure 1.22) and 7-methoxy-1-tetralone (Structure 1.21) in an arbitrary array

6. Which of the three structures (Structures 1.23–1.25) can be assigned to the mass spectrum presented in Fig. 1.21? List the main fragmentation reactions for the possible compounds *n*-hexyl-amine (Structure 1.23), N-methyl *n*-propyl amine (Structure 1.24), and dimethly *n*-butyl amine (Structure 1.25).

NH₂	NHCH₃	N(CH₃)₂
1.23	1.24	1.25

(continued)

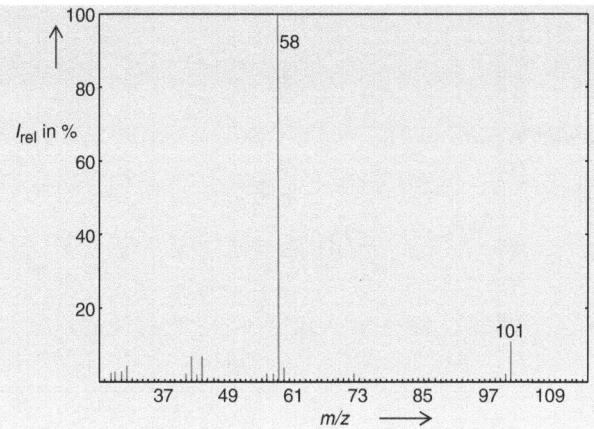

Fig. 1.21 Mass spectrum of an amine with the sum formula $C_6H_{15}N$

7. The mass spectrum of nicotine (Structure 1.47) shows an intensive [M-1] peak. Explain this fact.

1.5.3   McLafferty Rearrangement

The reaction named after McLafferty is a *specific* fragmentation that occurs in many chemical classes leading to peaks with abundant intensities. The McLafferty rearrangement often accounts for prominent and characteristic peaks and are consequently meaningful for structure analysis.

The McLafferty rearrangement is the migration of a γ hydrogen atom to a multiple bond via a six-ring transition state followed by cleavage of the β bond and the simultaneous elimination of a neutral molecule that contains the β and γ atoms. The denotation of the McLafferty rearrangement is mostly carried out by employing the symbol r_H.

The McLafferty rearrangement is demonstrated using the example of octane-2-one as in Eq. 1.29. Figure 1.22 displays the respective mass spectrum:

$$C_5H_{10} \ + \ CH_2{=}C(CH_3)OH^+ \qquad (1.29)$$

$$m/z = 58 \text{ amu}$$

1.26

The double bond necessary for the r_H reaction can be either a priori present in the molecule or it is generated by a previous fragmentation reaction, for example, by α

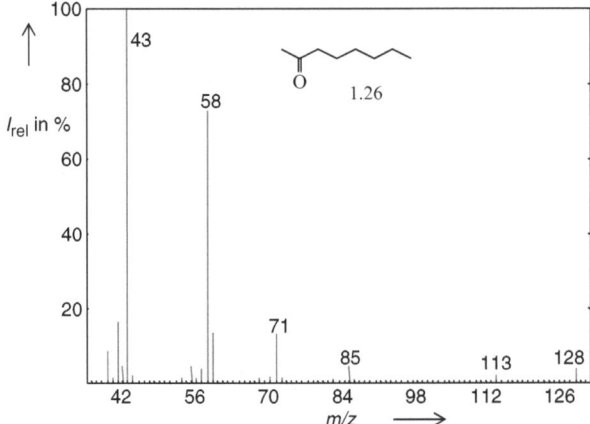

Fig. 1.22 McLafferty rearrangement demonstrated using the example of the 70-eV EI mass spectrum of octane-2-one (Structure 1.26)

Table 1.6 McLafferty product ions of some important chemical classes

Chemical classes	m/z in amu	McLafferty product ion
Aldehydes	44	$CH_2 = C(OH)H^{•+}$
Methylketones	58	$CH_2 = C(OH)CH_3^{•+}$
Ethylketones	72	$CH_2 = C(OH)C_2H_5^{•+}$
Hydrocarbon acids ($n \geq 4$)	60	$CH_2 = C(OH)_2^{•+}$
Methylesters	74	$CH_2 = C(OH)(OCH_3)^{•+}$
Amides	59	$CH_2 = C(OH)NH_2^{•+}$
Phenylketones	120	$CH_2 = C(OH)C_6H_5^{•+}$

cleavage in ethers. The McLafferty rearrangement can also be connected to an additional hydrogen transfer which results in a peak at $m/z = 60$ amu ($CH_3–C^+(OH)–NH_2$) in N-alkyl acetamides (see also Challenge 1.11).

The α, β, and γ positions are not limited to carbon atoms but they can also be occupied by nitrogen, oxygen, or sulfur. The only relevant condition is the presence of a γ hydrogen atom.

Some McLafferty products are listed in Table 1.6.

In addition, the McLafferty rearrangement can be induced by the influence of the aromatic group, providing that a γ hydrogen atom is present in the molecule which is demonstrated using the example of n-propylbenzene (Structure 1.27) as in Eq. 1.30:

$$(1.30)$$

The peak at $m/z = 92$ amu in the mass spectrum of n-propylbenzene (Fig. 1.14) is caused by r_H fragmentation. Thus, n-propylbenzene can be unambiguously distinguished by, for example, the methyl ethyl benzene isomer because a γ hydrogen atom is not present in this compound.

McLafferty fragment ions are produced with peak numbers other than $m/z = 92$ amu if the α and/or β position is substituted by heteroatoms. Thus, the characteristic peaks of the r_H fragmentation are $m/z = 94$ amu for phenylethers, and $m/z = 93$ amu for N-alkylamides.

The γ hydrogen atom can also migrate into a double bond of an alkene. Thus, the peak at $m/z = 42$ amu is the McLafferty product ion generated in alkenes possessing a γ hydrogen atom according to Eq. 1.31.

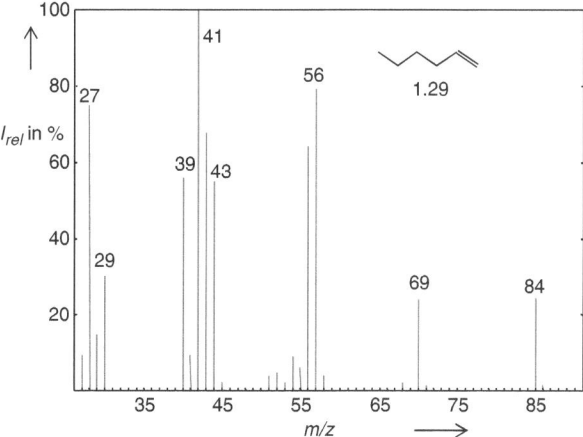

$$\tag{1.31}$$

The peak at $m/z = 56$ amu in the mass spectrum of hexene-1 (Structure 1.29) shown in Fig. 1.23 is the product ion of an r_H fragmentation generated by the isomeric 2-hexyl radical cation $CH_3–CH=CH^{+\cdot}–(CH_2)_2–CH_3$ (Structure 1.31).

Finally, the elimination of H_2O + olefin from acyclic alcohols proceeds via the McLafferty mechanism whereas a δ hydrogen atom occurs in the transition state.

The primarily formed alkene radical cation successively eliminates ethylene resulting in ion series [M-46] ($H_2O + C_2H_4$), ... [M − ($H_2O + C_nH_{2n}$)] which possess peaks at $m/z = 98$ amu, [M − H_2O], 70 amu [M − ($H_2O + C_2H_4$)], and 42 amu [M − ($H_2O + 2\,C_2H_4$)] in the mass spectrum of n-heptanol (Structure 1.32) shown in Fig. 1.24.

Fig. 1.23 70-eV EI mass spectrum of hexene-1 (Structure 1.29)

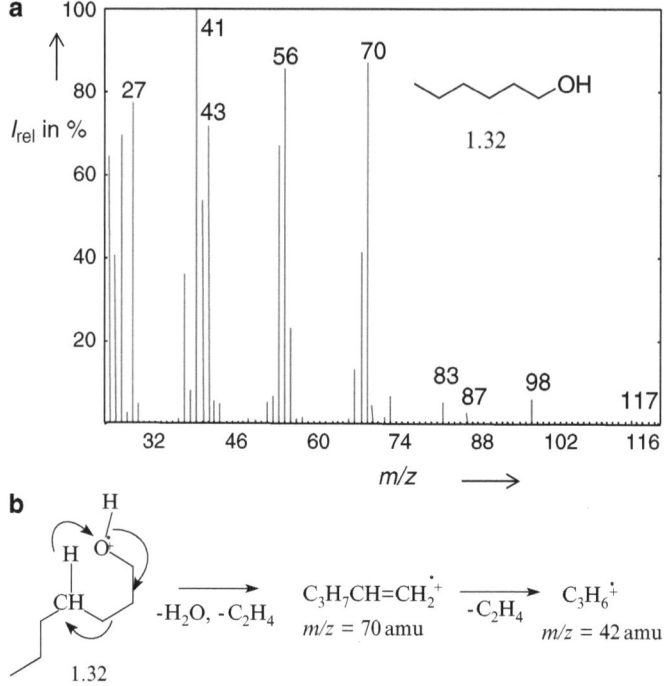

Fig. 1.24 70-eV EI mass spectrum of *n*-heptanol (Structure 1.32) and fragmentation reactions started by the McLafferty rearrangement

Challenge 1.10

For solutions to meet Challenge 1.10, see *extras.springer.com*.

1. Formulate the mass spectrometric fragmentation reactions for the given peaks:

a. Valerianic acid ($m/z = $ **60** amu)	b. *n*-Hexyl-benzene ($m/z = $ **92** amu)
c. Methyl *n*-pentyl ether ($m/z = $ **45** amu)	d. Methyl ethyl *iso*-propyl amine ($m/z = $ **72** amu)

2. Which fragmentation provides the peak at $m/z = 58$ amu in the mass spectrum of methyl *n*-butyl ketone (Fig. 1.17)?
3. Carboxylic acid methyl esters show an intensive [M-31] peak whereas in esters with C $\geq$ 2 the elimination of olefins dominates, for example, the loss of C_2H_4 at ethyl esters. Explain this fact.
4. Hexane-2-one, *n*-hexanale, methyl *tert*-butyl ketone are some isomers with the sum formula $C_6H_{12}O$. Formulate and name the most important fragmentation reactions for these three compounds. Which fragmentation reactions are relevant for the decision of the isomers?

(continued)

5. Assign the ethyl cyclo-hexanone isomers 1.33 and 1.34 to the mass spectra given in Fig. 1.25a, b and justify your decision. Formulate and name the relevant fragmentation reactions.

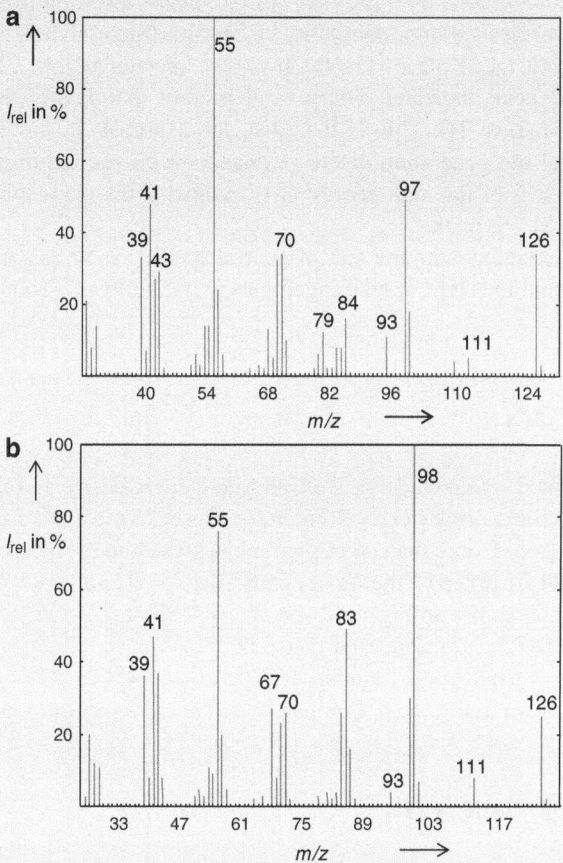

Fig. 1.25 Mass spectra of the ethyl cyclo-hexanone isomers shown in Structures 1.33 and 1.34

1.5.4   Loss of Neutral Molecules

There are some fragmentation steps associated with the loss of low-energy neutral species which may be either uncharged molecules such as CO, H_2O, HF, C_2H_2 and others or radicals, for example, $H^{\bullet}$, $CH_3{}^{\bullet}$, or $Cl^{\bullet}$. Peaks resulting from the elimination of neutral species are important hints for certain structural moieties, especially, if the elimination takes place directly from the molecular ion. Thus, the high-intensity [M-1] peak indicates a structural moiety that favors the loss of the highest-energy radical $H^{\bullet}$. This is the case, for example, in nicotine (Structure 1.47) because of the generation of the resonance-stabilized onium ion. The same is valid for the loss of the high-energy $CH_3{}^{\bullet}$ radical if the mass spectrum offers a high-intensity [M-15] peak.

Most fragmentations with the loss of neutral species occur as a consequence of primary fragmentation steps or rearrangements. In the following some examples are given.

1.5.4.1   Loss of CO

Cyclic unsaturated compounds as well as ions generated by an α cleavage of a carbonyl group eliminating CO resulting in the loss of 28 mass units. The loss of CO occurs step-by-step if more than one CO group is present in the molecule. Examples of the loss of CO are given by the fragmentation $m/z = 92$ amu $\rightarrow m/z = 64$ amu in the mass spectrum shown in Fig. 1.4d and $m/z = 119$ amu $\rightarrow m/z = 91$ amu in the mass spectrum which is presented in Fig. 1.18.

The loss of oxygen in phenol and its derivates also occurs as CO from a rearranged radical cation shown in Fig. 1.26. Compare with the mass spectrum in Fig. 1.4a.

1.5.4.2   Loss of Ethine, C_2H_2

Elimination of C_2H_2 (loss of the mass number 26) occurs, for example, in the case of the fragmentation of the tropylium ion by the following reaction steps: $m/z = 91$ amu $\rightarrow m/z = 65$ amu, $m/z = 65$ amu $\rightarrow m/z = 39$ amu; compare with Fig. 1.14.

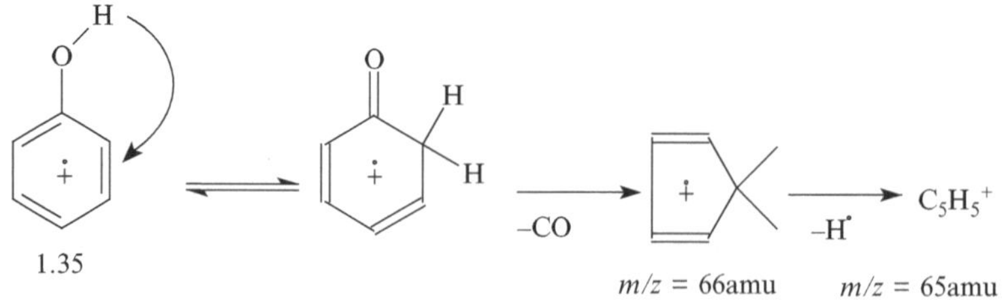

Fig. 1.26 Fragmentation reactions of phenol (Structure 1.35)

1.5.4.3   Loss of HCl and Cl˙

HCl (mass number = 36/38) and Cl˙ (mass number = 35/37) are eliminated from chlorine-substituted aromatic compounds. The loss of HCl is favored by the *ortho*-effect, see Sect. 1.4. Thus, the *ortho*-isomer can be easily recognized, but the mass spectra of the *meta*- and *para*-isomers are nearly equal.

1.5.4.4   Loss of HCN

N-heterocyclic compounds as well as arylamines eliminate nitrogen as the stable molecule HCN which is recognized by mass differences of $m/z = 27$ amu.

Thus, the fragmentation $m/z = 66$ amu $\rightarrow m/z = 39$ amu is caused by the loss of HCN in the three methyl-pyridines shown in Fig. 1.27 by the example of 4-methyl pyridine (Structure 1.36).

1.5.4.5   Loss of Ketene

The elimination of ketene with the mass number 42 is favored by the fragmentation of N- and O-acetyl compounds. The loss of ketene leads to the base peak in the mass spectrum of ethane acid benzyl ester (Structure 1.37) at $m/z = 108$ amu, see Fig. 1.28. Thus, the isomeric *ortho*-, *meta*-, or *para*-methyl benzoic acid methyl-esters (Structure 1.38) can be distinguished by the preferred α cleavage (elimination of $CH_3O˙$ at $m/z = 31$ amu) with the following loss of CO as it is pictured in Structure 1.38.

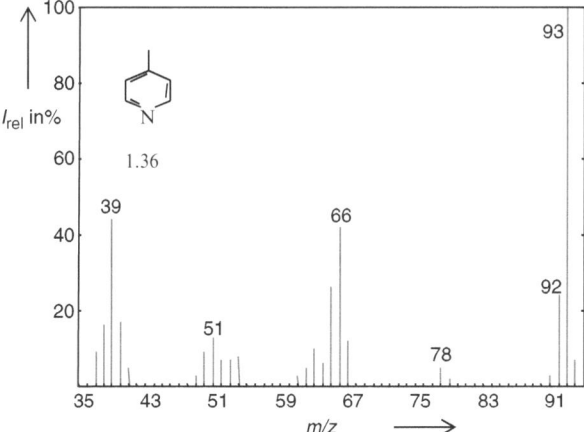

Fig. 1.27   70-eV EI mass spectrum of 4-methyl pyridine (Structure 1.36)

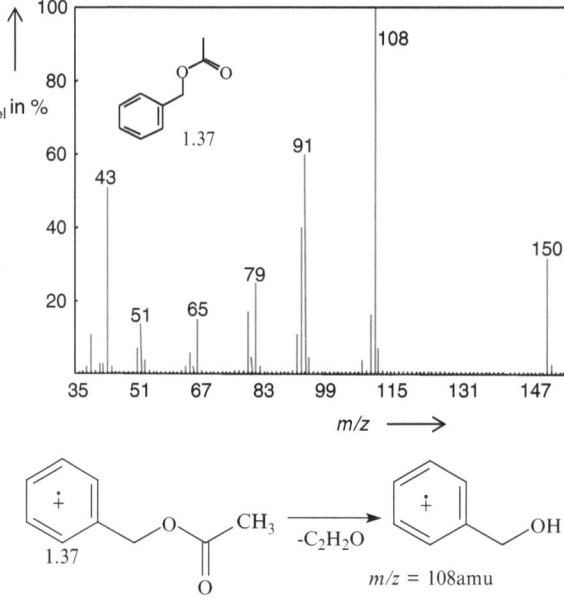

Fig. 1.28 70-eV EI mass spectrum of ethane acid benzyl ester (Structure 1.37) and the fragmentation reaction for the generation of the base peak

1.5.5 *Charge-induced Cleavage*

The charge-induced cleavage is caused by inductive effects and is generally denoted by the symbol **i**. The load-bearing atom pulls in an electron pair under cleavage of the neighboring bond and migration of the charge.

The charge-induced cleavage from the radical cation **OE$^{•+}$** occurs according to Eq. 1.32,

$$R - Y^{•+} - R' \rightarrow R^+ + Y^{•+} - R' \tag{1.32}$$

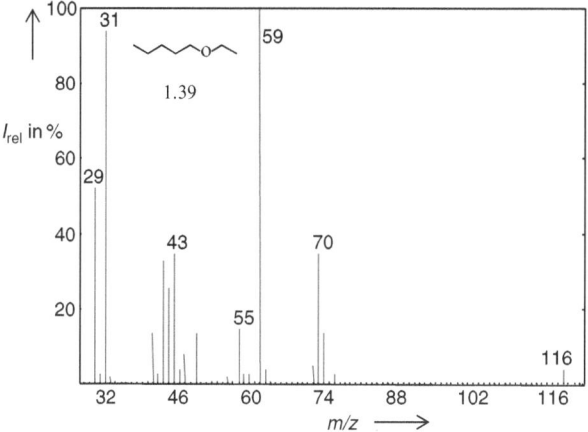

Fig. 1.29 Mass spectrum of ethyl *n*-pentyl ether (Structure 1.39)

and it is verified by the fragmentation reactions in Eqs. 1.33 and 1.34, in dialkyl-ethers

$$R-O^{\bullet+}-R' \longrightarrow R^+ + O^{\bullet+} - R' \tag{1.33}$$

$$\longrightarrow R - O^{\bullet} + R'^{+}. \tag{1.34}$$

The peaks at $m/z = 29$ and $m/z = 71$ amu in the mass spectrum of ethyl *n*-pentyl ether (Structure 1.39) shown in Fig. 1.29 are caused by the i cleavage.

The general reaction of a charge-induced cleavage started by the cation $\mathbf{EE^+}$ is given in Eq. 1.35.

$$R - YZ^+ \rightarrow R^+ + YZ \tag{1.35}$$

The peak at $m/z = 57$ amu ($C_4H_9^+$), for example, in the mass spectrum of methyl *n*-butyl ketone (Structure 1.15) shown in Fig. 1.17 is generated by such a fragmentation as Eq. 1.36 shows

$$C_4H_9 - C \equiv O^+ \rightarrow C_4H_9^+ + CO. \tag{1.36}$$

The frequency of the i cleavage diminishes in the series
Cl, Br, NO_2 > O, S >> N, C.

In the larger C_5 of alkyl-chlorides the intensity of the R^+ ions decreases because of the generation of cyclic products $C_nH_{2n}Cl$ with $n = 3, 4, 5$. The cyclic ion $C_4H_8Cl^+$ (with $m/z = 91$ amu) is the ion with the highest intensity of this series and mostly occurs as the base peak in the spectrum, which is shown in the mass spectrum of *n*-hexylchloride (Structure 1.40) in Fig. 1.30.

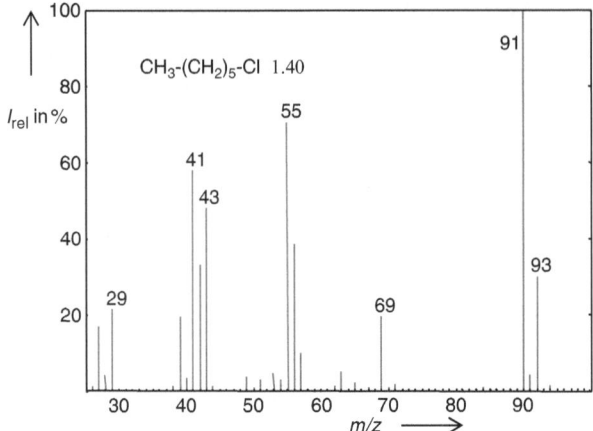

Fig. 1.30   70-eV EI mass spectrum of *n*-hexylchloride (Structure 1.40)

1.5.6   Onium Reactions

Onium reactions require cationic fragment ions with a heteroatom, such as onium ions, immonium ions, ammonium ions, sulfonium ions and others.

The onium reaction is characterized by hydrogen transfer from the heteroatom-containing alkyl residual under olefin elimination. Because the transferred hydrogen atom is unknown its source is symbolized by a bracket.

Amines, ethers, thioethers, and N-acetyl compounds are chemical classes in which onium reactions can occur.

The onium reaction is demonstrated by the example of the mass spectrometric fragmentation of ethyl *n*-pentyl ether (Structure 1.39). Its mass spectrum is shown in Fig. 1.29.

The essential fragmentations are summarized in the following four processes according to Eqs. 1.37, 1.38, 1.39, 1.40, 1.41, and 1.42:

1. α cleavage

$$CH_3-(CH_2)_2-CH_2-CH_2-O^{\bullet+}-CH_2-CH_3 \rightarrow \mathbf{CH_2=O^+-C_2H_5}+CH_3-(CH_2)_2-CH_2^{\bullet}$$
$$m/z = 59 \text{ amu}$$

$$(1.37)$$

$$CH_3-(CH_2)_2-CH_2-CH_2-O^{\bullet+}-CH_2-CH_3 \rightarrow \mathbf{CH_3-(CH_2)_3-CH_2-O^+=CH_2}+CH_3^{\bullet}$$
$$m/z = 101 \text{ amu}$$

$$(1.38)$$

According to the rule of repelling the largest radical the ion $m/z = 59$ amu is favorable resulting in the base peak. The relative intensity of the peak at $m/z = 101$ amu is smaller than 1%, therefore, it cannot be seen in the condensed mass spectrum given in Fig. 1.29.

2. McLafferty rearrangement of the first-formed onium ion at $m/z = 101$

$$CH_3-(CH_2)_2-CH_2-CH_2-O^+=CH_2 \rightarrow \mathbf{CH_3-O=CH_2^+}+CH_3-CH_2CH=CH_2$$

$$m/z = 45 \text{ amu}$$

$$(1.39)$$

3. Onium reaction

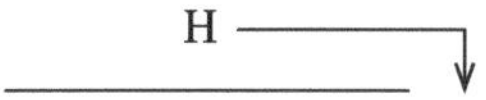

$$CH_3\text{–}(CH_2)_2\text{–}CH_2\text{–}CH_2\text{–}O^+\text{=}CH_2 \rightarrow \mathbf{HO=CH_2^+} + C_5H_{10} \qquad (1.40)$$

$$m/z = 31 \text{ amu}$$

4. Charge-induced cleavage

$$CH_3-(CH_2)_2-CH_2-CH_2-O^{\bullet+}-CH_2-CH_3 \rightarrow \mathbf{C_2H_5^+}+ CH_3-(CH_2)_2-CH_2-CH_2-O^\bullet$$

$$m/z = 29 \text{ amu}$$

$$(1.41)$$

$$CH_3-(CH_2)_2-CH_2-CH_2-O^{\bullet+}-CH_2-CH_3 \rightarrow \mathbf{C_5H_{11}^+}+CH_3-CH_2-O^\bullet$$

$$(1.42)$$

$$m/z = 71 \text{ amu}$$

1.5.7   Retro-*Diels-Alder Reaction*

The *retro*-Diels-Alder reaction (RDA), demonstrated by taking the example of cyclohexene, can proceed via two routes. Either the diene-component (Eq. 1.43) or the ene-component (Eq. 1.44) can bear the charge and can appear as a fragment ion in the mass spectrum.

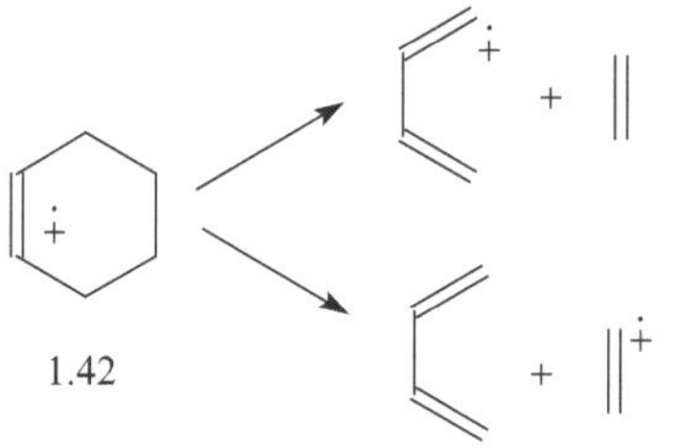

$$(1.43)$$

1.42

$$(1.44)$$

In general, the generation of the charged diene-component is the favorable route (Eq. 1.43), but also the charged ene-components can occur in the mass spectrum according to Eq. 1.44.

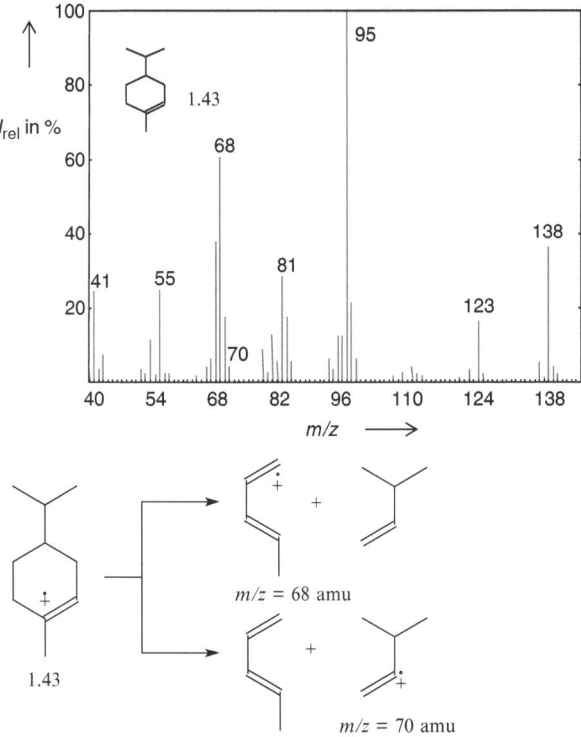

Fig. 1.31 70-ev EI mass spectrum and possible the RDA-reactions of menthene (Structure 1.43)

The RDA-reaction is only a preferred fragmentation if no other preferred reactions are possible. Thus, the presence of functional groups in the molecule can drastically diminish the intensity of RDA peaks.

The cyclohexane ring can also be substituted by one or more heteroatoms or it can be part of a larger ring system. The RDA reaction can be started from the molecule ion or from a fragment ion.

In the mass spectrum of menthene (Fig. 1.31) the diene-peak at $m/z = 68$ amu generated by the fragmentation reaction according to Eq. 1.43 is obviously more intensive than that of the ene-component at $m/z = 70$ amu according to Eq. 1.44. Both fragmentation reactions are presented in Fig. 1.31. But the RDA reactions do not dominate the mass spectrum because the loss of the *iso*-propyl group by a σ cleavage with the peak at $m/z = 95$ amu is the most favorable fragmentation. The elimination of the methyl radical is less probable according to the Stevenson rule and, hence, the peak at $m/z = 123$ amu ([M - 15]) has a lower intensity.

The RDA reaction takes place in many natural products such as retinol or flavonoxides. Note that the RDA fragmentation can also be caused by a *thermal* reaction in the inlet of the ionization chamber. In this case RDA product ions occur in the mass spectrum generated by thermal and/or mass spectrometric fragmentations which complicates the interpretation of the spectrum.

Challenge 1.11

For solutions to meet Challenge 1.11, *see extras.springer.com.*

1. Formulate and name the fragmentation reactions for the peaks in the mass spectrum of the following compounds:

   (a) Ethyl *sec*-butyl ether

   m/z in amu (I_{rel} in %): 31 (8), 45 (100), 59 (19), 73 (50), 87 (6)

   (b) N-acetyl hexyl amine

   m/z in amu (I_{rel} in %): 30 (100), 43 (31), 60 (10), 72 (28)

   (c) N-methyl ethyl *n*-propyl amine

   m/z in amu (I_{rel} in %): 44 (72), 58 (12), 72 (100), 86 (8)

2. Figure 1.32 shows the mass spectrum of a tertiary amine. Check whether the mass spectrum can be assigned to N-methyl ethyl *n*-propyl amine, $C_2H_5-N(CH_3)-CH_2-CH_2-CH_3$ (Structure 1.44). Formulate and name the relevant fragmentation reactions.

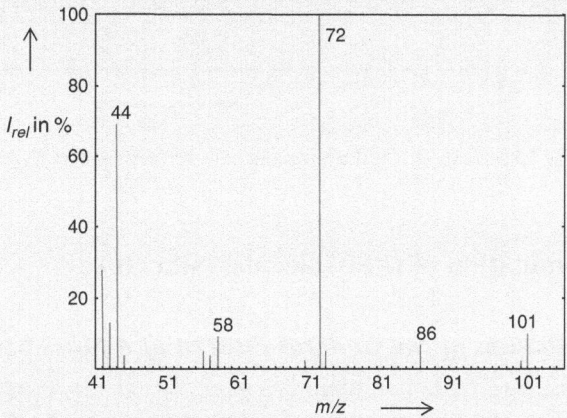

Fig. 1.32   70-eV EI mass spectrum of N-methyl ethyl *n*-propyl amine (Structure 1.44)?

3. The mass spectrum of a terpenoid with the sum formula $C_{10}H_{16}$ presented in Fig. 1.33 shows the base peak at $m/z = 68$ amu. Check whether the mass spectrum may be assigned to limonene (Structure 1.45). Formulate and name the fragmentation reaction for the base peak. Explain why Structure 1.46 can be excluded?

(continued)

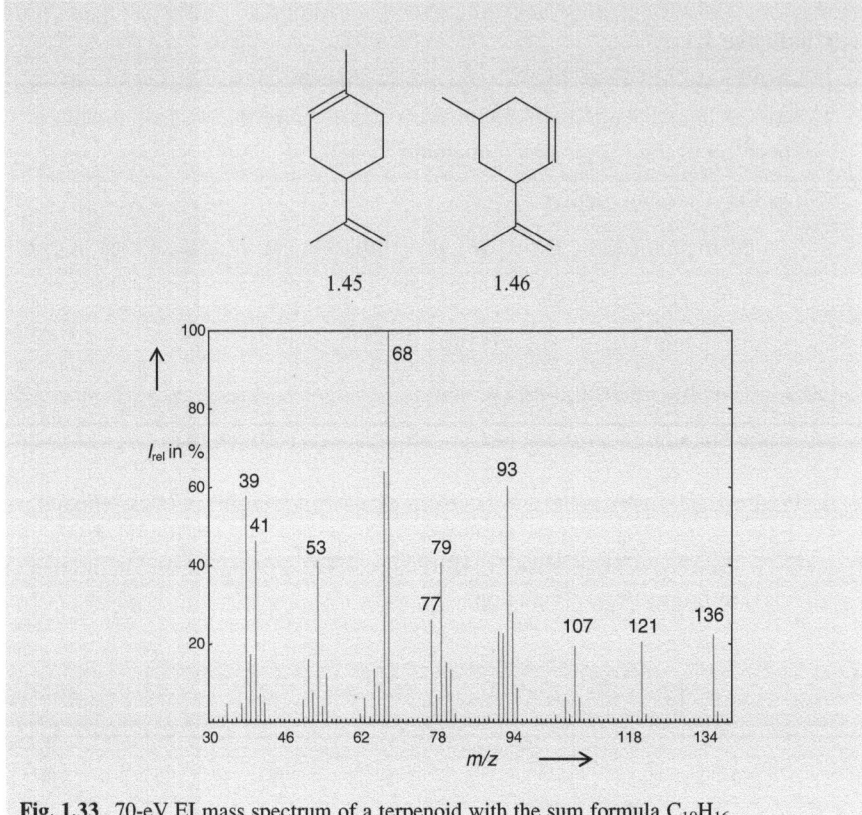

Fig. 1.33 70-eV EI mass spectrum of a terpenoid with the sum formula $C_{10}H_{16}$

1.6   Determination of the Molecular Structure

1.6.1   Evaluation of the General Pattern of a Mass Spectrum

The interpretation of a mass spectrum of an unknown compound should be started with its general evaluation which concerns, for example, the intensity of the molecular peak with its isotope peaks, the intensity of the [M-1] peak, the presence of few but very intense peaks, and the recognition of ion series. This will be illustrated by some examples.

Straight-chain, saturated aliphatic hydrocarbons provide a molecular ion peak, though one of low intensity; see for example the mass spectrum of decanes in Fig. 1.6a, b. The intensity of the molecular ion peak decreases at branched hydrocarbons and is very weak or missing.

The fragmentation pattern of saturated hydrocarbons or compounds possessing long-chain alkyl chains is characterized by clusters of peaks. They give rise to ion series of $m/z = 14\,n + 1$, accompanied by C_nH_{2n} and C_nH_{2n-1} fragments; see the mass spectrum in Fig. 1.7a. The most abundant fragments are obtained by C_3 and C_4 and the intensities decrease in a smooth curve to the molecular ion peak.

The mass spectra of branched hydrocarbons are grossly similar to those but the smooth curve of the decreasing intensities is broken by peaks of high intensity because of the preferred cleavage at the branched position; see the mass spectrum in Fig. 1.7b and the mass spectrum of octene-1 (Structure 1.20) in Fig. 1.19.

The mass spectra of alkenes are also dominated by peak clusters 14 mass units apart but the ion series C_nH_{2n} and C_nH_{2n-1} give more intense peaks than those of the C_nH_{2n+1} series.

Whereas the mass spectra of aliphatic compounds are characterized by weak, very weak, or missing molecular ion peaks (see Table 1.1) aromatic and heteroaromatic compounds show intense molecular peaks as long as groups with favorable fragmentation are not present, for example, long-chain alkyl groups. Furthermore, the aromatic pattern can be easily recognized by the presence of corresponding key fragments (see Table 6.4). The molecular ion peak is always the base peak in the unsubstituted aromatics (naphthalene, anthracene, and so on) with low intensity fragment peaks. A prominent peak at $m/z = 91$ amu indicates an alkyl-substituted benzene compound.

Few but very intense fragment peaks diagnose positions in the molecule which give rise to favorable fragmentations. Thus, for example, the intensities of all fragment peaks of nicotine (Structure 1.47) are smaller than 20% relevant to the base point $m/z = 84$ amu which is the resonance stabilized N-methyl tetrahydro pyrrole cation generated by an α cleavage.

Intense isotope peaks each separated from the next by 2 mass units indicate the presence of chlorine and/or bromine in the molecule. A weak [M + 2] peak is evidence for sulfur or silicon.

Unusually intense [M-1] peaks can be observed with alkines, nitriles, alkyl halogenides or compounds that possess resonance-stabilized onium ions by loss of the $H^{\bullet}$ radical, for example, aryl-aldehydes.

Relevant information important for structural analysis by mass spectrometry is summarized in Table 6.6 for some chemical classes.

1.6.2   Common Fragment Loss, [M-X] Peaks

[M-X] peaks originate from the loss of neutral species directly from the molecular ion peak. There are observed in the upper mass area. The loss of the neutral species X can only be indirectly recognized by the mass difference in relation to the molecular ion peak. The mass correlation table (Table 6.3) is a tool for the assignment of the eliminated species X. The [M-X] peaks enable the most simple and specific assignment of peaks in the mass spectrum and allow the recognition of *functional groups* which will be demonstrated by the example of the compounds shown in Structures 1.48 and 1.49.

The fragmentation of both molecule radical ions is started by α cleavage under generation of the resonance-stabilized benzoyl cation with $m/z = 105$ amu and leads to roughly the same fragmentation reactions. Thus, the fragment pattern of both molecules does not significantly differ. But the loss of $X = 15$ mass units in Structure 1.48 and the loss of $X = 31$ mass units in Structure 1.49 indicates the presence of a CH_3 and CH_3O group, respectively. The fragment peak at $m/z = 77$ amu generated by the following loss of CO (MN = 28) in both compounds indicates the functional groups $-C(=O)-CH_3$ and $-C(C=O)-OCH_3$, respectively. Note that the functional groups themselves cannot be directly recognized by corresponding fragment ions.

1.6.3   Common Specific Fragment Ions

Specific fragmentation ions in the lower mass range generated according to fragmentation rules are characteristic ions for the recognition of certain chemical classes and heteroatoms. Such characteristic ions (or clusters) are summarized as so-called key fragment ions in the mass correlation table (Table 6.4) and they are important for the recognition of structural moieties, heteroatoms, or functional groups.

Examples of characteristic ions and ion clusters are:

Aromatic compounds:
m/z in amu: 39 ($C_3H_3^+$), 51 ($C_4H_3^+$), 65 ($C_5H_5^+$), 91 ($C_7H_7^+$)
Monoalkylbenzenes: m/z in amu: 77 ($C_6H_5^+$), 78 ($C_6H_6^+$), 79 ($C_6H_7^+$)

Aliphatic amines	$R–CH_2–NH_2$:	$m/z = 30$ amu
	$R–CH_2–NHCH_3$:	$m/z = 44$ amu
	$R–CH_2–N(CH_3)_2$:	$m/z = 58$ amu

Thiols/thioethers: m/z in amu: 33 (SH^+ at thiols), 45 (CHS^+), 46 (CHS^+), 47 (CH_2SH^+)
CF_3^+: $m/z = 69$ amu
Phthalic esters (plasticizers) $m/z = 149$ ($C_8H_5O_3^+$) amu

---

1.6.4 General Steps for the Interpretation of Mass Spectra

– Consideration of all information about the sample: Chemical history, such as synthetic steps or other sources, for example, isolation from natural products, samples from the environmental analysis; possible by-products, such as plasticizers; the applied mass spectrometric techniques, such as EI, CI with a high- or low-energy reaction gas, quadrupole or ion-trap analyzer or others. Note that the ion-trap mass spectrometers provide incorrect intensities for the [M+X] peaks. Therefore, in comparison to quadrupole devices the determination of numbers of atoms is not possible with such spectrometers.
– Determination and protection of the molecular ion peak, if necessary, by applying the CI techniques or another soft ionization method. Bear in mind, the correct recognition of the molecular ion peak is an essential requirement in order to obtain proper information for the structural analysis.
– Determination of heteroatoms as well as the sum formula, if possible, under consideration of information obtained by other spectroscopic methods, such as the number of hydrogen and carbon atoms from NMR spectra or the presence of oxygen from infrared spectra and so on. Larger molecules require peak matching techniques through application of a high-resolution mass spectrometer for the determination of a correct sum formula. The sum formula enables the calculation of the double binding equivalents (DBE) which are important to determine the numbers of rings in the molecule.
– Evaluation of the general pattern of the mass spectrum according to aspects given in Sect. 1.6.1.
– Search for ionic series as well as characteristic ions (key fragment peaks) in the lower area and the [M-X] peaks in the upper area of the mass spectrum. Search for structural moieties and functional groups.
– Development of the molecular structure and checking whether the most intense peaks can be assigned to proper fragmentations. If possible, the molecular structure should be compared with a reference substance or a reference mass spectrum listed in the library of a modern spectrometer. Note that primary

thermal reactions or rearrangements of the primary generated molecular radical ion complicate the interpretation of mass spectra or can lead to false results. Thus, for example, cyclohexane shows an intense [M-15] peak though the molecule does not possess a methyl group.

Challenge 1.12
For solutions to Challenge 1.2, see *extras.springer.com*.

1. The computer-aided identification of mass spectra of unknown compounds by means of library mass spectra mostly offers more structural isomers. The decision as to the correct structure requires the interpretation of the corresponding mass spectrum. Some examples are given.
   Figure 1.34, 1.35, 1.36, and 1.37 present three mass spectra all with the same sum formula. Which molecular structure can be obtained from the mass spectra? Formulate the fragmentation reactions for the important fragmentations that are meaningful in terms of structure analysis.
2. Deduce a molecular structure from the mass spectra given in Fig. 1.38a, b, c, d, e, f, g, h, i, j, k, l, m, n. Determine the sum formula and check whether the molecular ion is an $OE^{+\cdot}$ ion. Calculate the double binding equivalents and assign the DBE to structural moieties. Evaluate the fragment peaks, formulate and name the relevant fragmentation reactions.

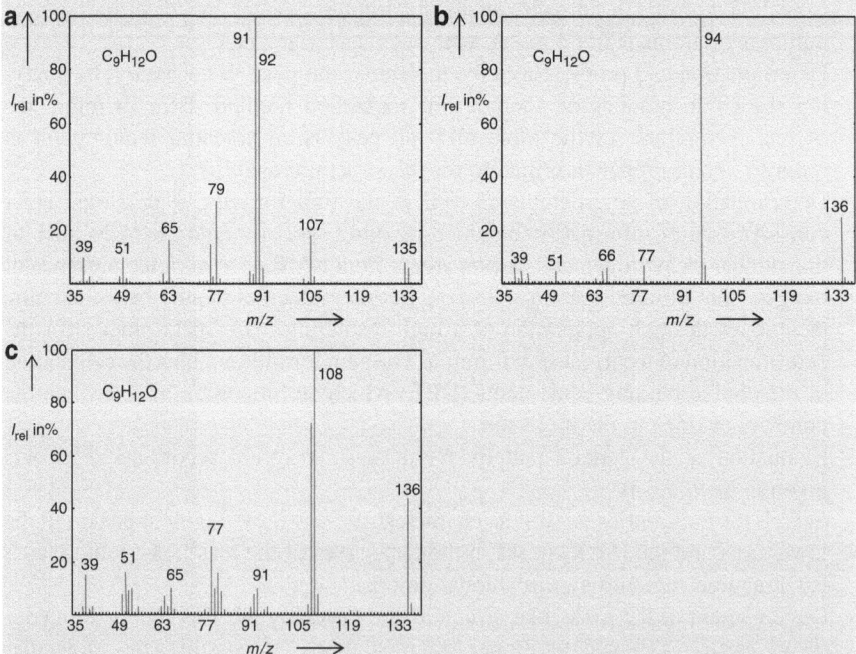

Fig. 1.34 (a–c) 70-eV EI mass spectra of unknown compounds with the sum formula $C_9H_{12}O$

(continued)

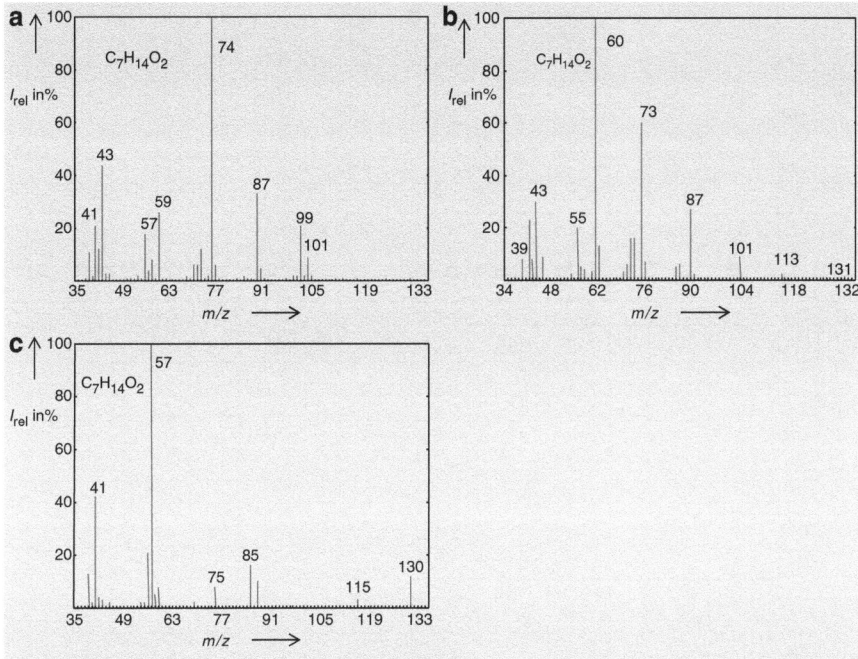

Fig. 1.35 (**a–c**) 70-eV EI mass spectra of unknown compounds with the formula $C_7H_{14}O_2$

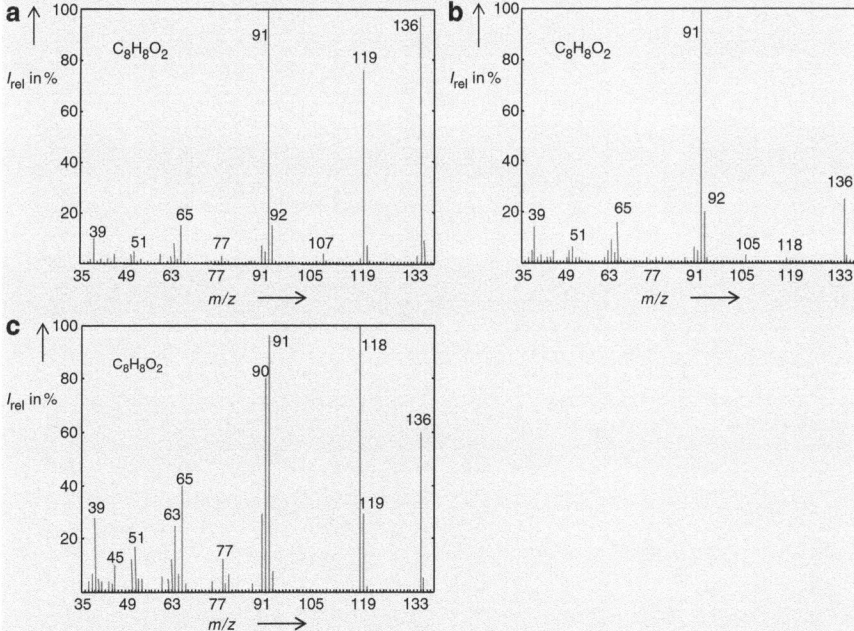

Fig. 1.36 (**a–c**) 70-eV EI mass spectra of unknown compounds with the formula $C_8H_8O_2$

(continued)

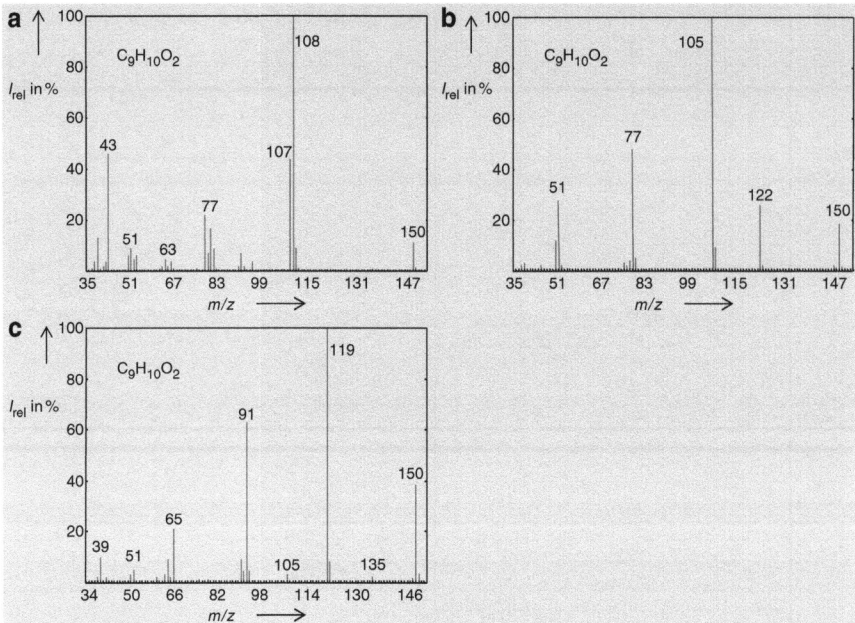

Fig. 1.37 (a–c) 70-eV EI mass spectra of unknown compounds with the formula $C_9H_{10}O_2$

(continued)

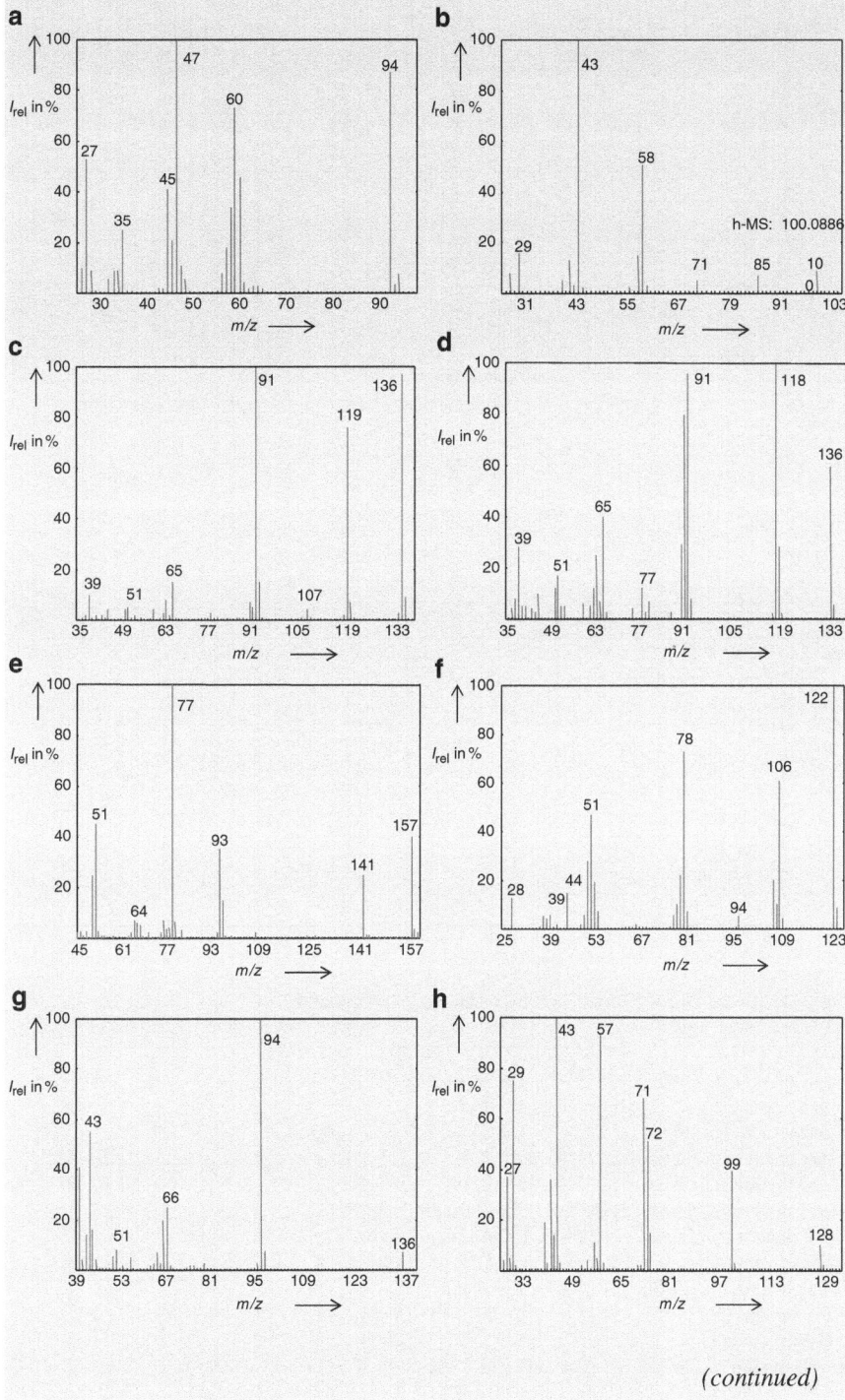

(continued)

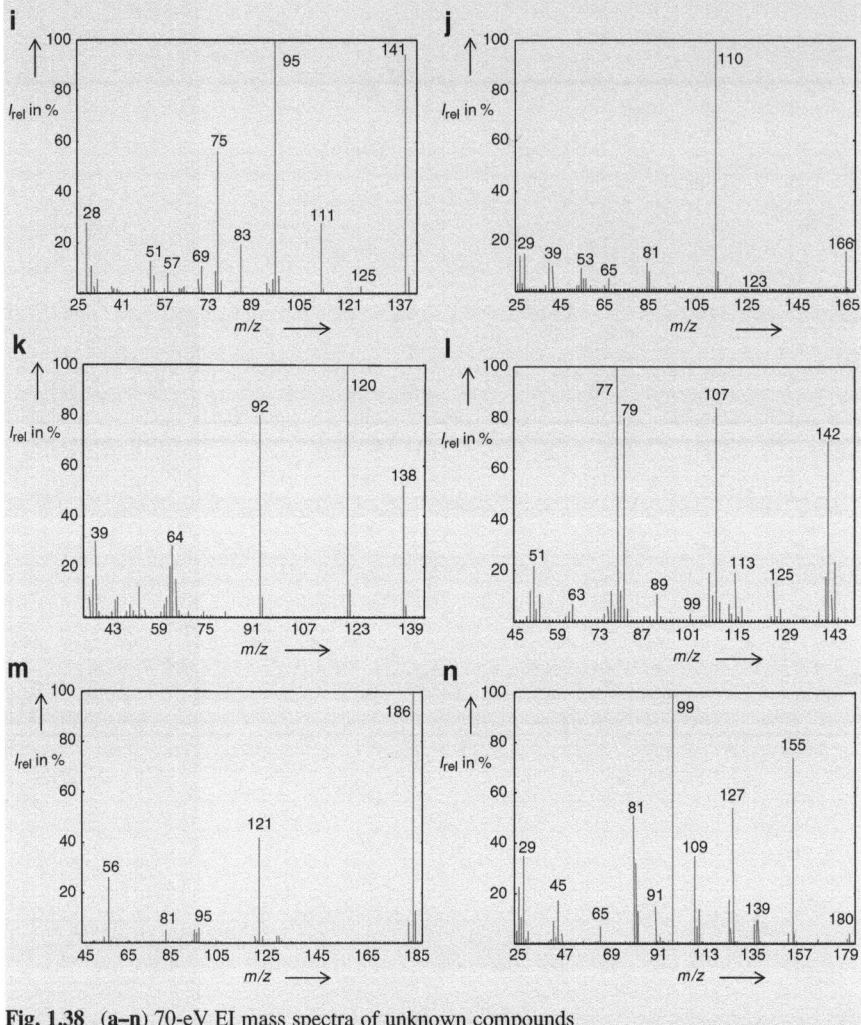

Fig. 1.38 (a–n) 70-eV EI mass spectra of unknown compounds

a. m/z in amu (I_{rel} in %): 94 (86.7), 95 (2.1), 96 (7.8)

c. m/z in amu (I_{rel} in %): 136 (97.2), 137 (8.8), 138 (0.7)

d. m/z in amu (I_{rel} in %): 136 (59.8), 137 (5.6)

e. m/z in amu (I_{rel} in %): 157 (39.8), 158 (3.7), 159 (2.1), 160 (0.2)

f. m/z in amu (I_{rel} in %): 122 (100), 123 (7.7)

g. m/z in amu (I_{rel} in %): 136 (7.0), 137 (0.6)

h. m/z in amu (I_{rel} in %): 128 (10.0) 129 (2.0), 130 (0.1)

i. m/z in amu (I_{rel} in %): 141 (94.1), 142 (6.7), 143 (0.6)

j. m/z in amu (I_{rel} in %): 166 (16.0), 167 (1.7), 168 (0.1)

k. m/z in amu (I_{rel} in %): 138 (52.3), 139 (4.5)

l. m/z in amu (I_{rel} in %): 139 (3.8), 140 (0.5), 141 (17.6), 142 (71.1), 143 (10.8),144 (23.4), 145 (1.8), 146 (0.1)

m. m/z in amu (I_{rel} in %): 182 (0.4), 183(0.1), 184 (8.3), 185 (1.6), 186 (100),187 (13.2), 188 (10.0)

Further Reading

1. Lee TA (1998) A Beginners Guide to Mass Spectral Interpretation, Wiley-VCH
2. McLafferty FW, Tureček F (2000) Interpretation von Massenspektren, Spektrum Akademischer Verlag Heidelberg
   Engl. Ed.: Interpretation of Mass Spectra, University Science Books, Mill Valley, CA, USA
3. deHoffmann E, Stroobant V (2002) Mass spectrometry – Principles and Application, J. Wiley
4. Budzikiewicz H, Schläfer M (2005) *Massenspektrometrie*, Wiley-VCH
5. Gross JH (2010) *Mass Spectrometry* Springer-Verlag

Chapter 2
Vibrational Spectroscopy

2.1 Introduction

2.1.1 Infrared (IR) Spectroscopy

The absorption of infrared (IR) radiation causes excitation of vibrations of the atoms of a molecule or the crystal lattice and causes bands in the spectra which are generally presented in the unit wave number $\tilde{v}$ in cm^{-1} (wavelength λ was used in the older literature). Commonly, the symbol v is used in vibrational spectroscopy instead of the symbol $\tilde{v}$ in the energy scale and they are sometimes named vibrational frequencies which are measured in the maxima of absorption bands whereas this is correct only for small oscillation strength.

The infrared range of the electromagnetic spectrum is divided into three regions, named after their relation to the visible spectrum:

- *Near-infrared* (NIR) (wave number ranges from 14,000 to 4,000 cm^{-1} and wavelength ranges from 0.8 to 2.5 μm) lying adjacent to the visible region excites so-called overtone or higher harmonic vibrations (*higher harmonics*).
- *Mid-infrared* (wave number ranges from approximately 4,000 to 400 cm^{-1} and wavelength ranges from 2.5 to 25 μm) excites mainly fundamental vibrations. This part of the infrared range may be used to study the structure of molecules that we shall be concerned with in this book. In general, the name "IR-spectroscopy" conventionally refers to the mid-IR region.
- *Far-infrared* (wave number ranges from approximately 400 to 10 cm^{-1} and wavelength ranges from 25 to 1,000 μm) is the lowest-energy region which excites mainly lattice vibrations or can be used for rotational spectroscopy.

A molecule can vibrate in many ways, and each way is called a *vibrational mode*. Let us start with a simple diatomic molecule A–B, such as CO or HCl. The wave number of absorbance v can be calculated by Eq. 2.1 derived by the model of the harmonic oscillator

M. Reichenbächer and J. Popp, *Challenges in Molecular Structure Determination*, DOI 10.1007/978-3-642-24390-5_2, © Springer-Verlag Berlin Heidelberg 2012

$$v = \frac{1}{2\pi c} \sqrt{\frac{f}{\mu}} \tag{2.1}$$

in which c is the light velocity, f is the force constant in the atomic scale and the spring constant for the macroscopic model, respectively, and μ is the reduced mass defined by Eq. 2.2

$$\mu = \frac{m_A \cdot m_B}{m_A + m_B}. \tag{2.2}$$

Note that the force constant is a criterion for the strength of the chemical bond in the molecule A−B. Therefore, the stronger the chemical bond (electronic effect) and the smaller the reduced mass μ (mass effect), the higher the wave number of the absorption band v.

The relation in Eq. 2.1 can be applied not only for the simple diatomic molecule but also for some structural moieties in more complex molecules as long as coupling between the atoms can be neglected, in other words, at so-called *characteristic vibrations* which are explained in Sect. 2.3.2. Such vibrations are provided, for example, by multiple bonds in the neighborhood of single bonds.

Electronic effects on the position of the absorption band

The force constant increases with the bonding order, for example, $f_{C-X} < f_{C=X} < f_{C\equiv X}$. Therefore, the vibrational frequencies increase for vibrations of equal atoms X according to Eq. 2.1:

$$v(C−C) \approx 1{,}000 \text{ cm}^{-1} < v(C=C) \approx 1{,}650 \text{ cm}^{-1} < v(C\equiv C) \approx 2{,}250 \text{ cm}^{-1}$$
$$v(C−O) \approx 1{,}100 \text{ cm}^{-1} < v(C=O) \approx 1{,}700 \text{ cm}^{-1}.$$

Note that Eq. 2.1 enables the recognition of the real strength of a chemical bond. For example, the hydrocarbon bond C−H is famously a *single* bond, but this is only a generalization. The neighborhood of oxygen at the carbon atom diminishes the electron density and, hence, the force constant and bonding strength, resulting in a smaller vibrational frequency according to Eq. 2.1. This is true for the following structural moieties: $v(\text{AlkylC}−H) \approx 3{,}000 \text{ cm}^{-1}$ and $v(O=C−H) \approx 2{,}750 \text{ cm}^{-1}$. Thus, the C−H-bond is slightly stronger in the alkyl group.

Let us now consider *I-* and *M-effects* on the vibrational frequencies.

The vibrational frequencies of the C=O double bond, $v(C=O)$ of compounds of the type $CH_3C(C=O)−X$ are: 1,742 cm^{-1} (for X = H), 1,750 cm^{-1} (for X = Cl), and 1,685 cm^{-1} (for X = phenyl). Thus, there are differences concerning the strength of the C=O bond due to the nature of the chemical bond: The −I-effect of the Cl atom diminishes the electron density at the C atom which is equalized by contribution of electron density from the free electron pair at the neighboring O atom. Thus, the bond order for the C=O bond is increased as is shown by the resonance structures $C−C(=O)Cl \leftrightarrow C−C\equiv O^+Cl^-$ which increases the force constant and, hence, a higher vibrational frequency is observed in carboxylic acid chlorides.

Otherwise, a substituent with a + M-effect (phenyl) causes conjugation of the C=O group as Structure 2.1 shows

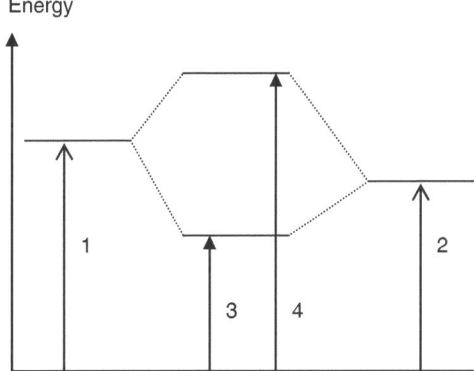

2.1

resulting in a smaller force constant and, hence, a smaller vibrational frequency.

Conjugation increases the thermodynamic stability of the molecule but decreases the bond strength and, hence, the force constant which results in smaller vibrational frequencies.

Conversely, the vibrational frequency obtained experimentally by the infrared spectrum provides hints to the strength of the chemical bond.

Note that coupling of various vibrations results in absorption bands of higher and lower frequency in the spectrum which are not caused by electronic effects. The transitions 3 and 4 caused by coupling of transitions 1 and 2 are shown in a simple energy diagram in Fig. 2.1.

Mass effects on the position of the absorption band

The smaller the mass of the atoms the higher the vibrational frequency, which is demonstrated by the C–H and C–D stretching frequencies,

$$v(C - H) \approx 3,000 \text{ cm}^{-1}; \ v(C - D) \approx 2,120 \text{ cm}^{-1}.$$

Fig. 2.1 Schematic energy diagram for the coupling of transitions 1 and 2 resulting in the transitions 3 and 4 with corresponding absorption bands in the infrared spectrum

The vibrational frequencies of heavier isotopes can markedly differ and they can occur with proper intensities for atoms with abundant isotopes.

As mentioned above, Eq. 2.1 is approximately valid for characteristic vibrations in multiatomic molecules. For such a structural moiety AB, for example, the C=O group of a carbonyl compound, the force constant f can be calculated according to the so-called "two-mass model" by Eq. 2.3

$$f \text{ in } N^{-1} = \frac{5.891 \cdot 10^{-5} \cdot v^2}{M_A^{-1} + M_B^{-1}} \qquad (2.3)$$

in which v is the wave number of the corresponding experimentally observed absorption band in cm^{-1}, and M_A and M_B are the relative atomic masses of the atoms A and B, respectively.

Challenge 2.1
Calculate the force constant of the P=O group in $OPCl_3$. The strong infrared absorption band of the P=O stretching vibration is observed at 1,290 cm^{-1}.

Solution to Challenge 2.1
The force constant is $f(P{=}O) = 1,035$ N m^{-1} calculated by Eq. 2.1 with $v = 1,290$ cm^{-1}, $M_P = 31$, and $M_O = 16$.

In the same manner the force constants can be calculated for carbon–carbon bonds and bonds in inorganic species resulting in the following rules:

– The relation of the force constants of the single, double, and triple bonds $f(C{-}C)$ ≈ 500 N m^{-1}, $f(C{=}C) \approx 1,000$ N m^{-1}, $f(C{\equiv}C) \approx 1,500$ N m^{-1} is 1:2:3.
– The force constant increases with increasing effective nuclear charge, for example, $f_{S\text{-}O}(SO_3^{2-}) = 552$ N m^{-1} $< f_{S\text{-}O}(SO_4^{2-}) = 715$ N m^{-1}.
– The force constant increases with increasing s-character of the hybrid orbital: $f_{C\text{-}H}(sp^3C{-}H) < f_{C\text{-}H}(sp^2C{-}H) < f_{C\text{-}H}(spC{-}H)$.

It is well known, however, that not all aspects of the atomic level can be described by the macroscopic ball-and-spring model. Instead the quantum mechanical formalism must be applied resulting in discrete energy levels

$$E_n = h \cdot v \cdot \left(v + \frac{1}{2} \right) \qquad (2.4)$$

in which v is the vibrational quantum number. According to this selection rule, only transitions are allowed with $\Delta v = \pm 1$. Because for most vibrations of organic molecules nearly exclusively the ground state $v = 0$ is occupied at ambient temperatures, the experimentally observed infrared absorption band of a two-atomic molecule is caused by the transition $v = 0 \rightarrow v = 1$ which is named the *fundamental vibration*.

Furthermore, the harmonic oscillator should be substituted by the *anharmonic oscillator model*. This substitution leads to a shape of the potential curve which is asymmetric and the right part of which converges to the dissociation energy. Therefore, the energy levels are no longer equidistant but decrease with increasing vibrational quantum number v. Moreover, the selection rule for the harmonic oscillator is no longer strongly valid; transitions with $\Delta v = \pm 2, \pm 3 \ldots$ can additionally occur, albeit with lower intensity, which are named *overtones* or *higher harmonics*. Thus, overtones are, in a first approximation, the whole number multiples of the fundamentals. Because of the anharmonicity of vibration the first overtone is somewhat smaller in frequency than the double value of the corresponding fundamental and so on.

The intensity of an absorption band is determined by the strength of the electro-magnetic absorption process which is mainly caused by the change of the electric dipole moment for a vibrational transition induced by infrared radiation. Note that the probability of a transition with $\Delta v = \pm 2$ is approximately tenfold smaller than that with $\Delta v = \pm 1$. The absorption coefficients α of fundamentals are in the range 1–100 L mol^{-1} cm^{-1}, therefore, overtones appear with low intensities or are missing in a routine spectrum.

In molecules with more than two atoms anharmonicity can additionally lead to *combination modes* which refer to the simultaneous excitation of two vibrational modes at one frequency which can lead to additional peaks and further complicate infrared spectra. Some of them are meaningful for structural analysis, for example, for the determination of aromatic substituent types (see Sect. 2.3).

Challenge 2.2

For solutions to Challenge 2.2, see *extras.springer.com*.

1. The experimentally observed N=O fundamentals $v(\text{N=O})$ of nitrosyl compounds O=NX are: 1,800 cm^{-1} (for X = Cl, gaseous state), 1,948 cm^{-1} (for X = Cl, solid state), and 2,387 cm^{-1} (for X = BF$_4^-$). Characterize the chemical bonding in these compounds and formulate the structure.

2. Calculate the force constants $f(\text{C=X})$ according to the two-mass model for the C=X bond for X = C and X = O, on the basis of the experimentally observed fundamentals: H$_2$C=O: $v(\text{C=O}) = 1,742$ cm^{-1}, H$_2$C=CH$_2$: $v(\text{C=C}) = 1,645$ cm^{-1}.

3. Calculate the force constant by means of the simple two-mass model for the C–H and C≡C bonds in H–C≡N with the experimentally observed fundamentals $v(\text{C–H}) = 3,311$ cm^{-1} and $v(\text{C≡C}) = 2,097$ cm^{-1}, respectively. The exact values obtained by a normal mode analysis are

(continued)

$f(C-H) = 582\,\text{N m}^{-1}$ and $f(C\equiv C) = 1{,}785\,\text{N m}^{-1}$. What is the percentage deviation between the exact and the approximate values? What knowledge can be gained from this result?

4. The stretching frequency of the ^{16}O-isotope is experimentally determined to be $\nu(^{16}O-^{16}O) = 832\,\text{cm}^{-1}$. At which wave number can the stretching frequency for the ^{18}O-isomer, $\nu(^{18}O-^{18}O)$ be expected?

2.1.2   Raman Spectroscopy and Some Applications

Raman spectra result from inelastic scattering of monochromatic light, usually from a laser in the visible, near-infrared, and near-ultraviolet range. Such lasers are for example: Ar ion with wavelengths at 514, 488, and 458 nm, He/Ne with wavelengths at 628, 578, and 442 nm, or Nd/YAG with wavelengths at 523 and 1,052 nm. Note that the last wavelength lies in the NIR range.

The Raman effect occurs when monochromatic light impinges upon a molecule and interacts with the electron cloud and the bonds of this molecule. For the spontaneous Raman effect to take place the molecule is excited by a photon from the ground state to a virtual energy state. The molecule can relax and return to the ground state by emission of a photon whose energy is the same as that of the exciting radiation resulting in elastic *Rayleigh* scattering. But a very small part of the scattered light can also have frequencies that are smaller than those of the elastically scattered part because a part of the energy of the incoming photons was employed to excite molecules to a higher vibrational state $v = 1$. As a result, the emitted photons will be shifted to lower energy, i.e., light with a lower frequency. This shift is designated the *Stokes shift*. The difference in energy between the original state and the final state corresponds to a vibrational mode far from the excitation wavelength.

If the process starts from a vibrationally excited state $v = 1$ and relaxes to the ground state $v = 0$, then the emitted photons will be shifted to higher frequency which is designated the *Anti-Stokes shift*. Compared with the Stokes-shifted light the Anti-Stokes shifted radiation has a lower intensity because of the small population of the vibrationally excited state compared to that of the ground state of the molecule at ambient temperatures. These processes are illustrated in a simple energy level diagram in Fig. 2.2. As Fig. 2.2 shows, the infrared absorption yields similar, but often complementary information. However, whereas the vibrational frequency directly corresponds to the absorption band in the infrared spectrum the vibrational frequency in the Raman spectra corresponds to the *difference* between Rayleigh and Stokes lines in the Raman spectrum. Note that the Raman effect is a scattering phenomenon and may not be confused with fluorescence whose emission starts from a discrete and not a virtual energy level (see Sect. 3.7).

Besides spontaneous Raman spectroscopy there are a number of advanced types of Raman spectroscopy, for example, surface-enhanced Raman (SERS), resonance

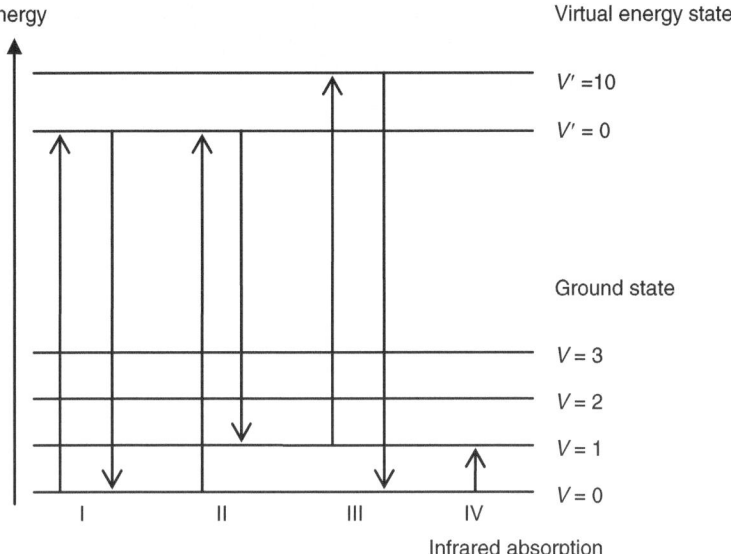

Fig. 2.2 Energy level diagram showing Rayleigh (**I**), Stokes (Raman) (**II**), and anti-Stokes (Raman) (**III**) scattering as well as IR-absorption (**IV**)

Raman spectroscopy (RRS), coherent anti-Stokes Raman spectroscopy (CARS), and others which are far beyond the scope of this book.

In principle, the registration of a Raman spectrum is simple: The sample is illuminated with an appropriate laser beam. The stray light is collected by a lens and sent through a monochromator. The elastic Rayleigh scattering light is filtered out while the rest of the collected light is dispersed onto a detector.

Although, in general, the organic chemist will prefer infrared spectroscopy for the registration of vibrational spectra, spontaneous Raman spectroscopy has a number of advantages; some of them will be given in the following text.

- The frequency of the excitation light can be chosen arbitrarily (within the limits of available laser frequencies). Thus, glass cuvettes can be applied for excitation with visible light instead of the expensive alkali discs used in infrared spectroscopy.
- Whereas the excitation of vibrations by infrared radiation is connected with changing the *electrical dipole moment*, in Raman spectroscopy only such vibrations are active that change the *induced dipole moment* (polarizability). Thus, polar bonds provide intense absorption bands whereas intense Raman lines are commonly observed from *nonpolar* bonds, such as C–C, C=C, C≡C, S–S, N=N, and others.
- Infrared and Raman spectroscopy are complementary for the investigation of symmetrical species, mostly present in inorganic chemistry, for the determination of *all* modes whereas there are also modes that cannot be excited either in IR or in Raman spectroscopy. Note that, according to the selection rules explained

in Sect. 2.1.3, symmetrical modes can only be observed in the Raman spectrum if the molecule possesses an inversion center.

- Because of the small intensities of the H–O stretching modes of water and alcohols, Raman spectroscopy is excellently suitable for alcoholic or aqueous solutions which is important, for example, in biological systems, such as the determination of the presence of S–S bridges in proteins, the symmetrical $O-P-O$ mode of nuclear acids, or for the in vivo denaturation of proteins by means of Raman microscopy.
- $C-C$ modes in rings can be better recognized and assigned because of their higher intensity in comparison to the infrared absorption bands.
- Raman spectra mostly show simpler patterns because of often-missing overtones and combination modes.
- Note that the Raman spectrum presents the *whole* range of molecular and lattice modes including those at low frequency $v < 200$ cm^{-1} for which an extra setup (far-infrared spectrometer with PE cuvette) may be necessary in infrared spectroscopy.

Examples for applications in the far-infrared range are investigations of:

- The low-frequency vibrations of metal–metal modes, such as, for example $v(Hg-Hg) = 169$ cm^{-1} in $Hg_2(NO_3)_2$, and $v(Mn-Mn) = 157$ cm^{-1} in $Mn_2(CO)_{10}$;
- Lattice modes in solids in order to characterize or to distinguish polymorphic forms;
- Stretching (v) and deformation (δ) modes of heavy atoms, for example, $[CdI_4]^{2-}$: 145 (v_{as}), 117 (v_s), 44 (δ_1), 23 (δ_2).

- The polarization of Raman scattered light also contains important information. This property can be measured using plane-polarized laser excitation and a polarization analyzer as shown in Fig. 2.3.

z-polarized excitation light induces a dipole moment in the molecule. Scattered light travelling in the x-direction is also z-polarized light $I_\parallel$ (**a**).

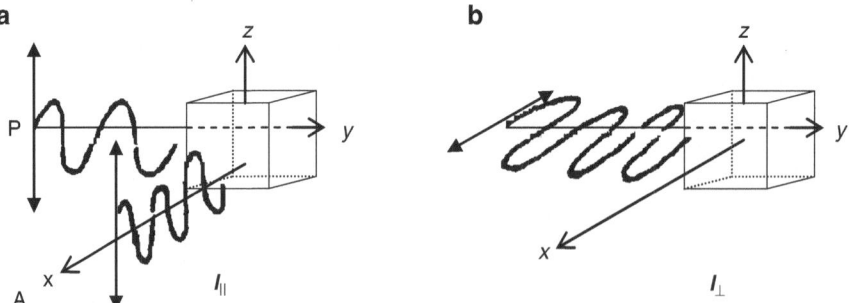

Fig. 2.3 Arrangement for measuring the polarization ratio of Raman scattered light. *P* polarizer, *A* analyzer, $I_\parallel$, $I_\perp$ intensity of Raman scattered light with the analyzer set parallel and perpendicular to the excitation plane, respectively

The polarization plane of the excitation laser light as well as the direction of the induced dipole is rotated around 90° in (**b**). Thus, in case of fully symmetric modes of optically isotropic substances, no z-polarized light is emitted travelling in the x-direction which means, the intensity $I_{\perp}$ is approximately zero.

Spectra acquired with an analyzer set both parallel (**a**) and perpendicular (**b**) to the excitation plane (P) can be used to calculate the *polarization ratio*, which is appropriate to experimentally recognize modes that are totally symmetrical (see Sect. 2.1.3). The degrees of polarization ρ are calculated by the ratio of the intensities measured employing the perpendicular and parallel analyzer set according to Eq. 2.5

$$\rho = \frac{I_{\perp}}{I_{\|}}. \qquad (2.5)$$

*Totally symmetrical modes provide polarized (**p**) Raman lines for which $0 < \rho < \frac{3}{4}$ is valid. All other modes are depolarized (**dp**) with $\rho \approx \frac{3}{4}$.*

Figure 2.4 shows the Raman spectra of carbon tetrachloride measured with an analyzer set both parallel ($I_{\|}$) and perpendicular ($I_{\perp}$) to the excitation plane and the polarization state is denoted by **p** and **dp**, respectively. As the degrees of polarization suggest, the Raman line at approximately 470 cm^{-1} is polarized. Thus, it must be assigned to the symmetric C–Cl stretching vibration. The connection between symmetry of modes, Raman activity, and the Raman spectra is the subject of the next section.

In addition to polarized IR spectroscopy, polarized Raman spectroscopy also can be used for example, in solid-state physics to determine the orientation of the crystallographic orientation of an anisotropic crystal and to understand macromolecular orientations in crystal lattices and liquid crystals. In most cases, however,

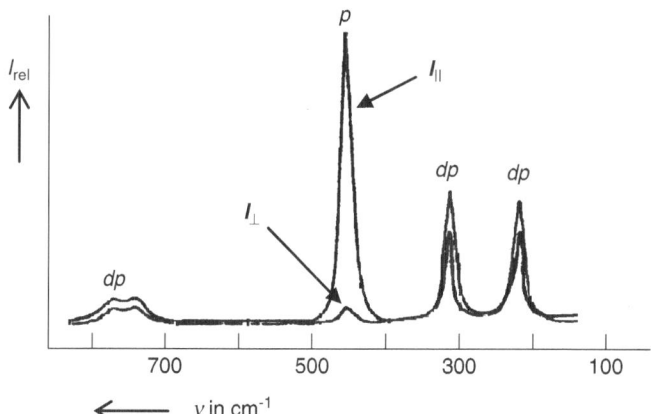

Fig. 2.4 The polarized Raman spectra of carbon tetrachloride measured with an analyzer set both parallel ($I_{\|}$) and perpendicular ($I_{\perp}$) to the excitation plane. p polarized Raman lines, dp depolarized Raman lines

the application is more complex than in IR spectroscopy. Both methods have their merits in polymorph analysis.

2.1.3   Symmetry, Selection Rules, and Applications

A nonlinear molecule with N atoms has $3 \cdot N - 6$ normal modes also called degrees of freedom in contrast to when the molecule is linear. Then only $3 \cdot N - 5$ normal modes are possible. As an example, the nonlinear molecule BF_3 has $3 \cdot 4 - 6 = 6$ degrees of freedom or normal modes while the linear CO_2 has $3 \cdot 3 - 5 = 4$ degrees of freedom.

There are two principal types of molecular vibrations: stretching modes and bending modes. A *stretching* vibration is characterized by movement along the bond axis with increasing and decreasing of the interatomic distance. Stretching vibrations are designated by the symbol ν. A *bending* vibration consists of a change of the bond angle between bonds or the movement of a group of atoms with respect to the remainder of the molecule with an accompanying change of bond angle. A movement *in-plane* is designated by the symbol δ and γ is used for *out-of-plane* movements (δ_{oop} vibrations). Note that normal mode vibrations may not change the center of gravity in the molecule.

The number of possible stretching vibrations is equal to the number of bonds and the number of bending vibrations is the difference of the total $3 \cdot N - 6(5)$ and the number of stretching modes.

Note that not only the "nonoscillating" molecule but also all modes of the molecule have a *symmetry* which means that they can be assigned to one of the symmetry classes of the point group to which every molecule belongs (for details, see the specialized literature on group theory). The algorithm for the determination of the point group of any molecule is presented in Table 6.7 and the symmetry classes of the normal modes are listed in Table 6.9 for the most important symmetry groups.

The principle of the determination of the symmetry properties of normal modes will be described by the example of a *planar* compound **AB₃** to which belongs, for example, the molecule BF_3 or the ion NO_3^-.

According to the procedure given in Table 6.7 the triangular planar species AB_3 belongs to the point group D_{3h}. In order to determine the symmetry classes of the fundamental modes, expediently, the base set of the so-called *inner coordinates* are used. In distinction from the Cartesian coordinates which include the total $3 \cdot N$ modes, inner coordinates only include the possible vibrations of the molecule. The inner coordinates of the stretching vibrations should be denoted by Δr_i, those of the bending in-plane by $\Delta \alpha_i$, and the bending out-of-plane by $\Delta \gamma_i$. The coordinates are defined in such a way that all possible movements of the molecule are included with the consequence that the bending movements are over-determined. Three angles cannot be simultaneously altered; the alteration of two angles fixes the third angle. Thus, more representations are obtained as alterations of angles are possible.

Therefore, without justification, a *totally symmetrical representation* Γ_1 must be eliminated.

Next, only the *character of the transformational matrix* has to be determined for the base set of the inner coordinates. The following rule should be applied:

Only that coordinate of the base set provides a contribution to the character of the transformational matrix that does not change its position at the execution of the respective symmetry operation.

By applying this procedure, one obtains a reducible base set of the transformation matrix which can be decomposed into the set of irreducible representations Z_i by means of the reduction formula (Eq. 2.6)

$$Z_i = 1/h \cdot \sum g_k \cdot \chi_r(K) \cdot \chi_{irr}(K) \qquad (2.6)$$

in which h is the order of the group (which is equal to the number of all symmetry operations), g_k is the number of the elements in each class K, and $\chi_r(K)$ and $\chi_{irr}(K)$ are the number of reducible and irreducible representations of the class K, respectively. The symmetry operations are listed in the *character table* which is given in Table 6.8 for some symmetry groups.

Let us now continue the problem for AB_3 given above. The molecule AB_3 has six normal modes: As the molecule has three bonds three stretching vibrations result. Therefore, three normal modes remain belonging to the bending modes which can be divided into in-plane (δ) and out-of-plane (γ) bending modes for a planar molecule. The base set of the inner coordinates divided as described are presented in Fig. 2.5.

The characters of the transformational matrix assigned to the three types of modes according to Fig. 2.5 are listed in Table 2.1. They are obtained by applying the rules given above.

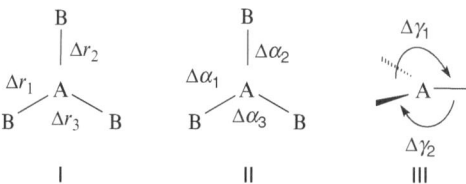

Fig. 2.5 Base set for the stretching (**I**), bending in-plane (**II**), and bending out-of-plane vibrations, for the triangular planar molecule AB_3 of the point group D_{3h}

Table 2.1 Characters of the transformational matrix assigned to the three types of modes (Fig. 2.5) for a triangular plane molecule AB_3 belonging to the point group D_{3h}

D_{3h}	E	$2\,C_3$	$3\,C'_2$	σ_h	$2\,S_3$	σ_v	Symmetry classes
$\chi(\Delta r_i)$	3	0	1	3	0	1	$\Rightarrow A_1' + E'$
$\chi(\Delta \alpha_i)$	3	0	1	3	0	1	$\Rightarrow A_1' + E'(-A_1')$
$\chi(\Delta \gamma_i)$	2	2	0	0	0	2	$\Rightarrow A_1'' + A_2''(-A_1')$

The last column contains the irreducible representations calculated by Eq. 2.6 with the reducible ones given in columns 2–7. A totally symmetrical class A_1' must be eliminated for the bending modes as explained above.

As Table 2.1 shows, the following six vibrations are obtained for a triangular planar molecule AB_3 with the symmetry D_{3h}. (Note that the symbol Γ is commonly used for the representation of a class).

$$\text{Stretching vibrations:} \qquad \Gamma_\nu = A_1' + E' \tag{2.7}$$

$$\text{Bending vibrations in - plane:} \qquad \Gamma_\delta = E' \tag{2.8}$$

$$\text{Bending vibrations out - of - plane:} \qquad \Gamma_\gamma = A_2''. \tag{2.9}$$

Although six vibrations are theoretically possible only four vibrations can be experimentally observed. The reason is that modes of the symmetry class E are twofold degenerated which means that both of the degenerated modes are observed as a *single* infrared absorption band or Raman line. Modes belonging to the symmetry class E consist of two energetic modes perpendicular to each other. Considering this fact, there are three stretching (Eq. 2.7), two bending in-plane (Eq. 2.8), and one bending out-of-plane (Eq. 2.9) vibrations, in summary six vibrations as calculated by $3 \cdot 4$–6.

The fundamental modes can be illustrated by so-called distortion vectors as is shown in Fig. 2.6 for the D_{3h} molecule AB_3. Note that the sum of these vectors may not change the center of gravity of the molecule as this would be caused by a translational motion.

Symmetrical stretching vibration, ν_s Asymmetrical stretching vibration, ν_{as}

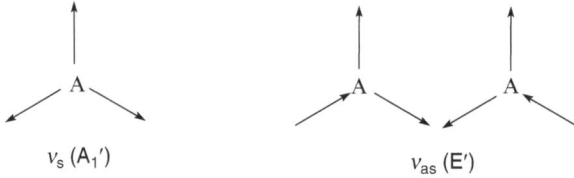

ν_s (A_1') ν_{as} (E')

Bending vibration in-plane   Bending vibration out-of-plane

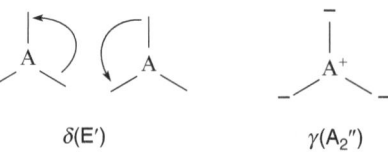

δ(E') γ(A_2'')

Fig. 2.6 Diagram of the fundamental modes of a molecule AB_3 with D_{3h} symmetry

The symmetry properties of molecules/ions of other symmetry groups can be obtained by the same procedure. The results for the most important structures are summarized in Table 6.9. Thus, the symmetry properties of the fundamental modes of a *pyramidal* molecule AB_3, for example NH_3, with the symmetry group C_{3v} are:

$$\text{Stretching vibrations:} \quad \Gamma_v = A_1 + E \qquad (2.10)$$

$$\text{Bending vibrations:} \quad \Gamma_\delta = A_1 + E. \qquad (2.11)$$

The question at hand is the following: Can one experimentally decide which of the two possible structures of the molecule AB_3 is present by means of the infrared and Raman spectroscopy: a planar structure with D_{3h} or pyramidal structure with C_{3v} symmetry? In order to answer this question the *selection rules* of vibrational spectroscopy must be employed.

2.1.3.1 Selection Rules in Vibrational Spectroscopy

A vibration is only infrared active, i.e., it can only be observed as an infrared absorption band if the corresponding vibration changes the electric dipole moment μ (note that a dipole moment need not exist in the equilibrium structure, but must then be built up by the vibration). For such a transition the intensity is determined by Einstein's transition moment $\vec{M}$ which is proportional to the transition moment integral in the volume $d\tau$

$$\vec{M} \sim \int \Psi_0 \, \vec{\mu} \, \Psi_e \, d\tau \qquad (2.12)$$

in which Ψ_0 and Ψ_e are the wave functions of the ground and the excited state, respectively and $\vec{\mu}$ is the dipole moment operator. The transition moment integral in Eq. 2.12 does not equal zero and hence, the transition has a certain probability if the integrant of the transition moment integral belongs to or contains the totally symmetric representation Γ_1.

In vibrational spectroscopy transitions are observed between different vibrational states. In a fundamental vibration the molecule is excited from its ground state ($v = 0$) to the first excited state ($v = 1$). Because the ground state belongs to the totally symmetrical representation the integrant of the transition moment integral contains only the totally symmetrical representation Γ_1 if the symmetry of the excited state wave function is the same as the symmetry of the transition moment operator. This fact gives rise to the selection rule in infrared spectroscopy.

2.1.3.2   Selection Rule for Infrared Spectroscopy

A vibration can be observed in the infrared spectrum, i.e., a vibration is infrared active, if the vibration is accompanied by a change in the electric dipole moment. This is the case if the symmetry of the respective vibration belongs to the same symmetry class as one of the dipole moment operators. Vibrations are infrared inactive if there is no change of the dipole moment with changing bond distance or angle.

The symmetry classes of the dipole moment operators are listed by x, y, and z, in the second to last column of the character table (Table 6.8) for the point group belonging to the molecule.

2.1.3.3   Selection Rule for Raman Spectroscopy and Alternative Forbiddance

The Raman effect requires a changing *induced dipole moment*, i.e., the electric polarizability which is the relative tendency of a charge distribution, like the electron cloud of an atom or molecule, to be distorted from its normal shape by an external electric field, while the molecule undergoes a vibration. The symmetry of the induced dipole moment components are listed as the binary products in the last column of the character table.

Raman active vibrations, i.e., vibrations which can be observed as Raman lines in the spectrum can only be such modes which are accompanied by a change of the induced dipole moment or the polarizability. Such vibrations belong to the same symmetry classes as the polarizability tensors listed in the last column of the character table as binary products.

Alternative Forbiddance

Vibrations of molecules or ions having an inversion center i are either infrared (asymmetric modes) or Raman active (symmetric modes).

2.1.3.4   Application of the Selection Rules

After having discussed the selection rules we can come back to the question if one can distinguish between the D_{3h} and C_{3v} structures for an AB_3 molecule using vibrational spectroscopy.

Table 2.2 shows the results of the application of the selection rules for both structures of AB_3-type molecules with D_{3h} and C_{3v} symmetry. The sign (+) means, that the vibration is allowed and (−) means that the vibration is forbidden.

According to Table 2.2, all six fundamental modes are active both in the IR- and the Raman spectrum for a molecule with pyramidal structure (C_{3v} symmetry). But

Table 2.2 Application of the selection rules to the vibrations of a molecule AB_3 having a triangular plane (D_{3h}) and a pyramidal structure (C_{3v}). The signs mean: (+): The transition is allowed and the vibration is observed, (–): the transition is forbidden and the vibration cannot be observed, and (p): the Raman line is polarized

D_{3h}	IR	Raman	C_{3v}	IR	Raman
$\nu_s(A_1')$	–	+ (p)	$\nu_s(A_1)$	+	+ (p)
$\nu_{as}(E')$	+	+	$\nu_{as}(E)$	+	+
$\delta(E')$	+	+	$\delta_s(A_1)$	+	+ (p)
$\gamma(A_2'')$	+	–	$\delta_{as}(E)$	+	+

only three bands and lines are observed in the infrared and Raman spectra, respectively for the planar structure D_{3h} where only two vibrations coincide. Thus, the application of vibrational spectroscopy can help to unambiguously distinguish between D_{3h} and C_{3v} symmetry of an AB_3 molecule or ion.

Note that sometimes the numbers of active vibrations are equal in the infrared and Raman spectra. In this case, the numbers of *polarized* Raman lines can be helpful in order to differentiate between possible structures. Bear in mind, only totally symmetrical vibrations of the symmetry classes A_1 or A_1' provide polarized Raman lines.

Challenge 2.3

For solutions to Challenge 2.3, see *extras.springer.com*

Determine the symmetry properties of the fundamental vibrations of molecules BAX_2 (for example, $O{=}CCl_2$ or $O{=}SCl_2$) with a planar and an angled structure. Apply the selection rules and answer whether both structures can be distinguished by means of vibrational spectroscopy.

2.1.4 Fermi Resonance

Most bands in the infrared spectrum arise from fundamental modes $v = 0 \rightarrow v = 1$, from overtones $v = 0 \rightarrow v = 2$, or from combination modes which involve more than one normal mode. There exist, however, absorption bands caused by a special combination, namely the Fermi resonance. The phenomenon of Fermi resonance can arise when two modes are similar in energy resulting in an unexpected shift in energy and intensity of absorption bands.

Two conditions must be satisfied for Fermi resonance:

- The two vibrational states of a molecule transform according to the same irreducible representation of the point group of the molecule.
- The transitions have accidentally almost the same energy.

Fermi resonance causes the high-energy mode to shift to higher energy and the low-energy mode to shift to even lower energy. Furthermore, the weaker mode gains intensity and the intensity of the more intense band decreases accordingly.

Thus, Fermi resonance results in a linear combination of parent modes and does not lead to additional bands in the spectrum.

Infrared absorption bands caused by Fermi resonance can be valuable for structural analysis. For example, two intense bands between 2,850 and 2,700 cm^{-1} are typical for the formyl group O=C–H caused by Fermi resonance of the overtone of the weak H–C=O bending mode at approximately 1,390 cm^{-1} with the H–C stretching mode. The occurrence of two (sometimes only one) bands in this region unambiguously indicates the presence of a formyl group in the molecule.

A further example is the double infrared absorption band of the C=O stretching vibration of aryl acid chlorides, Aryl–C(=O)Cl caused by coupling of the C=O stretching vibration with the overtone of the Cl–C=O bending modes, δ(Cl–C=O) which lies in the same range. Thus, aryl acid chlorides can be distinguished from alkyl acid chlorides by Fermi resonance. Further examples for Fermi resonances will be found in Sect. 2.3.

2.2 Interpretation of Vibrational Spectra of *Small* Molecules and Ions

The objective of the interpretation of vibrational spectra of small molecules and ions differs from that of organic molecules and can be mainly divided into three problems.

2.2.1 Determination of the Structure (Symmetry Group) of Molecules and Ions

The vibrational spectroscopic determination of the structure of a molecule or ion is realized by the following steps:

1. Determination of the symmetry (or point) group of a postulated structure by means of the algorithm given in Table 6.7.
2. Application of the selection rules on the fundamental modes the symmetries of which are taken from Table 6.9.
3. Assignment of the experimentally observed infrared absorption bands and Raman lines by means of the following rules:

   (a) The wave numbers of *stretching vibrations* are always higher than those of bending vibrations because the rhythmical movement along the bond axis requires higher energy than changing of the bond angle.
   (b) Consideration of the *force constants* and bond order, as well as the masses for the assignment of the stretching modes according to the relation in Eq. 2.1.

(c) Consideration of the *selection rules* and the *degrees of polarization* of the Raman lines.

(d) *Comparison of the vibrational spectra* of similar molecules or ions concerning structure, force constant, and masses. These are especially neighboring species in the periodic system of elements, for example, $AsCl_3/SeCl_3^+$.

(e) *Theoretical calculation* of the vibrational modes but this is outside the scope of this book.

When the experimentally observed data cannot be reasonably assigned to the possible fundamental modes a new structure must be postulated and the procedure described must be repeated.

Challenge 2.4

Let us turn again to a molecule of type AB_3. The infrared absorption bands and Raman lines listed in Table 2.3 are gathered from the respective spectra for the ions SO_3^{2-} and NO_3^-.

Table 2.3 Experimental IR-absorption bands and Raman lines of the ions SO_3^{2-} and NO_3^- given in cm^{-1} (p polarized Raman line)

SO_3^{2-}	IR	966	935	622	470
	Raman	967 (p)	933	620 (p)	469
NO_3^-	IR	1,370		828	695
	Raman	1,390	1,049 (p)		716

What structure do both ions possess?

Solution to Challenge 2.4

The ion of type AB_3 can possess a triangular planar structure D_{3h} or a pyramidal structure C_{3v}. The symmetry of the fundamental modes and the results of the selection rules are already listed in Table 2.2. As Table 2.2 shows the D_{3h} and C_{3v} structures can unambiguously be distinguished just on grounds of the number of the experimentally observed infrared absorption bands and Raman lines. All modes must be infrared and Raman active for the C_{3v} structure. This is realized for SO_3^{2-}. The ion NO_3^- belongs to the D_{3h} structure because only three infrared absorption bands and three Raman lines can be observed in the respective spectra.

But the justification of the proposed structures requires the assignment of *all* experimentally observed data. The two highest wave numbers must be assigned to the two stretching vibrations in the ions according to the rules given above. The polarized Raman line belongs to the symmetrical stretching vibration, thus the other one has to be assigned to the asymmetrical one. In the same manner, the symmetrical bending modes can be assigned by means of

(continued)

the polarization of the Raman lines. The reasonable assignment of all wave numbers summarized in Table 2.4 justifies the assumed structures for both ions.

Table 2.4 Selection rules and assignment of the infrared absorption bands and Raman lines of the experimentally observed wave numbers (cm^{-1}) to the theoretically possible vibrations of the ions SO_3^{2-} and NO_3^{-}

SO_3^{2-}				NO_3^{-}			
C_{3v}	IR	Raman	Assignment (IR/Raman)	D_{3h}	IR	Raman	Assignment (IR/Raman)
Stretching vibrations				Stretching vibrations			
$\nu_s(A_1)$	+	+	966/967	$\nu_s(A_1')$	−	+	−/1,049
$\nu_{as}(E)$	+	+	935/933	$\nu_{as}(E')$	+	+	1,390/1,390
Bending vibrations				Bending vibrations			
$\delta_s(A_1)$	+	+	622/620	$\gamma(A_2'')$	+	−	828/−
$\delta_{as}(E)$	+	+	470/469	$\delta(E')$	+	+	695/716

Vibrational spectra of materials in the *solid state* are determined by the symmetry of the crystal. When the symmetry of the crystal is lower than that of the free ion, then the following alterations are observed:

1. Splitting up of degenerate vibrations
2. Occurrence of modes that are symmetry-forbidden in the free ion.

Furthermore, interionical and intermolecular interactions in the crystal lattice cause shifts in comparison with free molecules and ions, respectively and additional translational and rotational modes of the lattice components appear in the lower frequency region ($\nu < 300$ cm^{-1}) and overtones as well as combination vibrations are observed in the general middle region. Thus, the spectra of solids are mostly characterized by more infrared absorption bands than observed for the free molecule and ion. The selection rules for vibrational spectra of solids are determined by the *site-symmetry* (see specialized literature). Band-rich infrared absorption spectra are provided by many inorganic salts with highly symmetrical ions, such as sulfates or carbonates.

Challenge 2.5

Note that the solutions can be found in *extras.springer.com*.

The sign "*p*" means that the respective Raman line is polarized and "Ra" is the abbreviation for Raman.

1. The structure of the molecules and ions is to be determined with the presented data (given in cm^{-1}) gathered from the infrared and Raman spectra following the steps explained above.

(continued)

(a)	BF_3	Ra	1,506	888 p	482			
		IR	1,505	719	480			
(b)	NO_2^-	Ra	1,323 p	1,296	827			
		IR	1,323	1,296	825			
(c)	NO_2^+	Ra	1,296 p					
		IR	2,360	570				
(d)	$VOCl_3$	Ra	1,035 p	504	409 p	249	164 p	131
		IR	All vibrations are infrared active					
(e)	$AlCl_4^-$	Ra	498	348 p	182	119		
		IR	496	FIR:	180			
(f)	$SO_3(g)$	Ra	1,390	1,065 p	530			
		IR	1,391	529	495			
(g)	AsF_5	Ra	809	733 p	642 p	388	366	123
		IR	810	784	400	365	FIR:	122
(h)	$S_2O_3^{2-}$	Ra	1,123	995 p	669 p	541	446 p	
		IR	1,122	995	667	543	445	
(i)	HSO_3^-	Ra	2,588 p	1,200	1,123	1,038 p	629 p	509
		IR	2,585	1,201	1,122	1,038	628	510

2. The stretching modes of $[PtCl_4]^{2-}$ are (given in cm^{-1}):

Ra		330 p		312
IR		313		

   Explain why this ion cannot have a tetrahedral structure? What structure can be proposed from these data?
3. Determine the structure of the ion $[AsF_4]^-$ and assign the infrared
   absorption bands and Raman lines (given in cm^{-1}):

Ra	829	745 p	272	213
IR	830	270		

4. Deduce the structure of PCl_5 in the gaseous and solid state and assign all
   infrared absorption bands and Raman lines (given in cm^{-1}).
   Solid state

Ra	662	458 p	360 p	283	255	238	178
IR	661	444	285	254			

   Gaseous state and in a nonaqueous solution

Ra	580	392 p	281 p	272	261	102
IR	583	443	300	272	101	

(continued)

5. Derive the structure of PF_5 from the data obtained by the infrared and Raman spectra (given in cm^{-1}). Assign all infrared absorption bands and Raman lines.

Ra	1,026	817 p	640 p	542	514	300
IR	1,025	944	575	533	302	

6. Develop a proposal for the structure of the ion $O_2PH_2{}^{2-}$ based on the data (given in cm^{-1}) gathered from the Raman spectrum. Assign the Raman lines to the theoretical modes if this is unambiguously possible. Note that the vibration that corresponds to the Raman line at 930 cm^{-1} is infrared inactive.

Ra	2,365 p	2,308	1,180	1,160 p	1,093	1,046 p	930	820	470 p

7. What structure can be derived from the data of the Raman spectrum (given in cm^{-1}) for NSF_3? Note that all vibrations are also infrared active. Assign the Raman lines to the corresponding modes of the molecule.

Ra (liquid)	1,523 p	815	773 p	525 p	432	346

8. The reaction of PCl_5 with $AlCl_3$ gives rise to a product for which the atomic ratio $Al : P : Cl = 1 : 1 : 8$ is analytically obtained. The structure is to be derived on the basis of the data gathered from the infrared and Raman spectra (given in cm^{-1}). Assign all bands and lines.

Ra	662	498	458 p	348 p	255	182	178	120
IR	665	497	258		FIR:	181		

9. Derive the structure of an arsenic compound with the sum formula $AsCl_2F_3$ and assign the infrared absorption bands and Raman lines (given in cm^{-1}).

Ra	689 p	573	500	422 p	375	187	157
IR	700	500	385		FIR:	186	

10. The addition of a solution of $SeCl_4$ in dioxane to a solution of $AlCl_3$ in dioxane gives rise to a solid product for which the atomic ratio $Al : Se : Cl = 1 : 1 : 7$ is analytically obtained. Derive the structure of this reaction product from the data of the infrared and Raman spectra (given in cm^{-1}). Assign the infrared bands and Raman lines. Note that $KAlCl_4$ shows lines at 182 and 119 cm^{-1} in the region $v < 200$ cm^{-1} in the Raman spectrum.

Ra	498	416 p	395	348 p	294 p	186	182	119
IR	497	415	396	296	FIR:	Not detected		

(continued)

11. A platinum compound has the analytical composition $PtCl_4 \cdot 2H_2O \cdot 2HCl$. Develop a proposal for the structure of this solid reaction product on the basis of the infrared absorption bands and the Raman lines (given in cm^{-1}). Assign the corresponding bands to the theoretical modes.

Ra	3,226 p	2,825	1,695	1,070 p	348 p	318	171
IR	3,225	2,825	1,694	1,068	342	183	

12. Compose a table with the results of the selection rules for the infrared absorption bands and Raman lines of the fundamental modes of the structural moiety $X–CH_3$. Assign the experimentally observed data for $X = Cl$ (given in cm^{-1}).

Ra	3,041	2,966 p	1,456	1,355 p	1,017	734 p
IR	3,042	2,967	1,455	1,354	1,015	732

Which modes will provide relatively constant values of the wave numbers at variation of X and which modes will show a great range for the observed wave numbers? Explain your findings.

13. Derive the structure of the two compounds $OXCl_2$ with $X = C$ and $X = S$, based on the data of the vibrational spectra. Assign the infrared absorption bands and Raman lines (given in cm^{-1}) to the respective modes of each molecule.

(a) X = C

Ra	1,827 p	849	580	569 p	440	285 p
IR	1,826	849	568	441	284	

(b) X = S

Ra	1,251 p	492 p	455	344 p	284	194 p	
IR	1,250	490	454	342	284	FIR:	Not detected

14. The totally symmetrical vibration v_1 in the fluoric complexes of the third main group of the type $[MeF_6]^{3-}$ with Me = Al, Ga, In, Tl is observed in the Raman spectrum at 478, 498, 535, and 541 cm^{-1}, respectively ordered according to increasing wave numbers. Assign v_1 to the respective Me–X mode. Why is the totally symmetrical vibration at best suitable in order to evaluate the bond strength via the force constant?

15. The totally symmetric vibration v_1 in the fluoric complexes of vanadium of the type $[VF_6]^{n-}$ with $n = 1$, 2, and 3 is observed at 533, 584, and 676 cm^{-1}, respectively. Assign v_1 to the respective complex ion and justify your decision.

16. Perovskites of the type A_2BMO_6 can be divided according to their $[MO_6]$-structure moiety into three groups:

(a) Perovskites with an undistorted $[MO_6]$-octahedron
(b) Perovskites with two different undistorted $[MO_6]$-octahedrons
(c) Perovskites with a distorted $[MO_6]$-octahedron.

(continued)

Fig. 2.7 Line diagram of the
Raman spectrum of three
perovskites. "*p*" means that
the respective Raman line is
polarized

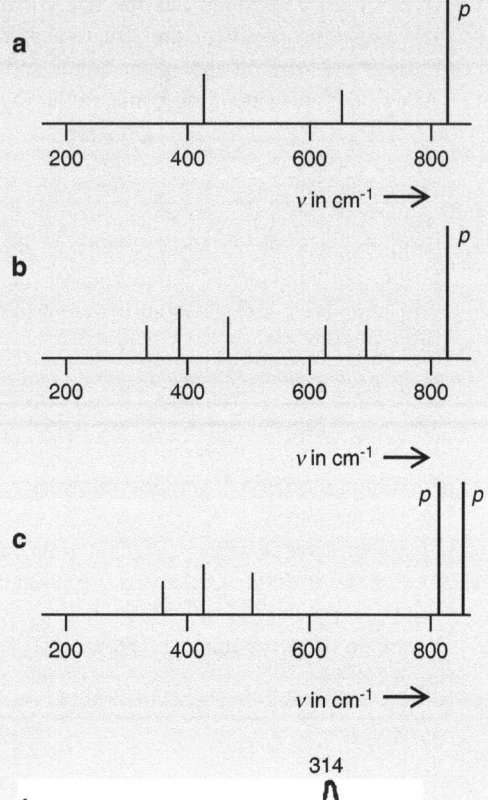

Fig. 2.8 Raman spectra of
the [FeS$_4$] structural moiety in
a peptide obtained by
polarized light

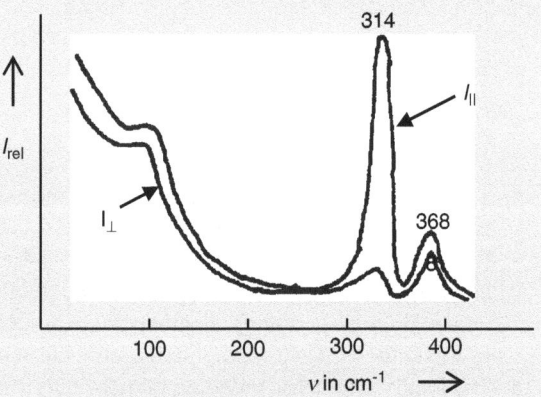

Assign the line diagrams of the Raman lines for three perovskites shown in
Fig. 2.7 to the three types of perovskites and justify your decision.

17. The [FeS$_4$] structure moiety of a ferric-containing peptide can be *planar* or
tetrahedral. Explain why *Raman* spectroscopy must be applied for solving this
problem. The Raman spectra obtained by polarized light are presented in
Fig. 2.8. What structure has the [FeS$_4$] moiety? Assign the stretching

(continued)

Table 2.5 Infrared absorption bands and Raman lines of the polymorphic forms of $CaCO_3$: Calcite and Aragonite

Calcite					
IR	≈1,460 (broad)	879	706		
Ra	1,432	1,087	714		
Aragonite					
IR	1,504	1,492	1,080	866	711	706
Ra	The lines correspond to the IR absorption bands.				

vibrations. Note the empirical fact that $v_s > v_{as}$ is valid for the D_{4h} structure, for the tetrahedral structure the reverse relation is observed.

18. The two degenerated bending vibrations of the *free* SCN ion are observed at 470 cm^{-1} in solutions as an asymmetrical absorption band. The same mode is split into two absorption bands at 486 and 471 cm^{-1} in the infrared spectrum of *solid* KSCN. What is the reason behind this?

19. There are two structures for $CaCO_3$: Calcite and aragonite. The data obtained from the vibrational spectra in the region above 400 cm^{-1} are listed in Table 2.5.

Assign the infrared absorption bands and Raman lines. Compare qualitatively both polymorphic forms of $CaCO_3$.

2.2.2   Coordination of Bifunctional Ligands at the Central Atom in Complex Compounds

The thiocyanate ion SCN$^-$ is a bifunctional ligand because it can be coordinated via the S and the N atom to a central atom in a complex compound. Vibrational spectroscopy can help to distinguish between both structures and this is explained in the following.

A linear three-atomic ion provides $3 \cdot 4-5 = 4$ fundamental modes, which are divided into two stretching and two bending vibrations. But the infrared spectrum of the free SCN$^-$ ion only has three absorption bands because the two bending modes at 470 cm^{-1} are degenerated. The infrared absorption bands at 2,053 cm^{-1} and that at 748 cm^{-1} can be roughly assigned to $v(C\equiv N)$ and $v(C-S)$, respectively. But the "real" $v(C\equiv N)$ obtained for nitriles lies in the range from 2,260 to 2,240 cm^{-1} and the "real" C–S stretching vibration is observed at approximately 680 cm^{-1}, see Sect. 2.3. This means, the C≡N bond is weaker than a threefold bond and the C–S bond is stronger than a single bond; thus, the chemical structure must be described by two resonance structures I and II, respectively (see Structure 2.1):

$$S\!\!=\!\!=\!\!C\!\!=\!\!=\!\!N(I) \quad \longleftrightarrow \quad S\!\!-\!\!C\!\!\equiv\!\!N(II)$$

2.1

Coordination to the *S atom* favors the resonance structure II in Structure 2.1:

$$\mathbf{Me} \leftarrow \mathbf{S - C \equiv N} \leftrightarrow \mathbf{Me} \leftarrow \mathbf{S = C = N}.$$

The wave numbers of v(C–S) should be smaller and that of v(C≡N) should be higher in a complex of the type Me←S–C≡N in comparison to the free ligand.

However, coordination to the *N atom* favors resonance structure I in Structure 2.1:

$$\mathbf{Me} \leftarrow \mathbf{N = C = S} \leftrightarrow \mathbf{Me} \leftarrow \mathbf{N \equiv C - S}.$$

The wave number of v(C–S) should be increased and the wave number of v(C≡N) should be decreased in a complex of the type Me←N=C=S in comparison to the free ligand.

Deviations from these rules can be caused by coupling of vibrations.

Challenge 2.6

The stretching modes of the SCN⁻ ligand in two thiocyanate complexes are:

$\left[\mathbf{Hg(SCN)_4}\right]^{2-}$: 2120   710 cm⁻¹ $\left[\mathbf{Fe(SCN)_6}\right]^{3-}$:   2055   828 cm⁻¹

Judge whether the ligand coordinates via S or N to the central atom.

Solution to Challenge 2.6

$$\left[\mathbf{Hg(SCN)_4}\right]^{2-}$$

Because the wave number of v(C≡N) for the complex is higher than that for the SCN⁻ ion and the wave number of v(C-S) for the complex is smaller than that for the SCN⁻ ion the coordination occurs via **S: Hg←SCN**.

$$\left[\mathbf{Fe(SCN)_6}\right]^{3-}$$

Because the following facts are valid: v(C≡N) (complex)<v(C≡N) (SCN⁻) and v(C-S) (complex)>v(C-S) (SCN⁻) the coordination occurs via **N: Fe←SCN**.

2.2.3   Investigations of the Strength of Chemical Bonds

Vibrational spectroscopy offers the possibility to investigate chemical bond strengths because the stretching modes are determined by the force constant according to Eq. 2.1 which is proportional to the bond order. This fact will be explained employing the example of nitrate complexes $[Me(NO_3)_n]^{m-}$.

The free ligand NO_3^- shows two stretching vibrations, the symmetrical $v_s(N-O)$ which is infrared inactive and the degenerated asymmetrical $v_{as}(N-O)$ which is observed at 1,390 cm^{-1} in the infrared spectrum. Coordination of the ligand to the metal ion via O diminishes the D_{3h} symmetry of the free NO_3^- ion with the result that the degenerated stretching mode is split. Furthermore, the symmetrical stretching mode is no longer forbidden. Thus, the following stretching modes are infrared active for a metal complex with a structure moiety shown in Structure 2.2

$$v(N-O) = v_1, v_s(NO_2) = v_2, \text{ and } v_{as}(NO_2) = v_3.$$

2.2

The equality of the three N–O bonds in the free NO_3^- ion is no longer present because of the metal-oxygen bond $Me\leftarrow O-NO_2$. Thus, the chemical bond in the ligand can be described by a (mainly) single N–O bond and two equal N=O bonds with conjugated double bonds which provide strongly coupled symmetrical, $v_s(NO_2)$ and asymmetrical, $v_{as}(NO_2)$ modes, respectively. Because the delocalization of the electrons is diminished to only two bonds the strength of the chemical N=O bond and hence, the wave numbers of the stretching modes of the NO_2 group increase more, the stronger the bond of the ligand to the metal ion is. In general, the mean value of the frequencies of the symmetrical and asymmetrical NO_2 stretching modes is used to qualitatively evaluate the strength of the chemical bond of the ligand to the metal ion. Furthermore, the splitting up of the degenerate N=O stretching mode will be all the greater the stronger the complex bond is. The maximum value is achieved at a "real" covalent bond, i.e., if the metal ion is substituted by H or an alkyl group; see, for example, the vibrational frequencies of HNO_3 in Table 2.6.

The comparison of stretching modes of the free ligand with those of a compound possessing a covalent bond of the ligand and those of the complex compound enables the evaluation of the strength of the chemical bond of the ligand to the central atom.

Challenge 2.7

Table 2.6 presents the stretching modes of two nitrate complexes and HNO_3.

Decide whether NO_3^- coordinates to each of the metal ions. If this is correct, evaluate qualitatively the strength of the $Me{\leftarrow}O{-}NO_2$ bond.

Table 2.6 Stretching modes of two nitrate complexes and HNO_3

	v_1	$v_2 = v_s(NO_2)$	$v_3 = v_{as}(NO_2)$	$\Delta v = v_3 - v_2$	$\bar{v}(v_2, v_3)$
$[Cu(NO_3)_4]^{2-}$	1,013	1,290	1,465	175	1,377.5
$[Zr(NO_3)_6]^{2-}$	1,016	1,294	1,672	378	1,428.5
$HONO_2$	920	1,294	1,672	378	1,483

Solution to Challenge 2.7

The degeneration of the asymmetrical stretching vibration of the free NO_3^- ion is no longer valid. Instead it is split into the symmetrical and asymmetrical NO_2 stretching vibrations. Furthermore, IR-forbidden symmetrical stretching is allowed in both complexes. Thus, the D_{3h} symmetry is no longer given because of coordination of the NO_3^- ligand to the metal ions.

To qualitatively evaluate the coordinative bond strength the mean values calculated by the wave number of the symmetrical and asymmetrical $N{-}O$ stretching modes are compared with that of HNO_3 in which the $H{-}O$ bond is a "true" covalent one. As the wave numbers in the last column of Table 2.6 show the coordinative bond is stronger in the Zr complex than in the Cu one because the mean value of the NO_2 stretching modes of the Zr complex is higher than that of the Cu complex and is closer to the value for a strong covalent bond.

Note that the procedure for the qualitative evaluation of the strength of a coordinative chemical bond described above can only be applied with true characteristic vibrations whose vibrational frequencies are influenced solely by electronic effects and not by couplings of vibrations.

Challenge 2.8

For solutions to Challenge 2.8, see *extras.springer.com*.

1. $K_3[Co(CN)_5(NCS)]$ can appear in two isomeric forms which give rise to the following infrared absorption bands:

Isomer I, v in cm^{-1}	2,065	810	Isomer II, v in cm^{-1}	2,110	718

Draw up and justify a proposal of the structure for the two isomeric forms of the Co(III) complex.

2. The following infrared absorption bands are observed for the complex Pd (bipyr)(NSC)$_2$ in the region of the NCS stretching modes:

2,117	2,095	842	700	cm^{-1}

Draw up and justify a proposal of the coordinative chemical bond in this complex compound.

3. Figures 2.9 and 2.10 show sections of the infrared spectra of the selenium compounds $(C_6H_5)_2Se(NO_3)_2$ and $(C_6H_5)_3SeNO_3$, respectively. Investigate whether nitrate is preferably covalently bonded to the selenium atom or whether an ionic structure with an NO_3^- ion is valid.

4. The mean values of the symmetrical and asymmetrical stretching vibrations of NO_2, NO_2^+, and NO_2^- ordered according to increasing wave numbers are: 1,307, 1,468, and 1,878 cm^{-1}. Assign and justify the wave numbers to the corresponding nitrogen compounds.

5. Evaluate the strength of the coordinative chemical bond of the oxalate ligand to the metal ions Pt and Cu on the basis of the C=O stretching vibrations. The observed infrared absorption bands of these complex compounds as well as those of the noncoordinate, free oxalate ion and covalently bonded oxalic acid dimethyl ester are summarized in Table 2.7.

6. The stretching frequencies of manganese compounds of the type K_nMnO_4 with $n = 1, 2$, and 3 are arbitrarily listed in Table 2.8. Assign the infrared absorption bands and Raman lines to the respective manganese compounds. Explain why one can recognize a relation between the wave

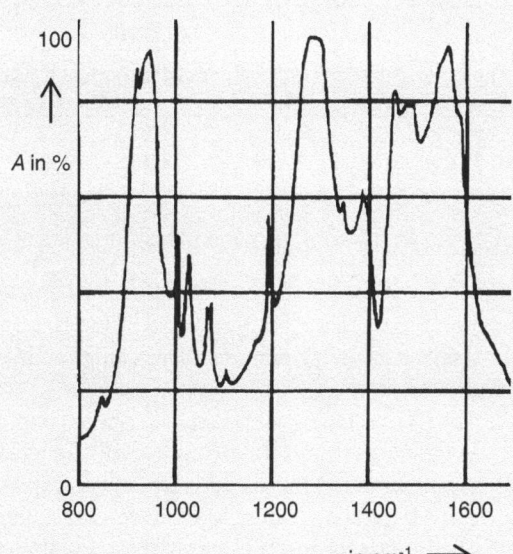

Fig. 2.9 Infrared spectrum in the range of the NO stretching modes of $(C_6H_5)_2Se(NO_3)_2$; sampling technique: Nujol mull

(continued)

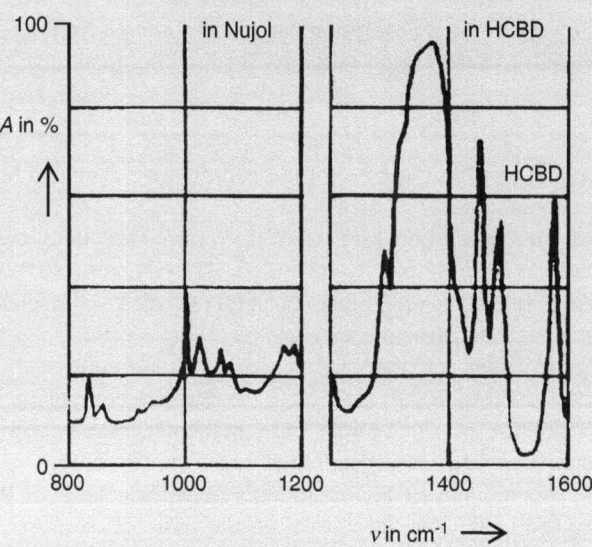

Fig. 2.10 Infrared spectrum in the range of the NO stretching modes of $(C_6H_5)_3Se(NO_3)$; sampling techniques used: Nujol mull (*left*) and hexachlor-butadiene (HCBD) (*right*)

Table 2.7 C=O stretching frequencies of the oxalate ion, oxalic acid dimethyl ester, and some oxalate complexes (given in cm^{-1})

$C_2O_4{}^{2-}$	1,640	1,650	$C_2O_4(CH_3)_2$	1,770	1,796
$[Pt(C_2O_4)_4]^{2-}$	1,674	1,709	$[Cu(C_2O_4)_4]^{2-}$	1,645	1,672

Table 2.8 Arbitrarily ordered infrared absorption bands and Raman lines of manganese compounds of the type K_nMnO_4

Compound I:	Raman:	820	812	332	325	cm^{-1}
	IR:	821	333			cm^{-1}
Compound II:	Raman:	902	834	386	346	cm^{-1}
	IR:	901	384			cm^{-1}
Compound III:	Raman:	789	778	332	308	cm^{-1}
	IR:	779	331			cm^{-1}

number of the symmetrical stretching modes and the effective charge of the Mn atom?

7. The CO stretching vibration of C≡O can be found at 2,155 cm^{-1} but it lies in the range from 2,100 to 1,800 cm^{-1} in metal carbonyl compounds. The chemical bond in metal carbonyl compounds is a superposition of a σ-Me-C bond and a Me-(CO)-π back donation. What information is obtained by the mean values of the C=O stretching modes according to the part of both contributions to the chemical bond in the following charged

(continued)

metal carbonyl ions? The wave number of $Ni(CO)_4$ may be used as the reference compound. All values are given in cm^{-1}.

$$Ni(CO)_4(2094), \left[Co(CO)_4\right]^{-}(1946), \left[Fe(CO)_4\right]^{2-}(1788), Pt(CO)_4]^{2+}(2258).$$

2.3 Interpretation of Vibrational Spectra of *Organic* Molecules

2.3.1 *Objectives and Special Features*

The interpretation of vibrational spectra mainly includes the following objectives

* Determination of *structural moieties* (alkyl, alkenyl-, alkin-, and aryl groups)
* Determination of *functional groups*
* Determination of *isomers*
* Investigation of *conformers*
* Investigation of *hydrogen bridges* (which is only possible by means of infrared spectroscopy)

In contrast to the interpretation of vibrational spectra of inorganic species symmetry properties of modes are, generally, not of interest.

The occurrence of overtones and combination vibrations increases whereas randomly degenerated vibrations decrease the number of the $3 \cdot N{-}6$ possible modes. Unlike the degeneration of modes on the basis of symmetry, *randomly degenerated* vibrations are caused if two or more atomic groups accidently possess the same vibrational frequency. This is the case, for example, for acetone, CH_3–C=O–CH_3. The *inner* vibrations of the two CH_3 groups accidently possess the same vibrational frequencies resulting in *one* symmetrical $v_s(CH_3)$, (degenerated) asymmetrical $v_{as}(CH_3)$ stretching modes, symmetrical $\delta_s(CH_3)$, and (degenerated) asymmetrical $\delta_{as}(CH_3)$ bending modes, respectively for both CH_3 groups. Whereas almost *all* infrared absorption bands and Raman lines can be assigned to the respective vibrations for small inorganic species the assignment is restricted only to a *few* for organic molecules. The extensive coupling of vibrations of the atoms of similar masses (C, O, N) linked with similar bond strength (C–C, C–O, C–N) give rise to many absorption bands whose source is mostly unknown. Thus, such absorption bands cannot be assigned to the vibration of a certain atomic group.

The following vibrations are meaningful for the interpretation of the vibrational spectra of organic molecules:

* Characteristic vibrations
* Structure-specific coupled vibrations
* The fingerprint region

2.3.2 Characteristic Vibrations

2.3.2.1 Definition

Characteristic vibrations of a multiatomic molecule are present if nearly the whole potential energy is localized in the movement of a certain atomic group of the molecule with the consequence that coupling of vibrations is negligible. Thus, characteristic vibrations give rise to infrared absorption bands at about the same wave numbers, regardless of the structure of the rest of the molecule.

These site-stable infrared absorption bands are called *group wave numbers* (also called *group frequencies*) because the absorption band can be unambiguously assigned to the vibration of certain atomic groups; therefore, they provide information about the absence or presence of specific atomic groups. Thus, they are valuable diagnostic markers for the recognition of functional groups. Group wave numbers are listed in tables and they are the basis for vibrational spectra interpretation.

2.3.2.2 Criteria for Characteristic Vibrations

Characteristic vibrations can occur if

1. The force constants of neighbored bonds are different (greater 25%).
2. The mass of neighbored atoms are strongly different (greater 100%).
3. The vibration is not accidently in the same region as another vibration.

The first and second criteria are very well fulfilled by, for example, H–C≡N. Infrared radiation at 3,311 cm^{-1} excites the vibration of only the H–C group because, according to calculations, 95% of the absorbed energy is utilized for the movement of this atomic group. However, 95% of the absorbed energy at 2,097 cm^{-1} is used for the vibration of the C≡N group. Because both stretching vibrations fulfill the criteria for characteristic vibrations the region between 3,300 and 3,320 cm^{-1} and at ≈2,100 cm^{-1} are the group wave numbers for the C–H stretching vibration with a *sp*-hybridized C atom and the C≡N stretching one, respectively. Note that the group wave numbers of the C≡N and C≡C groups are nearly the same because of the similar bond order and the approximately similar reduced masses.

According to the first criterion the stretching vibrations of nonconjugated multiple bonds provide characteristic vibrations whose group wave numbers are diagnostically valuable for the unambiguous recognition of the structural groups C=C, C≡C, C=O, C=N, or C≡N.

The first criterion is also fulfilled for the C=S group. Therefore, the C=S group should provide a characteristic vibration and the wave number range of this group should be consistent in analogy to the C=O group. Unfortunately this is not the case. Because of the greater mass of S the stretching frequency is diminished and lies in the range of the bending vibrations of the CH skeleton. Thus, criterion 3 is not fulfilled and the C=S stretching vibrations cannot be assumed to be a characteristic vibration with a small range of group wave numbers. According to calculations a large amount of the absorbed energy of the expected range of the C=S stretching vibration is distributed to bending vibrations of the CH skeleton, in other words, the C=S stretching vibration is a strongly coupled one and the denotation v(C=S) for an intense absorption band in the relatively large range of the C=S stretching vibrations is only approximately valid.

Because, in general, low-energy bending vibrations are strongly coupled the presence of characteristic vibrations is limited to stretching vibrations except for the *inner* bending vibrations of the CH_3 and CH_2 group.

The respective *inner* vibrations of the CH_3 and CH_2 groups are characterized by the fact that the vibration is mainly limited to the movement of CH atoms whereas outer vibrations also include movements of the neighbored atoms which are named *rocking* (ρ), *wagging* (ω), and *twisting* (τ). It is clear that these vibrations are not characteristic and do not provide group frequencies. Therefore, they are not of interest for structural analysis.

2.3.2.3   Examples of Characteristic Vibrations (Group Wave Numbers) According to Criterion 1 ($f_1 \neq f_2$)

Examples are listed in Table 2.9.

2.3.2.4   Examples of Characteristic Vibrations (Group Wave Numbers) According to Criterion 2 ($m_1 \neq m_2$)

Examples are listed in Table 2.10.

2.3.2.5   Internal and External Influences on Characteristic Vibrations

Because of negligible coupling of characteristic vibrations their group wave numbers should be observed in a narrow range. But this is not the case as is shown, for

Table 2.9 Examples of group wave numbers (in cm^{-1}) of compounds with multiple bonds

$\begin{array}{cc} f_1 & f_2 \end{array}$ —C—C=X	X = O	Carbonyls	v(C=O) $\approx$ 1,850–1,650
	X = N	Azomethine	v(C=N) $\approx$ 1,650
	X = C	Alkene	v(C=C) $\approx$ 1,640
$\begin{array}{cc} f_1 & f_2 \end{array}$ —C—C≡X	X = N	Nitrile	v(C≡N) $\approx$ 2,260–2,200
	X = C	Alkyne	v(C≡C) $\approx$ 2,260–2,100

Table 2.10 Examples of group wave numbers (in cm^{-1}) of compounds with the –C–X–H structural moiety

	m_1 m_2	X = O	Alcohol, phenol	$\nu(OH)_{free} \approx 3{,}650$
—C	—X——H		Not associated	$\nu(OH)_{ass} \approx 3{,}200$
			Associated	$\nu(NH) \approx 3{,}400$
		X = N	Amine, amide	2 IR-absorption bands
			Primary	1 IR-absorption band
			Secondary	$\nu(CH) \approx 3{,}300{-}2{,}850$
		X = C	Hydrocarbon	
			$\equiv C(sp)\text{–}H$	$\nu(CH) \approx 3{,}300$
			$=C(sp^2)\text{–}H$	$\nu(CH) \approx 3{,}100{-}3{,}000$
			$-C(sp^3)\text{–}H$	$\nu(CH) \approx 3{,}000{-}2{,}850$

example, for the group wave numbers of the C–H stretching vibrations (see Tables 2.10). The reason for this fact is that characteristic vibrations are influenced by internal and external effects. Knowledge of the correlation between peak position and these effects is valuable to recognize functional groups and structural moieties.

2.3.2.6 Inductive Effects

Increasing of the –I effect increases the bond order and, hence, the force constant in carbonyl compounds. Therefore, the wave number of the C=O stretching vibrations increases with the strength of the –I effect of X which is shown by the following examples:

$CH_3(C{=}O)\text{–}X$	X = H	$\nu(C{=}O)$: 1,740 cm^{-1}	X = Cl	$\nu(C{=}O)$: 1,800 cm^{-1}

2.3.2.7 Mesomeric Effects

Mesomerism diminishes the bond order and, hence, the force constant resulting in lower wave numbers of the stretching vibration. Examples are:

$CH_3(C{=}O)\text{–}X$	X = H	$\nu(C{=}O)$: 1,740 cm^{-1}
	X = Aryl	$\nu(C{=}O)$: 1,685 cm^{-1}
$-C{=}C\text{–}X$	X = H	$\nu(C{=}C)$: $\approx$ 1,645–1,640 cm^{-1}
	X = Aryl	$\nu(C{=}C)$: $\approx$ 1,600–1,450 cm^{-1}

2.3.2.8 Intermolecular Effects

Autoassociation, hydrogen bridges, interactions with the solvent and further effects diminish the strength of the chemical bond and, hence, the respective stretching vibrations are observed at lower wave numbers as is demonstrated by the experimentally observed wave numbers of the C=O stretching vibration of acetone: v(C=O), gaseous: 1,745 cm^{-1}, v(C=O), in CCl$_4$: 1,720 cm^{-1}. Therefore, information on the preparation techniques used for the registration of the vibrational spectrum of a sample should always be given.

2.3.3 Structure-Specific Coupled Vibrations

In contrast to characteristic vibrations structure-specific coupled vibrations cannot be assigned to the vibration of certain atomic groups but they appear as a known absorption band pattern in the infrared spectrum which is specific for a certain structural moiety.

Structure-specific coupled vibrations are accompanied by the presence of certain structural moieties in the molecule. Therefore, they diagnose a certain structural type rather than a functional group in a molecule.

Examples of structure-specific coupled vibrations

1. Overtones and combination vibrations of the CH out-of-plane bending modes and ring bending modes of the CC skeleton of aromatic compounds result in absorption bands in the range from 2,000 to 1,600 cm^{-1} whose pattern enables the recognition of the substitution type of the aromatic compound by comparison of the absorption band pattern with those of "ideal" ones shown in Fig. 6.1. Note that these absorption bands give rise to weak intensities. Hence, in general, they can be observed as more intense bands at higher concentrations or layer thickness of the cuvette.
2. The two intense (sometimes only one) infrared absorption bands in the range from 2,850 to 2,700 cm^{-1} diagnose the presence of the formyl group CHO in the molecule. As explained above these absorption bands specific for a formyl group are caused by Fermi resonance.
3. The weak infrared absorption band in the range from 1,840 to 1,800 cm^{-1} due to a combination of the two out-of-plane bending vibrations at ≈990 and ≈910 cm^{-1} diagnose the vinyl-group, –CH=CH$_2$.

2.3.4 The Fingerprint Region

The absorption bands in the range <1,500 cm^{-1} mostly result from strongly coupled vibrations that cannot be assigned to movements of certain atomic groups. Therefore, their position in the spectrum is determined by a certain structure of the molecule and only small structural changes in the molecule will markedly alter the pattern of these coupled vibrations. Whereas characteristic vibrations provide group wave numbers for functional groups and structure moieties regardless of the structure of the rest of the molecule the strongly coupled vibrations with absorption bands $<1,500$ cm^{-1} are specific for a *single* molecule. Therefore, this region is named the *fingerprint* region.

Whereas characteristic vibrations diagnose only functional groups and structure moieties the fingerprint region facilitates the identification of an individual compound.

The *identification of substances* is realized by visual or computer-aided comparison of the infrared absorption spectrum of the sample with that of a reference sample and a reference spectrum from a spectroscopic library, on condition that both spectra were measured under the same conditions. Especially, the pattern of the fingerprint region is valuable to recognize an individual compound crystallized in a special lattice. Bands of residual solvents or the habit of the crystal can give rise to additional, but weakly intense bands and slightly change the position of bands in the spectrum, respectively. But the most intense absorption bands may not differ more than ± 1 cm^{-1}.

Additional infrared absorption bands or Raman lines or changed band positions, especially observed in the low wave number range, can be caused by different crystal structures, i.e., *polymorphic forms*. Most of the organic compounds crystallize in polymorphic forms. Because vibrational spectra can be easily obtained from solids, vibrational spectroscopy is an excellent and simple method in order to determine polymorphic structures in routine analysis. Thus, such investigations are included, for example, in the regulations (Pharmacopeia) for pharmaceutical analysis. Note that polymorphic forms differ in physical properties such as solubility and stability. Therefore, the determination of the correct polymorphic form is an important condition to apply a compound as an active pharmaceutical ingredient. Besides vibrational spectroscopy, X-ray diffraction, thermal analytical methods, photoacoustic spectroscopy, and solid-state NMR spectroscopy are further methods to investigate polymorphism.

Because Raman spectroscopy, in contrast to routine infrared spectroscopy, also allows recording of the low-frequency range of lattice vibrations which are valuable for recognition of polymorphic forms, Raman spectroscopy is an excellent method for the investigation of polymorphism whereas Raman microscopy enables the investigation of single crystallites. But thermal transformation of polymorphic forms during the measurement must be considered.

2.3.5 Interpretation of Vibrational Spectra of Hydrocarbons

Vibrational spectra provide information about

1. The structural type of hydrocarbons (alkanes, alkenes, alkynes, aromatic hydrocarbons),
2. Branching of the carbon-carbon chains,
3. Structures of branch points,
4. Bonding geometry around the C=C-bond of alkenes,
5. Substitution pattern of aromatic compounds.

Table 2.11 Diagnostic information (wave numbers in cm^{-1}) for **alkanes** (sp^3C–H)

CH stretching vibrations, v_{CH} < **3,000**

CH₃	v_{as}(E): ≈ **2,965**	v_s: 2,875	$I(v_{as}) > I(v_s)$
CH₂	v_{as}: ≈ **2,925**	v_s: 2,855	$I(v_{as}) > I(v_s)$
CH	≈ 2,900 (weak intensity; mostly overlaid)		

Deviating position and intensity at CH₃–X and CH₂–X with X = O, N, Aryl

Bending vibrations

CH₃	δ_{as}(E) ≈ 1,470	δ_s ≈ **1,375** Splitting up at branching
CH₂	δ ≈ **1,455**	

Table 2.12 Diagnostic information (wave numbers in cm^{-1}) for **alkenes** and **aromatics** (sp^2C–H)

CH stretching vibration, v_{CH} < **3,000**

Alkenes		Aromatics	
ν(C=C)	**1,690–1,635**	ν(C=C)	**1,625–1,400**
>C=C < *types*		*Aryl substitution types*	
–CH=CH–(*E/Z*)	γ(CH)	γ(CH) + ring bending at mono and *meta* substitution	**850–650**
–CH=CH₂	**1,005–675**	Only γ(CH) at *ortho* and *para* substitutions	2,000–1,000 Band pattern, see Fig. 6.1
>C=CH₂		Overtones and combination vibrations	

Table 2.13 Diagnostic information (wave numbers in cm^{-1}) for **alkynes** (spC–H)

CH stretching vibration, v_{CH}	**3,340–3,250**
	Sharper than v(OH) and v(NH)
C≡C stretching vibration, $v_{C≡C}$	**2,260–2,100**
	Intensity is small to missing
Bending vibration, $\delta_{C≡CH}$	v ≈ **650** with an overtone at v ≈ 1,250

The most useful diagnostic bands to determine these structural alterations are the C–H, C=C, and C≡C stretching vibrations, the internal bending vibrations of the CH₃ and CH₂ groups, the out-of-plane bending modes of alkenes and aromatic compounds γ(CH) or δ_{oop}, the out-of-plane bending modes of the CC skeleton of the aromatics and their over and combination tones which are summarized in Tables 2.11, 2.12 and 2.13 and in detail in Table 6.15.I. Note that the $\delta_{in\text{-}plane}$ of alkenes and aromatics are meaningless for structural determinations.

Challenge 2.9

Figure 2.11 shows the infrared absorption spectrum of a synthetic product. Justify that the spectrum can be assigned to *para*-methyl styrene (Structure 2.3).

(continued)

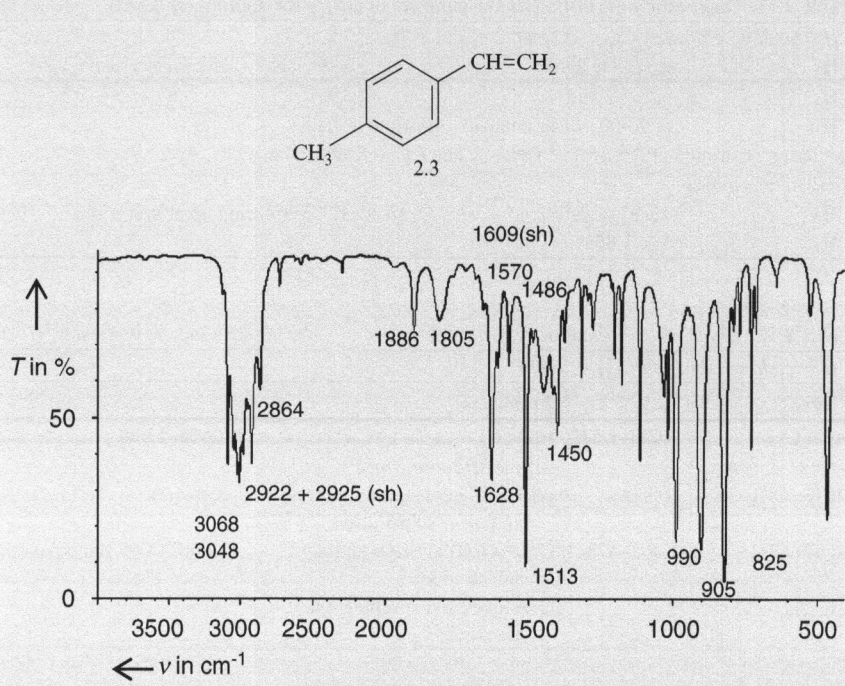

Fig. 2.11 Infrared spectrum of the synthetic product *para*-methyl styrene (Structure 2.3); sampling technique: capillary thin film

Solution to Challenge 2.9

The structural moieties that can be spectroscopically justified by infrared spectroscopy are summarized in Table 2.14.

Table 2.14 Assignment of the IR-absorption bands to *para*-methyl styrene (Structure 2.3)

Structural moiety	Wave numbers in cm^{-1}	Assignment of the infrared absorption bands
Aromatic compounds	3,070–3,010	$v(sp^2C–H)$
	1,570 w, 1,513 s, 1,486 w	$v(sp^2C=C)$, Aryl
–CH=CH$_2$	3,070–3,010	$v(sp^2C–H)$
	1,628 m	$v(C=C)$, in conjugation
	990 s + 905 s	$\gamma(CH)$ for –CH=CH$_2$
CH$_3$–aryl	2,922 m + 2,864 m	$v_s(CH_3)$ + Fermi resonance, Aryl–CH$_3$
	2,925 (sh), w	$v_{as}(CH_3)$ for aryl–CH$_3$
para-substitution	825 s	$\gamma(CH)$, Aryl
		(No ring bending!)
	2,000–1,650	Pattern of overtones and combination vibrations

Challenge 2.10
For solutions to Challenge 2.10, see *extras.springer.com*.

1. Figures 2.12, 2.13, 2.14 and 2.15 present infrared spectra of hydrocarbons. Assign the infrared absorption bands to the vibration of alkanes, alkenes, alkynes, and aromatic compounds. What structural moieties can and cannot be reliably recognized? The relevant wave numbers for the structure determination of the respective compounds are marked in bold.

a

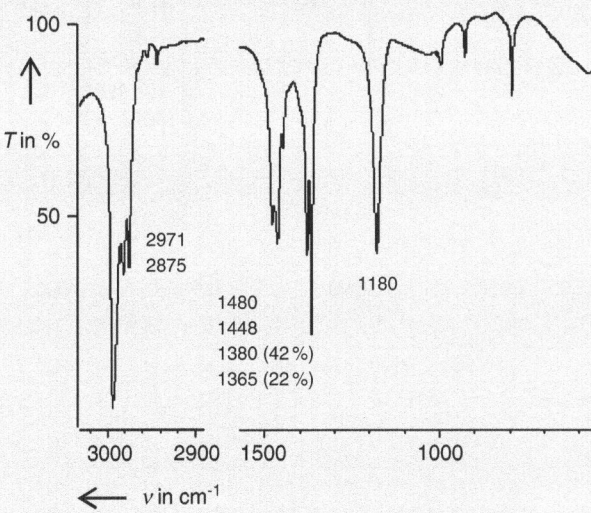

Assign the following infrared bands:
2,971; 2,875; 1,480; 1,448; 1,380/1,365; 1,180 cm^{-1}
What spectroscopic information gives rise to the evidence for

(1) The absence of a CH_2 group,
(2) The presence of the tertiary butyl group?

Fig. 2.12 Infrared spectrum of 2,2,3,3-tetra-methyl butane (Structure 2.4); sampling technique: capillary thin film

(continued)

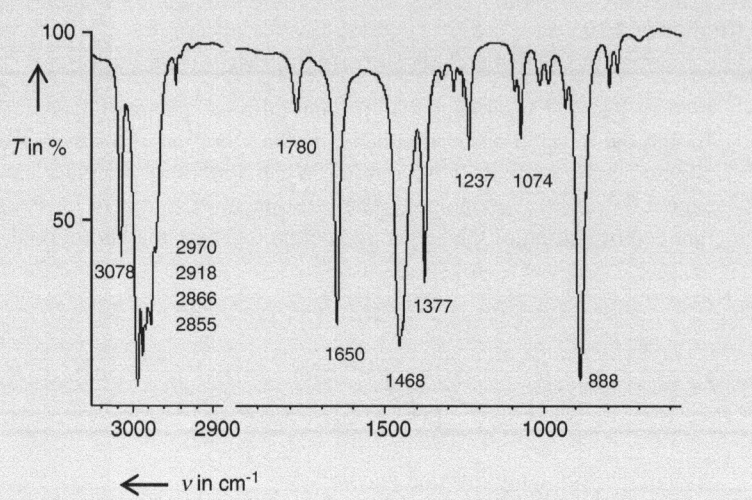

Fig. 2.13 Infrared spectrum of 2-methyl buten-1 (Structure 2.5); sampling technique: capillary thin film

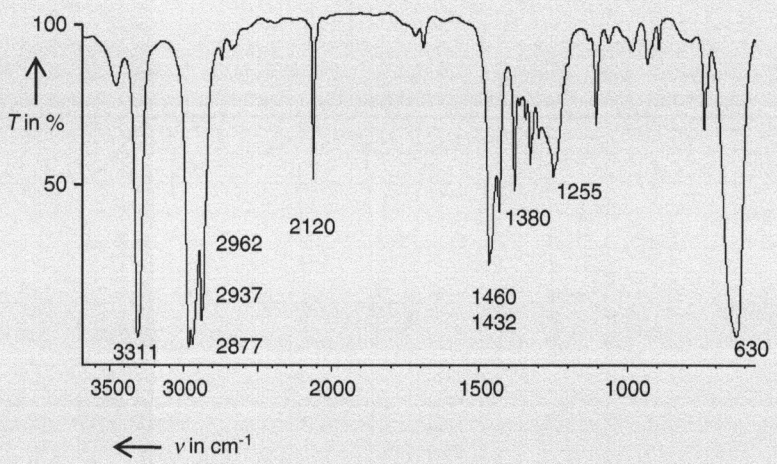

Fig. 2.14 Infrared spectrum of hexin-1 (Structure 2.6); sampling technique: capillary thin film

b

$$H_2C = C - CH_2 - CH_3$$
$$\qquad | \qquad\qquad 2.5$$
$$\qquad CH_3$$

(continued)

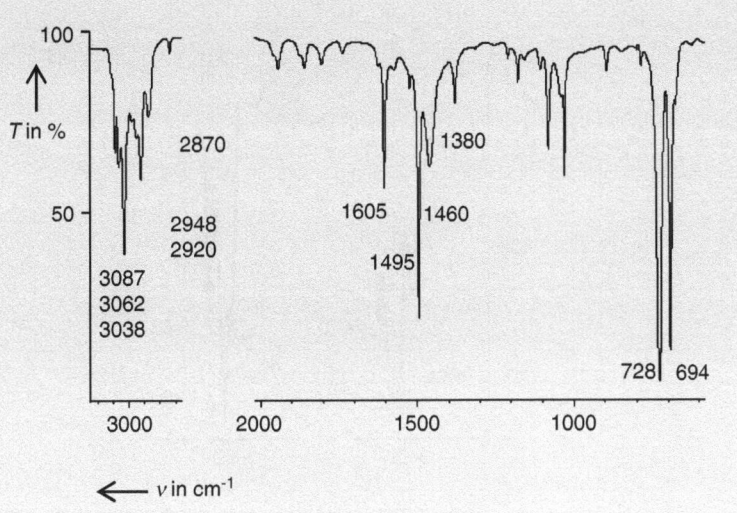

Fig. 2.15 Infrared spectrum of toluene (Structure 2.7); sampling technique: capillary thin film

Assign the following infrared bands:
3,078; 2,970; 2,918; 2,866; 2,855; **1,780**; **1,650**; 1,468; 1,377; **888** cm^{-1}.
The bold wave numbers belong to the alkene.
What spectroscopic information provides evidence for

(a) The presence of a $CH_2=CR_1R_2$ moiety,
(b) An unbranched C–C chain?

c

$$HC \equiv C \longrightarrow (CH_2)_3 \longrightarrow CH_3$$
$$2.6$$

Assign the following infrared bands:
3,311; 2,962; 2,937; 2,877; **2,120**; 1,460; 1,432; 1,380; **1,255**; **630** cm^{-1}
The bold wave numbers belong to the alkyne.
What spectroscopic information provides evidence for an unbranched C–C chain?

d

$$2.7$$

Assign the following infrared absorption bands:
3,087; **3,062**; **3,038**; 2,948; 2,920; 2,870; **2,000–1,650**; **1,605–1,495**;
1,460; 1,380; **728**; **694** cm^{-1}

(continued)

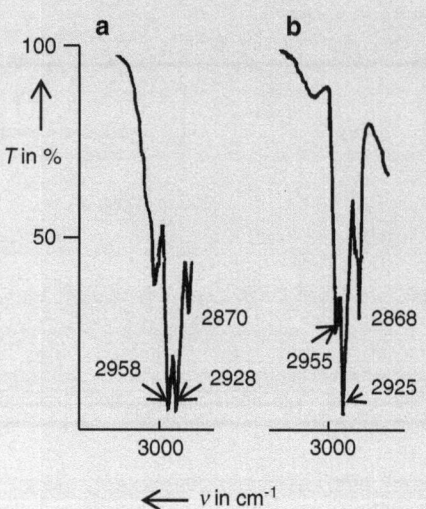

Fig. 2.16 (**a, b**) Infrared spectra of hydrocarbons extracted from seepage water; sampling technique: capillary thin film

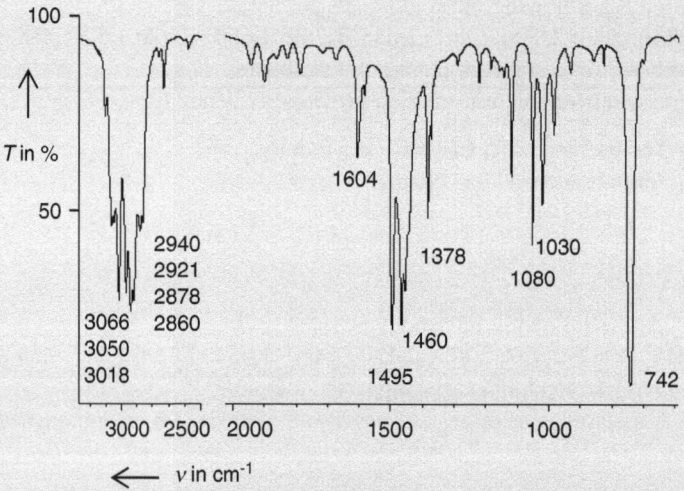

Fig. 2.17 Infrared spectrum of a liquid hydrocarbon compound; sampling technique: capillary thin film. **MS** : $m/z(I_{rel}$ in%) : 106(68)$M^{\bullet+}$; 107(6.0)

The bold wave numbers belong to the aromatic part of the molecule.

What spectroscopic information gives rise to the evidence for

(a) The presence of an aromatic and the absence of a C=C group,
(b) The *mono*-substitution of the aromatic compound.
(c) The presence of the CH_3–aryl structural moiety?

(continued)

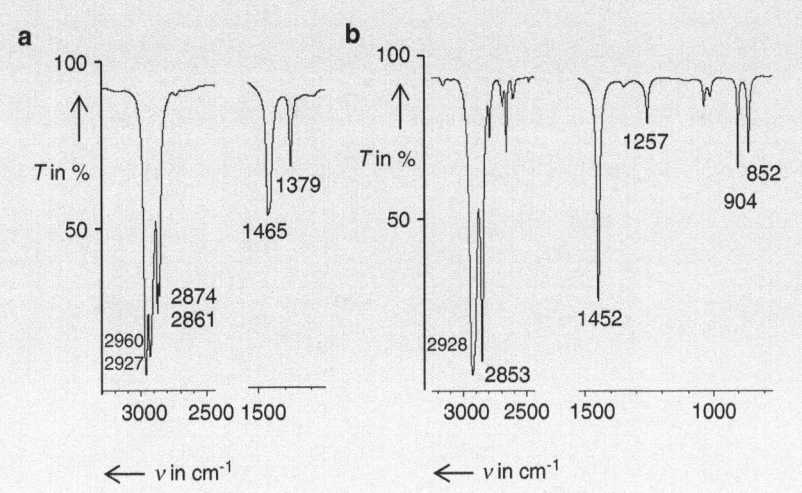

Fig. 2.18 (**a, b**) Infrared spectra of two C_6-hydrocarbon compounds; sampling technique: (**a**) in CCl_4; (**b**) capillary thin film

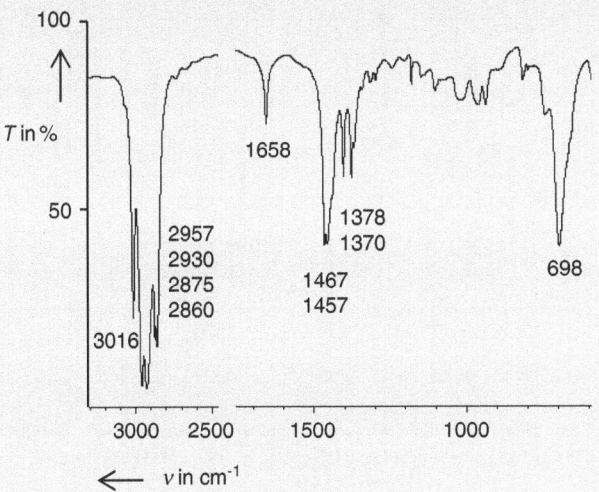

Fig. 2.19 Infrared spectrum of a liquid hydrocarbon compound; sampling technique: capillary thin film. **h** − **MS** : $m/z = 98.1097$ amu $M^{\bullet +}$

2. Seepage water of two retention basins at the interstate was extracted by $C_2Cl_3F_3$. The infrared spectra measured in the range of the C–H stretching vibrations after removal of the extraction solvent are shown in Fig. 2.16a, b.

(continued)

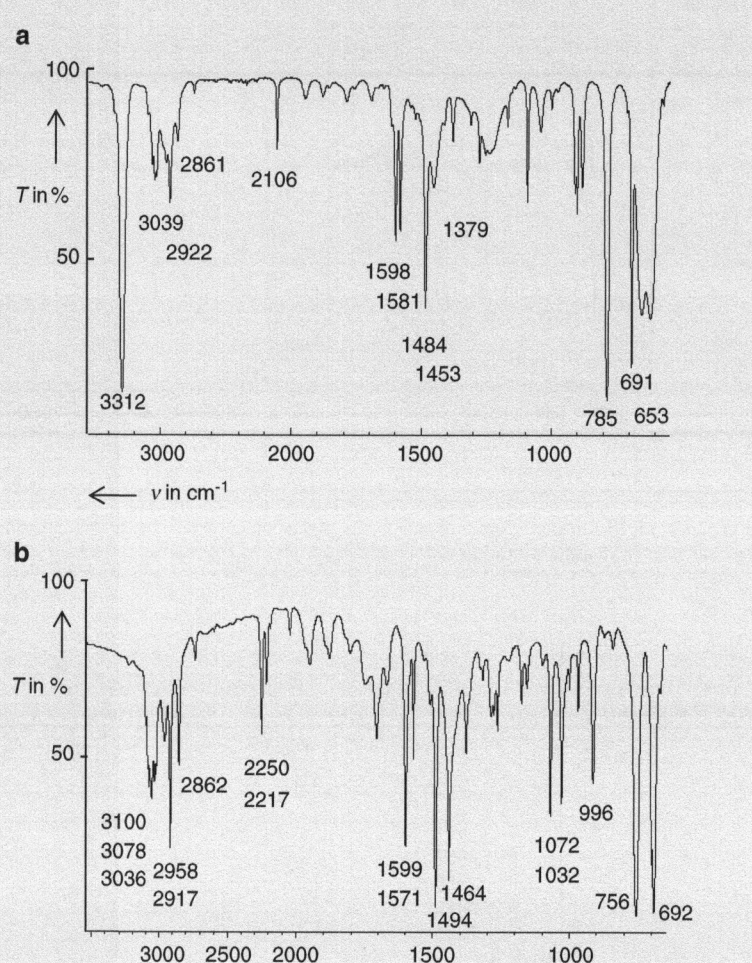

Fig. 2.20 (**a**, **b**) Infrared spectra of two structurally isomeric hydrocarbon compounds; sampling technique: capillary thin film. $\mathbf{h} - \mathbf{MS}$: $m/z = 116.0627$ amu $M^{\bullet +}$

Determine and justify whether the infrared spectra belong to gasoline or diesel oil.

3. Figures 2.17, 2.18, 2.19, and 2.20 present the infrared spectra of hydrocarbons. Determine structural moieties of these compounds and propose a structure for these with the aid of the additional information given.

(continued)

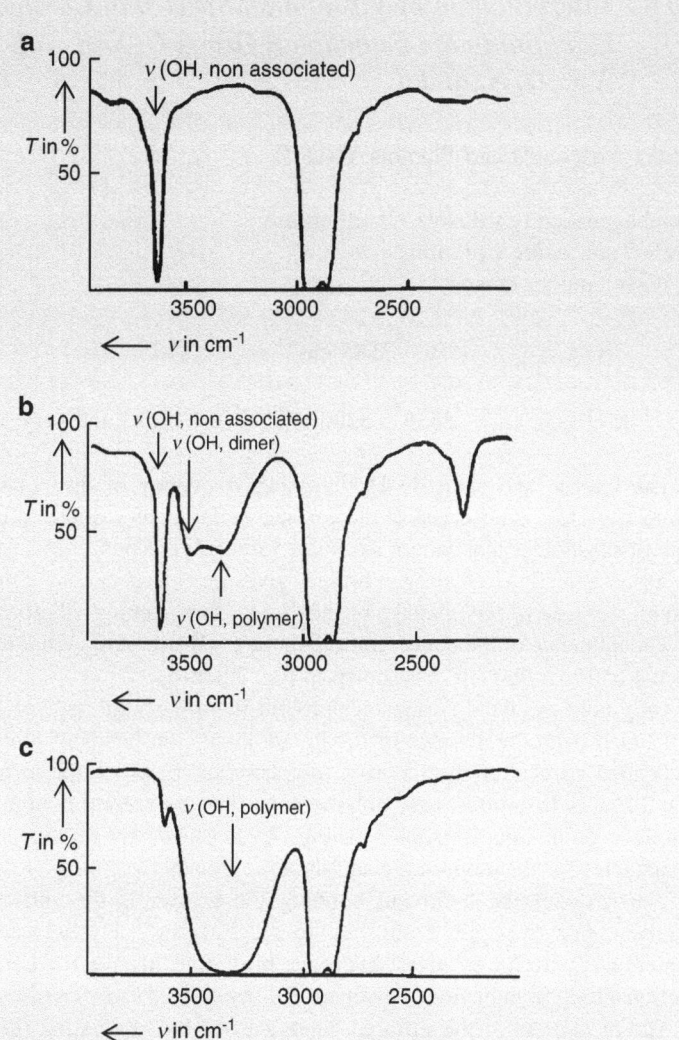

Fig. 2.21 (a–c) Infrared spectra of the O–H stretching vibrations of solutions of *n*-butanol in CCl$_4$ with various concentrations (in mol L^{-1}) and thicknesses of the cuvettes, l (in mm): (**a**) c = 0.005, l = 50; (**b**) c = 0.125, l = 2 mm; (**c**) c = 2, l = 0.05 mm

2.3.6 Interpretation of Vibrational Spectra of Compounds Containing the Functional Group C–X–R with X = O, N, and S

2.3.6.1 Alcohols and Phenols, C–O–H

The diagnostically valuable vibrations are:
H–O stretching vibrations
Wave number range:

$\nu(\text{OH})_{\text{free}}$ **3650 – 3585 cm^{-1}** (sharp and intense)

$\nu(\text{OH})_{\text{associated}}$ **3550 – 3200 cm^{-1}** (very broad and very intense).

The intense and sharp O–H stretching frequency of the nonassociated O–H group $\nu(\text{OH})_{\text{free}}$ is observed in the gaseous state, in very dilute solutions, or in the case of substances that cannot associate for steric reasons.

Association due to hydrogen bridges gives rise to a *decrease of the wave number* and an *increase of the intensity* of the O–H...O stretching vibration.

The influence of the concentration on the O–H stretching vibrations in the range from 3,650 to 3,000 cm^{-1} is shown in Fig. 2.21a–c.

Only nonassociated species are present in strongly diluted solutions as can be concluded from the absence of the broad absorption bands of dimeric and higher associated species at shorter wave numbers as shown in the infrared spectra of Fig. 2.21a, b. In contrast only polymeric species are present in highly concentrated solutions or in liquid samples as Fig. 2.21c shows by missing O–H stretching frequencies for the monomeric and dimeric species.

The stronger the hydrogen bonding the greater is the shift to lower wave numbers. For example, a shift up to $\approx$2,200 cm^{-1} is observed for inorganic compounds with the strongest hydrogen bridges. Thus, the *wave number shift* is a semiquantitative indicator to evaluate the *strength of hydrogen bridges*.

The *broadness* of the infrared bands caused by association through hydrogen bridges is determined by the degree of association. The broader the infrared bands the higher the degree of association. Because the hydrogen bridge in N–H compounds is weaker, only dimeric species are formed and the N–H stretching vibrations give rise to narrower bands which is evidence in order to distinguish N–H and O–H stretching vibrations in a given infrared spectrum.

*Intra*molecular hydrogen bridges are marked by a broad and flat pattern which is independent from the concentration; see, for example, the O–H stretching band for *o*-hydroxy acetophenone (Structure 2.8) in Fig. 2.22a, b. The *inter*molecular hydrogen bridge is caused by equilibrium; therefore, the pattern is determined by the concentration (see Fig. 2.21a–c).

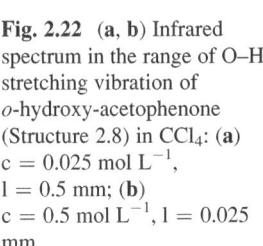

2.8

The strong intramolecular hydrogen bridges of diketones provide *very flat* and *very broad* bands in the range from approximately 3,200 to 2,000 cm^{-1} which is demonstrated by the very strong chelate bridge band of acetylacetone (Structure 2.9) in Fig. 2.23.

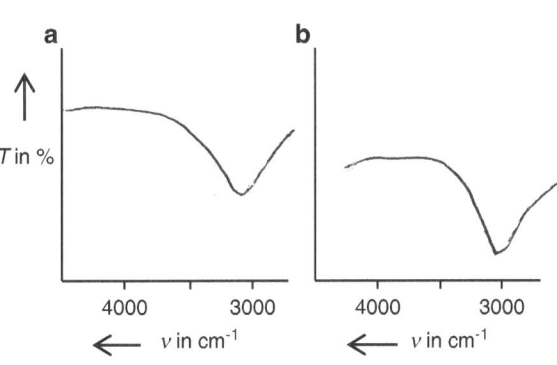

2.9

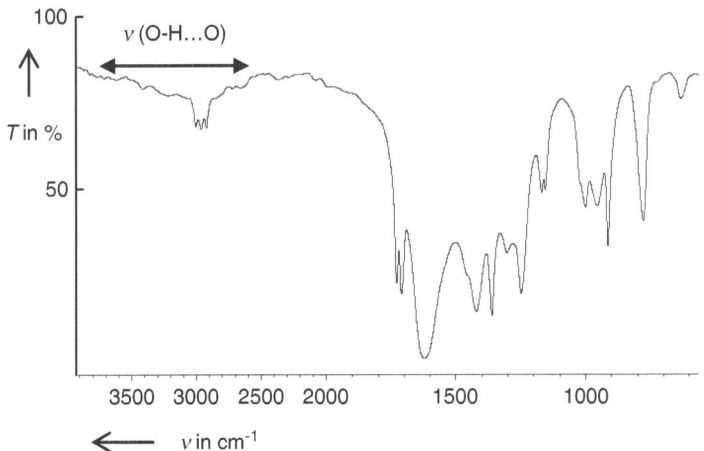

Fig. 2.22 (**a, b**) Infrared spectrum in the range of O–H stretching vibration of *o*-hydroxy-acetophenone (Structure 2.8) in CCl$_4$: (**a**) c = 0.025 mol L^{-1}, l = 0.5 mm; (**b**) c = 0.5 mol L^{-1}, l = 0.025 mm

Fig. 2.23 Infrared spectrum of acetylacetone (Structure 2.9) containing a very strong chelate bridge; cuvette: l = 0.01 mm

Bending deviations of the **C–O–H** *group*

Wave number range: $\delta_{\text{in-plane}}$: **1375 ± 45 cm^{-1}** $\delta_{\text{out-of-plane}}$: **650 ± 50 cm^{-1}**

The two types of bending deviations of the C–O–H group give rise to very broad absorption bands; thus, they can easily be distinguished by C–H bending modes of alkyl and aromatic structures. The in-plane bending vibration of primary and secondary alcohols couples with the C–H wagging mode resulting in two bands at $\approx$1,420 cm^{-1} and $\approx$1,330 cm^{-1}, respectively. Such coupling is not achievable for tertiary alcohols; therefore, in general, only one band is observed in this range. This evidence can be used to recognize tertiary alcohols.

The bending vibrations of phenols – $\delta_{\text{in-plane}}$ and $\delta_{\text{out-of-plane}}$ – are observed in the same range as those of the alcohols. The $\delta_{\text{s}}(CH_3)$ band, valuable for the recognition of the presence of a CH_3 group, can be distinguished from the $\delta_{\text{in-plane}}$ by its greater broadness.

"C–O" stretching vibrations

Wave number range: ν("C–O"): **1,280–980 cm^{-1}** (very intense bands)

The very intense bands in the range from 1,280 to 980 cm^{-1} are generally assigned to the stretching vibration of the C–O group, but these bands are caused by coupling with the neighboring C–C bond, especially in primary alcohols. Therefore, these bands should be named more precisely asymmetrical C–C–O stretching vibration ν_{as}(C–C–O). The bands are diagnostically very valuable because they enable the recognition of the alcohol type because of their various positions in the infrared spectrum. They are observed in the ranges from 1,075 to 1,000 cm^{-1} for primary (1°), from 1,150 to 1,075 cm^{-1} for secondary (2°), and from 1,210 to 1,100 cm^{-1} for tertiary (3°) alcohols, respectively. As one can see, the wave numbers of secondary and tertiary alcohols partially overlap which prevents the recognition of the alcoholic type. Sometimes, the *symmetrical* C–C–O stretching vibration ν_{s}(C–C–O) can be helpful which is observed between 900 and 800 cm^{-1} in 1° and 2° alcohols and lies at $\approx$1,000 cm^{-1} in 3° alcohols, respectively.

The C–O stretching vibration of phenols lies in the range from 1,260 to 1,200 cm^{-1} which is approximately the same range as for 3° alcohols. However, the distinction between both structures should not be difficult because of the characteristic pattern of the aromatic moiety.

Let us now turn to the problem of the occurrence of absorption bands of *water*. Water is contained in many compounds, even if only in traces, or it is incorporated during the sample preparation. Water provides stretching vibrations that overlap with those of alcohols. Therefore, the recognition of the O–H group of an alcohol can be complicated. However, in contrast to alcohols water provides a bending vibration at $\approx$1,650 cm^{-1} which is demonstrated by the infrared spectrum of a hydrocarbon which is contaminated with water, shown in Fig. 2.24.

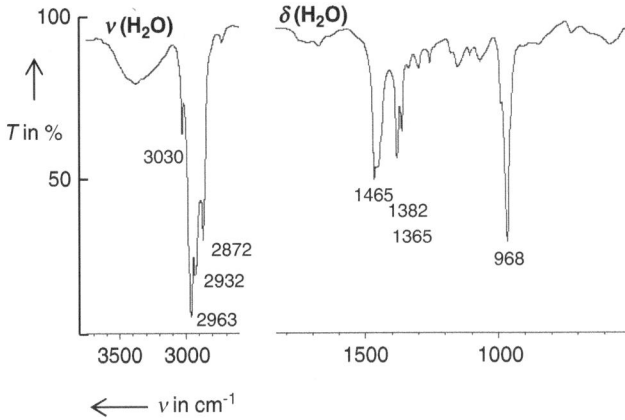

Fig. 2.24 Infrared spectrum of a hydrocarbon compound which is contaminated with water

2.3.6.2 Ethers, C–O–C(α)

Saturated ethers or such containing a *nonbranched* α-C atom provide a strongly asymmetric C–O–C stretching vibration in the range from 1,150 to 1,080 cm^{-1} (mostly at $\approx$1,125 cm^{-1}) and a symmetric C–O–C stretching vibration with medium intensity which lies in the range from 890 to 820 cm^{-1}. Ethers with a *branched* α-C atom show two or more absorption bands with approximately equal intensity in the range from 1,210 to 1,070 cm^{-1} because of coupling of vibrations of the C–O with the neighboring C–C group.

Note that the range of the stretching vibrations of ethers overlaps with those of alcohols (and also carboxylic acid esters, see below), but the presence of an alcohol requires additionally a very intense O–H stretching vibration as explained above (and esters are characterized by intense C=O stretching vibrations).

For other types of ethers intense absorption bands are observed in the following ranges:

Aryl alkyl ether: v_{as}(C–O–C): $1{,}275 - 1{,}200$ cm^{-1}; v_{s}(C–O–C) $:\approx 1{,}210 \pm 10$ cm^{-1}; *Di-aryl ethers*: v_{as}(C–O–C): $1{,}300 - 1{,}200$ cm^{-1}.

A representative infrared spectrum employing the example di-*sec*-butyl ether (Structure 2.10) is shown in Fig. 2.25. The bands assigned to the vibration of the ether group are marked in bold.

2.10

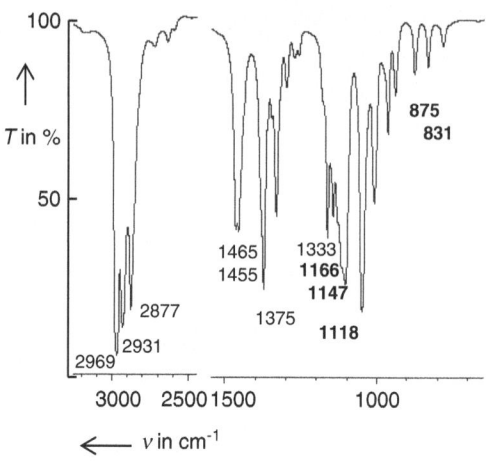

Fig. 2.25 Infrared spectrum of di-*sec*-butyl ether (Structure 2.10); sampling technique: capillary thin film

2.3.6.3   Amines, C–NHR (R = H, Alkyl, Aryl)

N–H stretching vibrations, v(N–H)

Wave number range: **3500 – 3280 cm^{-1}**

Primary (1°) amines show two IR-absorption bands: v_{as}(N–H) at 3,380–3,350 cm^{-1} and 3,500–3,420 cm^{-1} for *saturated* and *aromatic* amines, respectively, and v_s(N–H) at lower wave numbers. The intensities of v(N–H) are weaker than those of v(O–H) because of the weaker hydrogen bridges in amines in contrast to alcohols and phenols. Thus, only dimer species are formed resulting in narrower bands as is shown in the infrared spectrum of benzyl amine (Structure 2.11) presented in Fig. 2.26.

$$\text{C}_6\text{H}_5-\text{CH}_2-\text{NH}_2$$
2.11

Secondary (2°) amines give rise to only *one* N–H stretching vibration which lies in the range of 3,320–3,280 cm^{-1} and at $\approx$3,400 cm^{-1} for saturated and aromatic amines, respectively.

Tertiary (3°) amines do not show any N–H stretching vibration.

Bending vibrations

Wave number range: $\delta_{in-plane}$(NH$_2$): **1650 – 1580**; δ_{oop}(NH$_2$): **850 – 700 cm^{-1}**

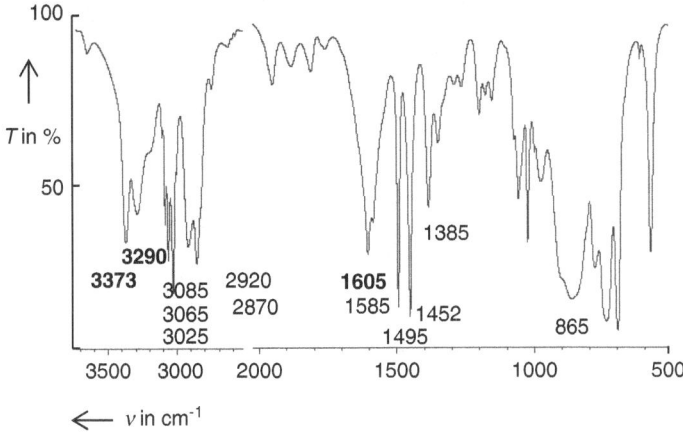

Fig. 2.26 Infrared spectrum of benzyl amine (Structure 2.11); sampling technique: capillary thin film

The medium intense scissoring $\delta_{\text{in-plane}}(NH_2)$ is observed as a relatively broad band. Thus, it can be distinguished from other vibrations in this range and is important in order to distinguish primary amines and alcohols.

The out-of-plane bending vibration lies in the range from 850 to 750 cm^{-1} and from 750 to 700 cm^{-1} for primary and secondary amines, respectively. Sometimes, this difference may be used to distinguish both types of amines.

C–N stretching vibrations, ν(C–N)

Wave number range: $\mathbf{1350 - 1020\,cm^{-1}}$

The medium intense ν(C–N) lies in the range from 1,250 to 1,020 cm^{-1} and from 1,350 to 1,250 cm^{-1} for saturated and aromatic amines, respectively. Because of its lower intensity it can be easily distinguished from the very intense C–O bending vibration provided it is not superimposed by a ν(C–O). Primary amines containing a branched αC atom show new bands due to coupling of vibrations but they cannot be reasonably assigned.

The asymmetrical C–N–C stretching vibration of secondary amines can be used to distinguish saturated and aromatic secondary amines: Saturated secondary amines show bands at 1,180–1,130 cm^{-1} and those of aromatic ones are observed in the range from 1,350 to 1,250 cm^{-1}.

2.3.6.4 Thiols, C–S–H

Wave number range: ν(S − H): $\mathbf{2600 - 2550\ cm^{-1}}$

The S–H stretching frequency lies in a range which is not occupied by other fundamentals; hence, the functional group S–H can be well recognized using infrared spectroscopy. Note that the intensity of v(S–H) is low because of the small dipole moment change during vibration of this group.

Challenge 2.11
Justify the structure of N-methyl p-toluidine (Structure 2.12) by assignment of the marked bold infrared absorption bands in Fig. 2.27.

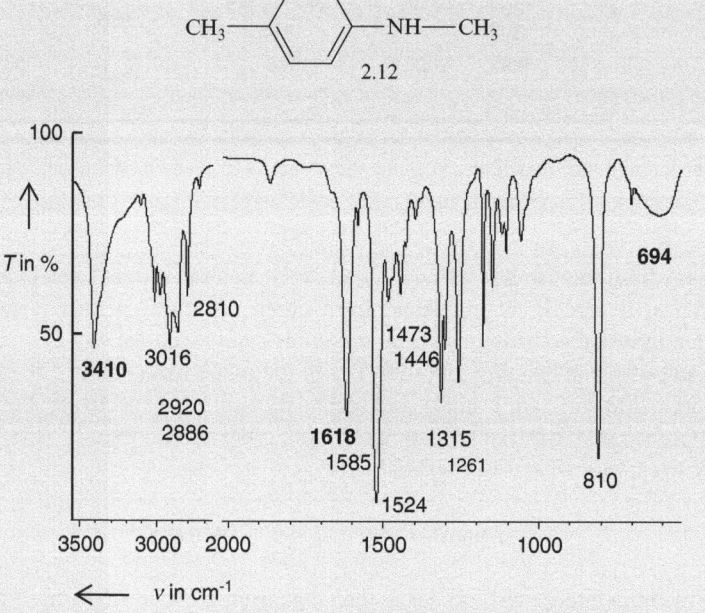

Fig. 2.27 Infrared spectrum of N-methyl p-toluidine (Structure 2.12); sampling technique: capillary thin film

Solution to Challenge 2.11
The assignment of the absorption bands of the infrared spectrum in Fig. 2.27 is summarized in Table 2.15.

(continued)

Table 2.15 Assignment of infrared absorption bands of N-methyl *p*-toluidine (Structure 2.12)

v in cm^{-1}	Assignment	v in cm^{-1}	Assignment
3,410 vs	v(N–H),2° Amine	1,473 m	$\delta_{as}(CH_3)$
		1,446 m	
3,016 m	$v(sp^2C–H)$, Aryl	1,315 w	$\delta_s(CH_3)$
2,920 m	$v_s(CH_3)$ + Fermi	1,261 m	v(C–N–C), N–aryl
2,886 m	resonance		
2,810 m	$v_s(CH_3)$, N-aryl	810 s	γ(C–H), 1,4-disubstituted
1,618 s–,524	v(C=C), Aryl	694 m	δ_{oop} 2° Amine
		2,000–1,650 w	Overtones and combination vibrations for aryl 1,4-disubstituted

The following structural moieties can be justified according to results in Table 2.15:

1. The *aromatic* structure is recognized by $v(sp^2C–H)$ in the range from 3,100 to 3,000 cm^{-1} in addition to v(C=C) at $\approx$1,600–1,490 cm^{-1}. Note that there are no hints of an alkene because of the missing absorption band $v > 1,600$ cm^{-1}.

   The *para*-substitution is proposed by γ(C–H) at 810 cm^{-1} in combination with the missing ring bending vibration. Note that the latter fact is also valid for a 1,2-disubstitution, but the pattern of the weak overtone and combination vibrations in the range from 2,000 to 1,650 cm^{-1} indicate the *para*-substitution rather than the *ortho*-one (see Fig. 6.1).

2. The presence of the secondary amine structure is indicated by the intense single band at 3,410 cm^{-1} which belongs to v(N–H) and, additionally, by δ_{oop} at $\approx$700 cm^{-1}.

3. Alkyl groups are indicated by $v(sp^3C–H)$ in the range from 3,000 to 2,800 cm^{-1} as well as the asymmetric and symmetric bending vibrations at 1,473, 1,446, and 1,315 cm^{-1}.

   The stretching vibrations of a CH_3 group which is directly linked to aryl or N are diminished and are observed for the N-aryl structure at 2,925 $\pm$ 5, 2,865 $\pm$ 5, and 2,815 $\pm$ 5 cm^{-1}, which is experimentally justified by the respective absorption band.

   The CH_2 group cannot be recognized by its respective vibrations because the position of its sole indication, namely the presence of v_{as}(C–H) at $\approx$2,925 cm^{-1} (see Table 2.11), is overlaid by the shifted stretching vibrations of the N-CH_3 group. Note that recognition number and types of alkyl groups are not the subject of infrared spectroscopy.

Challenge 2.12
For solutions to Challenge 2.12, see *extras.springer.com*.

1. Assign the given absorption bands of the infrared spectrum of di-*sec*-butyl ether (Structure 2.10) in Fig. 2.25. The wave numbers that are important for the ether group are marked in bold.
2. Assign the given absorption bands of the infrared spectrum of benzylamine (Structure 2.11) in Fig. 2.26. The wave numbers that are important for the amine group are marked in bold.
3. An unknown compound with the sum formula $C_4H_{11}N$ give rise to the infrared spectrum shown in Fig. 2.28. Derive the structure.
4. The infrared spectra of three structurally isomeric compounds are presented in Fig. 2.29a–c. Derive the structure of each compound. The sum formula can be determined by the spectrometric data of the isomer **B**; m/z (I_{rel} in%): 108 (83.5) $M^{\bullet+}$, 109 (6.8).

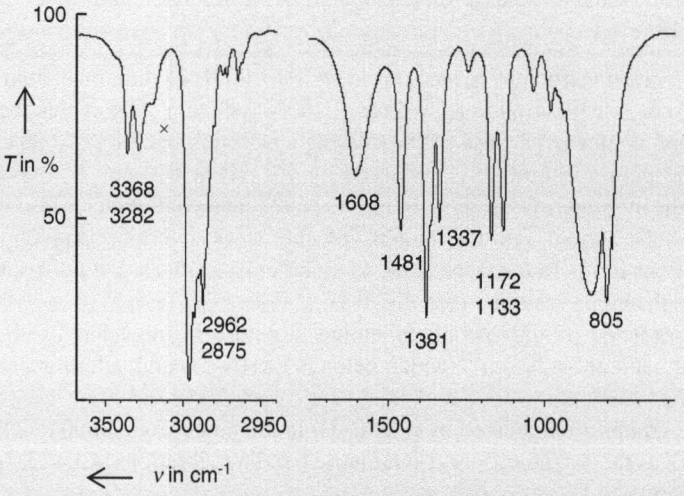

Fig. 2.28 Infrared spectrum of an unknown liquid compound; sampling technique: capillary thin film; × - Fermi resonance: Overtone of $\delta(NH_2) + \nu(NH)$

(continued)

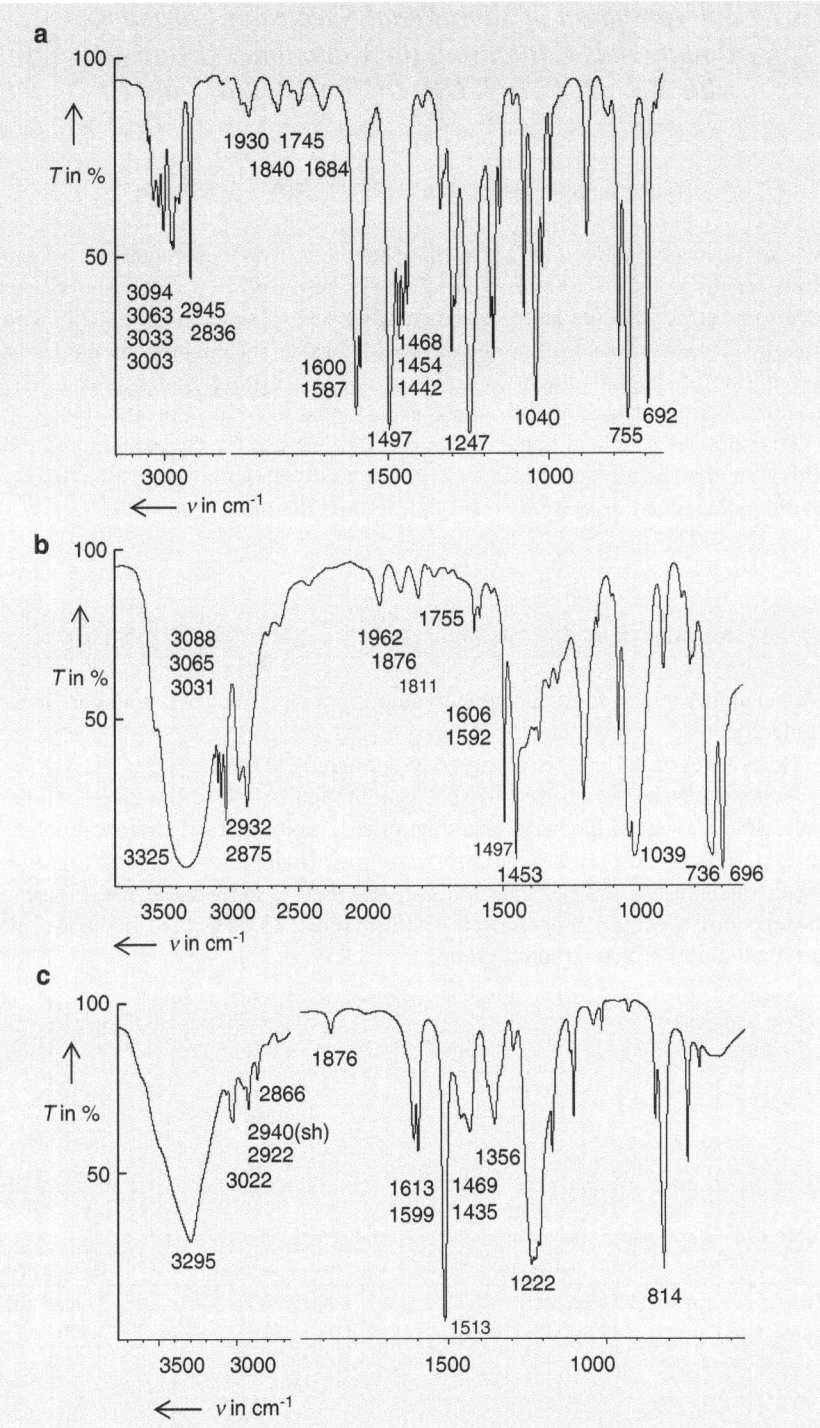

Fig. 2.29 (**a**–**c**) Infrared spectra of unknown liquid compounds; sampling technique: capillary thin film

2.3.7   Interpretation of Vibrational Spectra of Carbonyl Compounds Containing the Functional Group –(C=O)–X with X = H, C, OH, OR, O(C=O)R, NR₂, and Cl

Wave number range: $v(C = O)$: **1,850 − 1,650 cm^{-1}**

As explained above, the C=O stretching mode is a very characteristic vibration. However, its range of 200 cm^{-1} is relatively large for a characteristic vibration. Electronic effects are the main reasons for this, as explained in Sect. 2.3.2. Knowledge of these relations, in combination with further information, is the basis for recognition of the substituent X linked at the C=O group which is summarized in this section. Note that a weak band mostly occurs in the range $v \approx$ 3,400–3,650 cm^{-1} which is the overtone of the intense C=O stretching vibration. Although this overtone coincides with the O–H stretching vibration it can be unambiguously assigned by its low intensity and the missing broadness.

2.3.7.1   Ketones, X = C

Wave number ranges for diagnostically important vibrations of ketones are listed in Table 2.16.

Decreasing of $v(C=O)$ is observed by branching at the αC atom.

Ketones can be recognized by the combination of $v(C=O)$ and $v(C-C-C)$. In contrast to $v(C-C)$ of the alkyl chain the intense asymmetrical stretching vibration of the C–C(=O)–C skeleton indicates a ketone provided that functional groups of the aldehyde, ester, and carbonic acids are not present in the molecule. Figure 2.30 shows the IR-spectrum of acetophenone (Structure 2.13). The bold IR-bands belong to the vibration of the carbonyl group.

2.13

Table 2.16 Wave number ranges of important vibrations of ketones in cm^{-1}

	Saturated ketones	Aryl-alkyl-ketones	Di-aryl ketones
$v(C=O)$	1,715 ± 10	1,700−1,670	1,680−1,650
$v(C-C-C)$	1,230−1,100	1,300−1,230	1,300−1,230

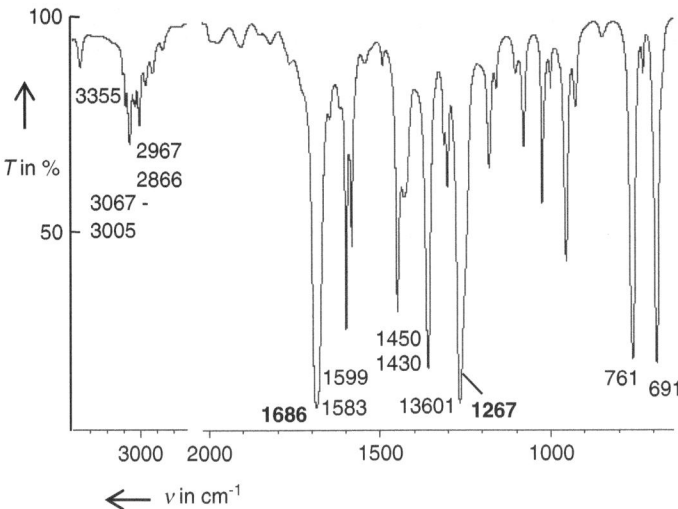

Fig. 2.30   Infrared spectrum of acetophenone (Structure 2.13); sampling technique: capillary thin film

Table 2.17   Wave number ranges of important vibrations of aldehydes in cm^{-1}

v(CH) + Fermi resonance	2,850–2,700, 1 or 2 bands
v(CH), saturated, non α-C branched	2,730–2,715
v(CH), saturated, α-C branched	2,715–2,700
v(C=O), saturated	1,730 ± 10
v(C=O), aryl	1,700 ± 10
δ(CHO)	≈1,390

2.3.7.2   Aldehydes, X = H

Wave number ranges for diagnostically important vibrations of aldehydes are listed in Table 2.17.

Although the range of the very intense C=O stretching vibration lies admittedly somewhat higher, it overlaps partially with that of the corresponding ketones. Thus, this band is not an evidence to distinguish both functional groups. *However, the aldehyde can be unambiguously distinguished from the corresponding ketone by two or one intense absorption bands at 2,850 − 2,700 cm^{-1} combined with δ(CHO) at ≈1,390 ± 10 cm^{-1}.* This is shown in the infrared spectrum of *n*-hexanal (Structure 2.14) presented in Fig. 2.31. Note that only the band of the Fermi resonance is well separated, whereas v(CH) is developed only as a shoulder because of overlap with the CH-stretching vibration. This fact is also to be expected for a saturated, non α-C branched aldehyde. In the case of α-branched saturated aldehydes, however, v(CH) is decreased (see Table 2.17) and hence, two separate bands will be observed in this range.

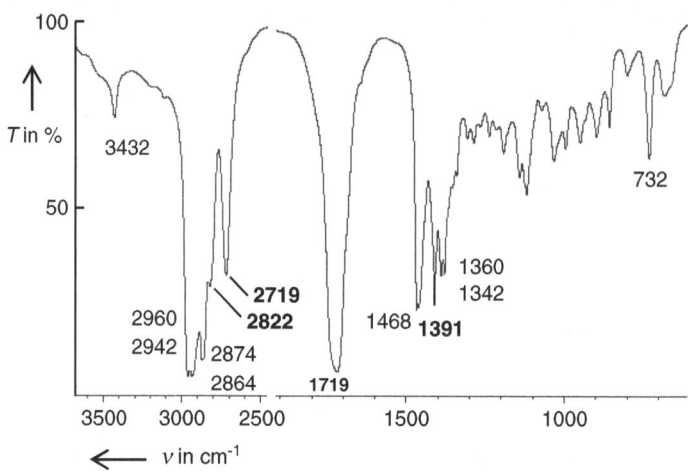

Fig. 2.31 Infrared spectrum of *n*-hexanal, $CH_3-(CH_2)_4-COH$ (Structure 2.14); sampling technique: capillary thin film

2.3.7.3 Carboxylic acids, X = OH

Since carboxylic acid contains a carbonyl and a hydroxyl group, bands of both functional groups can be seen in the corresponding spectra. A representative spectrum is shown in Fig. 2.32 for propionic acid.

Typical bands for the carboxylic acid groups are marked in bold.

O—H stretching vibrations

The very broad and intense infrared band in the range from 3,500 to 2,500 cm^{-1} is an unambiguous evidence for the carboxylic group. No other functional group shows such a broad and intense band, which is caused by the strong hydrogen bonding. Even in the gas state dimeric species linked by hydrogen bridges are still present. Additionally, overtone and combination modes feature bands of medium intensity on the low side of the O—H stretching up to 2,500 cm^{-1}. As shown in Fig. 2.32, the sharper C—H stretching bands are superimposed upon the broad O—H stretching band. Note that the free $v(OH)$ stretching mode appears only in very strongly diluted solutions at $\approx 3,520$ cm^{-1}. The broad O—H stretching bands of β-diketones caused by the strong chelate hydrogen bridge lie in the same range as those of the carboxylic group, however, the intensity is markedly lower (compare the infrared spectra of Figs. 2.23 and 2.32).

C=O stretching vibrations

The intensity of the C=O stretching $v(C=O)$ is higher than that of ketones, compare Figs. 2.30 and 2.32. Monomeric saturated carboxylic acids absorb at 1,760 cm^{-1}. The common dimeric structures feature only the asymmetrical C=O stretching vibration, which lies at 1,710 ± 10 cm^{-1} for aliphatic carboxylic acids. Conjugation diminishes the band position of $v(C=O)$ which is found at 1,700 ± 15 cm^{-1}. Intramolecular hydrogen bonding can be recognized by bands

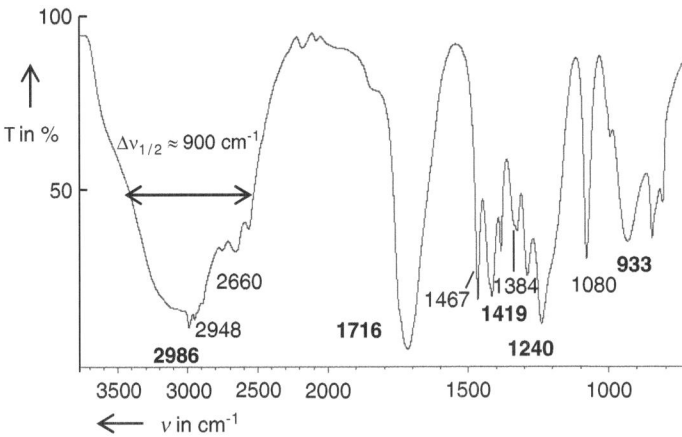

Fig. 2.32 IR-spectrum of propionic acid CH_3-CH_2-COOH (Structure 2.15); sampling technique: capillary thin film

markedly shifted to lower wave numbers. Thus, the C=O stretching bands of *o*-hydroxy benzoic acid and *p*-hydroxy benzoic acid are observed in KBr at 1,667 and 1,682 cm^{-1}, respectively.

C—O and O—H bending vibrations

Coupling of the C—O stretching vibration v(C—O) with the in-plane bending $\delta_{in-plane}$(C—O—H) provides two bands. The bending named by "$\delta_{in-plane}$(O—H)" lies in the range from 1,440 to 1,395 cm^{-1} and the more intense "v(C—O)" lies at 1,315–1,280 cm^{-1}. The latter is split into more bands for long-chained carboxylic acids which is an evidence to recognize *fatty acids*.

Carboxylic acid salts (carboxylates)

Infrared spectra of carboxylates differ from the carboxylic acids by the lack of broad stretching and bending bands. The carboxylic group provides two bands in the range from 1,650 to 1,350 cm^{-1}. The more intense band at the higher wave number is assigned to the asymmetric CO_2 stretching vibration and that in the range from 1,450 to 1,350 cm^{-1} to the symmetric CO_2 stretching vibration, respectively.

2.3.7.4   Carboxylic Acid Esters, X = OR

*Aliphatic and aromatic carboxylic acid esters are characterized by three intense infrared bands which lie in the ranges $v \approx 1,700\ cm^{-1}$, $\approx 1,200\ cm^{-1}$, and $\approx 1,100\ cm^{-1}$. This so-called "**three band rule**" is diagnostic for the carboxylic acid ester group.*

The band with the highest wave number belongs to the C=O stretching vibration v (C=O). The medium band is caused by asymmetric C—C and C—O attached to the carbonyl carbon and is called v_{as}(C—C—O). The third of these bands includes the asymmetrical vibration of the ester oxygen and the next two carbon atoms in the

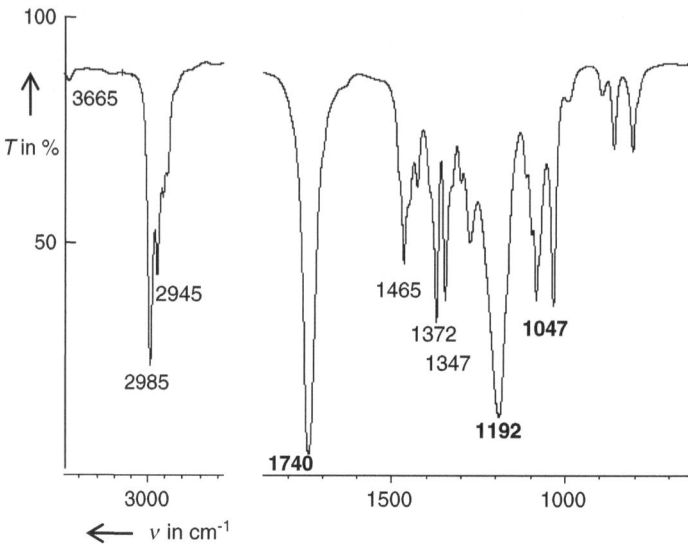

Fig. 2.33 Infrared spectrum of propionic acid ethyl ester (Structure 2.16); sampling technique: capillary thin film

Table 2.18 Wave number ranges for various types of carboxylic esters

	Saturated esters	Aromatic or α,β-nonsaturated esters
ν(C=O)	1,750–1,735	1,730–1,710
ν(C–C–O)	1,210–1,165	1,330–1,250
ν(C–C–O), acetates	≈1,240	
ν(O–C–C)	1,100–1,030	1,130–1,000

hydrocarbon chain ν_{as}(O–C–C). All of these bands can be found in the infrared spectrum of propionic acid ethyl ester C_2H_5–COO–C_2H_5 (Structure 2.16) shown in Fig. 2.33. The bands typical for the carboxylic acid groups are marked in bold.

Wave number ranges for various types of esters are summarized in Table 2.18.

2.3.7.5   Carboxylic Acid Anhydrides, X=O–(C=O)–OR

C=O stretching vibrations

The coupling of the two C=O groups results in two intense C=O stretching vibrations of which the symmetrical vibration is observed at higher wave numbers than the asymmetrical one. The symmetrical stretching vibration provides more intense bands than the asymmetric one for noncyclic anhydrides; the reverse intensity ratio is true for cyclic anhydrides.

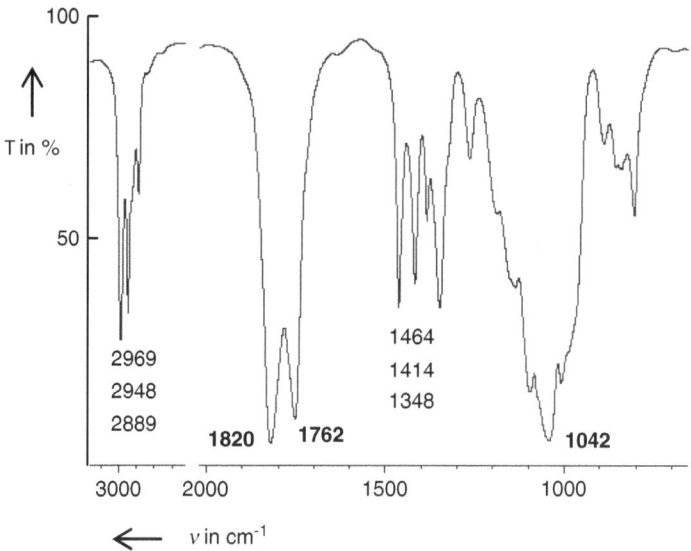

Fig. 2.34 Infrared spectrum of noncyclic propionic acid anhydride (Structure 2.17); sampling technique: capillary thin film

The intensity ratio of $v_s(C=O)$ and $v_{as}(C=O)$ enables us to distinguish between cyclic and noncyclic anhydrides, respectively.

In the infrared spectrum of noncyclic propionic acid anhydride (Structure 2.17) shown in Fig. 2.34 the stretching vibrations lie at 1,820 cm^{-1} [the more intense $v_s(C=O)$] and 1,762 cm^{-1} [the weaker $v_{as}(C=O)$]. The respective C=O stretching vibrations in the infrared spectrum of cyclic 3-methyl glutaric acid anhydride (Structure 2.18) presented in Fig. 2.35 at $v_s(C=O) = 1,809$ cm^{-1} and $v_{as}(C=O)$ = 1,762 cm^{-1} have the reverse intensity ratio. The same result is true for five-ring anhydrides.

2.17

The infrared bands due to the anhydride group are marked in bold.

2.18

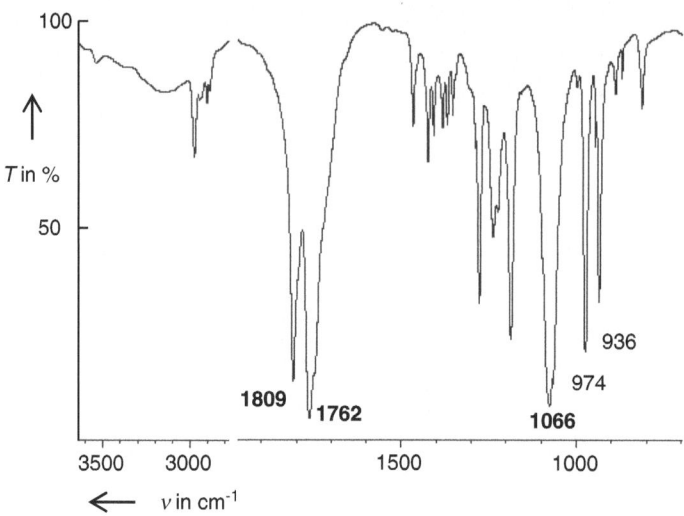

Fig. 2.35 Infrared spectrum of 3-methyl glutaric acid anhydride (Structure 2.18); sampling technique: capillary thin film

Table 2.19 Wave number ranges of the C=O and C–O stretching vibrations for some anhydride types

		Noncyclic anhydrides	Cyclic anhydrides
v_s(C=O)	Saturated	$1{,}820 \pm 5$ (stronger)	$1{,}860 \pm 10$ (weaker)
	Nonsaturated	$1{,}775 \pm 5$ (stronger)	$1{,}850 \pm 10$ (weaker)
v_{as}(C=O)	Saturated	$1{,}750 \pm 5$ (weaker)	$1{,}785 \pm 15$ (stronger)
	Nonsaturated	$1{,}720 \pm 5$ (weaker)	$1{,}870 \pm 10$ (stronger)
v(C–O)		$1{,}050 \pm 10$	$1{,}300$–$1{,}175$ and 950–880

C–O stretching vibrations

The C–O and C–C–O stretching vibrations, in most cases more intense than v(C=O), lie at $v \approx 1{,}050$ cm^{-1}.

The wave number ranges of the C=O and C–O stretching vibrations are listed in Table 2.19 for some anhydride types.

2.3.7.6 Carboxylic Acid Amides, X = NR$_2$

N–H stretching vibrations

The number of N–H stretching vibrations is equal to the number of N–H bonds.

The two bands of *primary* (1°) *amides* R–(C=O)–NH$_2$ are seen at $v \approx 3{,}520$ and $\approx 3{,}400$ cm^{-1} with medium intensity in the spectrum which is obtained by employing diluted solutions. In the solid state the bands are shifted to smaller wave numbers because of hydrogen bonding.

Secondary (2°) *amides* R–(C=O)–NRH show only one band in the range from 3,500 to 3,400 cm^{-1} in diluted solutions. In the solid state generally several bands appear caused by various conformeric structures that are formed by hydrogen bonding.

Tertiary (3°) *amides* R–(C=O)–NR$_2$ do not show any bands in the range from 3,500 to 3,300 cm^{-1}.

C=O stretching vibrations

v(C=O) is shifted to smaller wave numbers due to mesomerism as is shown by the two resonance structures:

Furthermore, the C=O stretching vibration cannot be assumed to be purely v(C=O) because of coupling with other vibrations. Therefore, "v(C=O)" is named the *amide I* band. It is seen in the spectrum always below 1,700 cm^{-1} and its position is determined by the sampling technique and the amide type. The amide I band of tertiary amides is nearly independent of the physical state because of missing hydrogen bridges.

N–H bending vibrations

The NH$_2$ scissoring δ(NH$_2$) couples with the N–C=O stretching vibration, because both vibrations lie in nearly the same wave number range, result in a band which is named *amide II*. In 1° amides the amide II bands mostly coincides with the amide I band or it is developed only as a shoulder but in solution both bands are separate.

Coupling of the amide II band with the δ(NH$_2$) band gives rise to the *amide III* band which appears at $v \approx 1{,}250$ cm^{-1}.

Other vibrations

1° amides show a band at $v \approx 1{,}400$ cm^{-1} which is assigned to the C–N stretching vibration.

The out-of-plane bending of the NH$_2$ group δ_{oop}(NH$_2$), also named "wagging NH$_2$," appears as a broad band in the range from 750 to 600 cm^{-1} whose intensity is smaller for 2° amides. Typical wave numbers of various types of amides are summarized in Table 2.20.

Figure 2.36 presents the infrared spectrum of benzamide (Structure 2.19) as an example of a primary amide prepared in KBr. Typical wave numbers for the aryl-benzamides are marked in bold.

2.19

Table 2.20 Typical wave numbers (in cm^{-1}) of various types of amides

	1° amide	2° amide	3° amide
$v(NH)_{nonassociated}$	v_s: 3,520	3,500–3,400	No
	v_{as}: 3,400		
$v(NH)_{associated}$	v_s: 3,350	3,300–3,060	No
	v_{as}: 3,180	Mostly several	
Amide I$_{nonassociated}$	1,690	1,700–1,680	1,680–1,630
Amide I$_{associated}$	1,680–1,630	1,680–1,630	No
Amide II	1,655–1,620	1,570–1,515	No
	Separated in diluted solutions	High intensity	
$v(C–N)$	1,430–1,390	1,310–1,230	$\approx$1,505
			R=CH$_3$
$\delta_{oop}(O{=}C{-}NH_2)$	750–600 (broad)	750–680 broad	No

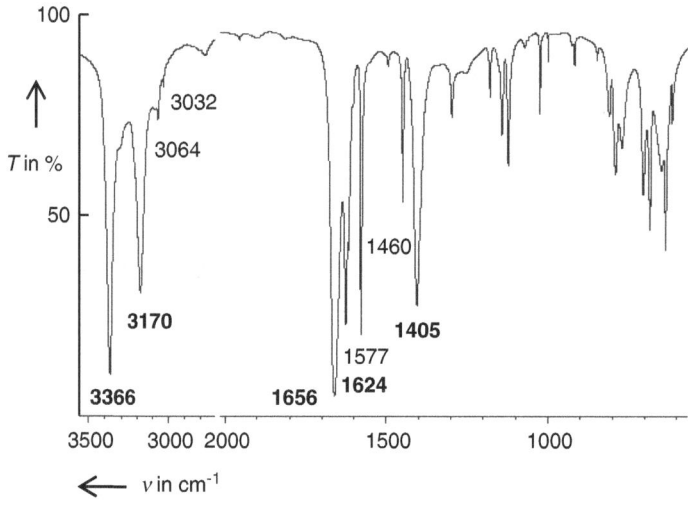

Fig. 2.36 Infrared spectrum of benzamide (Structure 2.19) as an example of a primary amide; sampling technique: KBr pellet

2.3.7.7 Carboxylic Acid Chlorides, X = Cl

Recall from the discussion in Sect. 2.3.1 that a substituent with an –I effect linked at the carbonyl group decreases the wave number of $v(C{=}O)$ bands. Because of the strong –I effect of the Cl atom, the shift of $v(C{=}O)$ towards higher wave numbers is large. Thus, $v(C{=}O)$ of nonconjugated carboxylic acid chlorides appears in the high wave number range from 1,815 to 1,785 cm^{-1}. Conjugation, as discussed above, diminishes the $v(C{=}O)$ band position to 1,800–1,770 cm^{-1}.

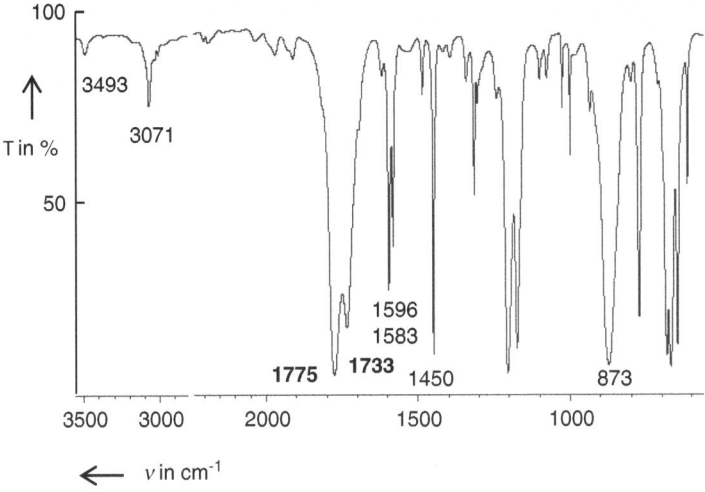

Fig. 2.37 Infrared spectrum of benzoyl chloride (Structure 2.20); sampling technique: capillary thin film using AgCl discs

The (C=O) stretching vibration of *aryl-carboxylic acid chlorides* is split into two absorption bands due to Fermi resonance as discussed in Sect. 2.1.4. This is an evidence to distinguish alkyl and aryl carboxylic acid chlorides. In general, the band at lower wave number has a lower intensity.

The infrared spectrum of benzoyl chloride (Structure 2.20) is shown in Fig. 2.37. The typical wave numbers for carboxylic acid chlorides are marked in bold.

2.20

2.3.8 Interpretation of Vibrational Spectra of N-Containing Compounds

2.3.8.1 NO$_2$ Compounds

The NO$_2$ group gives rise to two intense infrared bands that are easy to identify and this is demonstrated in the infrared spectrum of *p*-ethyl nitro benzene (Structure 2.21, R=C$_2$H$_5$) shown in Fig. 2.38. The group wave numbers of the NO$_2$ group are marked in bold in the spectrum.

R —⟨ ⟩— NO$_2$

2.21

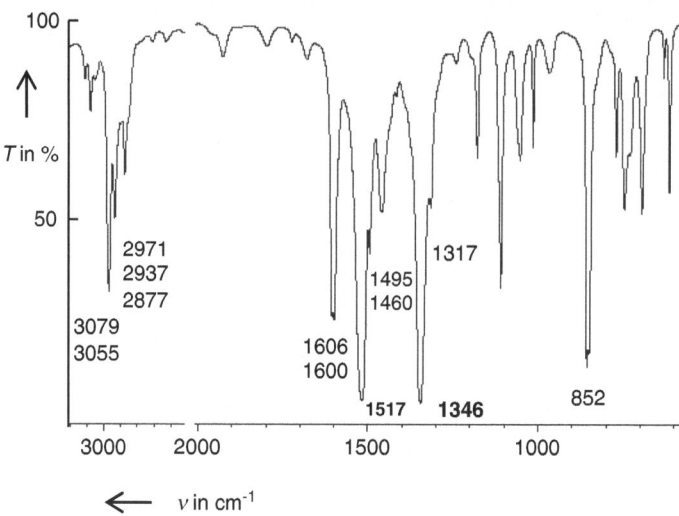

Fig. 2.38 Infrared spectrum of *p*-ethyl nitro benzene (Structure 2.21); sampling technique: capillary thin film

Table 2.21 Group wave numbers of the functional group NO_2 (in cm^{-1})

	$v_{as}(NO_2)$	$v_s(NO_2)$
Alkyl-NO_2	1,570–1,540 (very intense)	1,390–1,340 (intense)
Aryl-NO_2	1,560–1,500 (very intense)	1,360–1,300 (intense)

The most intense asymmetrical N–O stretching vibration $v_{as}(NO_2)$ appears at higher wave numbers than the less intense symmetrical one $v_s(NO_2)$. Conjugation in aryl nitro compounds shifts the wave number to somewhat smaller values, see Table 2.21.

Electron donor substituents decrease the group wave numbers and increase their intensity. Thus, $v_{as}(NO_2)$ and $v_s(NO_2)$ of *p*-nitro aniline (Structure 2.21, R=NH_2) appear at 1,475 and 1,310 cm^{-1}, respectively as very intense infrared bands.

Coupling between $\delta_{oop}(NO_2)$ and $\gamma(C–H)$ of the aromatic group gives rise to new bands which alter the pattern of the infrared bands in the range from 2,000 to 1,650 cm^{-1}. Hence, the pattern cannot be used to recognize the aromatic substitution type.

2.3.8.2 N=O Compounds

Because of similar mass and bond order the N=O stretching vibration should give rise to infrared bands similar to those of the C=O group. Thus, $v(N=O)$ is observed as an intense infrared band at 1,585–1,540 and 1,510–1,490 cm^{-1} for alkyl–N=O and aryl–N=O compounds, respectively.

2.3.8.3   Nitriles

The C≡N group gives rise to characteristic vibrations in a unique wave number range. The C≡N stretching vibration is observed from 2,260 to 2,240 cm^{-1} for saturated nitriles and from 2,250 to 2,200 cm^{-1} for aromatic ones, respectively. Thus, the position of v(C≡N) enables us to distinguish between saturated and aromatic nitriles.

2.3.9   Interpretation of Vibrational Spectra of S–O-Containing Compounds

2.3.9.1   Structure Types RR'S=O, RO–(S=O)–O–R'

Because of the larger mass of S the bands of the S=O stretching vibrations are shifted to lower wave numbers compared to the respective C=O ones and appear in the range from 1,225 to 980 cm^{-1} as intense infrared bands. For further information, see Sect. XIII in Table 6.15.

2.3.9.2   Structure Types –SO₂–

Like nitro compounds the SO_2 group gives rise to an asymmetric $v_{as}(SO_2)$ and a symmetric $v_s(SO_2)$ stretching vibration which can be well recognized in the infrared spectrum in the range from 1,420 to 1,000 cm^{-1} because of their high intensity. $v_{as}(SO_2)$ lies at a higher wave number than $v_s(SO_2)$. Note that the range of the group wave numbers of the SO_2 compounds overlap with those of other functional groups.

Challenge 2.13
Mass and infrared spectra are given for an unknown compound in Figs. 2.39 and 2.40, respectively. Create a proposal for the structure of this compound as far as it is possible. The peaks and wave numbers given in the mass and infrared spectra, respectively have to be assigned.

(continued)

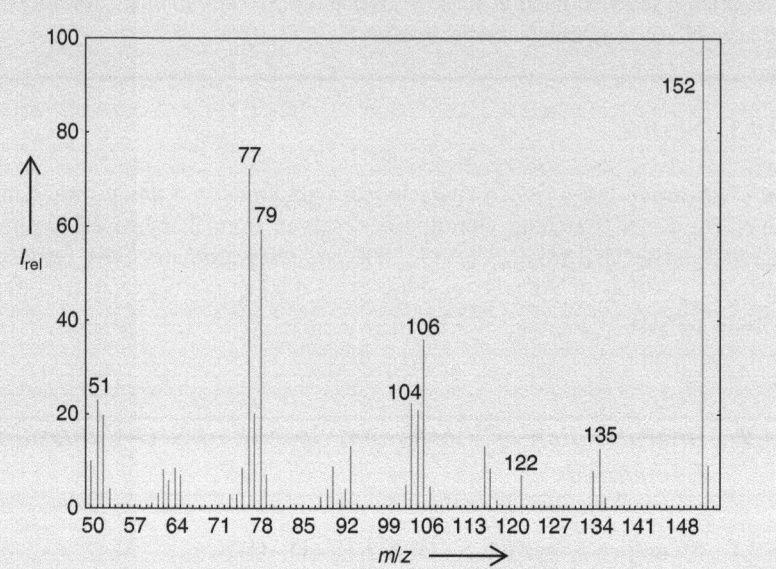

Fig. 2.39 70-eV mass spectrum of an unknown compound. m/z (I_{rel} in%): 152 (100), 153 (8.8)

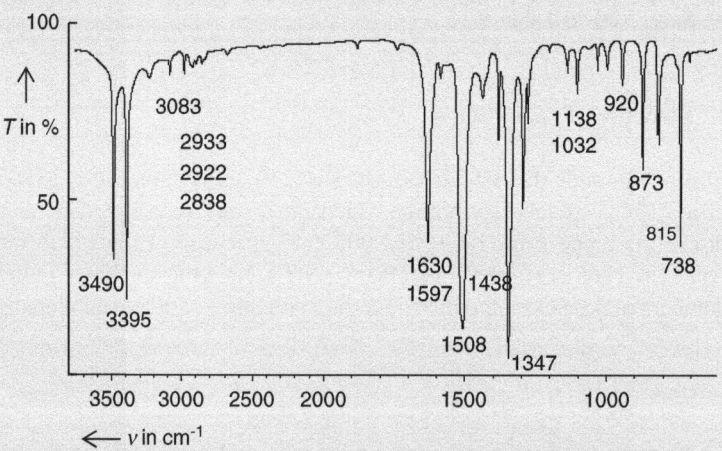

Fig. 2.40 Infrared spectrum of an unknown compound; sampling technique: KBr pellet

Solution to Challenge 2.13
Mass spectrum

- Derivation of the *sum formula*
  The results for the derivation of the sum formula are summarized in Table 2.22.
- Calculation and assignment of the *double bond equivalents* (DBE)
  Transformation of the sum formula into C_nH_x according to the rules given in Sect. 1.3:
  $C_7H_8N_2O_2$ (-2 O–2 N + 2 CH) $\Rightarrow C_9H_{10}$
  DBE $= 5$ calculated by Eq. 1.19.
  Assignment: Aryl (4 DBE) + NO_2 (1 DBE),
  $\Rightarrow$ No further rings can be present in the molecule.
- Structural moieties determined by mass spectroscopy

Structural moiety/functional group	Recognized by
NO_2, NH_2	Molpeak (N rule), (M − X) fragment peaks
Aromatic compound	DBE, specific fragment peaks: $m/z = 51$, 65, 77 amu
CH_3	Difference to the sum formula

Table 2.22 Results for the development of the sum formula. The m/z values and differences are given in amu

Number of C and N atoms n_C, n_N				
m/z (I_{rel} in%)	Assignment	Information		
152 (100)	$M^{\bullet+}$	$n_N = 0$ or 2		
153 (8.8)	$[M + 1]^{\bullet+}$	$n_C = 8 \pm 1$		
Evidence of N in the molecule				
Search for N-containing fragment peaks				
m/z (start ion)	m/z (end ion)	Difference	Loss of	Hints for
152	122	30	N=O	Aryl–NO_2
152	106	46	NO_2	Aryl–NO_2
106	79	27	HCN	Aryl–NH_2 or pyridine
104	77	27	HCN	Aryl–NH_2 or pyridine
Result	$n_N = 2$			
Search for O-containing fragment peaks				
m/z (start ion)	m/z (end ion)	Difference	Loss of	Hints for
152	136	16	O	Aryl–NO_2
152	135	17	OH	*Ortho*-effect?
Development of the sum formula				
Summation of the atoms recognized as yet		8 ± 1 C + 2 N + 2 O		
Proposals	Mass number	Difference to M	n_H	
$C_7N_2O_2 + H_x$	**144** + xH	152–144 = 8	8	
$C_8N_2O_2 + H_y$	**156**	Not possible, because larger than M		
Result	$C_7H_8N_2O_2$	$OE^{\bullet+}$ proved according to Eq. 1.11		

(continued)

Table 2.23 The assignment of the infrared bands obtained by the infrared spectrum shown in Fig. 2.40

v in cm^{-1}	I	Assignment	Hints for structural moieties
3,490	vs	$v_{as}(NH)$	
3,395	vs	$v_s(NH)$	1° amine, R–**NH₂**
3,180	w	OT $\delta_{as}(NH_2) + v(NH)$	Fermi resonance
3,083	w	$v(sp^2CH)$	Benzene ring/alkene
2,933	w	$v_{as}(Aryl\text{-}CH_3)$	
2,922	w	$v_s(Aryl\text{-}CH_3)$	**Aryl–CH₃**
2,861	w	OT $\delta_{as}(CH_3) + v(CH)$	Fermi resonance
1,950–1,700	vw	OT + CV	Pattern for 1,2,4-trisubstitution (?)
1,630	s	$\delta(NH_2)$	1° amine, **NH₂**
1,597	w	$v(C=C)$ aryl	Aromatic compound
1,508	vs	$v_{as}(NO_2)$	
1,347	s	$v_s(NO_2)$	**Aryl–NO₂**
1,438	m	$\delta_{as}(CH_3)$	CH₃
1,371	m	$\delta_s(CH_3)$	
1,032	m	$v(C–N)$?	
873	m	$\delta(NO_2)$	
920	m	$\gamma(Aryl\text{-}CH)$	Aryl, 1 neighboring H atom?
815	m	$\gamma(Aryl\text{-}CH)$	Aryl, 2 neighboring H atoms?

OT overtone, *CW* combination vibration, *I* estimated intensity

- The compound is a nitro amino toluene (Structure 2.22) whose substitution type is unknown. Remember that, in general, the recognition of aromatic isomers is not the subject of mass spectroscopy.

2.22

Infrared spectrum

The assignment of the infrared bands is listed in Table 2.23.

Structural moieties and functional groups determined by infrared spectroscopy:

Aryl–NO₂, R–NH₂, Aryl–CH₃, 1,2,4-trisubstituted (?) aromatic compound.

Proposal for structures

The benzene ring must be trisubstituted, the substituents are NO₂, NH₂, and CH₃. *Ortho*-substitution of NO₂ and CH₃ should be excluded because of the missing [M − 17] peak due to the elimination of O–H via the *ortho*-effect in the mass spectrum. Thus, Structures 2.23–2.25 come into consideration. Note that the true structure can only be established by comparison of the

(continued)

infrared spectrum with those of the respective reference spectra. But this is not the subject of this book. With the employment of further methods presented in the next chapters the evidence of the final structure will be adduced.

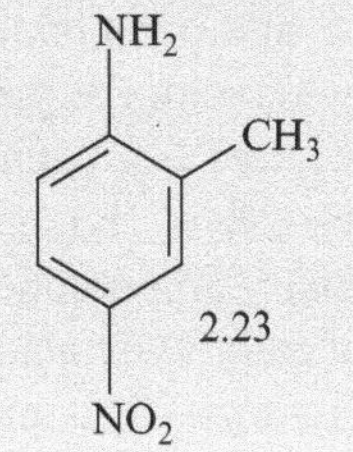

2.23

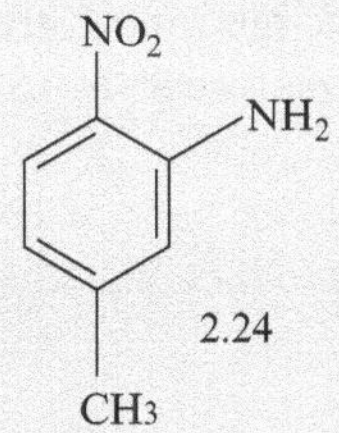

2.24

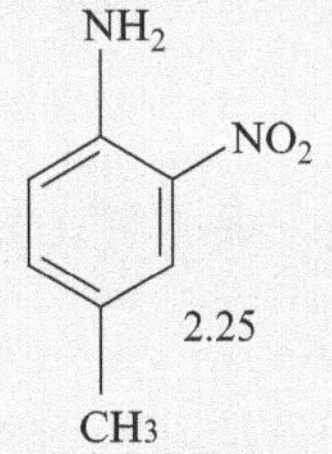

2.25

Challenge 2.14

For solutions to Challenge 2.14, see *extras.springer.com*.

1. Assign the given wave numbers in the infrared spectra of Figs. 2.30, 2.31, 2.32, 2.33, 2.34, 2.35, 2.36, 2.37 and 2.38 to the respective vibrations. The wave numbers distinctive for a certain substance class are marked in bold. Assemble certain and uncertain structural moieties.

2. Carbonyl compounds can be easily recognized by the intense infrared band in the range from 1,800 to 1,650 cm^{-1} but there are many substance classes containing the C=O group. Show how the following substance

(continued)

classes can be evidenced and distinguished from the others by infrared
spectroscopy: aldehydes, ketones, carboxylic acids, carboxylic acid esters,
amides, and anhydrides.

3. Show how the following structures can be distinguished:

   (a) Intra- and intermolecular hydrogen bonding
   (b) *E/Z* isomers for compounds with the general structure RHC=CHR
   (c) Substitution types of aromatic compounds
   (d) Distinction between aryl–CH_2–CHO and CH_3–aryl–CHO
   (e) Distinction between primary and secondary amines
   (f) Recognition of a tertiary carboxylic acid amide
   (g) Distinction between CH_3–aryl–CHO and aryl–(C=O)–CH_3
   (h) Distinction between CH_3–aryl–(C=O)–Cl and Cl–CH_2–aryl–CHO
   (i) Distinction between R–S–H, R–S–R´, and R–S–S–R´.

4. The oxidation of 3-methoxy-tetraline (Structure 2.26) gives rise to the
   main product whose molecular peak is 16 amu greater than that of the
   starting compound. The infrared spectrum of a solution of the reaction
   product in CCl_4 shows an intense band at 1,686 cm^{-1}. What structures can
   be assigned to the reaction product and what can be excluded?

2.26

5. The following wave numbers for the C=O stretching vibrations are
   obtained in the infrared spectrum of compounds of the structural type
   shown in Structure 2.27 with R = H, OH, N(CH$_3$)$_2$, NO$_2$: 1,737, 1,727,
   1,715, and 1,685 cm^{-1}. Assign the infrared bands to the respective
   carbonyl compounds and justify your decision.

2.27

6. An intense absorption band is observed at 1,766 cm^{-1} in the infrared
   spectrum. Decide whether the infrared spectrum belongs to a compound
   of Structures 2.28 or 2.29.

(continued)

2.28

2.29

7. The range of the characteristic C=O stretching vibrations from 1,850 to 1,650 cm^{-1} is smaller than that observed for the C=S stretching vibration (1,275–1,030 cm^{-1}). Explain this fact.

8. Figure 2.41 shows the infrared spectrum of a liquid compound. Develop the structure and assign the infrared bands. The molecular peak is $m/z = 128$ amu.

9. Figure 2.42 shows the infrared spectrum of a liquid compound. Develop the structure and assign the infrared bands. The molecular peak is $m/z = 78$ amu.

10. Figure 2.43 shows the infrared spectrum of a solid compound. Derive the structure and assign the infrared bands. The following information is obtained by the mass spectrum given in m/z (I_{rel} in%): 157 amu (40) M^{+}, 158 amu (3.3), 159 amu (2).

11. The infrared spectrum of a di-*tert*-butylphenol (Structure 2.30) is presented in Fig. 2.44. Develop the structure and assign the infrared bands.

2.30

(continued)

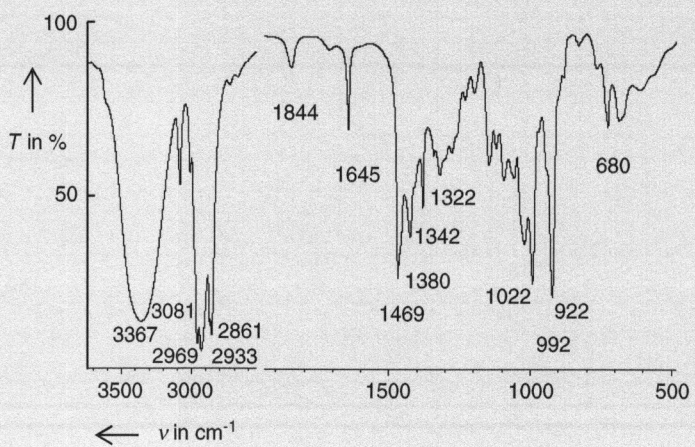

Fig. 2.41 Infrared spectrum of a liquid compound; sampling technique: capillary thin film; **MS:** $m/z = 128$ amu $M^{\bullet+}$

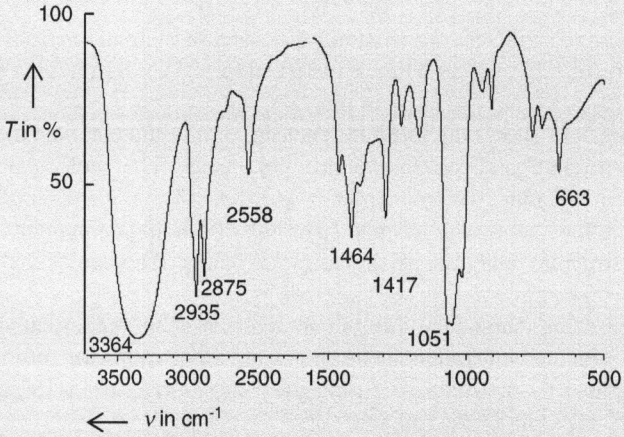

Fig. 2.42 Infrared spectrum of a liquid compound; sampling technique: capillary thin film; **MS:** $m/z = 78$ amu $M^{\bullet+}$

12. Derive the structure and assign the given wave numbers in the infrared spectrum shown in Fig. 2.44. The CI-mass spectrum (reaction gas: *iso*-butylene) shows an intense peak at $m/z = 85$ amu.

13. Dependent on the solvents the thermic decomposition of phenyl azo cyclo hexene provides the two products whose structure is given in Structure 2.31A, B. The infrared spectra feature the following bands in the range $v > 3,000$ cm^{-1}:

(continued)

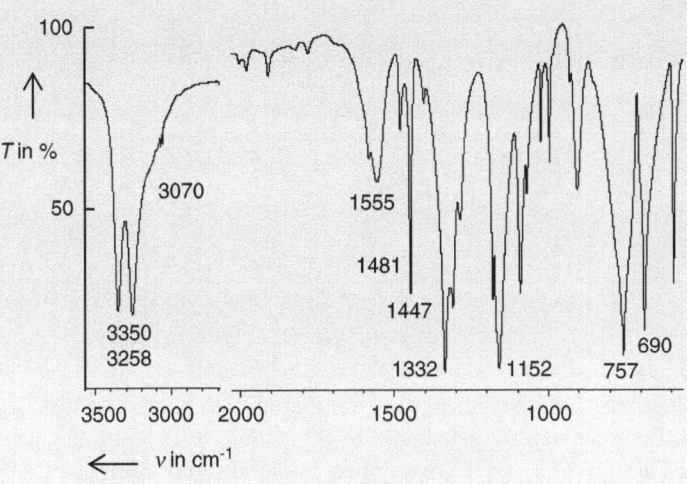

Fig. 2.43 Infrared spectrum of a solid compound; sampling technique: KBr pellet; **MS**: m/z (I_{rel} in%): 157 amu (40) $M^{+\cdot}$, 158 amu (3.3), 159 amu (2)

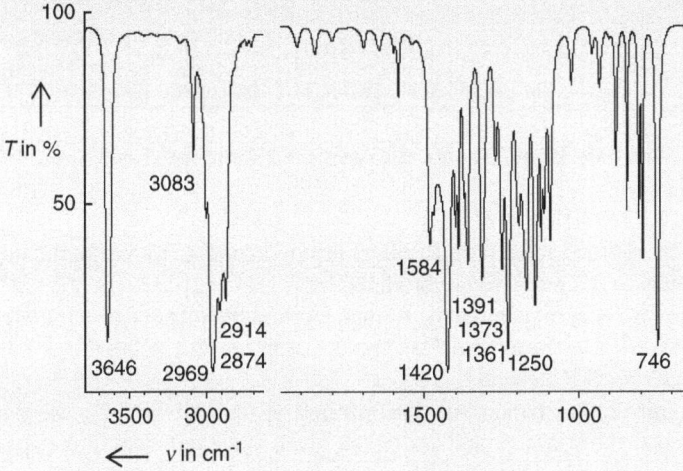

Fig. 2.44 Infrared spectrum of a di-*tert*-butylphenol (Structure 2.30); sampling technique: KBr pellet

Product I: sharp band at 3,360 cm^{-1},
Product II: broad band centered around 3,150 cm^{-1}.
What structures have the reaction products?

(continued)

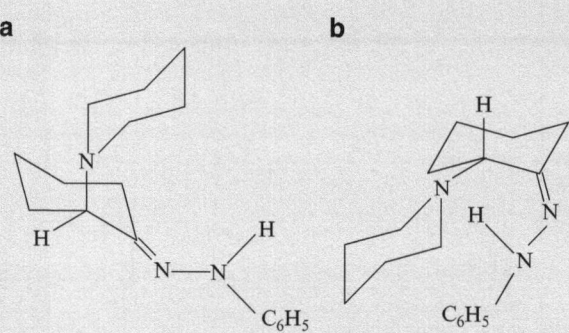

14. Some mass spectrometric data and the infrared spectra of two isomeric liquid compounds are featured in Fig. 2.45a, b. Develop the structures and assign the given wave numbers in the infrared spectra.

15. The following bands listed in arbitrary order are observed in the infrared spectra of four isomeric pentenes possessing the structures

$$H_2C = CHC_3H_7 (\text{Structure } 2.32) \quad H_2C = C(CH_3)C_2H_5 (\text{Structure } 2.33)$$

$$trans - C_2H_5CH = CHCH_3 (\text{Structure } 2.34)$$
$$cis - C_2H_5CH = CHCH_3 (\text{Structure } 2.35)$$

A: 966 (s); **B:** 1,780 (w), 887 (s); **C:** 1,826 (w), 1,643 (m), 993 (s), 912 (s);
D: 1,659 (m), 689 (m).
Which structure belongs to which infrared spectrum? Assign the infrared bands to the respective vibrations.

16. Figure 2.46a, b features the Raman (A) and the infrared spectra (B) of an aromatic C_{14} compound. Develop the structure and assign the given wave numbers in the spectra.

17. Figure 2.47a, b presents the infrared spectra of two C_8 compounds. Derive the structure and assign the given wave numbers in the spectra.

18. Some mass spectrometric data and the infrared spectra of four structurally isomeric compounds are given in Fig. 2.48a–d. Derive the structure, assign the given wave numbers in the spectra and formulate the relevant mass spectrometric fragmentation for each compound.

19. A section of an infrared spectrum in the range 1,800–1,600 cm^{-1} obtained by a reaction product of succinic acid is shown in Fig. 2.49. Which reaction product was obtained?

(continued)

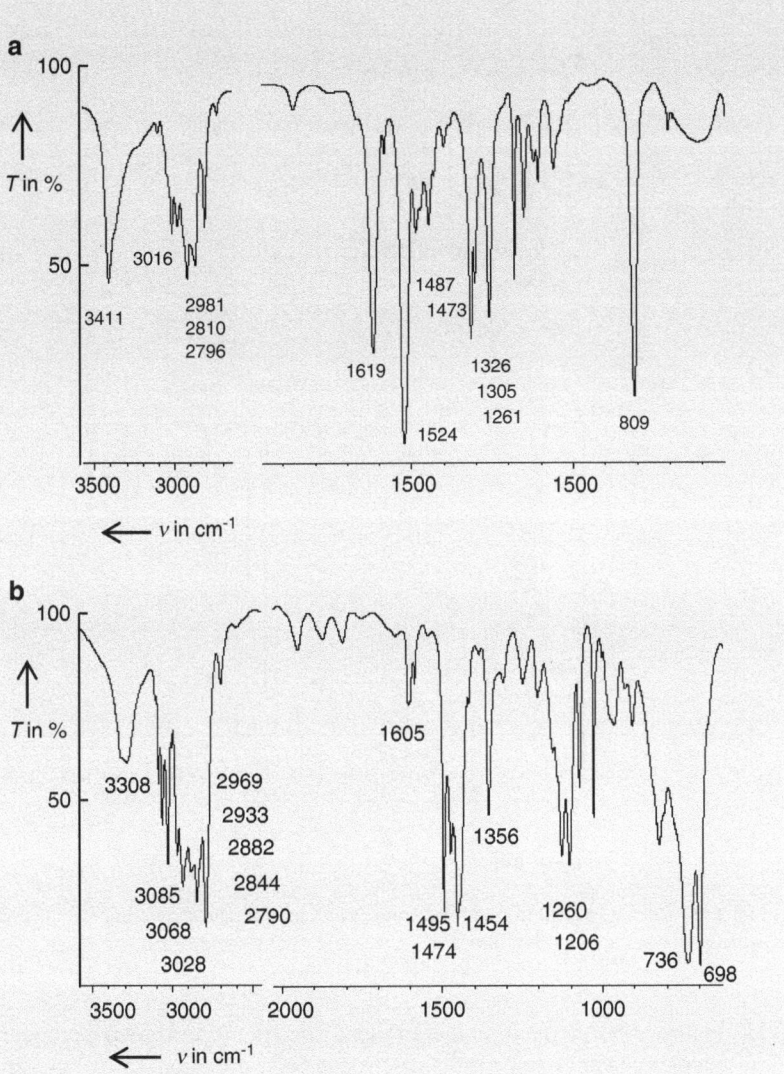

Fig. 2.45 (**a, b**) Infrared spectra of a two isomeric compounds; sampling technique: capillary thin film **a: h-Ms:** $[M]^{+}$: $m/z = 121.0891$ amu; base peak: $m/z = 120$ amu. **b: MS:** Base peak: $m/z = 120$ amu; intense peaks: m/z (I_{rel} in%): 44 (93), 91 (68)

20. In general, the asymmetric C–O stretching vibration is observed at higher wave numbers than the corresponding symmetrical one. This order is reversed in carboxylic anhydrides. How can this fact be confirmed experimentally?

(continued)

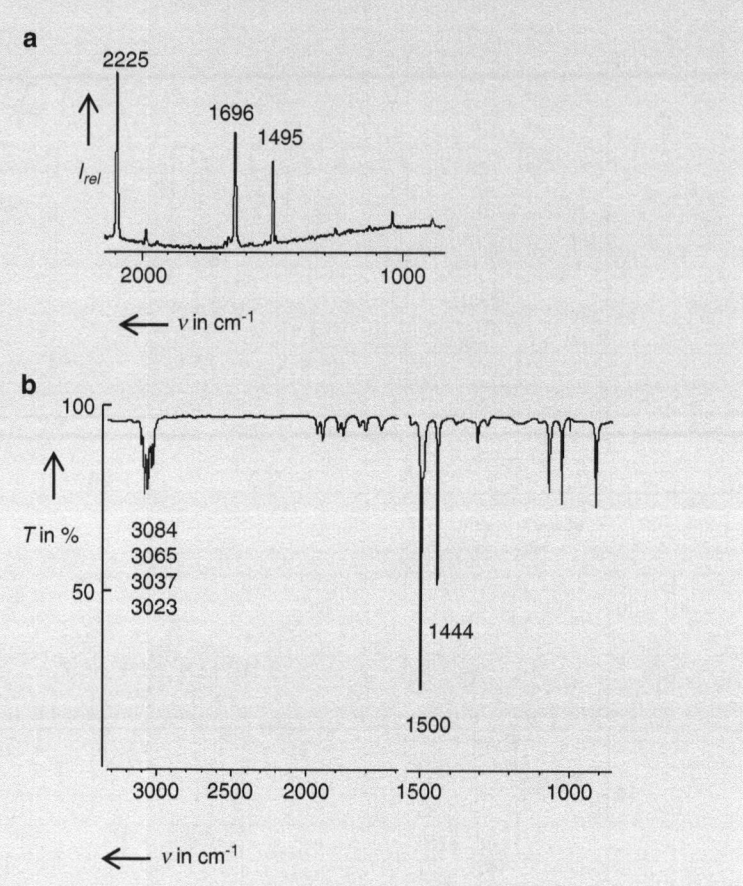

Fig. 2.46 Raman (**a**) and infrared spectrum (**b**) of an aromatic C_{14} compound; sampling technique (IR): capillary thin film

21. Figure 2.50a–e feature the infrared spectra of unknown compounds supplemented by some mass spectrometric information. Derive the structure and assign the given wave numbers in the spectra.

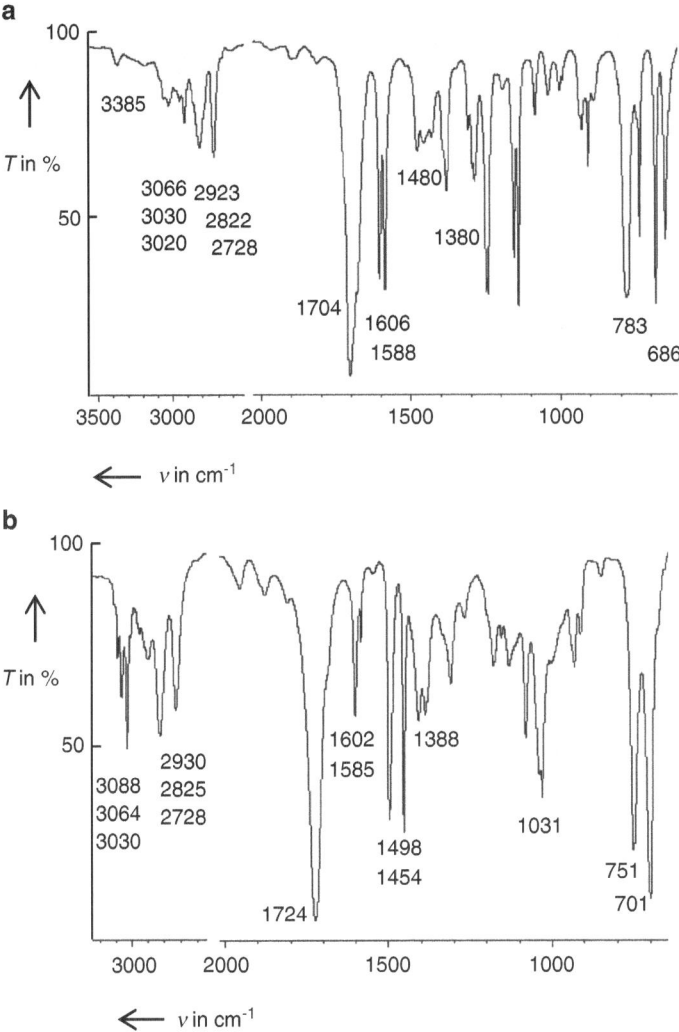

Fig. 2.47 (a, b) Infrared spectra of two C_8 compounds; sampling technique: capillary thin film.
a: MS: intense $[M - 1]^{+\bullet}$ peak. **b: MS**: Base peak: $m/z = 91$ amu

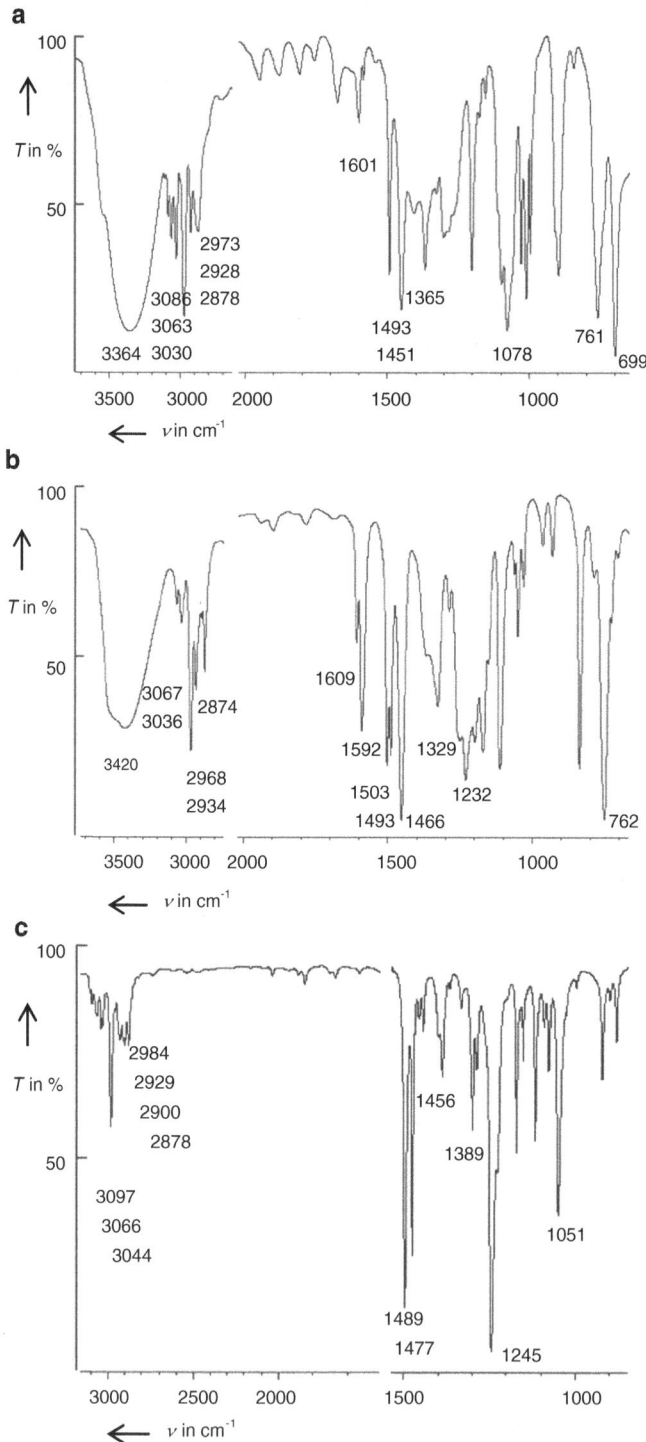

Fig. 2.48 (continued)

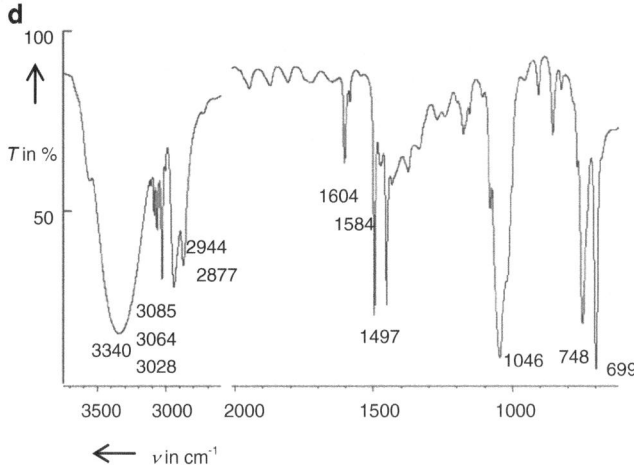

Fig. 2.48 (**a–d**) Infrared spectra of four structurally isomeric compounds; sampling technique for all samples: capillary thin film. **a: MS**: m/z in amu (I_{rel} in%): 107 (100), 122 M⁺⁺, intense peaks: 51, 77, 79. **b: MS**: m/z (I_{rel} in%): 107 (100), 122 (40) M⁺⁺, peaks with $I_{rel} > 10\%$: 39, 51, 77, 79, 91. **c: MS**: Base peak: $m/z = 94$ amu; peaks with $I_{rel} > 10\%$: m/z in amu = 39, 51, 65. **d: MS**: Base peak: $m/z = 91$ amu

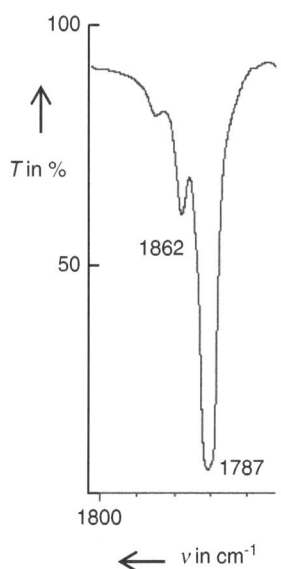

Fig. 2.49 Part of the infrared spectra of a reaction product of succinic acid; sampling technique: solution in carbon tetrachloride

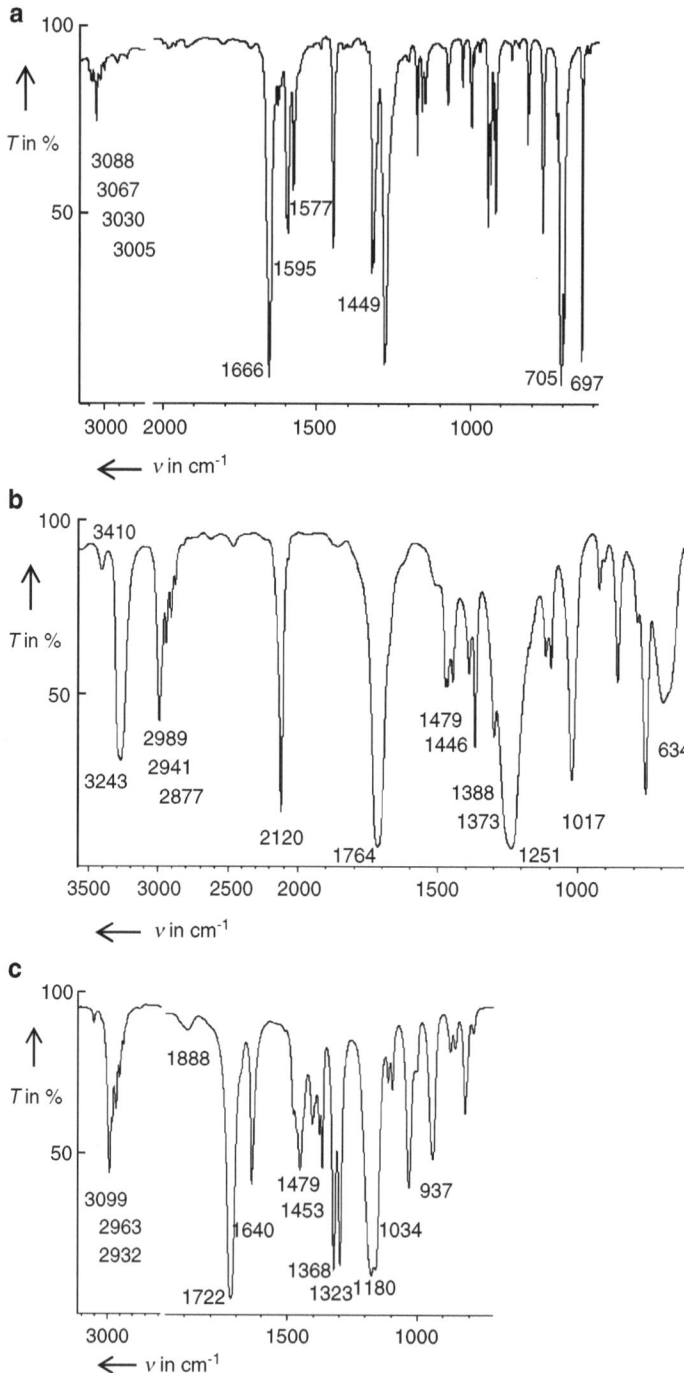

Fig. 2.50 (continued)

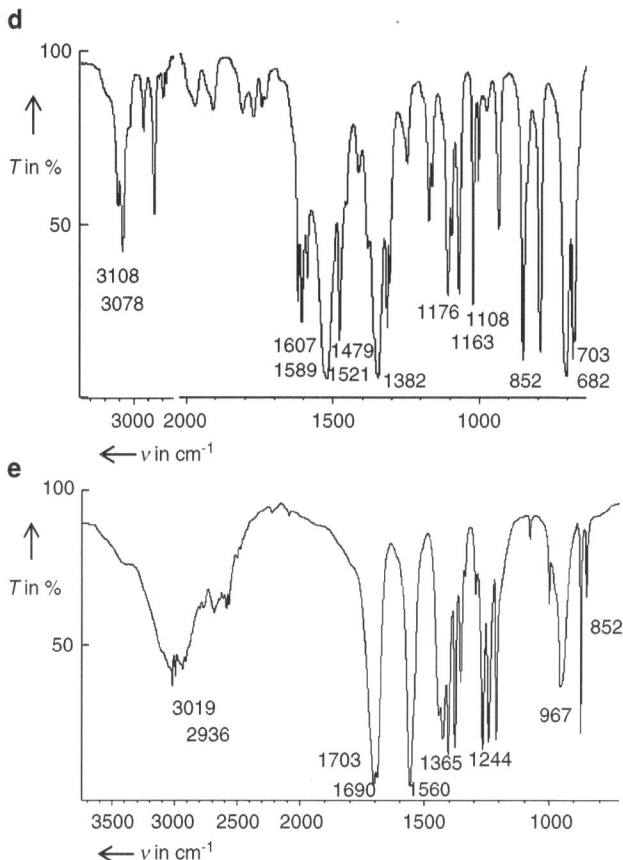

Fig. 2.50 (a–e) Infrared spectra and some mass spectrometric data of unknown compounds. **a:** Sampling technique: KBr pellet, **MS:** m/z in amu (I_{rel} in%): 51 (15), 77 (49), 105 (100), 182 (32) $M^{\bullet+}$, 183 (4.2). **b:** Sampling technique: capillary thin film, **MS:** m/z in amu (I_{rel} in%): 53 (100), 97 (4.8), 98 (0.4) $M^{\bullet+}$. **c:** Sampling technique: capillary thin film, **MS:** $m/z = 114$ amu [M-1]$^{\bullet+}$; C_6 compound. **d:** Sampling technique: KBr pellet, **MS:** m/z in amu (I_{rel} in%): 123 (85) $M^{\bullet+}$, 124 (6). **e:** Sampling technique: KBr pellet, **MS:** m/z in amu (I_{rel} in%): 119 (< 0.1) $M^{\bullet+}$, 45 (100), 55 (43), 73 (82)

Further Reading

1. Nakamodo K (1997) Infrared and Raman spectra of inorganic and coordination compounds. Wiley, New York
2. Smith B (1999) Infrared spectral interpretation. CRC Press, Boca Raton
3. Long DL (2001) The Raman effect. Wiley, New York
4. Colthup ND, Daly LH, Wiberley SE (1990) Introduction to infrared and Raman spectroscopy. Academic, New York

Chapter 3
Electronic Absorption Spectroscopy

3.1 Introduction

3.1.1 Range Within the Electromagnetic Spectrum

Electronic absorption spectroscopy is based on electron transitions from the ground to excited states caused by the absorption of electromagnetic radiation in the *ultraviolet* (UV) and *visible* (Vis) range. The electromagnetic waves are typically characterized by any of the following physical properties: frequency f in Hertz, wavelength λ in nm, wave number $\tilde{v}$ in cm^{-1}, or photon energy E in eV. The relations are illustrated by Eqs. 3.1, 3.2, 3.3 and 3.4,

$$f = \frac{c}{\lambda} \tag{3.1}$$

$$f = \frac{E}{h} \tag{3.2}$$

$$E = \frac{h \cdot c}{\lambda} \tag{3.3}$$

$$\tilde{v} = \frac{1}{\lambda} \tag{3.4}$$

where c is the speed of light in a vacuum ($c = 299{,}792{,}458$ m s^{-1}) and h is Planck's constant ($h = 6.6261 \ 10^{-34}$ J s $= 4.1357 \ 10^{-15}$ eV s). Generally, the unit wavelength λ in nm and wave number $\tilde{v}$ in cm^{-1} are used in connection with this spectroscopic method in the following ranges:

M. Reichenbächer and J. Popp, *Challenges in Molecular Structure Determination*, 145
DOI 10.1007/978-3-642-24390-5_3, © Springer-Verlag Berlin Heidelberg 2012

Vacuum UV ←	Ultraviolet (UV)	Visible (Vis)	→ NIR
λ in nm	200–400	400–800	
$\tilde{\nu}$ in cm^{-1}	50,000–25,000	25,000–12,500	

The energetically higher vacuum UV is the adjacent range of UV and the near infrared (NIR) is adjacent to the Vis range at the lower energy side of the electromagnetic spectrum. Note that only the unit wave number, but not wavelength, is directly proportional to the energy.

Absorption of electromagnetic radiation in the UV/Vis range can only take place if two conditions are fulfilled: The matter (molecules, crystals) must possess a *chromophore* and the interaction of the electromagnetic radiation with matter must *change the electronic dipole moment* which is the subject of the next sections.

3.1.2   Chromophore

A chromophore is a chemical group capable of absorbing selectively light resulting in absorption bands in the UV/Vis spectrum. Historically, the use of the term "chromophore" was limited to the absorption of visible light resulting in coloration of certain organic or bioorganic compounds. Nowadays, the "nonvisible" UV range is also included. Because the energetically very stable σ electrons cannot be excited by UV radiation beyond 200 nm the chromophore is bound to π and n electron systems. The chromophore can be localized to only a *part* of the molecule but it can also be extended into the *whole* molecule which will be demonstrated by some examples:

– Molecules whose chromophores are extended by the π system over the whole molecule (Structures 3.1, 3.2).

3.2

3.1

– Molecules whose π system is induced by substituents (Structure 3.3).

3.3

– Molecules possessing two or more separated π systems, which do not couple or couple only weakly with another (Structure 3.4).

The sp^3 hybridized spiro-C atom causes a "conjugation barrier" resulting in two subchromophore systems I and II.

The *addition rule* is valid for separated chromophores: The spectrum is the addition of the subchromophores.

Molecules having solely σ bonds or, additionally, one or more free n electron pairs (alkanes, aliphatic saturated alcohols, or ethers) do not possess a chromophore and, hence, do not give rise to absorption bands in the UV/Vis range because the possible $\sigma \rightarrow \sigma^*$ or $n \rightarrow \sigma^*$ transitions lie in the vacuum UV range. Therefore, such compounds can be used as solvents for the measurement of UV spectra.

The following information obtained by electronic absorption spectra is meaningful for molecular structure determination:

1. *Number* of absorption bands in the UV/Vis range
2. *Position* of the absorption bands, generally, given as the wavelength in the center of the absorption band (λ_{max} value in nm)
3. *Intensity* of the absorption bands (expressed as the molar absorptivity α in $\text{mol L}^{-1} \text{ cm}^{-1}$, also named extinction coefficient)
4. *Structural pattern* of the absorption bands.

The answer to these questions is generally the subject of quantum mechanics. For practical purposes, let us go the following route:

1. Reducing the molecule to the *base chromophore* which is, in general, well known from theoretical studies for most chemical classes
2. Application of *empiric rules* with regard to interactions between the base chromophore and substituents.

The first step, generally, enables the recognition of chemical classes and functional groups can be recognized by step two.

The *base chromophore* is the part of the molecule that is left if all substituents without contribution to the spectral behavior are removed.

Some examples to explain the term "base chromophore" follow:

– The spectral behavior of *substituted aromatic compounds* is determined by the base chromophore of *benzene* (Structure 3.1). Thus, the ultraviolet spectrum of acetophenone (Structure 3.3) should look like that of benzene. Deviations of the spectroscopic data are caused by the $CH_3–C=O$ substituent.
– The spectral behavior of a $\Delta^{4,6}$-*diene steroid* (Structure 3.5), *trans*-stilbene (Structure 3.6), or indigo (Structure 3.7)

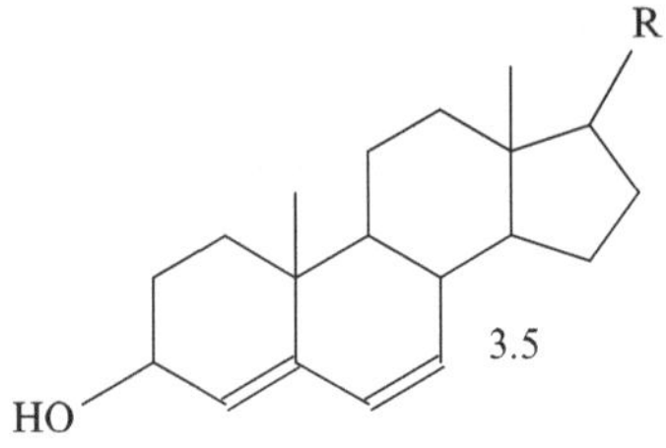

3.5

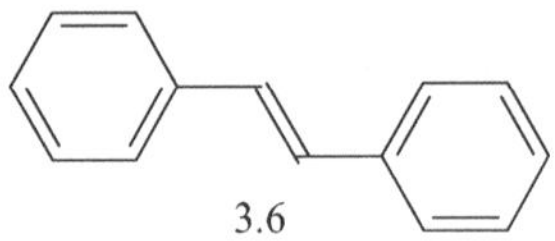

3.6

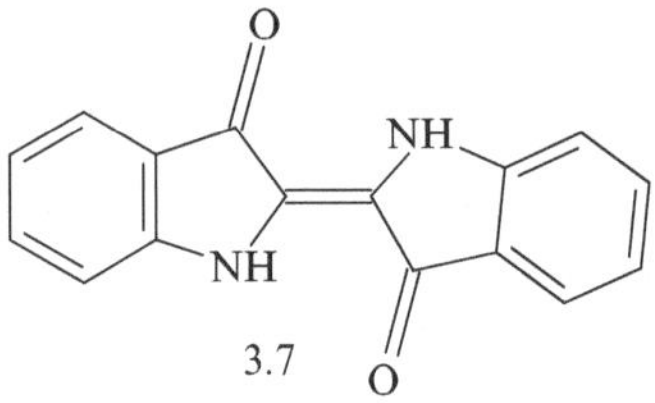

3.7

can be reduced to that of the base chromophores of Structures 3.8, 3.9, and 3.10, respectively.

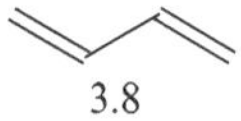

3.8

$-(CH{=}CH)_n-$

3.9

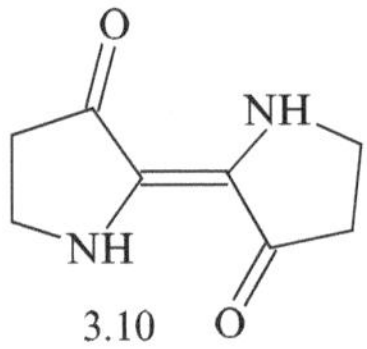

3.10

Note that, for example, the loss of a C=O group in Structure 3.7 leads to another spectrum which does not show any similarity; thus, this compound will be colorless whereas the molecule of Structure 3.10 is a colorant whose λ_{max} value and absorptivity α are like those of indigo (Structure 3.7).

3.1.3   Term System of Molecules

Molecular structure determination by means of electron absorption spectroscopy only requires a *qualitative* answer to the four questions given above which can be achieved by the *term system* of the respective base chromophore of the molecule.

The term system represents the *real existing electronic states* of the molecule. Therefore, the name "*spectroscopic states*" is used also because the electronic transitions within the term system caused by the absorption of electromagnetic radiation in the UV/Vis range provide the respective absorption bands.

The term system can be qualitatively obtained on the basis of the respective molecular orbital diagrams applying the so-called π approximation. The π

approximation considers alone the (energetically higher) free electron pairs (*n*) as well as the *p* orbitals generating the bonding (π) and antibonding (π^*) molecular orbitals in order to construct the molecular orbital (MO) diagram starting with the respective atomic orbitals (AO). The energetically lower *s* orbitals which provide the σ pattern of the molecule are not considered because, as mentioned above, such electrons will not be excited by UV light with a wavelength greater than 200 nm.

The number of possible transitions in the term system corresponds to the *number* of absorption bands. However, the "*position*" of the absorption bands (expressed as the λ_{max} values) cannot be obtained by this simple procedure because the energy of the terms is not known. Therefore, we will make recourse to empirical rules and laws provided by results from quantum mechanics.

Let us start with the development of the term system of formaldehyde $H_2C=O$. $H_2C=O$ is the base chromophore of all saturated aldehydes and ketones.

3.1.3.1 Molecular Orbital (MO) Diagram and the Symmetry of the Wave Function of the MOs

Figure 3.1 shows the base set of atomic orbitals (AO) to construct the molecular orbital (MO) diagram of formaldehyde $H_2C=O$. As shown in Fig. 3.1, solely the p_x orbital of the carbon and those of the oxygen atoms are employed in the π bond as well as the free *n* electron pair of oxygen. The second free electron pair of oxygen lies energetically too deep, so that it is not excited by UV light over 200 nm. The σ skeleton is pictured by lines.

The molecular orbitals are obtained by linear combination of the respective atomic orbitals. The following rule is valid for the energetic order of the MOs:

The energy of a molecular orbital lies higher the greater is the number of nodal planes.

As will be shown later, the application of the selection rules needs the symmetry of the wave functions of the electronic states and, hence, the symmetry of the wave functions of the molecular orbitals from which they are obtained. As in the procedure explained above for the determination of the symmetry of normal

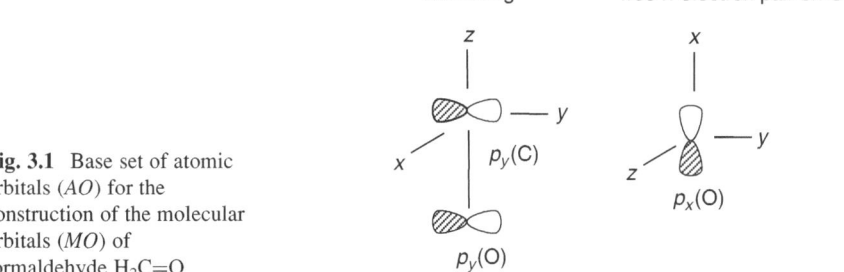

Fig. 3.1 Base set of atomic orbitals (*AO*) for the construction of the molecular orbitals (*MO*) of formaldehyde $H_2C=O$

modes just the character of the transformation matrix is necessary. Therefore, the following rules are valid:

1. All symmetry operations of the respective symmetry group of the molecule must be acted upon the base set of the atomic orbitals.
2. Only that atomic orbital provides a contribution to the character of the transformation matrix that does not change its position by the execution of the respective symmetry operation.
3. The (most) reducible base set of the character of the transformation matrix must be transferred into the set of the respective irreducible base sets applying the formula given in Eq. 2.6.
4. The symmetry classes of the irreducible representations are assigned to the respective molecular orbitals. Note that, generally, the symmetry classes of the MOs are denoted in small type because molecular orbitals do not possess physical reality.
5. The number of MOs and, hence, the number of symmetry classes must be the same as the number of AO used for the linear combination.

Let us now apply these rules to formaldehyde which belongs to the symmetry group C_{2v}. The transformation matrix for the π bond—$\chi(\pi)$—and that of the free electron pair—$\chi(n)$—are summarized in Table 3.1. The latter is just an irreducible representation because only one AO was used, but the transformation matrix of the π bond must be reduced by application of Eq. 2.6. The result is given in the last column of Table 3.1.

The MO diagram for $H_2C=O$ is presented in Fig. 3.2.

The π^* orbital has a higher energy than that of the π orbital because it possesses an additional nodal plane. The assignment of the symmetry classes is simple in this case because both orbitals belong to the same symmetry class b_2.

3.1.3.2   Development of the Term System and the Symmetry of the Electron States

The molecular orbitals are occupied with the respective electrons. The representation of the electronic configuration shown in Fig. 3.2 belongs to the energetic *ground state* of the molecule. All four electrons are arranged in a way that the energy is minimized taking into account Pauli's exclusion principle and Hund's rule.

Electronically excited states are created by the occupation of molecular orbitals with higher energy. Thus, the *first* electronically excited state is obtained if an

Table 3.1 Symmetry of the molecular orbitals of $H_2C=O$

C_{2v}	E	$C_2(z)$	$\sigma_v(x,z)$	$\sigma_v(x,z)$	Symmetry classes of the MO
$\chi(\pi)$	2	−2	−2	2	$b_2 + b_2$
$\chi(n)$	1	−1	1	−1	b_1

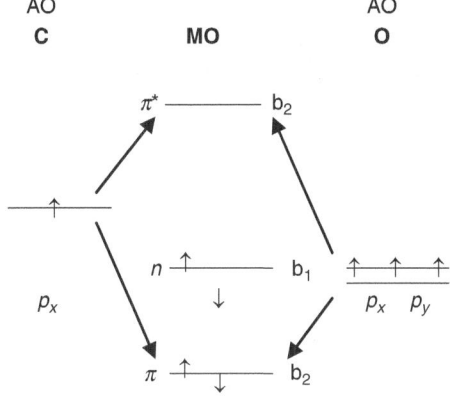

Fig. 3.2 MO diagram of $H_2C=O$ in the ground state with designation of the symmetry of the MOs

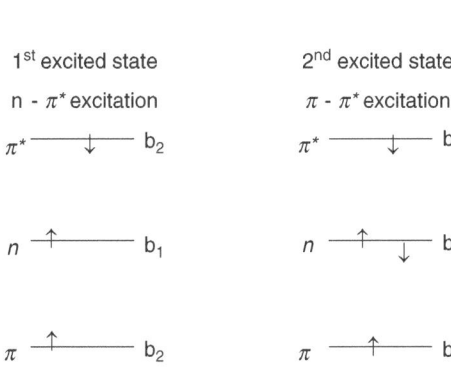

Fig. 3.3 Electronic configuration for the first and second excited state

n electron is transferred to the π^* orbital. The electron configuration $b_2{}^2b_1{}^1b_2{}^1$ corresponds to the $n \rightarrow \pi^*$ excitation. The *second* electronically excited state is created by transfer of a π electron into the π^* orbital resulting in the electron configuration $b_2{}^1b_1{}^2b_2{}^1$ which corresponds to the $\pi \rightarrow \pi^*$ excitation. Both electronic configurations are presented in Fig. 3.3.

The electronic configurations shown in Figs. 3.2 and 3.3 represent the three terms of the molecule: ground state, first and second excited states, respectively. The real energy belonging to these terms can admittedly not be determined by this simple procedure, but the second excited state has a higher energy than that of the first one.

Though the wave functions of the electronic states are unknown we can obtain their symmetry by applying the following rules:

1. Molecular orbitals that are *fully* occupied belong to the *totally symmetric representation*.
2. The symmetry of occupied molecular orbitals that are *half-full* is obtained by the *direct product* of the MO's irreducible representations, i.e., by a multiplication of the respective character sets.

The symmetry classes of the terms are denoted by capital letters because terms are real existing electronic states.

Because the molecular orbitals of the electronic *ground state* of organic molecules are always fully occupied, they belong to the totally symmetric class Γ_1.

The symmetry of the *excited electronic states* is obtained by the direct product ($\times$) of the respective character sets:

First excited state:

C_{2v}	E	$C_2(z)$	$\sigma_v(x,z)$	$\sigma_v(y,z)$	Symmetry of the wave function of the first excited state
b_1	1	-1	1	-1	
b_2	1	-1	-1	1	
$b_1 \times b_2$	1	1	-1	-1	$\Rightarrow \mathbf{A_2}$

Second excited state:

C_{2v}	E	$C_2(z)$	$\sigma_v(x,z)$	$\sigma_v(y,z)$	Symmetry of the wave function of the second excited state
$b_2 \times b_2$	1	1	1	1	$\Rightarrow \mathbf{A_1}$

Thus, the wave function of the first and second excited electronic states belong to the symmetry classes A_2 and A_1, respectively.

3.1.3.3 Spin Multiplicity

Remember that the electron spin $s = 1/2$ is an intrinsic property of electrons which determines the *spin multiplicity M* of the terms. The spin multiplicity of organic molecules consisting mostly of light atoms is given by Eq. 3.5

$$M = 2S + 1 \tag{3.5}$$

S is total spin angular momentum which is obtained for such molecules by the sum of the spin s_i of the single electrons.

If two electrons are arranged with antiparallel spin the total spin angular momentum S is equal to zero. For such a term the spin multiplicity M is 1 according to Eq. 3.5, and the term is named *singlet state S*. This is always valid for the ground state of any organic molecule.

However, if the excitation of an electron gives rise to a reversal of the spin, then both electrons are arranged with parallel spin. In this case, the total spin angular momentum quantum number S is equal to 1 and the spin multiplicity is 3. Such a term is named *triplet state T*. Note that a triplet state T results whenever two electrons are arranged with parallel spin.

The rule is valid according to the energetic position of the triplet state:

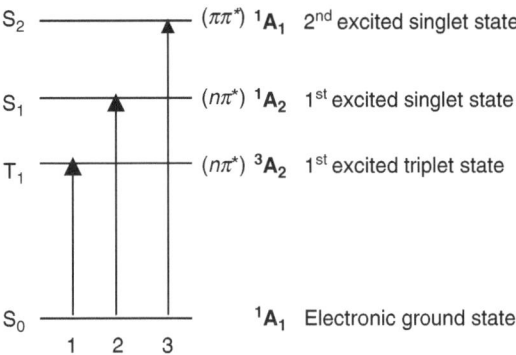

Fig. 3.4 Term system, term symbols, and possible electronic transitions of $H_2C=O$

The triplet state is always characterized by a lower energy than the respective singlet state.

Now, we can picture the term system, i.e., the total electronic states of formaldehyde $H_2C=O$ (Fig. 3.4). The three arrows correspond to the three possible electronic transitions which results in three absorption bands in the UV range at $\lambda > 200$ nm: $S_0 \rightarrow T_1$, $S_0 \rightarrow S_1$ ($n \rightarrow \pi^*$ transition), and $S_0 \rightarrow S_1$ ($\pi \rightarrow \pi^*$ transition).

3.1.4 Selection Rules

According to Fig. 3.4, the UV spectrum of $H_2C=O$ should possess three absorption bands. But the question is how intense are these bands? The intensity is determined by the transition strength which can be qualitatively evaluated by the selection rules.

3.1.4.1 Spin Selection Rule

Electronic transitions between states of different multiplicity are forbidden.

Thus, singlet-triplet transitions are forbidden. Note that this rule can be lifted by spin-orbit coupling. But the spin-orbit coupling constant and, hence, the transition strength is small for molecules with light atoms. Therefore, the intensity of singlet-triplet absorption bands is very small and such transitions are meaningless in structural determination of organic compounds.

3.1.4.2 Symmetry Selection Rule

The electron transition is accompanied by a change of the dipole moment; hence, as has already been explained in Sect. 2.1.3, the strength of an absorption band is determined by the transition moment which is proportional to the transition moment integral according to Eq. 3.6.

$$\vec{M} \sim \int \Psi_0 \, \vec{\mu} \, \Psi_n \, d\tau \qquad (3.6)$$

Ψ_0 and Ψ_n are the wave functions of the ground and excited state, respectively, and $\vec{\mu}$ is the dipole moment operator.

When the transition moment integral is zero the intensity will be zero and thus, the transition will be forbidden. Symmetry consideration can again help us to determine if the electron dipole change is zero (forbidden) or not (allowed).

An integrand can be nonzero if the direct product of the integrand belongs to the totally symmetric representation A_1 or if it is contained in the symmetry species set. Because for organic molecules the wave function of the ground state Ψ_0 is always totally symmetric (A_1), the integrand belongs to the symmetry species A_1, and only then, if the symmetry species of the wave function of the respective excited state is equal to that for the dipole moment, i.e., $\Gamma(\mu) \times \Gamma(\Psi_n) \in \Gamma_1$ if $\Gamma(\mu) = \Gamma(\Psi_n)$. This condition is the basis of the symmetry selection rule:

> Electron transitions are only allowed if the symmetry of the respective excited state belongs to the same symmetry species as one of the dipole moment components which are given in the second to last column of the character table of the symmetry group of the molecule.

Note that electron transitions of molecules are accompanied by the excitation of vibrations of atoms. Thus, the symmetry selection rule can be partially lifted by contribution of suitable vibrations accompanying the electron transition. Examples are asymmetric vibrations of the C–C skeleton in benzene because such vibrations distort the D_{6h} symmetry which is responsible for rendering the electron transfer allowed.

Electron transitions that are not limited by the selection rules give rise to intense absorption bands. Table 3.2 presents an overview of the molar absorptivity of the allowed and forbidden electron transitions.

A special symmetry selection rule is the *Laporte* rule which is relevant, in particular, for the electronic spectroscopy of transition metals (see Sect. 3.6). The Laporte rule states that electronic transitions conserving either symmetry or asymmetry with respect to an inversion center are forbidden. Thus, the transitions g (gerade) $\rightarrow g$ or u (ungerade) $\rightarrow u$, respectively are forbidden. Because d and p orbitals are symmetric and asymmetric, respectively in relation to the inversion center, $d \rightarrow d$ and $p \rightarrow p$ transitions respectively are Laporte forbidden.

Let us now apply the selection rules to the electron transitions in Fig. 3.4.

Table 3.2 Molar absorptivity α of possible electron transitions	Electron transition	α in L mol^{-1}
	Spin forbidden	$\approx 10^{-3}$
	Symmetry forbidden	$\approx 10^{2}$
	Allowed	10^{4}–10^{5}

Clearly, transition 1 is a spin-forbidden electron transfer but all other transitions are spin allowed because the promotion of electrons does not include a change in their spin. Thus, transition 1 will not be visible in the routine absorption spectrum of formaldehyde because of its very small absorptivity (see Table 3.2).

From the character table it follows that the symmetry class A_2 does not contain one of the dipole moment components. Hence, transition 2, though spin allowed, is symmetry forbidden. The intensity of the absorption bands caused by this electron transition will be $\alpha \approx 50-100 \text{ L mol}^{-1} \text{ cm}^{-1}$. Note that this electron transition is of the type $n \rightarrow \pi^*$.

Both the dipole moment component in the z-direction as well as the symmetry of the wave function of the second singlet electron transfer belong to the same symmetry species A_1. Hence, the transition $S_0 \rightarrow S_2 \, (\pi \rightarrow \pi^*)$ is allowed providing an intense absorption band with $\alpha \approx 10^4 \text{ L mol}^{-1} \text{ cm}^{-1}$.

Note that the results obtained for the simple molecule $H_2C{=}O$ are valid for all saturated aldehydes and ketones, because transitions between σ electrons are not considered.

Furthermore, these results can be generalized in the following rules:

$n \rightarrow \pi^*$ *electron transitions* are *forbidden* because of the symmetry selection rule. They will give rise to only weak absorption bands; hence, they are, generally, meaningless for structure determination.

$\pi \rightarrow \pi^*$ *electron transitions* are, apart from some highly symmetric molecules like benzene, *allowed* and provide intense absorption bands that can be used for structure determination.

Challenge 3.1
Develop the term system of *trans*-butadiene and estimate the intensity for the possible electron transitions in the range $\lambda > 200$ nm according to the procedure outlined for formaldehyde.

Solution to Challenge 3.1
The procedure is similar to that of formaldehyde.

–   Symmetry group according to the algorithm in Table 6.2.1: C_{2h}
–   Base set of atomic orbitals:

4 p_z orbitals:  z

(continued)

- Transformation matrix obtained by the action of all symmetry operations of the point group C_{2h} on the base set of atomic orbitals as well as the symmetry classes after reduction of the reducible character set by means of Eq. 2.6:

C_{2h}	E	$C_2(z)$	I	σ_h	Symmetry classes of the molecular orbitals (MO)
$\chi(\pi)$	4	0	0	-4	$\Rightarrow 2\,a_u + 2\,b_g$

- The four molecular orbitals must be energetically ordered according to the number of nodal planes as Fig. 3.5 shows.

The symmetry property with regard to the inversion center i may be helpful to assign the four symmetry classes to the respective molecular orbitals. As is shown in Fig. 3.5, the two molecular orbitals, for example, π_1 and π_3^* are antisymmetric in respect to i because the symmetry operation i transforms, for example, the hatched lobe into a non-hatched one. Therefore, such an MO belongs to a symmetry class whose character $\chi(i) = -1$, but there is only one symmetry class for it, namely a_u. Thus, the molecular orbitals π_1 and π_3^* are assigned to a_u and, hence, the molecular orbitals π_2 and π_4^* belong to the symmetry class b_{2g}.

The electron configurations for the first and second excited states and the resulting term system are presented in Fig. 3.6.

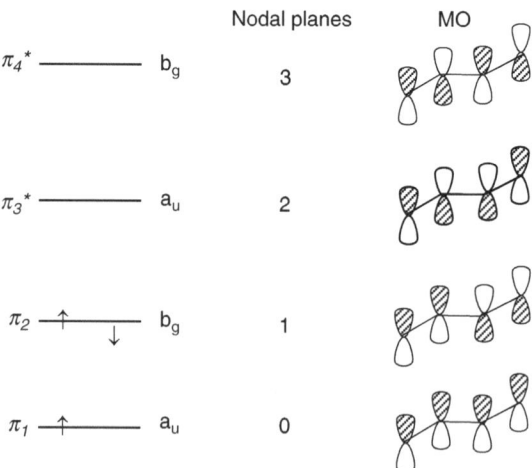

		Nodal planes	MO
π_4^* ———— b_g		3	
π_3^* ———— a_u		2	
π_2 ———— b_g		1	
π_1 ———— a_u		0	

Fig. 3.5 MO diagram for the electron ground state of *trans*-butadiene

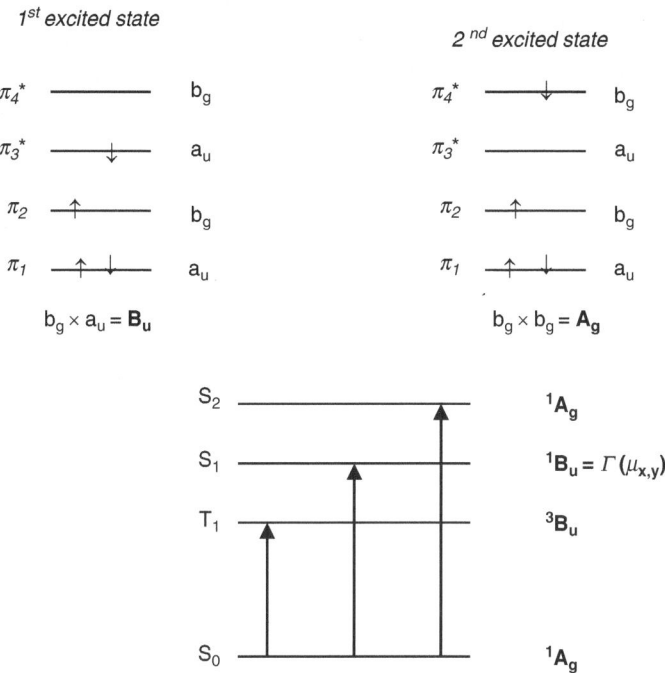

Fig. 3.6 Electron configurations for the first and second excited states of *trans*-butadiene and the resulting term system

The following three electron transitions are possible for *trans*-butadiene: $S_0 \rightarrow T_1$, $S_0 \rightarrow S_1$, and $S_0 \rightarrow S_2$. The transition $S_0 \rightarrow T_1$ is spin forbidden and, hence, it is not observed in the absorption spectrum. The transition $S_0 \rightarrow S_1$ is spin and symmetry allowed because the wave function of the first excited singlet state belongs to the symmetry class that contains a dipole moment component. But $S_0 \rightarrow S_2$ is a symmetry forbidden transition because there is no dipole moment component contained in symmetry class A_1. Therefore, the absorption spectrum will possess an intense absorption band ($\alpha \approx 10^4$ L mol^{-1} cm^{-1}) and a band with lower intensity at shorter wavelengths which will lie in the vacuum UV range.

Note that *trans-butadiene is the base chromophore of dienes and polyenes* and compounds of theses chemical classes show electron absorption spectra like that of *trans*-butadiene.

Challenge 3.2

For solutions to Challenge 3.2, see extras.springer.com.

1. Derive the term system of benzene which is the base chromophore of the aromatic compounds. The six wave functions of the molecular orbitals obtained by the base set of the 6 p_z atomic orbitals belong to the following symmetry classes ordered according to increasing energy: a_{1u}, e_{1g}, e_{2u}, b_{2g}.

(continued)

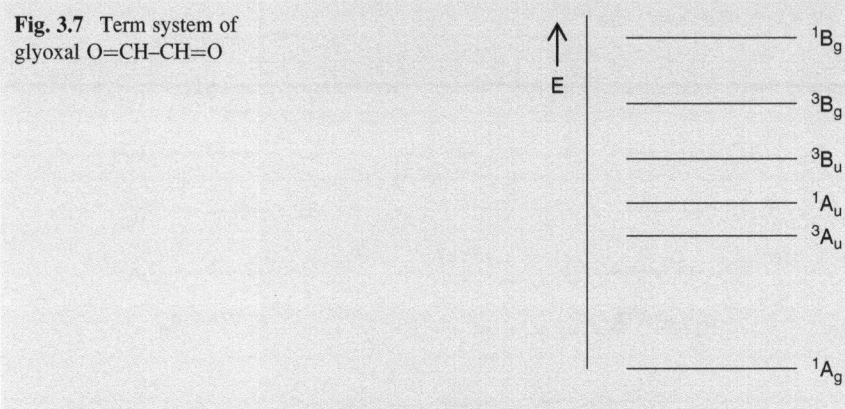

Fig. 3.7 Term system of glyoxal O=CH–CH=O

Which of the possible electron transitions are allowed and forbidden, respectively?

2. Figure 3.7 shows the term system of glyoxal. Which of the possible electron transitions are allowed and forbidden, respectively? Evaluate the respective absorptivity of the absorption bands.

3. Explain why electron transitions of the same azimuthal quantum number are forbidden.

4. Which solution (concentration; solvent) is necessary to create the $S_0 \rightarrow S_1$ absorption band for a. benzene, b. glyoxal?

3.1.5   The Structural Pattern of Electron Absorption Bands

The radiation-induced excitation of electron transitions is accompanied by the excitation of vibrations of the atoms. Therefore, these absorption bands are characterized by their relatively large half-width. The absorption bands can show an overlying vibrational pattern (⇒ *vibrationally structured electron absorption bands*) or they can be observed as *structureless bands* caused by the time of the deactivation steps. Some rules, meaningful in structural determination, will be given:

– The gaseous state as well as nonpolar solvents favors the occurrence of a vibrational structure which can disappear in polar solvents.

– A vibrational pattern is only observed in "rigid" molecules, i.e., in molecules that possess only few low-energetic deactivation channels. Thus, unsubstituted aromatic classes like benzene, acenes, phenes, or polyenes show a well-formed vibrational pattern. However, an alkyl substituent diminishes the vibrational pattern with increasing C–C chain length. But if a "rigid" molecule is produced by linking the alkyl substituent at the aromatic ring the vibrational pattern can again be observed.

– Dienes –CH=CH–CH=CH– show, generally, a well-pronounced vibrational pattern, in contrast to enones –CH=CH–CH=O.

– Charge transfer (CT) transitions always provide structureless absorption bands.

The vibrational pattern of the absorption bands is determined by the *Franck-Condon principle* which states:

> The electron transition is so fast ($\approx 10^{-16}$ s) that there is no change of the atomic positions in a molecule; hence, a vertical transition will be the one with the highest intensity. All other transitions will be less intense.

Let us assume a molecule that has approximately the same C–C distance in the ground and excited states, then the (vertical) $0 \rightarrow 0$ transition is the most favorable one. The intensity of the transitions $0 \rightarrow 1$, $0 \rightarrow 2$, $0 \rightarrow 3$ and so on will be progressively diminished resulting in an asymmetrically structured vibrational pattern. Such an absorption band is, for example, observed for anthracene (see Fig. 3.13c). Note that the 14 π system of anthracene and, hence, the bond strength is not markedly weakened by loss of one electron after electron excitation.

However, the excitation of one of the 6 π electrons in benzene will diminish the bond strength in the excited state resulting in a greater C–C distance. As a consequence, the vertical electron transition no longer transfers the molecule into the ground vibrational state of the electron excited state but into a higher one. Thus, the $0 \rightarrow 3$ transition is the most favored one for benzene and the lower und higher energetic transitions possess a lower intensity resulting in a symmetric vibrational pattern (see Fig. 3.13a).

The vibrational pattern gives rise to information about the structure of the molecule in the only short-lived first electron excited state. The distance of the sub-bands corresponds to the vibrational spectrum of the molecule in the excited state.

For example, the distance between the sub-bands in benzene is approximately 1,300 cm^{-1} which can be assigned to the C=C stretching wave number of the S_1 excited benzene molecule. Note that the corresponding vibrations lie at 1,600–1,500 cm^{-1} for the S_0 state which corresponds to a higher CC bond strength. Calculation of the force constant can provide more insight into the bond strength of the molecule in the excited states which can be of interest in the photochemistry.

Challenge 3.3

For solutions to Challenge 3.3, see *extras.springer.com*.

1. The electron absorption band of *n*-propylbenzene (Structure 3.11) lacks a fine structure. In contrast, hydrindene (Structure 3.12) shows a well-structured vibrational pattern. Explain this fact.

3.11   3.12

(continued)

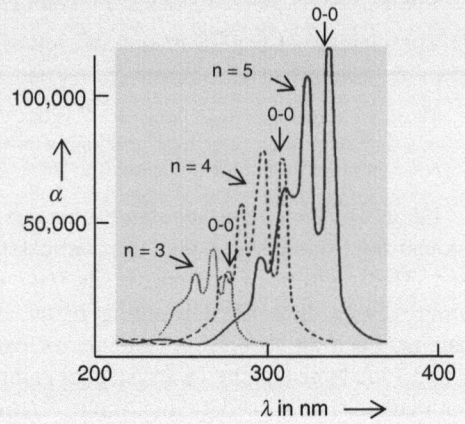

Fig. 3.8 Electron absorption spectra of polyenes $CH_3(CH=CH)_nCH_3$ with $n = 3, 4,$ and 5 in n-hexane with the marked 0–0 transitions. The absorptivity α is given in L mol^{-1} cm^{-1}

2. Figure 3.8 presents the electron absorption spectra of polyenes $CH_3(CH=CH)_nCH_3$ with $n = 3, 4,$ and 5. Explain the change in the vibrational pattern of these three bands. What information is obtained with respect to the structure of these compounds?

3.2   Electron Absorption Spectra of Organic Chemical Classes

3.2.1   Small Nonconjugated Chromophore

1. *Alkenes*
   Ethene represents the base chromophore of all saturated or nonconjugated alkenes. The allowed $\pi \rightarrow \pi^*$ transition lies in the vacuum-UV range at $\lambda_{max} \approx 170$–180 nm. Thus, only the longest wavelength tail of the absorption band is observed in the range $\lambda > 200$ nm.

   Increasing alkyl substitution of the base chromophore achieves a *bathochromic shift*, i.e., the shift of the absorption band to longer wavelength, also called a *red shift*. Thus, the spectroscopic data of three examples from very different chemical classes shown in Table 3.3 possess similar spectroscopic behavior because they belong to the same base chromophore.

2. *Nonconjugated carbonyl compounds and derivates*
   The base chromophore of these chemical classes is formaldehyde whose term system is shown in Fig. 3.4. The allowed $\pi \rightarrow \pi^*$ transition possesses intense absorption bands at wavelengths $\lambda < 200$ nm. The longest absorption band is caused by the forbidden $n \rightarrow \pi^*$ transition. Thus, the molar absorptivity α is

Table 3.3 Absorption spectra of nonconjugated alkenes

Compound	n_C	λ_{max} in nm	α in L mol^{-1} cm^{-1}
Cyclohexene Structure 3.13 with R = H	2	182	7,500

3.13

Cholesterol Structure 3.14	3	190	7,500

3.14

Dimethyl-cyclohexene Structure 3.13 with R = CH$_3$	4	194	9,000

n_C – number of C substituents attached at the base chromophore

Table 3.4 Spectroscopic data of the $n \rightarrow \pi^*$ transition of acetyl derivatives CH$_3$C(O)–X

X	λ_{max} in nm	α in L mol^{-1} cm^{-1}	X	λ_{max} in nm	α in L mol^{-1} cm^{-1}
H	290	15	OC$_2$H$_5$	204	50
CH$_3$	280	50	NH$_2$	214	
OH	204	50	Cl	235	53

small which is shown in the example of some acetyl derivatives listed in Table 3.4.

The interaction of the free electron pair of the substituent OH with the π system of the carbonyl group gives rise to a *hypsochromic shift* (or blue shift, i.e., shift into shorter wavelengths) for the $n \rightarrow \pi^*$ transition for carboxylic acids. On the basis of this hint it is possible to distinguish by means of UV spectroscopy carboxylic acids from aldehydes and ketones.

3. *α-Dicarbonyl and nitroso compounds*

The interaction of the n and π^* orbitals of neighbored carbonyl groups in α-dicarbonyl compounds gives rise to a split resulting in two weak bands belonging to $n \rightarrow \pi^*$ transitions. One of these lies in the visible range ($\lambda > 400$ nm). Thus, glyoxal HOC–CHO or biacetyl CH$_3$CO–COCH$_3$ are "colorants" (see Table 3.5) though both compounds do not possess any conjugated π system.

Table 3.5 $n \rightarrow \pi^*$ transitions for α-dicarbonyl compounds as well as acrolein

Compound	λ_{max} in nm	α in L mol^{-1} cm^{-1}
CH₃CO–COCH₃ (biacetyl)	224	29
	274	18
	422 (yellow-orange!)	22
HOC–CHO (glyoxal)	**450** (orange!)	35
H₂C=C–CH=O (acrolein)	333 (colorless)	28

Fig. 3.9 Simple MO diagram for the N=O group

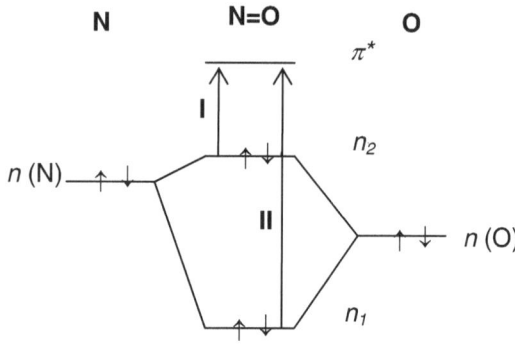

Table 3.6 Spectroscopic data for alkyl-nitroso compounds

Transition	λ_{max} in nm	α in L mol^{-1} cm^{-1}
$n \rightarrow \pi^*$	I: 685	≈30
	II: 280	80
$\pi \rightarrow \pi^*$	220	≈5,000

If one of the two interacting carbonyl groups is removed the orbital interaction can no longer take place and the additional $n \rightarrow \pi^*$ transitions vanish; compare the spectroscopic data of acrolein H₂C=C–CH=O with glyoxal HOC–CHO in Table 3.5.

The especially strong interaction of the n orbitals of nitrogen and oxygen of the nitroso group –N=O, outlined in Fig. 3.9, gives rise to two $n \rightarrow \pi^*$ transitions and one of them (transition I) absorbs in the low-energy Vis range.

The spectroscopic data for alkyl-nitroso compounds are given in Table 3.6.

3.2.2 Dienes, Enones

The electron absorption spectra of dienes –CR=CR′–CR″=CR‴ and enones –CR=CR′–C=O can be traced into the base chromophore of *trans*-butadiene which is explained above. The allowed $S_0 \rightarrow S_1$ transition is observed in the range from $\lambda = 250$–350 nm as an intense absorption band ($\alpha \approx 10{,}000$–15,000 L mol^{-1} cm^{-1}).

The additional weak $n \rightarrow \pi^*$ transition for enones is superimposed by the intense $\pi \rightarrow \pi^*$ transition. The position of the λ_{max} value of the longest wavelength can be evaluated by the empirical *Woodward diene and enone rules* (see Table 6.16) providing important information for structure determination. Note that only microgram amounts are necessary for the simple registration of the UV spectrum.

Challenge 3.4

The longest wavelength of an unknown compound having the sum formula $C_9H_{12}O$ is observed at $\lambda_{max} = 307$ nm ($\alpha = 10{,}500$ L mol^{-1} cm^{-1}). Deduce a proposal for the structure of this compound?

Solution to Challenge 3.4

The number of double bond equivalents (DBE) is 4 calculated according to Eq. 1.19 which can be assigned to three double bonds and one ring. (Note that aromatic compounds with DBE = 4 must be excluded because of the high absorptivity of the absorption bands observed, see next section.)

The spectroscopic data are well in accordance with the diene or enone chromophore. Structure 3.15 can be suggested by DBE (three double bonds, one ring), and the sum formula. This structure can be confirmed by the very good correspondence of the experimental ($\lambda_{exp} = 307$ nm) and the theoretical value calculated according to the Woodward rules listed in Table 3.7.

3.15

Base chromophore	Enone λ_{max} in nm
Base value	215
1 additional double bond	+30
Homoannular order	+39
1 exocyclic double bond	+5
1 δ-C substituent	+18
Solvent correction	0
Sum	307

Table 3.7 Calculation of the λ_{max} value for structure 3.15 according to the Woodward rules

However, this good correspondence is not always given because this simple estimation does not consider many influences on the absorption spectra, including for example steric factors among others. Thus, Structures 3.16–3.18 possess the same chromophore as Structure 3.15 with the same theoretical value $\lambda_{max} = 307$ nm; however the experimental λ_{max} values differ considerably:

3.16

exp. 318 nm

3.17

exp. 322 nm

3.18

exp. 328 nm

Challenge 3.5

For solutions to Challenge 3.5, see *extras.springer.com*.

1. A compound with the sum formula C_4H_9NO provides the following spectroscopic data: λ_{max} in nm (α_{max} in L mol^{-1} cm^{-1}): 220 (5,500), 289 (80), 685 (22). What base chromophore is present? Order the electron transition according to the electronic configuration. What color may possess a solution of

   (a) 10^{-5} mol L^{-1} and
   (b) 0.1 mol L^{-1}?

   (Note that the absorbance must approximately be at least 0.05 to be visible to the human eye.)

2. A solution of 7.4 mg of an unknown compound having the sum formula $C_{14}H_{10}O_2$ in 5-mL *n*-hexane possesses the UV spectrum shown in Fig. 3.10. Determine the kind of electron transitions, the number of double bond equivalents (DBE), and propose a structure for this compound.

3. The main product of a reaction mixture obtained by removal of the respective spot from the thin layer chromatogram is presented in Fig. 3.11. Structures 3.19–3.21 come into question. Which of them possesses the UV spectrum in Fig. 3.11? Explain your decision.

3.19

(continued)

Fig. 3.10 Absorption spectrum of a compound having the sum formula $C_{14}H_{10}O_2$. Sample: 7.4 mg in 5-mL n-hexane (UV grade); length of the cuvette: l = 1 cm

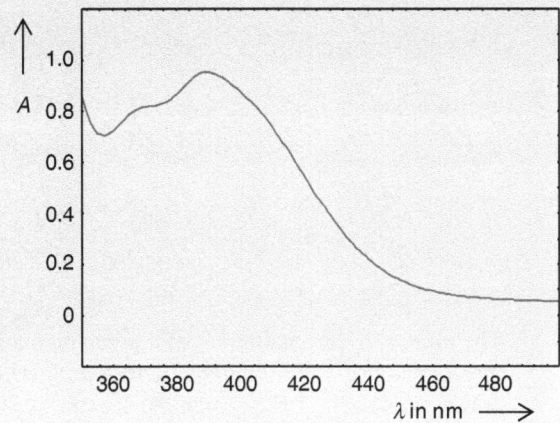

Fig. 3.11 UV spectrum of a steroidal reaction product. Sample: 175 μg in 5-mL acetonitrile (UV grade); length of the cuvette: l = 1 cm

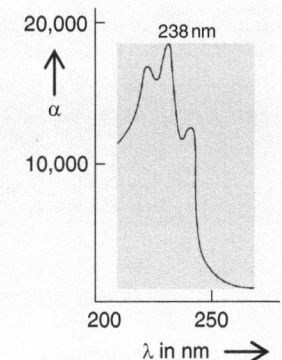

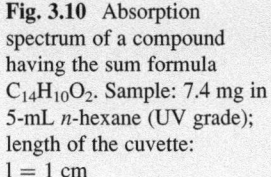

3.20

(continued)

4. The thin layer chromatogram of a photochemical reaction mixture shows three spots that were removed by methanol (UV grade). The UV spectra are shown in Fig. 3.12. Assign the spectra to Structures 3.22–3.24 which could be possible products.

5. Which of the two alternative structures belongs to the experimentally measured λ_{max} values of the longest wavelength absorption band? All spectra were measured with ethanol as solvent.

(continued)

(a) $CH_3-CH_2-CH=C(CH_3)-CH=CH-C(=O)-CH_3$ (I) $\lambda_{max} = 280$ nm
    $CH_3-CH=CH-CH_2-C(CH_3)=CH-C(=O)-CH_3$ (II)

(b) $CH_3-CH=CH-CH=CH-CH=CH-CH=O$ (I) $\lambda_{max} = 285$ nm
    $CH_3-CH=CH-C(O)-CH=CH-CH=CH_2$ (II)

(c)

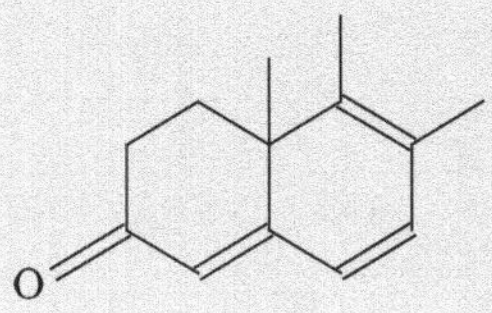

3.25

$\lambda_{max} = 328$ nm

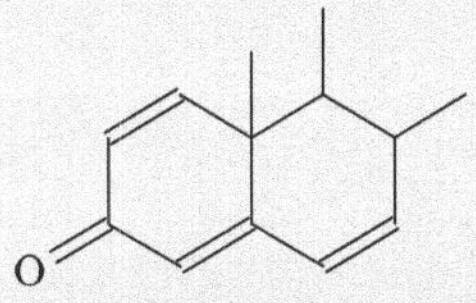

3.26

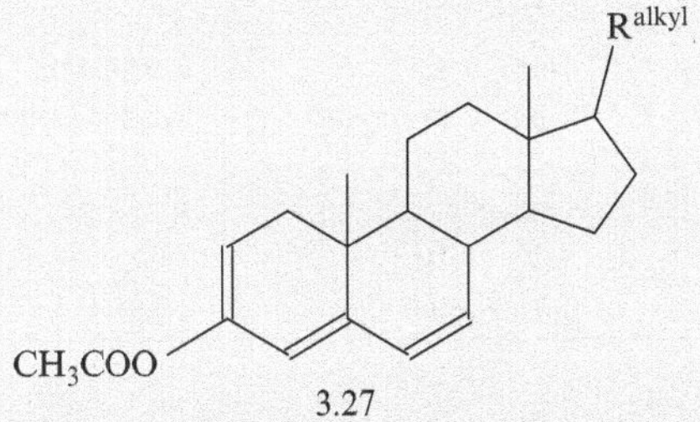

3.27

$\lambda_{max} = 306$ nm

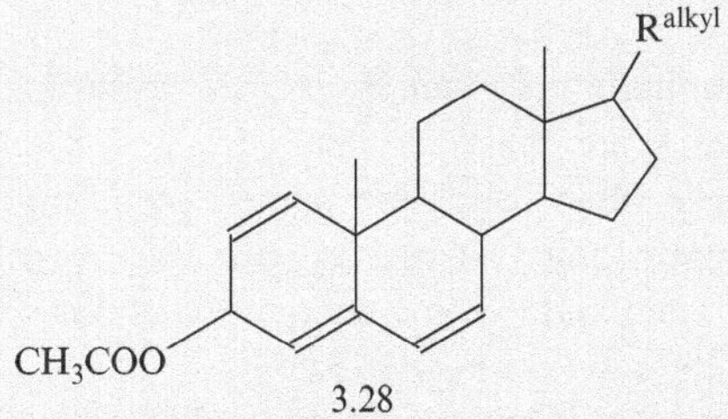

3.28

(continued)

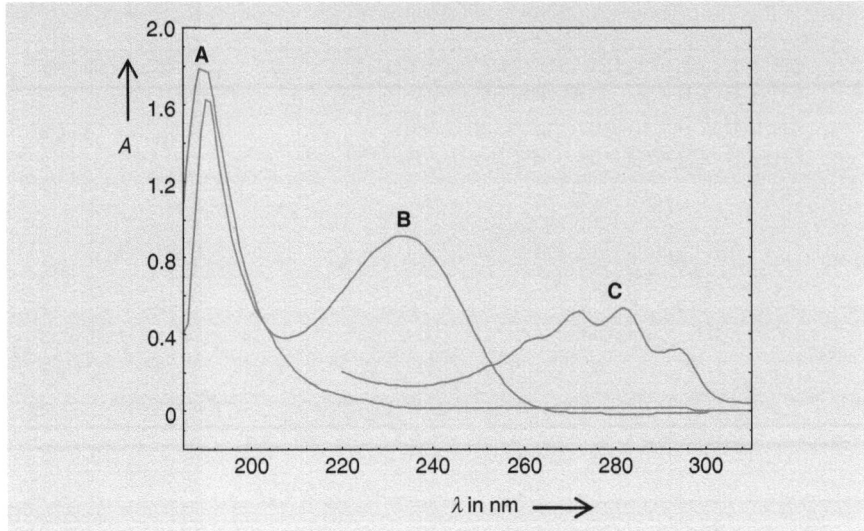

Fig. 3.12 UV spectra of steroidal reaction products (Structures 3.22–3.24 with R=$C_8H_{17}O$).
Samples: **A, B**: 275 μg in 5-mL acetonitrile (UV grade); length of the cuvette: $1 = 1$ cm.
C: 108 μg 5-mL methanol (UV grade); length of the cuvette: $1 = 1$ cm

Table 3.8 $S_0 \rightarrow S_n$ electron absorption bands of benzene and the corresponding term designation

ν in cm^{-1}	54,300	49,000	39,300
λ in nm	184	201	254
α in L mol^{-1} cm^{-1}	68,000	8,500	250
Selection rule	Allowed	Forbidden	Forbidden
Pattern of the band	Nonstructured	Weakly structured	Well structured
Nomenclature			
According to multiplicity	$S_0 \rightarrow S_3$	$S_0 \rightarrow S_2$	$S_0 \rightarrow S_1$
According to symmetry	$^1A_{1g} \rightarrow {}^1E_{1u}$	$^1A_{1g} \rightarrow {}^1B_{1u}$	$^1A_{1g} \rightarrow {}^1B_{2u}$
According to Clar	β band	*para* band	α band

3.2.3   Aromatic Compounds

3.2.3.1   Benzene and derivatives

Benzene is the base chromophore of aromatic compounds. Its UV spectrum is characterized by three absorption bands that are caused by the π (e_{1g}) $\rightarrow \pi^*(e_{2u})$ transition, see Table 3.8. In general, these three bands are named according to Clar's nomenclature for condensed aromatic ring systems: α, *para*, and β band. However, only the longest wavelength α band and the low-energy tail of the *para* band are observed in the UV range $\lambda > 200$ nm (see Fig. 3.13a).

The α band is well symmetrically patterned as is expected for a "rigid" molecule. The most intense sub-band is caused by the $0 \rightarrow 3$ transition which means,

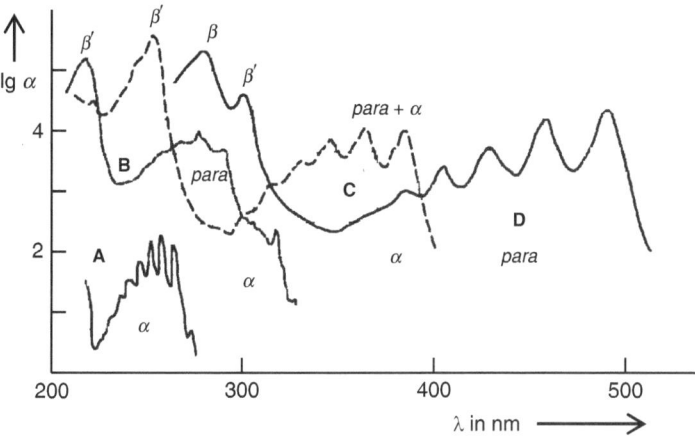

Fig. 3.13 Electron absorption spectra of benzene (**A**) and the acenes naphthalene (**B**), anthracene (**C**), and tetracene (**D**)

Table 3.9 Spectroscopic data of benzene derivatives R–C$_6$H$_5$ and R–C$_4$H$_4$–R$'$ *without* conjugation effects of the substituents R and R$'$, respectively

R; R$'$	*para* band		α band	
	λ_{max} in nm	α_{max} in L mol^{-1} cm^{-1}	λ_{max} in nm	α_{max} in L mol^{-1} cm^{-1}
H	201	8,500	254	250
CH$_3$	206	7,000	261	225
OH	211	6.200	270	1,450
OCH$_3$	217	6,400	269	1,500
Cl	210	7,500	257	170
NH$_3^+$	203	7,500	254	160
o-(CH$_3$)$_2$	210	8,500	263	300
m-(CH$_3$)$_2$	212	7,200	265	300
p-(CH$_3$)$_2$	216	7,500	269	750

according to the Franck-Condon rule, a considerable change of the geometry of the molecule in the excited state.

Nonconjugated substituents have only a weak influence on the base chromophore; hence, the absorption spectra of such derivatives hardly differ from that of benzene as is shown for some examples in Table 3.9.

Let us turn to Challenge 3.4 once more. The four double bond equivalents can be assigned also to a benzene derivative without further double bonds. But an absorption band at λ_{max} =307nm possessing an absorptivity of 10,500 L mol^{-1} cm^{-1} cannot be assigned to a benzene derivative with a nonconjugated substituent.

Conjugating substituents give rise to the intense additional *K band* which lies, generally, between the *para* and the α band. The *K* band is caused by an allowed $\pi \rightarrow \pi^*$ transition; hence, its absorptivity is approximately 10,000–10,500 L mol^{-1} cm^{-1}. The α band is red shifted in comparison to benzene, but it is mostly

Table 3.10 The spectroscopic data of some examples of benzene derivatives with conjugating substituents R^{conj}–C_6H_5. Wavelength in the maximum of the band λ_{max} in nm; absorptivity α in L mol^{-1} cm^{-1}

R^{conj}	para band		K band		α band		$n \rightarrow \pi^*$ band	
	λ_{max}	α	λ_{max}	α	λ_{max}	α	λ_{max}	α
$CH=CH_2$			248	14,000	282	280		
CHO	200	28,500	240	13,600	278	1,100	336	25
			250					
$COCH_3$			245	13,000	279	1,200	315	55
NO_2	208	9,800	251	9,000	292	1,200	322	150
CN			221	12,000	269	830		
COOH			230	10,000	270	800		
$CH=CHO$	218	12,400	298	25,000			351	100

superimposed by the high intensity of the neighboring K band and is observed only as a shoulder.

Conjugating substituents that possess a free electron pair give rise to an additional weak $n \rightarrow \pi^*$ transition being the longest wavelength absorption band.

The spectroscopic data of some examples of benzene derivatives with conjugating substituents are listed in Table 3.10.

Multiple-substituted benzene derivatives possessing *donor* and *acceptor* substituents in *1,4 position* show a considerable deviation from the spectroscopic behavior of the aromatic chromophore. The push-pull system gives rise to an intense charge transfer transition which should be demonstrated by the spectroscopic data of p-nitro aniline (Structure 3.29): $\lambda_{max} = 375$ nm, $\alpha = 56,000$ L mol^{-1} cm^{-1}. Such push-pull systems cannot be assigned from a spectroscopic viewpoint to the aromatic chromophore.

3.29

The absorption band of the large number of aromatic α *carbonyl compounds* (Structure 3.30) can be evaluated by means of the *Scott rules*, see Table 6.17.

3.30

Challenge 3.6

The UV spectrum of N-dimethylamino benzoic acid (Structure 3.31) shows a structureless absorption band at $\lambda_{max} = 322$ nm ($\alpha \approx 24,000$ L mol^{-1} cm^{-1}). Which isomer is present?

N(CH$_3$)$_2$

COOH

3.31

Solution to Challenge 3.6

Structure 3.31 is a α-dicarbonyl compound whose longest wavelength band can be estimated by the Scott rules (Table 6.17). The results are summarized in Table 3.11 for the three possible isomers.

Result: The UV spectrum must be assigned to the *para*-isomer. Note that differences between the experimental and calculated values are certainly possible due to the simplicity of the applied procedure, but the two other isomers can be excluded unambiguously.

Table 3.11 λ_{max} values estimated by the Scott rules for the *ortho*-, *meta*-, and *para*-N-dimethylamino benzoic acid (Structure 3.11)

	ortho	*meta*-	*para*-
Base value for X = OH	230 nm	230 nm	230 nm
N(CH$_3$)$_2$	+85 nm	+20 nm	+20 nm
Total	315 nm	250 nm	250 nm

3.2.3.2   N-Heterocyclic compounds

The electron absorption spectra of *five-membered N-heterocyclic* compounds resemble rather those of dienes but the UV spectra of *six-membered N-heterocyclic* compounds can be derived from the base chromophore benzene as is shown employing the spectral data of some examples presented in Table 3.12.

The influence of *aza* substitution (substitution of CH by N) upon the base chromophore of aromatic compounds can be summarized as follows:

– No significant changes of the position of the absorption bands in comparison to benzene

Table 3.12 Spectral data of six-membered N-heterocyclic compounds λ_{max} is given in nm and α in L mol^{-1} cm^{-1}

Compound	β band		*para* band		α band		$n \rightarrow \pi^*$	
	λ_{max}	α	λ_{max}	α	λ_{max}	α	λ_{max}	α
(benzene)	184	68,000	201	8,500	254	250		
(pyridine)	176	70,000	198	6,000	251	2,000	270	450
(pyridazine)	170		192	5,400	251	1,400	340	315
(pyrimidine)	168		189	10,000	244	2,050	298	325

- Increasing of the intensity of the forbidden α band because of perturbation of the D_{6h} symmetry
- Addition of a weakly intense $n \rightarrow \pi^*$ transition as the longest wavelength absorption band
- Two nitrogen atoms in *ortho* position give rise to an extraordinarily red-shifted $n \rightarrow \pi^*$ transition because of the interaction of the free electron pairs of the neighbored N atoms as already explained above for the N=O group and α-dicarbonyl compounds.

3.2.3.3   Polycyclic aromatic compounds (acenes, phenes)

Acenes are *linear* annulated aromatic compounds starting with benzene: naphthalene, anthracene, and tetracene. The electron absorption spectra of acenes are shown in Fig. 3.13.

The following spectral changes are observed in the row of the acenes in comparison with the base chromophore benzene:

- Splitting up and red shifting of the degenerated β band with increasing annulation. Thus, the longer wavelength β' band lies for naphthalene and anthracene in the UV range $\lambda > 200$ nm and both bands are observed for tetracene.
- The *para* as well as the α band are red shifted with increasing annulation.
- The bathochromic effect is larger for the *para* band with the consequence that *para* and α possess the same position in the spectrum of anthracene and their order is reversed in the case of tetracene.

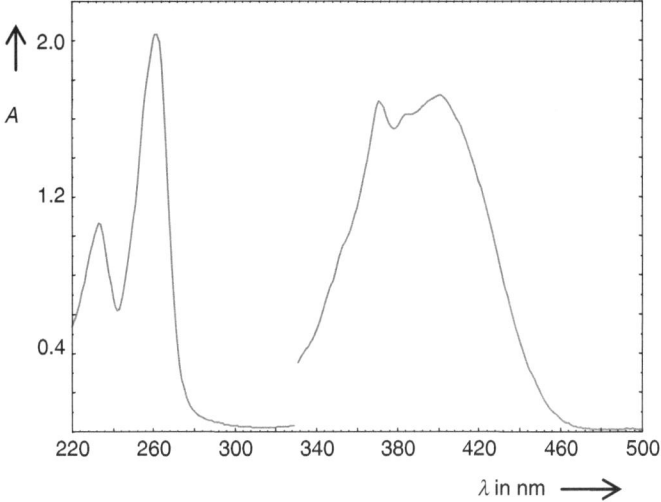

Fig. 3.14 Electron absorption spectrum of 9-anthracene aldehyde. Samples: Range 220–330 nm: 10 μg in 10-mL ethanol. Range 330–500 nm: 100 μg in 10-mL ethanol. Cuvette: l = 1 cm

The crossed annulated *phenes* also show the typical spectral behavior of the base chromophore but their absorption bands lie at shorter wavelengths than that of the respective acenes.

Substituent effects upon the chromophore of acenes and phenes are smaller than those for benzene; hence, the electron absorption spectra of substituted compounds do not strikingly differ from that of the nonsubstituted acenes and phenes. Compare, for example, the absorption spectra of 9-anthracene aldehyde (Structure 3.32; R=CHO) in Fig. 3.14 with spectrum C in Fig. 3.13. Similar spectra are observed for the derivatives azomethine (R=CHNCH$_3$) and hydrazone (R=CHN–NH$_2$).

3.32

Challenge 3.7

For solutions to Challenge 3.7, see *extras.springer*.com.

1. The longest wavelength absorption band of a compound with the sum formula C$_6$H$_5$O$_3$N lies at λ_{max} = 400 nm ($\alpha \approx$ 18,000 L mol^{-1} cm^{-1}) measured in a 0.1-molar NaOH solution. To which chromophore can the absorption band be assigned? Propose a structure for this compound.

(continued)

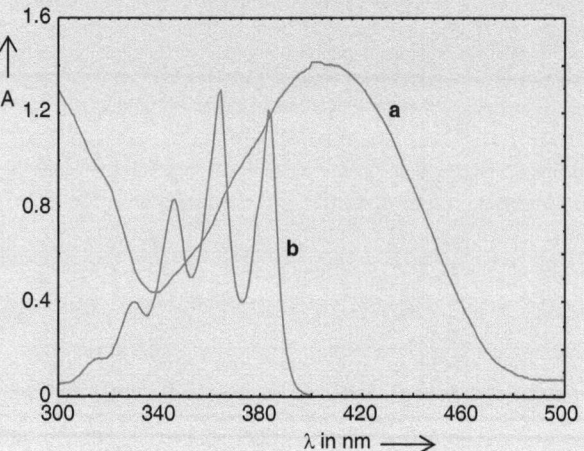

Fig. 3.15 Absorption spectra of (**a**) the thermodynamically stable benzophenone 9-anthraldazine (Structure 3.33) and (**b**) the photoproduct. Samples: **a**: 0.175 mg in 10-mL toluene; **b**: (without concentration) the respective spot in the TLC was eluted by methanol

2. After irradiation of a solution of benzophenone 9-anthraldazine (Structure 3.33) with UV light a photoisomer can be isolated by thin-layer chromatography (TLC) which thermally converts into the starting product. This procedure which can be repeated several times is based on photochromism. The absorption spectra of the starting and the photoproduct are shown in Fig. 3.15. Derive the structure for the photoisomer. On which principle is the photochromism of this system based? Help: Consider also the spectrum in Fig. 3.14.

$$CH=N-N=C(C_6H_5)_2$$

3.33

3. The maximum of a UV spectrum is observed at $\lambda_{max} = 272$ nm (lg $\alpha = 4.1$). Which of the two compounds shown in Structures 3.34 and 3.35 can be assigned to the spectrum?

3.34

3.35

(continued)

4. The oxidation of 3-methoxy-tetralone (Structure 3.36) gives rise to a main product possessing the sum formula $C_{11}H_{12}O_2$. A solution of the reaction product in n-hexane shows an intense infrared band at $v = 1{,}685$ cm^{-1} and a UV band at $\lambda_{max} = 275$ nm ($\alpha \approx 12{,}500$ L mol^{-1} cm^{-1}). Develop a proposal for the structure of the oxidation product.

CH_3O 3.36

3.2.4   Polyenes

The following chemical classes possess the chromophore of polyenes:

R–(CH=CH)$_n$–R	dialkyl, diaryl polyenes
O=CR–(CH=CH)$_n$–C(R)=O	polyene-diketones, -dialdehydes

Polyenes are characterized by the following features:

– Conjugated double bonds must be present.
– The number of π electrons is equal to the number of the π centers.
– The bond order of the conjugated system is alternating.
– The electron density is equal at each carbon atom included in the conjugation system.
– The very intense absorption bands are mostly well structured. The absorptivity increases with n, i.e., the length of the conjugated system.
– The longest wavelength absorption band (λ_{max}) is red shifted with the increasing conjugation. The relation between λ_{max} and the number of the conjugated double bonds n is given by Eq. 3.7.

$$\lambda_{max} = a + b \cdot \sqrt{n}. \tag{3.7}$$

The limit value for polyenes lies approximately at $\lambda_{max} = 550$ nm which gives rise to red-orange solutions. (Note that, for example, blue solutions cannot be assigned to the polyene chromophore!)

The chromophore of polyenes can be reduced to that of $trans$-butadiene (see above) and it is present in many natural compounds. The best known compound possessing the polyene chromophore is β carotene (Structure 3.37) with the spectroscopic data λ_{max} (0 → 1) = 451 nm, ($\alpha \approx 142{,}000$ L mol^{-1} cm^{-1}).

3.37

3.2.5   Polymethines

Now let us turn to colorants. One of the chemical classes of colorants is that of the polymethines.

The general chemical structures of the polymethines are outlined in Structure 3.38:

$$X - (CH = CH)_n - CH = Y \leftrightarrow X = (CH - CH)_n = CH - Y$$

Structure 3.38

$X = NR_2$	$Y = NR_2$	(kationic) **cyanine** (3.8)
$X = NR_2$	$Y = O$	(neutral) **merocyanine**
$X = O$	$Y = O$	(anionic) **oxonole**

Polymethines are characterized by the following features:

- Conjugated double bonds with two heteroatomic end groups (N or O) must be present.
- Equation 3.9 is valid for the number of the π electrons [$n(\pi$ electron)] for Structure 3.38

$$n(\pi\text{-electron}) = n(\pi\text{-centers}) \pm 1 \qquad (3.9)$$

Thus, the polymethine chromophore can be easily recognized taking into account the following observations:

The number of conjugated methine (CH) groups ordered between the two heteroatoms must be odd.

- The bond order of the conjugated system is approximately the same caused by resonance as shown in Eq. 3.8. The better the similarity of the order of the conjugated bonds the stronger the polymethinic behavior.
- The very intense longest wavelength band is accompanied by a shorter wavelength shoulder. Note that this fact is helpful to recognize the polymethine chromophore.
- The relation between the longest wavelength λ_{max} and the number of conjugated methine groups n is linear according to Eq. 3.10:

$$\lambda_{max} = a + b \cdot n. \qquad (3.10)$$

Table 3.13 Effects of substitution of =CH groups and the hydrogen atom of the polymethine chain upon the λ_{max} values

Polymethine chain	X –	CH –	CH –	CH –	...
Substitution of =**CH** by					
More electronegative atom		Red shift	Blue shift	Red shift .	...
Less electronegative atom		Blue shift	Red shift	Blue shift	...
Substitution of **H** by					
A donor		Red shift	Blue shift	Red shift	...
An acceptor		Blue shift	Red shift	Blue shift	...

Note that "*red shift*" means bathochromism (shift to larger wavelengths) and "*blue shift*" hypsochromism (shift to smaller wavelengths), respectively

Table 3.14 Spectral data λ_{max} in nm (lg α) of polymethines

n	Cyanine	Merocyanine	Oxonole
1	312 (4.81)	283 (4.57)	276 (4.83)
2	416 (5.09)	361.5 (4.70)	362.5 (4.75)
3	519 (5.32)	421.5 (4.75)	455 (4.88)
4	625 (5.47)	462.5 (4.81)	≈ 550
5	735 (5.55)	491.5 (4.83)	
6	848 (5.34) (NIR!)	512.5 (4.86)	
Solvent	CHCl$_3$	CHCl$_3$	DMFA

– The vinylene shift b is approximately 100 nm for symmetrically substituted cyanines. Thus, cyanines with $n = 2$ absorb in the visible range and, hence, they are colored compounds. NIR (!) absorbing cyanines are existent with $n = 6$. The constant a corresponds to a λ_{max} value for $n = 0$ of $\lambda_{max} \approx 230$ nm. Thus, the length of the conjugated chain can be estimated by application of Eq. 3.10 converted to n.

– Substituents at the conjugated carbon atoms and substitution of CH groups by heteroatoms, respectively shift the λ_{max} values according to the known polymethine rules given in Table 3.13.

– The polymethinic behavior is less pronounced for oxonoles and merocyanines, therefore, the λ_{max} values are observed at shorter wavelengths in comparison to the respective cyanines. Table 3.14 presents the spectral data of some examples of the three polymethine classes.

Challenge 3.8

Mark the chromophore in the compounds of Structures 3.39 and 3.40 and estimate the respective λ_{max} value.

3.39

(continued)

$$CH_3 — N^+ \langle\rangle — CH=CH — \langle\rangle — O^-$$

3.40

Solution to Challenge 3.8

The polymethine chromophore is present in both compounds because the number of methine groups between the heteroatoms nitrogen and oxygen, respectively is odd. Structure 3.39 can be assigned to a nonamethine-aza-cyanine chromophore, which is marked in bold in Structure 3.41 for both resonance forms.

3.41

The λ_{max} value can be estimated by Eq. 3.10: $\lambda_{max} = 230$ nm $+ 100$ nm $\times$ $4 = 630$ nm. But the λ_{max} value should be somewhat larger because one methine group is substituted by the more electronegative nitrogen at a position which will cause a red shift according to the polymethine rules given in Table 3.13. Thus, the λ_{max} value for the longest wavelength absorption band can be estimated to appear approximately at 680 nm which is in good agreement with the experimentally determined value: $\lambda_{max} = 685$ nm ($\alpha = 65,000$ L mol^{-1} cm^{-1}).

The compound of Structure 3.40 also possesses the polymethine chromophore which can be named nonamethine-merocyanine. The λ_{max} value is the same as was calculated for Structure 3.39 because $n = 4$. Remember that the merocyanines absorb at somewhat shorter wavelength than the respective cyanines, hence, the estimated λ_{max} value should be $\lambda < 600$ nm; 558 nm ($\alpha = 55,000$ L mol^{-1} cm^{-1}) was determined experimentally. Thus, each of the two structures is justified by its electron absorption spectrum.

Merocyanines form resonance between a polar benzoic and an nonpolar quinoid structure as is shown in Structure 3.42.

3.42

3.2.5.1   Di- and Triarylmethane Dyes

The technically important di- and triarylmethane dyes possess the polymethine chromophore.

Diaryl- and *symmetrically substituted triarylmethane* dyes provide electron absorption spectra of type I: Only *one* absorption band typical for polymethines is observed in the Vis range. A representative absorption spectrum of type I is shown in Fig. 3.16 for crystal violet (Structure 3.43 with R=H).

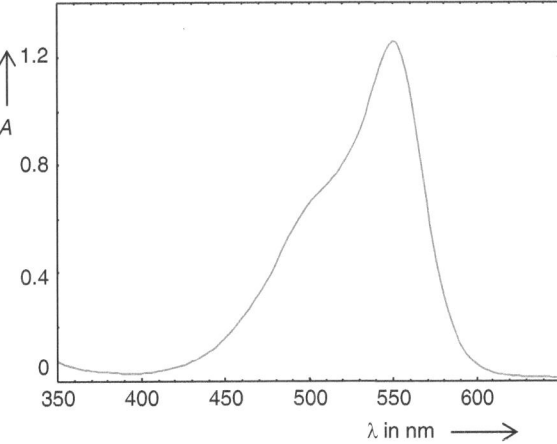

The absorption spectrum presented in Fig. 3.17 for malachite green (Structure 3.44) is typical for the asymmetrically substituted triarylmethane chromophore. The spectrum of type II is characterized by an additional absorption band at shorter wavelengths caused by an electron transition whose polarization direction is marked in Structure 3.44. This additional band lies in the blue range of the visible, and, in combination with the absorption band at approximately 600 nm, it is responsible for the *green* color of the compound.

Fig. 3.16 Vis spectrum of crystal violet (Structure 3.43 with R=H) as an example for the *symmetrically* substituted triarylmethane chromophore. Sample: 90 μg in 10-mL methanol

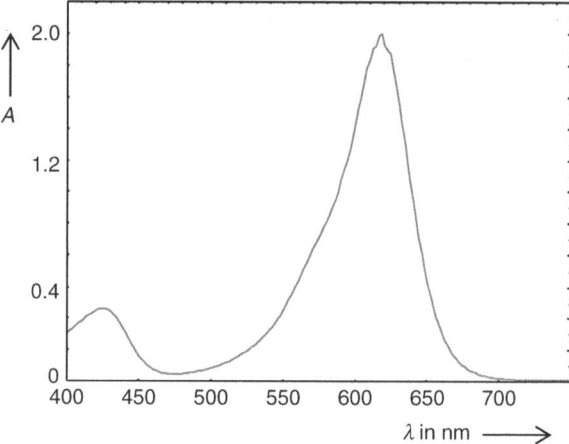

Fig. 3.17 Vis spectrum of malachite green (Structure 3.44) as an example for the *asymmetrically* substituted triarylmethane chromophore. Sample: 130 µg in 10-mL methanol

3.44

Asymmetrically substituted triarylmethane compounds whose substituents hardly influence the electron system provide also absorption spectra of type I. Thus, the Vis spectrum of fuchsine (Structure 3.43 with R=CH$_3$) hardly differs from that of crystal violet (Structure 3.43 with R=H).

Most of the electron absorption spectra cannot be assigned unambiguously to one single chromophore system.

3.45

Thus, for example, the aromatic chromophore for 4′-dimethylamino benzophenone (Structure 3.45) must be rejected because the spectral data $\lambda_{max} = 330$ nm ($\alpha = 24{,}000$ L mol^{-1} cm^{-1}) do not belong to an aromatic chromophore. Conversely, the absorption spectrum of isomer 3′-dimethylamino-benzophenone (Structure 3.45) hardly differs from that of the unsubstituted benzophenone which possesses a typical aromatic chromophore. The strong difference with regard to the spectroscopic behavior of both structurally isomeric compounds is understandable because *para* substitution of the dimethylamino group in 4′-dimethylamino benzophenone provides a polymethinic chromophore system. Furthermore, an odd number of methine groups is distributed between the heteroatoms oxygen and nitrogen resulting in a pentamethine-merocyanine chromophore. This is confirmed by the good correspondence of the spectral data of 4′-dimethylamino-benzophenone and that of the merocyanines for $n = 2$ in Table 3.14. The difference between the λ_{max} values of the "true" merocyanine and 4′-dimethylamino benzophenone can be caused by the fact that the true chromophore of the benzophenone derivative is superimposed by the polymethinic and aromatic chromophore.

The 3′-isomer must be assigned to a purely aromatic chromophore, because the conditions for a polymethine chromophore are not given. Thus, knowledge of the chromophore systems can be helpful in structural determination.

3.2.6 Azo Compounds

Azo compounds contain one or more azo groups –N=N–. The stable configuration is the *trans*-form.

The base chromophore of the aromatic azo compounds is given by *trans*-azobenzene (Structure 3.46 with X=N and R=H) whose absorption spectrum hardly differs from that of *trans*-stilbene (Structure 3.46 with X=CH and R=H), see Fig. 3.18. But as opposed to *trans*-stilbene azo benzene shows additionally a weak $n \rightarrow \pi^{*}$ transition in the visible spectral range, which is responsible for its yellow-orange color.

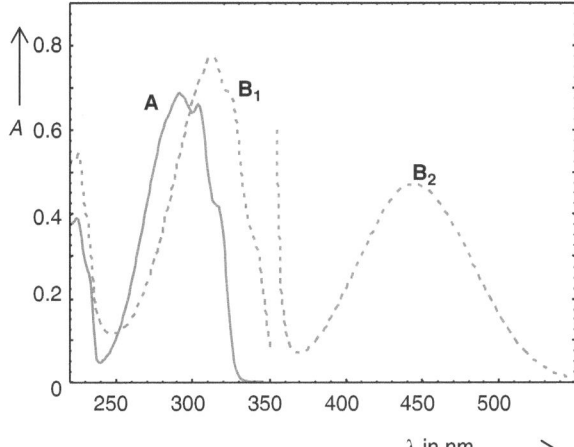

3.46

Fig. 3.18 UV–vis spectra of (**a**) *trans*-stilbene (Structure 3.46 with X=CH and R=H) and (**b**) *trans*-azobenzene (Structure 3.46 with X=N and R=H). Samples: **A**: 44 μg in 10-mL *n*-hexane; **B$_1$**: 67 μg in 10-mL *n*-hexane; **B$_2$**: 1.88 mg in 10-mL *n*-hexane

According to quantum mechanics the longest wavelength for azo compounds is caused by a charge transfer from the π system of the aromatic moiety to the π^* system of the azo group $-N=N-$. Thus, the spectral behavior of the large number of azo compounds can be understood with knowledge of the kind of electron transfer which is important to recognize the azo chromophore for structural determination. The following facts are valid:

– *Donor* substituents attached to the aromatic moiety favor the electron transfer resulting in a red shift of the longest wavelength absorption band. Such donor substituents, for example, can be amino groups ($-NR_2$) or polycyclic aromatics. For example, the spectral data for $R=N(CH_3)_2$ in Structure 3.46 are: $\lambda_{max} = 400$ nm (log $\alpha = 4.49$).
– An additional red shift is provided if an *acceptor* is attached at the azo group which is realized by protonation. Thus, all azo compounds absorb at longer wavelengths in acid solutions which should be kept in mind for the recognition of azo compounds within the scope of structural determination. Furthermore, this spectroscopic behavior of azo compounds is the base of their application as pH indicators, for example, methyl orange, see Structure 3.47 with λ_{max} (basic form) $= 405$ nm and λ_{max} (acid form) $= 485$ nm.

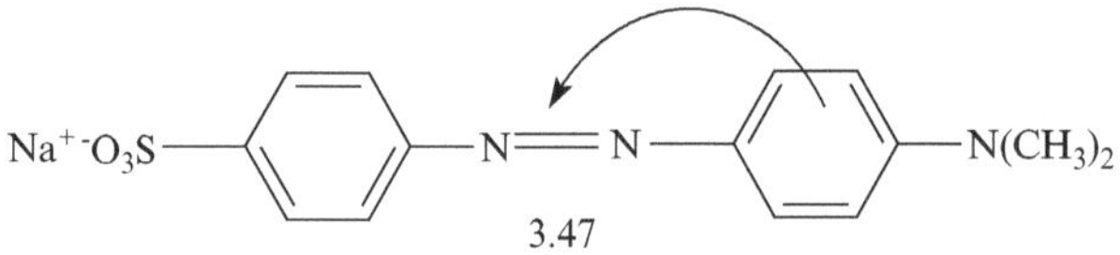

3.47

3.2.7 Quinoid Compounds

The base chromophore of quinoid compounds is given by *ortho-* and *para-*benzoquinone (Structures 3.48 and 3.49, respectively), and their UV–vis spectra are presented in Fig. 3.19.

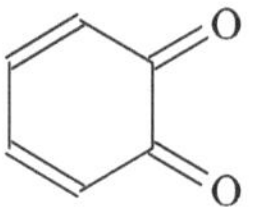

3.48

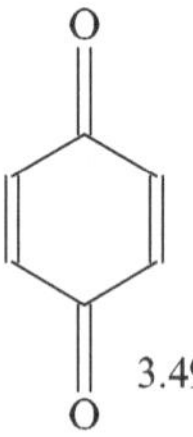

3.49

The longest wavelength absorption band ($S_0 \rightarrow S_1$) is caused by an $n \rightarrow \pi^*$ transition in both compounds. Because of the term interaction of the neighbored

Fig. 3.19 UV–vis spectra of
(**A**) *para*-benzoquinone
(Structure 3.49) and
(**B**) *ortho*-benzoquinone
(Structure 3.48)

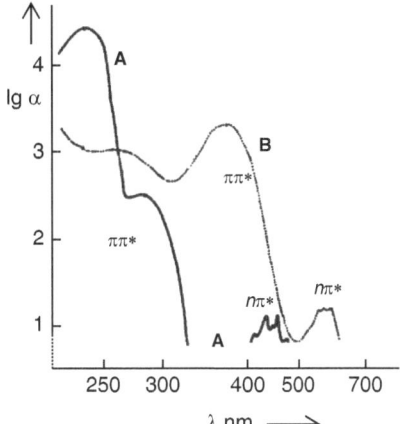

free electron pairs, like in α-dicarbonyl compounds, the $n\pi^*$ absorption band is red shifted for *ortho*-benzoquinone in comparison to the *para*-isomer.

According to quantum mechanics, the longest wavelength $\pi \rightarrow \pi^*$ ($S_0 \rightarrow S_2$) transition is forbidden for *para*-benzoquinone, only the ($S_0 \rightarrow S_3$) transition is allowed which is reflected in the intensities of the respective absorption bands.

The $\pi \rightarrow \pi^*$ transitions of substituted *para*-benzoquinones are red shifted in the order Cl $<$ Br $<$ I $<$ and $OCH_3 <$ $SCH_3 <$ $N(CH_3)_2$, respectively, see, for example, the spectral data for a *para*-benzoquinone derivative shown in Structure 3.50: λ_{max} (log α): $S_0 \rightarrow S_2$: 380 nm (4.3), $S_0 \rightarrow S_3$: 500 nm (2.6).

A similar absorption spectrum is observed for 9,10-anthroquinone (Structure 3.51 with $R_1 = R_2 = H$), see Fig. 3.20.

Fig. 3.20 Absorption spectra
of 9,10-anthroquinone
(Structure 3.51):
(A) $R_1=R_2=H$; (B) $R_1=H$,
$R_2=NH_2$

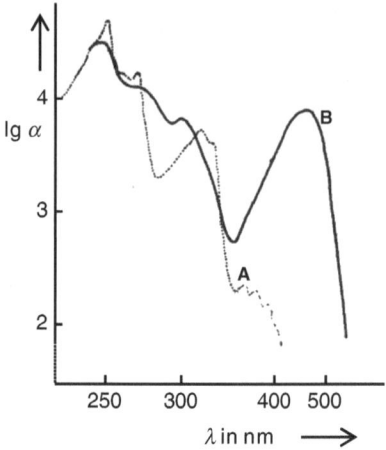

Whereas acceptor substituents like NO_2, Cl, or CN do not or hardly influence the chromophore system, donor substituents give rise to a strong red shift in the absorption spectrum of 9,10-anthroquinone derivatives and superimpose the weak $n \rightarrow \pi^*$ transition; see, for example, spectrum **B** in Fig. 3.20. An especially strong red shift is caused by donors like O^- or $N(CH_3)_2$ which manifests itself in the spectral data of Structure 3.51 with $R_1 = R_2 = N(CH_3)_2$: $\lambda_{max} = 592$ nm, log $\alpha = 4.19$ (measured in ethanol).

3.3 Stereochemical Influence on the Electron Absorption Spectra of Olefins

According to quantum mechanics, the double bonds in conjugated systems are weakened. Conversely single bonds are stronger in the S_1 state resulting in changed energy barriers for the rotation around the respective bond in the ground (S_0) and first excited state (S_1) of a conjugated system; see Fig. 3.21 for the structure moiety aryl–C=C–.

As Fig. 3.21 shows, the transition energy which is reflected in the length of the arrow, at any torsion angle ϕ is larger for a molecule which is twisted around the single bond than that for a molecule which is twisted around the double bond. Thus, the following rules are valid for conjugated systems:

Distortion caused by a rotation around the C=C double bond gives rise to a red shift, distortion caused by a rotation around the C–C single bond shifts the absorption band to shorter wavelengths.

Distortion is always coupled with a decrease of the absorptivity.

Note that very detailed structural information can be obtained by means of these rules.

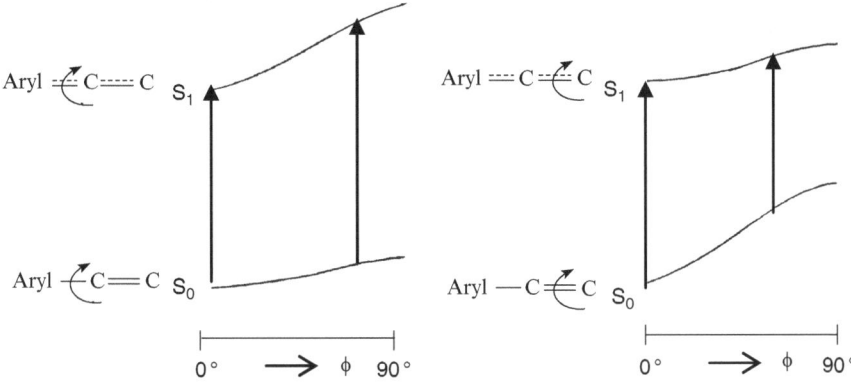

Fig. 3.21 Qualitative diagram of the energy barriers for the rotation from $\phi = 0°$ to $\phi = 90°$ around the *aryl*–CC– single and the aryl–C=C– double bond, respectively in the ground (S_0) and S_1 excited state

Challenge 3.9

There is no steric reason why *trans-stilbene* (Structure 3.52) should not be planar resulting in a most effective overlapping of the *p* orbitals.

The UV spectrum (Fig. 3.18a) is characterized by $\lambda_{max} = 289$ nm ($\alpha = 28,500$ L mol^{-1} cm^{-1}).

The spectral data of *cis-stilbene* and *bis-fluorenylidene* are: $\lambda_{max} = 280$ nm ($\alpha = 22,500$ L mol^{-1} cm^{-1}) and $\lambda_{max} = 450$ nm, respectively. (Note that the last compound is colored *red!*) Which structural information can be obtained from these experimental data for both compounds?

Solution to Challenge 3.9

In comparison to the planar *trans*-stilbene (Structure 3.52), the *cis* isomer must be twisted because of the repulsion of the *ortho*-hydrogen atoms of the benzene rings. The question follows in which manner does the molecule prevent the steric hindrance. The electron absorption spectrum can provide

(continued)

an answer. In comparison to *trans*-stilbene the *cis* isomer absorbs at shorter wavelengths with a smaller absorptivity. According to the rules given above, *cis*-stilbene must be *twisted around the C–C single* bond which is outlined in Structure 3.53.

The planarity of bis-fluorenylidene is sterically prevented by the interaction of the aromatic hydrogen atoms. But the distortion around the energetically lower C–C single bond cannot be realized. Thus, the molecule is *twisted around the central double bond* as is shown in Structure 3.54 resulting in a strongly red shift of the longest wavelength absorption band which is in accordance with the rules given above. Thus, bis-fluorenylidene is a red compound without any chromophore group responsible for an electron transfer in the visible range.

3.4   Solvent Effects on Electron Absorption Spectra

Knowledge of the effect of solvents on electron absorption bands is important not only for structure determination but also for gaining an understanding of some photochemical and photophysical processes. The electron absorption spectra of all molecules show a more or less strongly pronounced dependence on the polarity of the solvent used. This effect is named *solvatochromism*. Thus, some chemical classes like polymethines or azo compounds can be easily recognized because the absorption bands are strongly influenced by the polarity and pH of the solvent, respectively.

To understand solvatochromism let us restrict our discussion to solvent effects upon the $n \rightarrow \pi^*$ and $\pi \rightarrow \pi^*$ transitions in the C=O group. The qualitative energy diagram for the change of the ground (S_0) and excited states (S_n) of the C=O group with crossing from the nonpolar solvent n-hexane to the strongly polar methanol is shown in Fig. 3.22.

The main interaction between the C=O group of a carbonyl compound and the solvent will be caused on the basis of dipole–dipole interactions. Generally, the higher the dipole moment of the respective state the stronger it is energetically stabilized by the dipole interactions with the solvent. The polar C=O group is characterized by a considerable dipole moment. Therefore, increasing solvent polarity should increasingly stabilize the energetic ground state (S_0) which is shown by the stabilized S_0 states for the polar solvent methanol in comparison to the nonpolar n-hexane shown in Fig. 3.22. But to understand the spectral changes caused by solvent polarity knowledge is required of the energy of the respective *excited* states.

The relative amount of dipole moment in the $n\pi^*$ and $\pi\pi^*$ states can be easily estimated. Thus, the electron density at the oxygen atom is diminished by $n \rightarrow \pi^*$ excitation resulting in a smaller dipole moment in the $n\pi^*$ state. Therefore, the effect of the energetic stabilization by the polar solvent methanol should be smaller and, hence, the energy gap in Fig. 3.22 is smaller. Otherwise, the $\pi \rightarrow \pi^*$ excitation is coupled by increasing the dipole moment in the excited $\pi\pi^*$ state. Hence, the respective state is strongly stabilized. The comparison of the energy differences illustrated by the length of the arrows in Fig. 3.22 provides the following rules: (Note that the longer the arrow the shorter is the wavelength observed in the spectrum.)

$n \rightarrow \pi^*$ *transitions cause a hypsochromic effect with increasing solvent polarity, i.e., shift to shorter wavelengths (blue shift).*
$\pi \rightarrow \pi^*$ *transitions can provoke a bathochromic effect with increasing solvent polarity, i.e., shift to longer wavelengths (red shift).*

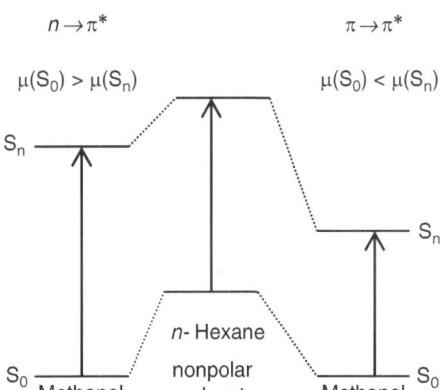

Fig. 3.22 Qualitative energy diagram for the ground (S_0) and excited states (S_n) of the C=O group in nonpolar (n-hexane) and strongly polar (methanol) solvents

The condition for a solvent effect for a $\pi \to \pi^*$ excitation is that the solvent is localized at the center of the molecule in which the respective $\pi \to \pi^*$ excitation takes place. Thus, for example, the K band ($\pi \to \pi^*$) in acetophenone (Structure 3.3) is red shifted with increasing solvent polarity but the α band (also a $\pi \to \pi^*$ transition) is not affected by the solvent. Note the α band is caused by an electron transition in the aromatic moiety, whereas the K band involves the conjugated C=O substituent. Thus, the solvatization of a molecule by the polar solvent does not take place at the aromatic moiety (source of the α band) but at the C=O group. This kind of structural analytical problem can only be solved by electron spectroscopy.

Challenge 3.10

For solutions to Challenge 3.10, see *extras.springer.com*.

The UV spectrum of a compound with the sum formula $C_6H_{10}O$ shows the following absorption bands (λ_{max} in nm):

Solvent	Absorption band 1	Absorption band 2
Hexane	231	335
Ethanol	237	320
Water	245	300

Assign the absorption band to the respective electron transitions and justify the decision. Which structural moieties can be recognized by this experimental fact?

3.5 Some General Remarks About Molecular Structure Determination by Means of UV–Vis Spectroscopy

In general, electron absorption spectroscopy does not provide detailed information with respect to molecular structure determination which may be the reason why in most of the excellent books on this subject UV–vis spectroscopy is not considered or only λ_{max} values of possible absorption bands for $\pi\pi^*$ transitions in some chemical classes are listed in tables. Clearly, the main information is doubtless obtained, nowadays, by NMR methods and the potential of the electron absorption spectra lies in *quantitative* analysis (spectrophotometry). However, the bare listing of λ_{max} values for chromophoric groups is of little use, because further experimental data must be taken into account like the absorptivity, number of absorption bands, or their structural pattern. For example, the information $\lambda_{max} = 265$ nm can be assigned to an $n \to \pi^*$ transition of a carbonyl group, to the $\pi \to \pi^*$ transition in a diene or in an enone, or it can belong to the α band of a benzene derivate with one or more nonconjugated substituents. The latter can be valid if the absorption band is well patterned and its absorptivity lies in the range $\alpha \approx 500$ L mol^{-1} cm^{-1}. If the diene or enone chromophore is present, the absorptivity must be $\alpha > 10,000$ L mol^{-1} cm^{-1} and the absorption band of diene is, generally, well patterned, whereas enones provide structureless bands. Finally, the absorption band the position of which is given above may be assigned to a carbonyl group of a nonconjugated system (aliphatic aldehyde or ketone) if the absorption band does not show a vibrational pattern and its absorptivity is approximately 50 L mol^{-1} cm^{-1}. Thus, evaluation of not only the λ_{max} value but also intensity and pattern can provide the chromophore system which can be important information at the starting point of molecular structure determination.

Note that electron absorption spectroscopy is characterized by a high sensitivity, i.e., it needs only very small sample amounts; in general, a few micrograms are sufficient which can be provided, for example, from spots on the TLC plate.

Furthermore, this method works nondestructively which means that the sample can be used, after removing the solvent, for further spectroscopic methods, for example, for FT-IR-microscopy and, afterwards, for mass spectrometry. The results obtained by these methods can provide important structural information which will be demonstrated by some examples in the next challenge.

Note that the structural determination of *short-lived intermediates* and *thermally unstable products* can only be obtained by UV–vis spectroscopic data. In general, such investigations start with a known compound, so that the structure of such products can be well derived; see, for example, Challenges 3.7.2 and 3.11.8.

For some problems, the photochemist may be interested in the structure of molecules in the short-lived excited state. Such questions can only be answered by electron absorption spectroscopy and, as shown later, by emission spectroscopy.

The analytical chemist who deals with the HPLC equipped with a standard DAD (diode array detector) will obtain routine electron absorption spectra with the help of which one can answer structural questions (see Challenge 3.11.17).

Last but not least, electron absorption spectroscopy is fascinating in its experimental simplicity in connection with its high sensitivity. Thus, it can be applied, especially in combination with thin layer chromatography, as a simple and fast method for *controlling chemical reactions*.

Many of the structural problems given in the next challenges can be solved in the best way or generally only by UV–vis spectroscopy.

Challenge 3.11

For solutions to Challenge 3.11, see *extras.springer.com*.

1. An ethanolic solution of the compound [RR′N–(CH=CH)$_n$–CH=NRR′]$^+$Cl$^-$ with R=phenyl and R′=CH$_3$ absorbs at $\lambda_{max} = 550$ nm (log $\alpha = 5.1$). What is the base chromophore of this compound? Determine the length of the conjugated chain, i.e., n in the formula.

2. A solution of the compound of Structure 3.55 in a water-alcohol mixture absorbs only in the UV range. The solution becomes red after addition of some drops of diluted sodium hydroxide solution.

(continued)

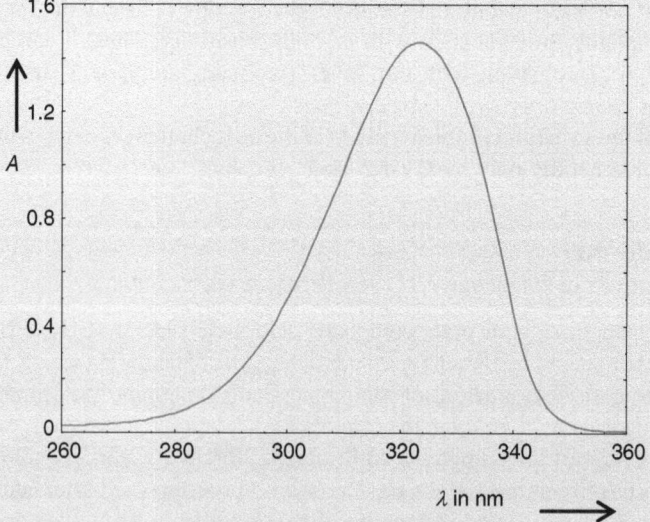

3.55

The absorption band with $\lambda_{max} = 550$ nm shows a shoulder at shorter wavelengths. Which structure and which base chromophore is present in aqueous alkaline solution?

3. A compound with the sum formula $C_7H_{13}N_2{}^+Cl^-$ provides the absorption spectrum shown in Fig. 3.23. Derive a structure for this compound. Which additional data are necessary to justify the true structure?

4. After addition of sulfanilamide (para-H_2N–C_6H_4–SO_2–NH_2) and N(1-naphthyl)-ethylene-diamine·2HCl ($C_{12}H_{16}Cl_2N$) to a weakly acid solution of a wastewater sample an absorption spectrum is measured which is

Fig. 3.23 Absorption spectrum of an unknown compound with the sum formula $C_7H_{13}N_2{}^+$ $ClO_4{}^-$

(continued)

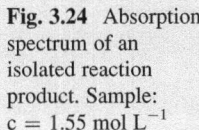

Fig. 3.24 Absorption spectrum of an isolated reaction product. Sample: $c = 1.55 \ \text{mol L}^{-1}$

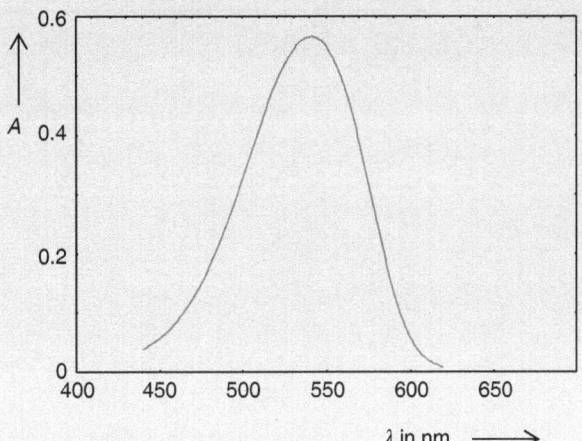

presented in Fig. 3.24. The λ_{max} value is dependent on the pH of the solution. Which is the base chromophore for the reaction product? Propose a structure. Which compound is present in the wastewater sample?

5. The four compounds **a–d** show the following absorption bands in the visible spectral range, λ_{max} (α in L mol^{-1} cm^{-1}): **a**: 650 nm (60), **b**: 450 nm (150,000), **c**: 445 nm (1,500) and **d**: 630 nm (58,000). Which compound can be assigned to a polyene and which can be a polymethine? Justify your decision. Why can azo compounds be excluded?

6. For an unknown compound the molecular peak is determined to be $m/z = 120.0576$ and its absorption spectrum is given in Fig. 3.25. Assign the individual bands to the respective electron transitions and propose a structure for the compound.

7. The spectral data are given for the compounds with the Structures 3.56 and 3.57: $\lambda_{max} = 680$ (log $\alpha = 4.86$) and $\lambda_{max} = 580$ nm (log $\alpha = 4.76$), respectively. Assign the spectra to the respective compounds and justify your decision. Which is the base chromophore for each compound?

$(CH_3)_2N$ O

3.56

$(CH_3)_2N$ $N(CH_3)_2Cl$

3.57

8. An approximately 10^{-4} molar solution of the spiropyrane in Structure 3.58 does not absorb in the Vis range. If the solution is irradiated by UV

(continued)

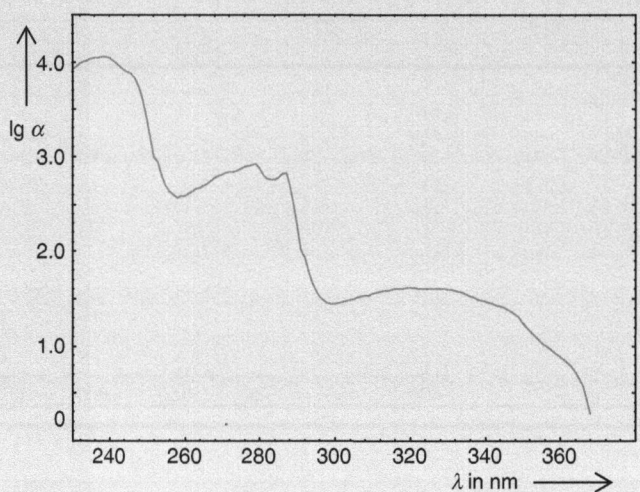

Fig. 3.25 Absorption spectrum of a compound with the molecular mass $m/z = 120.0576$ amu

light at $0°C$ in a cuvette with $l = 1$ cm, a deep-blue color develops. The measured absorption band absorbs at $\lambda_{max} = 610$ nm ($A = 1.5$) and possesses a shoulder at shorter wavelengths. When the irradiation is interrupted the solution becomes colorless again or will stay blue if the irradiation is continued. Such a procedure is named photochromism if it can be repeatedly reproduced. From additional investigations it is known that only approximately 20% of the sample is converted to the reaction product. To which base chromophore can the reaction product be assigned? Propose a structure for it.

3.58

9. Assign the two substituents $N(CH_3)_2$ to the structural moiety in Structure 3.59 in a way, that the compound with the sum formula $C_{16}H_{18}ClN_3S$ can be assigned to the absorption spectrum presented in Fig. 3.26. Complete the Structure 3.59. Which spectral changes are to be expected if a. N is substituted by CH and b. S is substituted by O?

(continued)

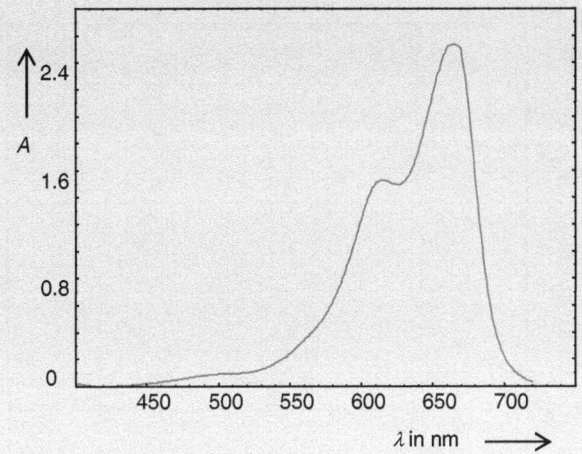

Fig. 3.26 Absorption spectrum of a compound with the skeleton of Structure 3.59. Sample: 0.128 mg in 10-mL ethanol

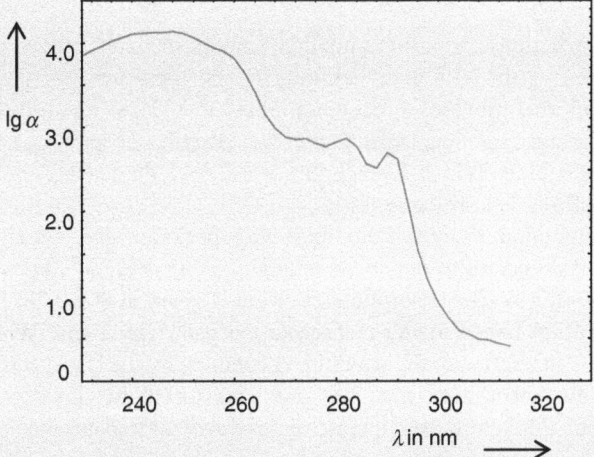

Fig. 3.27 Absorption spectrum of a compound with the molecular mass $m/z = 104.0625$ amu

3.59

10. The molecular peak of an unknown compound was determined to be $m/z = 104.0625$ amu. Its absorption spectrum is shown in Fig. 3.27. Propose a structure for the compound.

(continued)

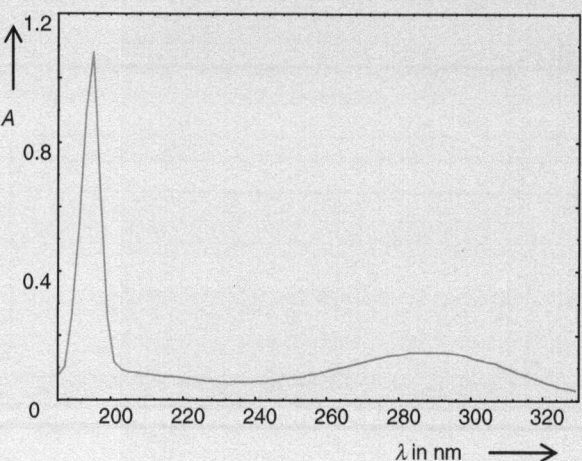

Fig. 3.28 Absorption spectrum of an unknown compound. Sample: 0.008 mol L^{-1} (aceto-nitrile); l = 1 cm

11. The reaction of 4,4′-tetramethyl-diamino-benzophenone (Michler's ketone) with N-dimethylaniline in the presence of phosphorous oxychloride provides a reaction product that shows an absorption band possessing a shoulder at the shorter wavelength side with $\lambda_{max} = 570$ nm (log α = 4.5). Which base chromophore does the reaction product possess? Propose a structure for it.

12. A compound provides the following peaks in the upper range of the mass spectrum: $m/z = 98$ amu ($I_{rel} = 45.4\%$) M$^{\bullet+}$, $m/z = 99$ amu ($I_{rel} = 3.0\%$). The absorption spectrum is given in Fig. 3.28. Assign the absorption bands to the respective electron transitions. Which information in relation to the structure is obtained by the UV spectrum? Note that the absorbance $\lambda < 200$ nm is distorted by absorption of the solvent. Therefore, this intense short-wavelength band manifests itself only in the range of its slope at the long-wavelength side.

13. Figure 3.29a and b shows the mass and UV–vis spectra of an unknown compound, respectively. Derive the structure of this compound.

14. Propose a structure for the compound which provides the mass and UV–vis spectra in Fig. 3.30. Confirm the structure by the respective mass spectrometric fragmentation routes.

15. A steroid (Structure 3.60) reacts with chloranile (Structure 3.61). The HPLC chromatogram of a reaction mixture as well as the DAD spectra of the three by-products are shown in Fig. 3.31. The main product P gives rise to a structureless absorption band with $\lambda_{max} = 287$ nm. The measured absorbance of a solution of 0.116-mg P in 10-mL acetonitrile is $A = 0.42$. Which structure does the main reaction product P possess?

(continued)

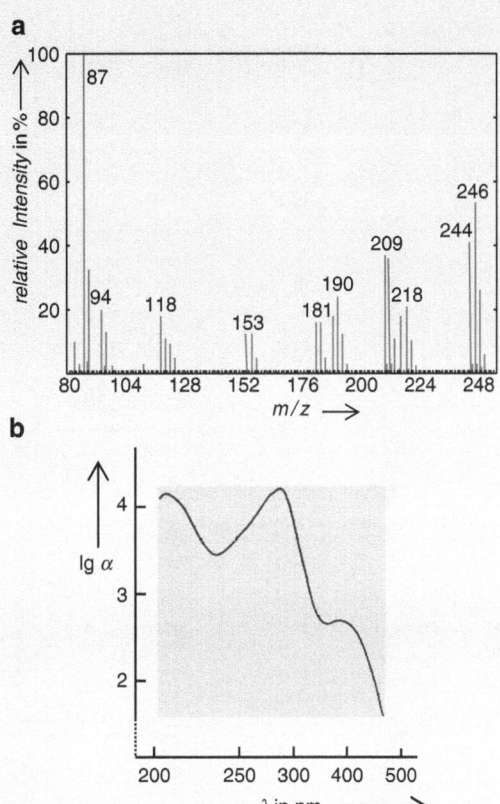

Fig. 3.29 Mass and absorption spectra of an unknown compound MS data: m/z in amu (I_{rel} in %): 244 (40.8), 245(2.7), 246(53.4), 247(3.1), 248(26.0), 249(1.6), 250(5.7), 251(0.3), 252(0.3)

Propose a structure of the by-products and state which additional spectroscopic data are necessary to justify the structures

3.60

3.61

16. The skeleton (Structure 3.62) and the Vis spectrum (Fig. 3.32) are given for an ionic dye salt. To which base chromophore can the compound be assigned? Attach the two substituents $N(CH_3)_2$ to the skeleton of Structure 3.62 and decide whether X can be CH, N, O, or S.

(continued)

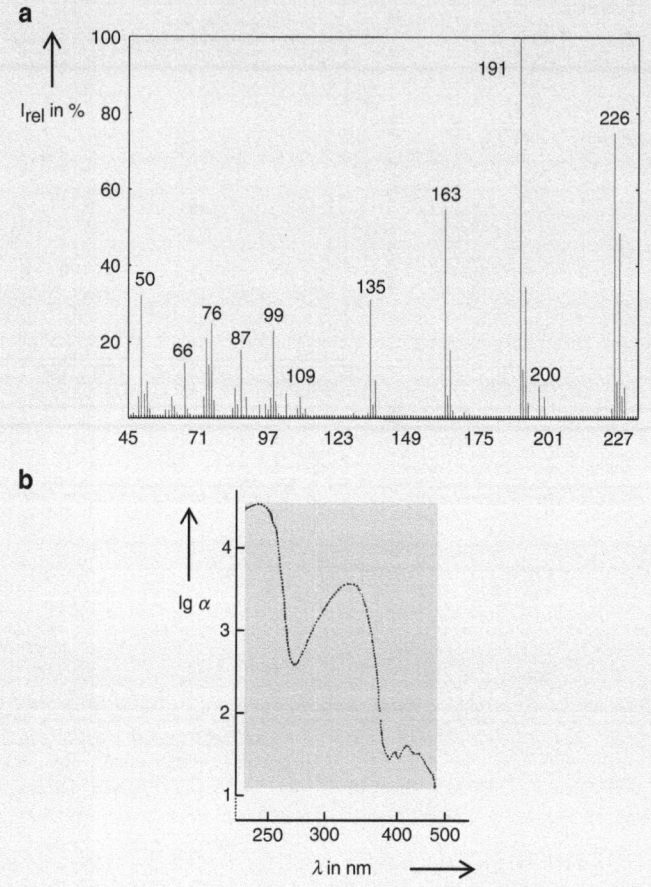

Fig. 3.30 Mass and absorption spectra of an unknown compound

3.62

17. An unknown compound with the mass spectrum in Fig. 1.1 provides the absorption spectrum shown in Fig. 3.33. Note that the determination of its structure was started with the help of mass spectrometry in Challenge 1.5. Which structural information can be obtained by the UV–vis spectrum? Which structural questions cannot be answered yet?

(continued)

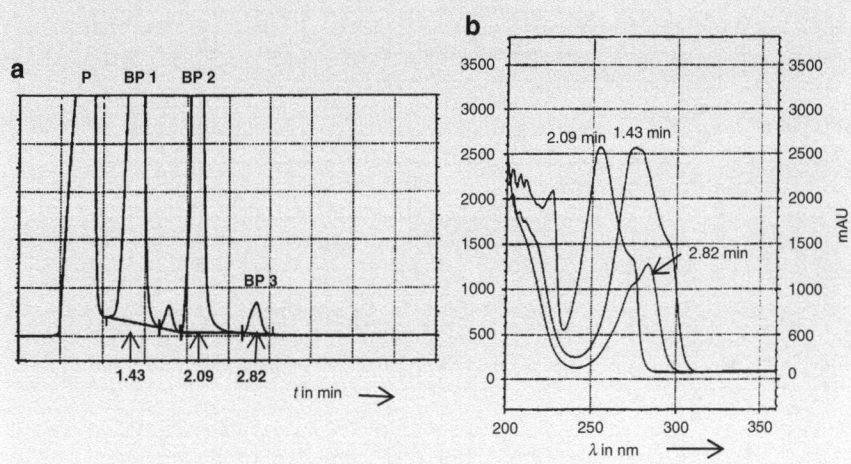

Fig. 3.31 HPLC chromatogram (**a**) and DAD spectra (**b**) of the by-products (BP) of a reaction mixture; P – main product

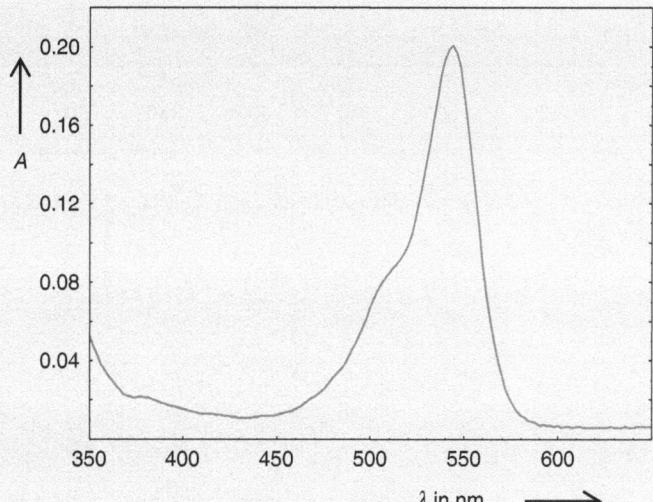

Fig. 3.32 Absorption spectra of an ionic dye

18. Structure 2.22 could be determined from the mass and infrared spectra. Which information can be obtained from the absorption spectrum in Fig 3.34 for this unknown compound?

19. The absorption spectrum shown in Fig. 3.35 belongs to a certain isomer of Structure 3.63. Which isomer is present? Why can the two other isomers be unambiguously excluded? Name the chromophore of the compound?

(continued)

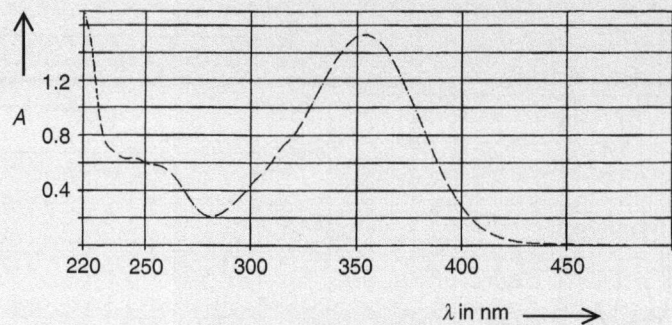

Fig. 3.33 Absorption spectra of the unknown compound; its mass spectrum is presented in Fig. 1.1

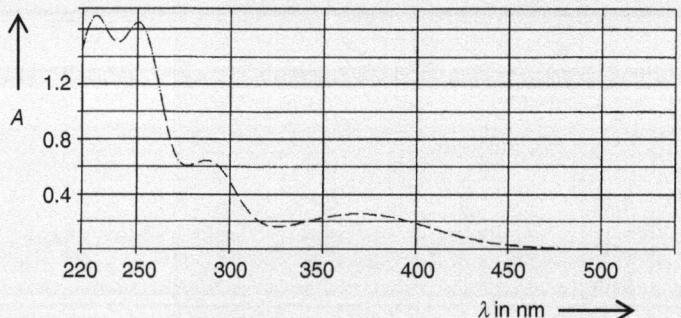

Fig. 3.34 Absorption spectra of the unknown compound of Challenge 2.12. Sample: 0.20 mg in 10-mL methanol

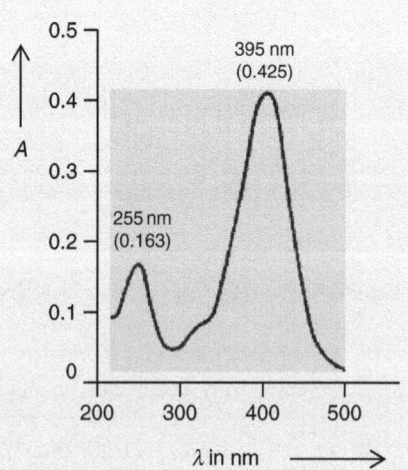

Fig. 3.35 Absorption spectra of an isomer of Structure 3.63. Sample: 0.0462 mg in 25-mL ethanol; 1 = 1 cm

(continued)

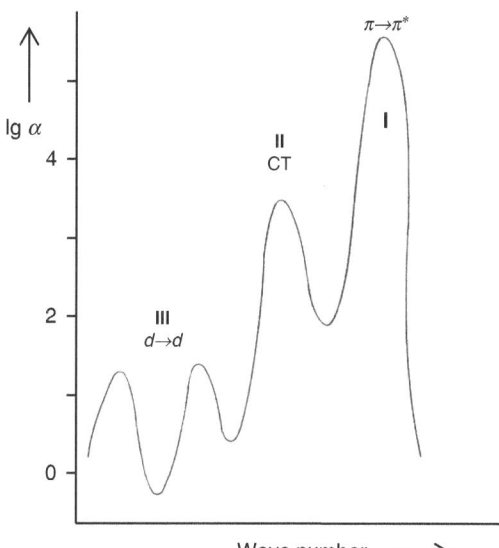

$$(CH_3)_2N \qquad \overset{H}{\underset{O}{\diagdown}}$$

3.63

3.6   Electron Absorption Spectra of Inorganic Complex Compounds

Inorganic complex compounds are characterized by three types of absorption bands, which are outlined in Fig. 3.36.

1. *Inner ligand transitions* (log $\alpha \approx$ 4–5); absorption band **I** in Fig. 3.36
   The absorption spectra of electron transitions within the ligand conform substantially to those of the free ligand. The absorption spectrum shows only the absorption band of the ligand if the central atom does not possess d electrons. Thus, for example, the absorption spectrum of $[Tl(pyridine)_3]Cl_3$ does not differ from that of pyridine.
2. *Charge transfer* (CT) (log $\alpha \approx$ 3–4)
   Absorption bands of type **II** in Fig. 3.36 are caused by electron transitions from a π^* level of the ligand to the central atom or vice versa. Such electron transitions determine substantially the absorption spectra of the complex

Fig. 3.36 Pattern of a general absorption spectrum of an inorganic complex compound

compounds of electron-rich metal ions in low valence state or those of heavy d metal ions, for example, [Mo(pyridine)$_3$]Cl$_3$. Unfortunately such transitions are meaningless for structural determination.

3. *d-d transitions* (log $\alpha \approx$ 1–2)

Electron transitions within the same orbital (here: the d orbital) are forbidden according to Laporte's rule (see Sect. 3.1.4). Such transitions can be observed, in spite of Laporte's rule, because the actual transitions are coupled to vibrations that are antisymmetric and have the same symmetry as the dipole moment operator thereby lifting Laporte's rule. Their intensities, however, remain weak (see type **III** in Fig. 3.36).

   d-d transitions that are additionally spin-forbidden are even much weaker than spin-allowed transitions.

   Although *d-d* transitions are only weakly allowed and, accordingly, give rise to only weak absorption bands, they are meaningful for structural determination. However, the relation between structure of the complexes and their term systems must be known as already explained for organic compounds above. The term systems caused by *d-d* transitions are best obtained by the *ligand field theory* according to the *method of the weak ligand field* which primarily considers the electron interactions within the central ion and secondarily the effect of the ligand field upon the terms of the free ions. Therefore, the basis of this theory will be shortly outlined in the following.

Step 1: Consideration of the splitting-up of the terms of the 3 d^N system resulting from the interactions of the N 3 d electrons:

Remember, a quantum mechanical approach to determine the energy of electrons in an element or ion is based on the results obtained by solving the Schrödinger Wave Equation for the H-atom. The various solutions for the different energy states are characterized by the three quantum numbers, n, l, and m_l. Note that m_l is a subset of l, with allowable values ranging from $m_l = -l$ to $m_l = +l$. Thus, there are (2 l + 1) values of m_l for each l value. There is a fourth quantum number m_s that identifies the orientation of the spin of one electron relative to those of other electrons in the system. The value of m_s is either + ½ or −½. The manner in which the angular momenta associated with the orbital and spin motions in many-electron atoms can be combined together are versatile. But *Russell–Saunders coupling* is a good approximation for the light 3 d elements. According to this theory,

– The spin-orbital interaction (J) can generally by ignored. The overall spin **S** arises from adding the individual m_s together and is a result of coupling of spin quantum numbers for the separate electrons.
– The total orbital angular momentum quantum number **L** defines the energy state for a system of electrons.

The Russell–Saunders term symbol that results from these considerations is given by ^{2S+1}L:

Table 3.15 Term symbols and symmetry-caused splitting-up of the terms of the free ion resulting from the influence of the cubic-ordered ligands (octahedron O_h or tetrahedron T_d)

L	Symbol	Orbit multiplicity	Ligand field components for cubic symmetry
0	S	1	A_1
1	P	3	T_1
2	D	5	$E + T_2$
3	F	7	$A_2 + T_1 + T_2$
4	G	9	$A_1 + E + T_1 + T_2$

Note that, for simplicity, spin multiplicities are not included in the symbols since they remain the same for each term. Furthermore, the subscripts g and u must be added for the octahedral environment which means symmetric and antisymmetric with respect to center of inversion, respectively.

$$d^N \xrightarrow{\quad \text{Electron interaction} \quad} \overset{\text{Term symbolic of the free ion}}{{}^{2S+1}\mathbf{L}}$$

The states or term letters are represented as given in the second column of Table 3.15. The multiplicity M of the term is given by Eq. 3.11

$$M = 2S + 1 \tag{3.11}$$

The multiplicity states how many levels are energetically degenerated, see column three in Table 3.15.

Note that, for example, the d^2 system is 45-fold degenerated employing two triplet terms ${}^3F$, ${}^3P$ and three singlet terms ${}^1G$, ${}^1D$, ${}^1S$, and the d^3 system is 120-fold degenerated. However, generally, only the *ground state* is important to understand the absorption spectra in the UV–vis range. The *ground terms* are deduced by using Hund's and Pauli's rules:

1. The ground term will have the maximum multiplicity S.
2. If there is more than one term with maximum multiplicity, then the ground term will have the largest value of L.

Thus, for a d^2 system, the ground term must be a triplet term according to rule 1, i.e., only the terms ${}^3F$ or ${}^3P$ are to be considered. Because of the greater orbit moment L in comparison to P the ${}^3F$ must be the ground term of the 3 d^2 system. Note that this term is 21-fold degenerated. The energetic position of the higher terms can be seen on the left side of the Tanabe-Sugano term diagrams (Fig. 6.2).

Step 2: Crystal field splitting of Russell-Saunders terms upon being surrounded by an array of point charges in octahedral or tetrahedral geometry consisting of the ligands.

All Russell-Saunders terms, higher than threefold orbit-degenerated, split up under the strength of a ligand field. The terms resulting from the effect of a crystal field on the Russell-Saunders terms in an octahedral or tetrahedral environment are listed in the last column of Table 3.15. The x-axis of a Tanabe-Sugano diagram is expressed in terms of the ligand field splitting parameter Δ and the y-axis is detailed in terms of energy. Both axes are scaled by the Racah parameter B. The plots of the

Fig. 3.37 Term system of a d^2 system in octahedral symmetry and all possible electron transitions

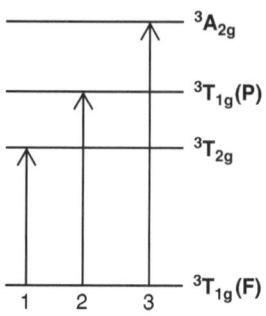

energies calculated for the electronic states of each electron configuration are known as Tanabe-Sugano diagrams, which are presented in Fig. 6.2 for the d^N system in an octahedral environment. Note that these diagrams are also valid for the tetrahedral symmetry. However, the terms switch their position. In a Tanabe-Sugano diagram the energy of the ground state is taken to be zero for all field strengths, and the energies of all other terms and their components are plotted with respect to the ground term.

The term diagram for the d^2 system at a certain ligand field strength Δ is shown in Fig. 3.37. Note that only such terms are considered which are responsible for the spectrum in the UV–vis range. These are the ground state (^3F) and the first excited state with the same spin multiplicity ^3P. Because electron transitions accompanied by a change in the spin multiplicity are strictly forbidden (see selection rules above), singlet terms are not considered.

Note that, if two or more of the same terms appear, the source of the Russell-Saunders terms are added in brackets. Thus, one $^3T_{1g}$ term arises from the ^3F term and the other from the excited ^3P term. The subscript g is necessary because of the octahedral symmetry.

The three possible electron transitions give rise to three absorption bands. This is shown in Fig. 3.38 employing $[V(H_2O)_6]^{3+}$ as an example.

Band 1:	$^3T_{1g} \rightarrow {}^3T_{2g}$	$\tilde{v} = 17{,}000 \text{ cm}^{-1}$ ($\lambda = 588$ nm)
Band 2:	$^3T_{1g} \rightarrow {}^3T_{1g}$	$\tilde{v} = 25{,}000 \text{ cm}^{-1}$ ($\lambda = 400$ nm)
Band 3:	$^3T_{1g} \rightarrow {}^3A_{2g}$	$\tilde{v} = 38{,}000 \text{ cm}^{-1}$ ($\lambda = 263$ nm)

The number of absorption bands corresponds to that of the electron transitions for O_h symmetry and, furthermore, none of the bands shows asymmetry which could be caused by a disturbance of the symmetry, hence the octahedral symmetry can be justified by the electron absorption spectroscopy.

The values of the absorptivity of spin-allowed but symmetrically forbidden transitions of octahedral complexes lie in the range from 10 to 50 L mol^{-1} cm^{-1}, those of tetrahedral ones are tenfold higher. This is important to distinguish between both structures.

Because higher ordered terms for the electron configuration $d^4 - d^7$ drop sharply, *term crossing* is observed at certain ligand field strength. Thus, for example, the $^5T_2$

Fig. 3.38 UV–vis spectrum of $[V(H_2O)_6]^{3+}$ in aqueous solution

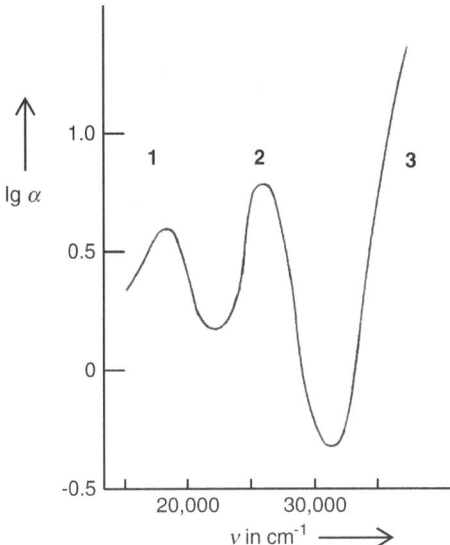

term of the d^6 system (Fe^{2+} or Co^{3+}) is the ground state at low ligand field strength. For stronger ligand fields, however, the $^1A_1$ term becomes the ground term. In the first case a so-called *high-spin complex* with four unpaired electrons is present and in the second case a *low-spin complex* exists in which all electrons are paired. Because for the d^6 system the excited $^1A_1$ terms drop very sharply and become the ground term at low Δ values already, most of the Co(III) complexes are diamagnetic. The absorption spectra of octahedral Co(III) complexes are caused by the transitions $^1A_1 \rightarrow {}^1T_{1g}$ and $^1A_1 \rightarrow {}^1T_{2g}$ as shown in Fig. 3.39 II. Thus, for example, both absorption bands are observed for $[Co(NH_3)_6]^{3+}$ at $\tilde{\nu} \approx 20,000$ cm^{-1} ($\lambda \approx 500$ nm) and $\tilde{\nu} \approx 20,000$ cm^{-1} ($\lambda \approx 357$ nm).

The change from $[CoX_6]$ to *trans*-$[CoX_4Y_2]$ and *cis*-$[CoX_4Y_2]$ reduces the symmetry to D_{4h} and C_{2v}, respectively. The reduction of the octahedral symmetry also takes place in distorted octahedrons. These normally result from elongation of the bonds to the ligands lying along the z-axis, but occasionally also a shortening of these bonds occurs. The symmetry degradation results in a splitting-up of threefold-degenerated T terms which is outlined in Fig. 3.39 I and II. This has the consequence that, instead of two, three bands are observed in the absorption spectra, sometimes only as asymmetrical bands if they are not resolved. This is shown in Fig. 3.40 for the absorption spectra of *trans*-$[Coen_2F_2]$ and *cis*-$[Coen_2F_2]$. The term splitting is smaller in the *cis* isomer, therefore, only an asymmetric band is observed instead of two bands as in the case of the *trans* isomer.

The distortion of the octahedral geometry can also have the causes outlined by the *Jahn–Teller theorem*. It essentially states that any nonlinear molecule with degenerated electronic ground state will undergo a geometrical distortion that

Fig. 3.39 Term diagrams of the electron transitions in the Vis range for the Co(III) complexes $[CoX_6]$ (O_h) I, trans-$[CoX_4Y_2]$ (D_{4h}) II, and trans-$[CoX_4Y_2]$ (C_{2v}) III

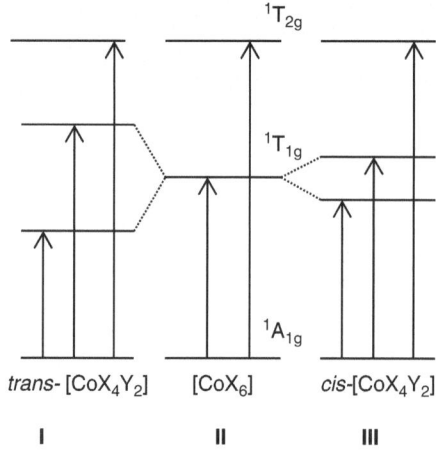

Fig. 3.40 Absorption spectra of trans-$[Coen_2F_2]$ and cis-$[Coen_2F_2]$. The arrows point at the asymmetry of the bands for the cis isomer

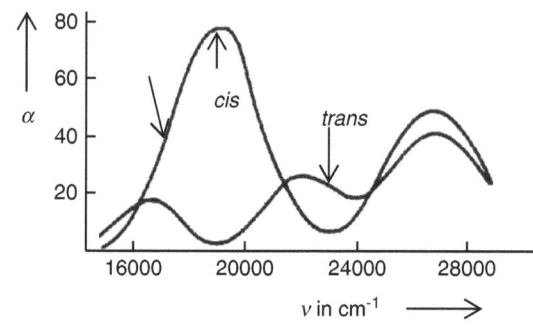

Fig. 3.41 Qualitative term diagram for the electron transitions in the Vis range for Ti(III) complexes (d^1 system) with octahedral (O_h) and tetragonal D_{4h} symmetry

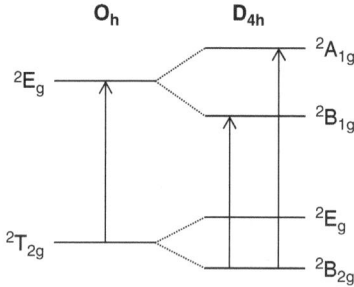

removes the degeneracy. Unfortunately it is not able to predict the direction of the distortion, only the presence of an unstable geometry.

As shown in the left part of Fig. 3.41 the ground term of the octahedral d^1 system is the threefold-degenerated $^3T_{2g}$ term. However, this degenerated term cannot form the stable ground state according to the Jahn-Teller theorem. Instead it will split into the terms $^2B_{2g}$ and ^2E_g, see the right part of Fig. 3.41. Hence, the octahedral geometry is distorted into a geometry having D_{4h} symmetry which can be seen in

Fig. 3.42 Absorption spectrum of the distorted [Ti(H$_2$O)$_6$]$^{3+}$ ion measured in diluted sulfuric acid. The arrow points at the asymmetry of the longest wavelength absorption band caused by a structural distortion following the Jahn–Teller theorem

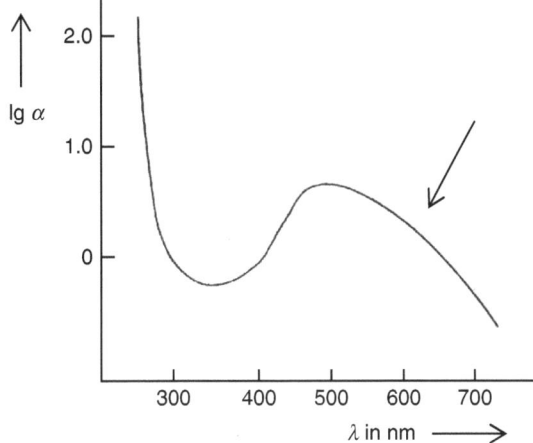

the absorption spectrum of the [Ti(H$_2$O)$_6$]$^{3+}$ ion measured in diluted sulfuric acid (Fig. 3.42).

Note that the reduction of the symmetry of the coordination polyhedron is connected with an increase of the absorptivity. This increase is higher for the *cis* form than the *trans* form.

Challenge 3.12

Figure 3.43 shows the absorption spectrum of Ruby (Cr^{3+}/Al$_2$O$_3$). Which structure does the Cr^{3+} complex possess? Assign the electronic transitions to the respective absorption band.

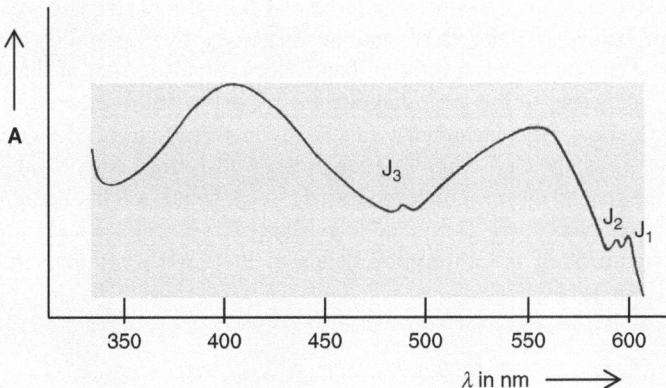

Fig. 3.43 Absorption spectrum of Ruby. J$_i$ are intercombination bands

Solution to Challenge 3.12

Let us start with the assumption that the Cr^{3+} ion is octahedral coordinated by oxygen atoms in Ruby. The section of the term diagram for the octahedral Cr^3+ ion (d^3 system of Fig. 6.2) responsible for the possible electron transitions in the UV–vis range is shown in Fig. 3.44.

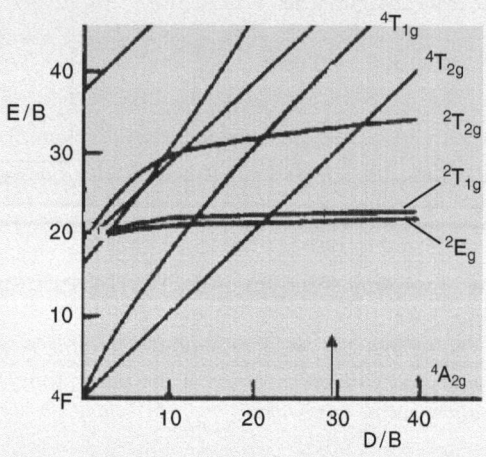

Fig. 3.44 Section of the Tanabe-Sugano diagram for a d^3 system obtained by Fig. 6.2. The arrow approximately indicates the ligand field strength for the possible electron transitions. Note that, for simplicity, the Russell-Saunders terms of the excited states are not included in the symbols

The absorption spectrum can be easily assigned to the *octahedral* structure of the Cr^{3+} ion. The three weak intercombination bands caused by a transition accompanied by a change of the spin multiplicity can be helpful to find the position of the ligand field strength expressed as the Δ/B value. This position is marked by an arrow in the diagram. Two intercombination bands provide the two weak longest wavelength bands and the third one lies between two spin-allowed electron transitions. The number of absorption bands corresponds to that of the electronic transitions according to the term diagram for O_h geometry. Furthermore, none of these bands show any asymmetry caused by a distortion of the coordination polyhedron. Thus, the Cr^{3+} complex possesses a well-formed octahedral geometry. Note that the ground term is nondegenerated ($^4A_{2g}$), hence, a Jahn-Teller distortion should not take place.

The assignment of the absorption bands of the Cr(III) complex in Rubin is summarized in Table 3.16.

Table 3.16 Assignment of the absorption bands of the Cr(III) complex in Ruby

Transition for O_h	Comment	λ in nm
$^4A_{2g} \rightarrow {}^2E_g$	Weak spin-forbidden intercombination transition	600
$^4A_{2g} \rightarrow {}^2T_{1g}$	Weak spin-forbidden intercombination transition	592
$^4A_{2g} \rightarrow {}^4T_{2g}$	Spin-allowed transition	558
$^4A_{2g} \rightarrow {}^2T_{2g}$	Weak spin-forbidden intercombination transition	487
$^4A_{2g} \rightarrow {}^4T_{1g} \, (^4F)$	Spin-allowed transition	405
$^4A_{2g} \rightarrow {}^4T_{1g} \, (^4P)$	Spin-allowed transition	<350

Challenge 3.13
For solutions to Challenge 3.13, see *extras.springer.com*.

1. For Nickel complexes of the type $[NiX_2Y_2]$ a tetrahedral or, alternatively, a planar-quadratic structure can be assumed. The absorption spectrum of a nickel complex of such type is shown in Fig. 3.45 in the Vis and NIR range. Which structure can be derived from the absorption spectrum? Assign and name the absorption bands to the respective electron transitions.

2. Which structure can be derived from the absorption spectra (Fig. 3.46) for the nickel complexes $[Ni(H_2O)_6]^{2+}$ and $[Ni(en)_3]^{2+}$? Assign and name the absorption bands to the respective electron transitions. Which color can be attributed to a 0.1-molar solution of each of these complexes and what is the difference based on?

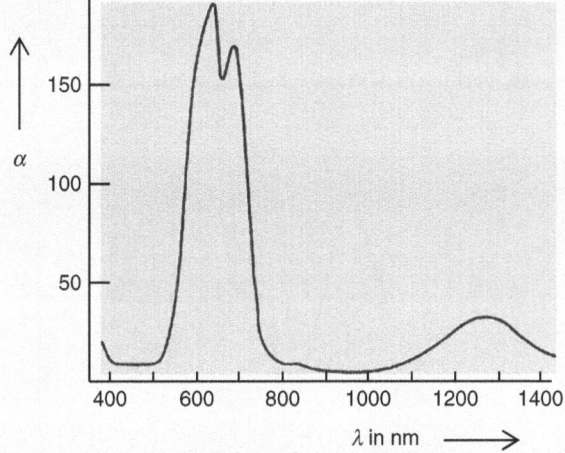

Fig. 3.45 Absorption spectrum of a nickel complex of the type $[NiX_2Y_2]$ in the Vis and NIR range

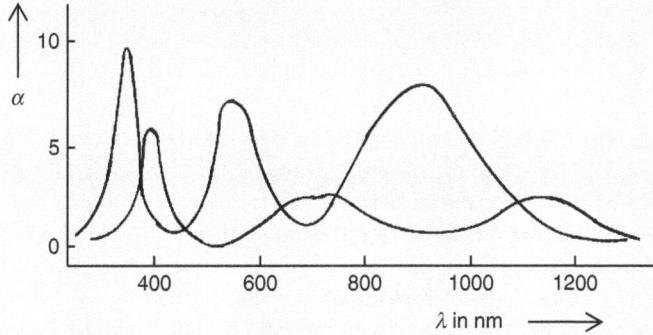

Fig. 3.46 Absorption spectra of the complexes $[Ni(H_2O)_6]^{2+}$ and $[Ni(en)_3]^{2+}$ in the Vis and NIR range

(continued)

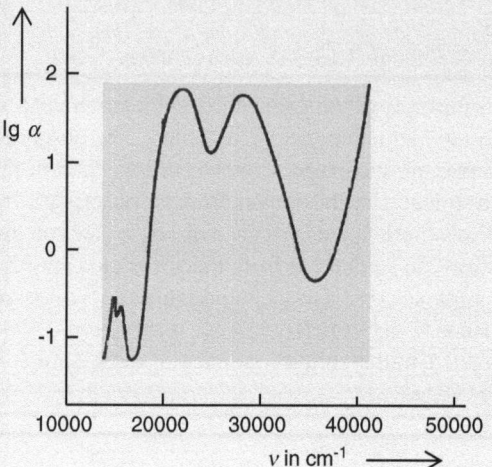

Fig. 3.47 Absorption spectrum of a complex of the type $[Me(H_2O)_6]^{n+}$, V_i αin L mol cm^{-1}

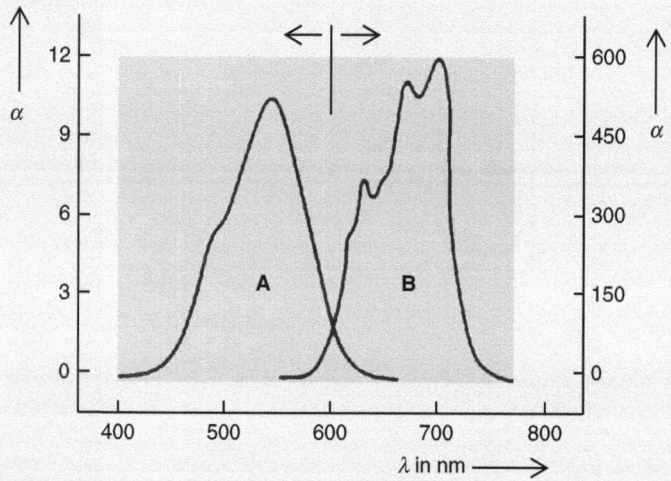

Fig. 3.48 Absorption spectrum of complex ions $[CoCl_4]^{2-}$ and $[Co(H_2O)_6]^{2+}$

3. Decide which complex compound of the type $[Me(H_2O)_6]^{n+}$ with Me=Cr^{3+}, Fe^{2+}, Co^{3+}, Ni^{2+} can be assigned to the absorption spectrum shown in Fig. 3.47 and justify your decision.

4. Figure 3.48 shows the absorption spectra of $[CoCl_4]^{2-}$ and $[Co(H_2O)_6]^{2+}$. Assign the absorption spectra to the respective cobalt complex. Which structure can be derived from the absorption spectra?

5. Which complex compound of the type $[Me(H_2O)_6]^{n+}$ with Me=V^{3+}, Cr^{3+}, Mn^{2+}, Fe^{2+}, or Fe^{3+} can be assigned to the absorption spectrum presented

(continued)

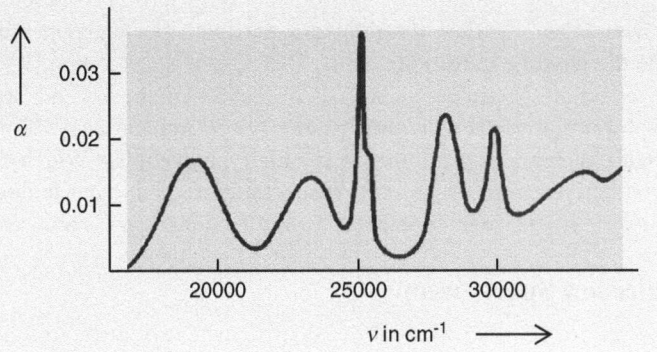

Fig. 3.49 Absorption spectrum of a complex of the type $[Me(H_2O)_6]^{n+}$; α in $L\ mol^{-1}\ cm^{-1}$

Fig. 3.50 Absorption
spectrum of the type [Fe
$(H_2O)_6]^{2+}$ as well as $[CoF_6]^{3-}$
(**a**) and all other Co(III)
complexes as well as [Fe
$(CN)_6]^{4-}$ (**b**)

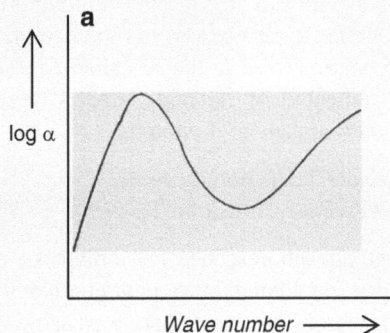

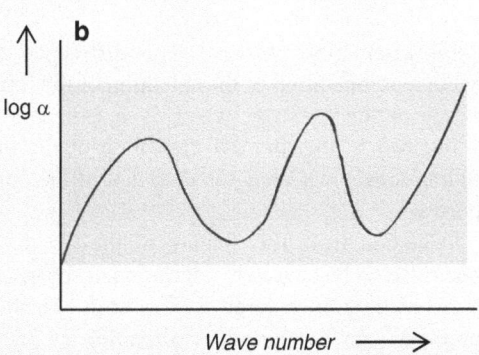

in Fig. 3.49? What is the reason for the very weak intensity of the
absorption bands?

6. The absorption spectrum shown in Fig. 3.50a is characteristic for both [Fe
$(H_2O)_6]^{2+}$ as well as $[CoF_6]^{3-}$. In contrast, all other Co(III) complexes as
well as $[Fe(CN)_6]^{4-}$ provide absorption spectra like the one shown in
Fig. 3.50b. Which geometry and magnetic behavior can be derived for

(continued)

the complexes? Outline the term diagram with the relevant terms and name the possible transitions.

7. The reflection spectrum of solid $K_3[CoF_6]$ shows in the Vis and NIR range two absorption bands with maxima at $\lambda \approx 680$ nm and $\lambda \approx 880$ nm. The complex is magnetically normal, i. e. a high spin complex. Which structure can be derived from the absorption spectrum? Explain your findings.

3.7 Emission Spectroscopy

Up to now, only electron transfer from the ground state to excited states has been considered resulting in *absorption* spectra. However, the reverse process resulting in *emission* spectra can also provide structural information. There are some routes for the molecule to relax from the excited state into the ground state which, in general, are summarized in the so-called *Jablonski* diagram. Besides nonradiative transitions indicated by squiggly arrows there are the following two radiative transitions with emission of photons:

– *Fluorescence* Transition $S_1 \rightarrow S_0$
– *Phosphorescence* Transition $T_1 \rightarrow S_0$.

Note that phosphorescence can only take place if the triplet state T_1 of the molecule was populated by a nonradiative intersystem crossing process (ISC) from the S_1 state. The two radiative transitions indicated by straight arrows and the resulting emission spectra are outlined in Fig. 3.51.

In general, the fluorescence spectrum is the mirror image of the respective absorption spectrum as shown in Fig. 3.51 and the difference between the 0 and 0 transitions is small. This difference, the so-called *Stokes shift* is caused by the reorganization of the solvation of the molecule in the excited state. Thus, the energy of the S_1 state, which is responsible for the fluorescence, is somewhat diminished so that the 0–0 transition of the fluorescence spectra are observed at slightly longer wavelengths than that of the corresponding transition in absorption spectra.

Deviation from this feature of the "in the usual way" obtained fluorescence spectrum can be caused by alteration of the geometry of the molecule in the excited state resulting in a large Stokes shift. Furthermore, deviation from the mirror symmetry can be caused by alteration of the flexibility of the excited molecule because of stronger bonds. Thus, besides the analysis of the vibrational fine structure of the electron absorption bands as explained in Sect. 3.1.5, fluorescence spectroscopy (deviation from the mirror symmetry, Stokes shift) also gives rise to information on the structure of the molecule in the short-lived excited state.

Note that phosphorescence spectra are meaningless for structure determination.

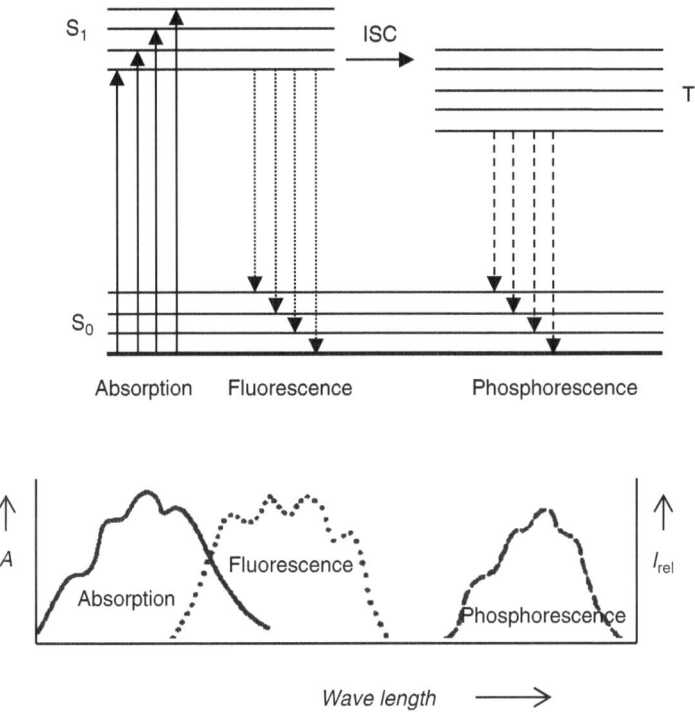

Fig. 3.51 Simple term diagram for the radiative deactivation processes of a molecule excited by absorption of electromagnetic radiation. ISC – nonradiative intersystem crossing

Challenge 3.14
Figure 3.52 presents the absorption and fluorescence spectra of biphenyl (Structure 3.64).

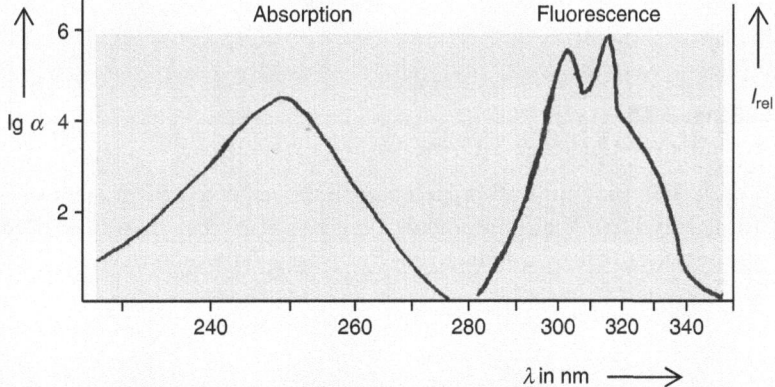

Fig. 3.52 Absorption and fluorescence spectra of biphenyl (Structure 3.64)

(continued)

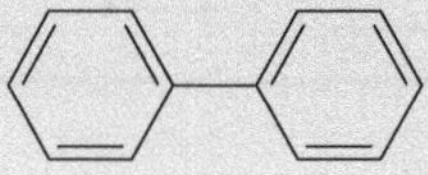

3.64

Which information can be obtained concerning the structure of the molecule in the excited singlet state?

Solution to Challenge 3.14

The very large Stokes shift $(3,310 \text{ cm}^{-1})$ reveals that the molecule possesses another structure in the excited state. According to the results obtained by quantum mechanics, biphenyl is distorted in the ground state by approximately $23°$ and should be planar in the excited state. This fact explains the large Stokes shift. Furthermore, the central C–C bond should be transformed into a double bond in the excited state. This theoretical finding can be justified by the fluorescence spectrum. The double bond in the excited state diminishes the "flexibility" of the molecule, thus, the fluorescence spectrum shows a vibrational pattern in contrast to the absorption spectrum. The structure of biphenyl in the ground and excited state obtained by fluorescence spectroscopy is outlined in Structure. 3.64A.

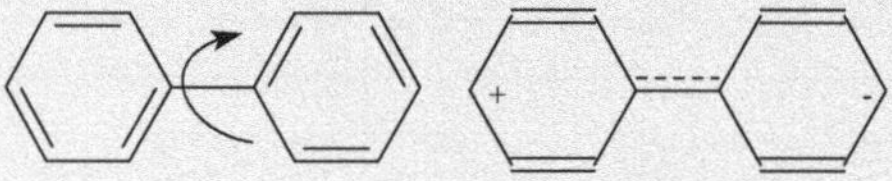

S_0 S_1
          3.64.A

Challenge 3.15

For solutions to Challenge 3.15, see *extras.springer.com*.

1. Figure 3.53 shows the absorption and fluorescence spectra of anthracene (Structure 3.65). Which information can be obtained by these spectra for the geometry of the excited state? Why do the 0–0 transitions differ in both spectra?

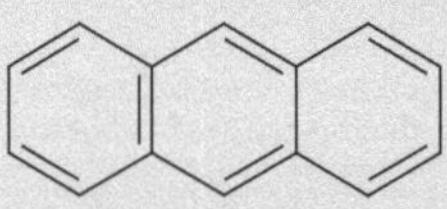

3.65

(continued)

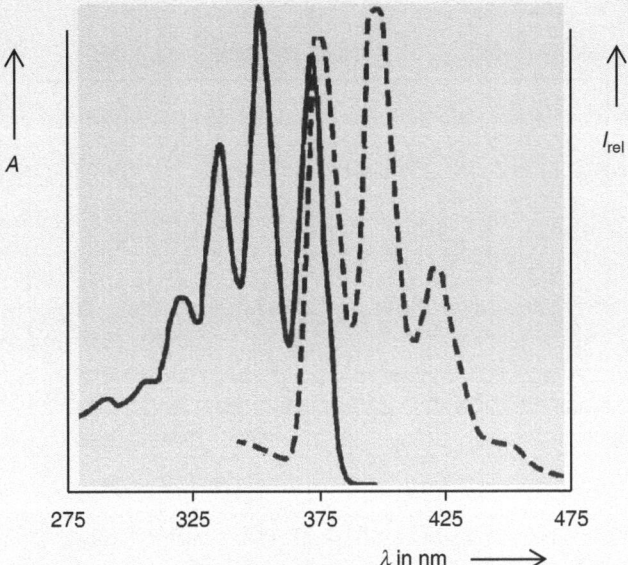

Fig. 3.53 Absorption and fluorescence spectra of anthracene

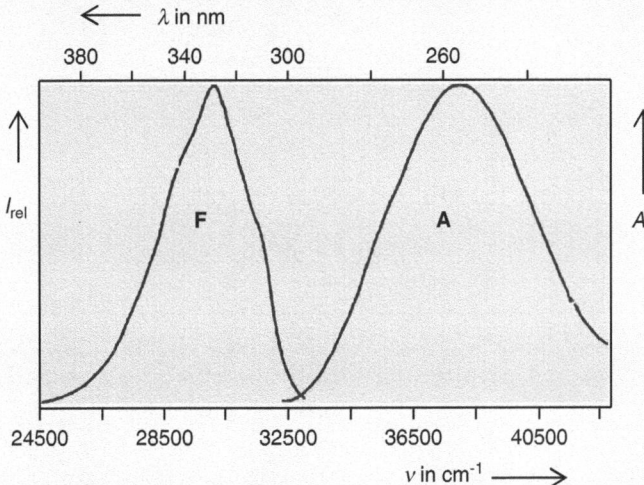

Fig. 3.54 Absorption and fluorescence spectra of 4-methoxy biphenyl (Structure 3.66) $\lambda_{exc} = 265$ nm; Stokes shift: 3,320 cm^{-1}

2. The absorption and fluorescence spectra of the two biphenyl derivatives shown in Structures 3.66 and 3.67 are depicted in Figs. 3.54 and 3.55, respectively. Which geometry can be derived for their S$_1$ state?

(continued)

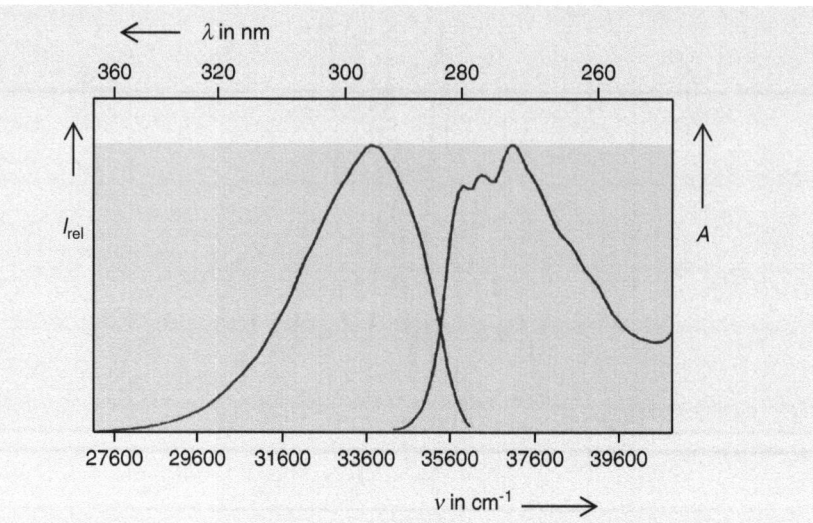

Fig. 3.55 Absorption and fluorescence spectra of octamethyl biphenyl (Structure 3.67) $\lambda_{exc} = 265$ nm; Stokes shift: 1,950 cm^{-1}

3.66

3.67

Further Reading

1. Perkampus H-H (1985) UV/VIS-Spektroskopie und ihre Anwendung. Springer, Berlin
2. Fabian J, Hartmann H (1985) Light absorption of organic colorants. Springer, Berlin
3. Schläfer HL, Gliemann G (1967) Einführung in die Ligandenfeldtheorie. Akad. Verlagsgesellschaft, Leipzig

Chapter 4
Nuclear Magnetic Resonance Spectroscopy (NMR)

4.1 Introduction

In contrast to electron spectroscopy, there are very well-written introductions to NMR spectroscopy in the literature (for example, see Sect. 4.5). Therefore, we will only summarize the basic theory behind the technique. Note that we are performing experiments on the nuclei of atoms, not the electrons.

Nuclear magnetic resonance spectroscopy is based on the interaction of the *magnetic moment* $\vec{\mu}$ with an outer homogeneous magnetic field with the strength B_0. The magnetic moment is caused by the spin of the atomic nuclei. The nuclear magnetic moment is determined by the nuclear spin $\vec{J}$ and the *magnetogyric ratio* γ:

$$\vec{\mu} = \gamma \cdot \vec{J}. \tag{4.1}$$

According to the laws of quantum mechanics the absolute value of the nuclear magnetic spin is given by Eq. 4.2

$$\left| \vec{J} \right| = \frac{h}{2\pi} \cdot \sqrt{I(I+1)} \tag{4.2}$$

in which I is the nuclear spin quantum number, which is determined by the protons and neutrons of the atomic nucleus, and h is Planck's constant.

The subatomic particles protons and neutrons can be figuratively seen as spinning around their axis with a frequency v called the *Lamor frequency*. If these spins are paired against each other (such as in ^{12}C or ^{16}O), the nucleus of the atom has no overall spin ($I = 0$) and, hence, no magnetic moment according to Eqs. 4.1 and 4.2. Such nuclei cannot be detected in NMR spectroscopy, based on the fact that the number of neutrons and the number of protons are both even.

M. Reichenbächer and J. Popp, *Challenges in Molecular Structure Determination*,
DOI 10.1007/978-3-642-24390-5_4, © Springer-Verlag Berlin Heidelberg 2012

If the number of neutrons plus the number of protons is odd, then the nucleus has a half-integer spin, i.e., $1/2, 3/2, \ldots$ Nuclei with $I = \frac{1}{2}$ are ^{1}H, ^{13}C, ^{19}F, and ^{31}P and they are important for structure determination in organic and bioorganic chemistry.

If the number of neutrons and the number of protons are both odd, then the nucleus has an integer spin, i.e., $1, 2, 3, \ldots$ Nuclei with $I = 1$ are, for example, ^{2}H and ^{14}N. Note that all nuclei with $I \geq 1$, i.e., nuclei with more than one unpaired nuclear particle (protons or neutrons) will have a charge distribution that results in an *electric quadrupole moment*.

According to quantum mechanics, magnetic nuclei of spin I will have $2 \cdot I + 1$ orientations in an outer homogeneous static magnetic field of the strength B_0. Thus, nuclei with spin $I = \frac{1}{2}$ (^{1}H, ^{13}C, ^{19}F, ^{31}P) will have two possible orientations. If the magnetic field is absent, these orientations are degenerate, i.e., of equal energy. If the magnetic field is applied, then the energy levels split.

Each level is given by a *magnetic quantum number*, m whose values reach from $-I$ to $+I$. Hence, for nuclei with $I = \frac{1}{2}$ two states result with $m = -\frac{1}{2}$ and $m = +\frac{1}{2}$ corresponding to two energy levels. The energy is given by

$$E = m \cdot \hbar \cdot \gamma \cdot B_{\text{eff}} \tag{4.3}$$

with $\hbar = \frac{h}{2\pi}$

B_{eff} is the strength of the magnetic field *at the nucleus* which is somewhat smaller than the applied magnetic flux density B_0:

$$B_{\text{eff}} = (1 - \sigma) \cdot B_0 \tag{4.4}$$

Because the *shielding constant* σ is very small ($\approx 10^{-5}$) $B_{\text{eff}} \approx B_0$ is valid in a first approximation.

The energy diagram for nuclei with $I = \frac{1}{2}$ and $m = \pm\frac{1}{2}$, respectively is outlined in Fig. 4.1.

The difference of both energy levels is given by

$$\Delta E = \hbar \cdot \gamma \cdot B_{\text{eff}} \tag{4.5}$$

Equation 4.5 reveals the difference between NMR spectroscopy and optical spectroscopy:

The energy gap is determined by the applied magnetic field B_0. The higher B_0 the wider is the energy gap ΔE. It also means that if a nucleus has a relatively large magnetogyric ratio γ, then ΔE is correspondingly large.

When nuclei are in an applied magnetic field, the initial populations of the energy levels are determined by thermodynamics, as described by the Boltzmann distribution

$$\frac{N_\beta}{N_\alpha} = e^{-\frac{\Delta E}{k_B \cdot T}} \tag{4.6}$$

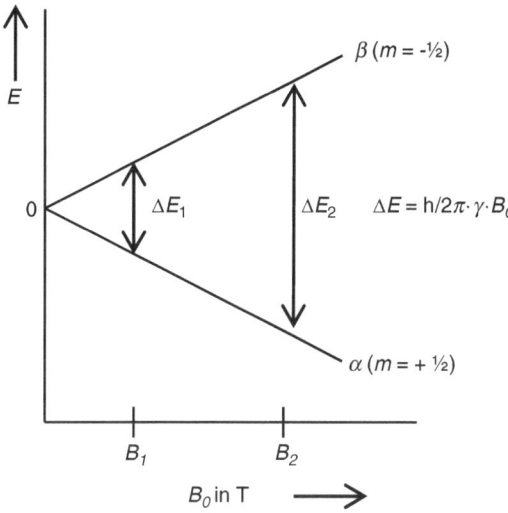

Fig. 4.1 Energy diagram levels and energy differences in relation to the applied magnetic field strength B_0

where N_β and N_α are the number of spins in the spin upper and down states, respectively, the number k_B is the Boltzmann constant, and T is the absolute temperature.

Because the energy gap ΔE is significantly smaller than the thermal energy $k_B \cdot T$ the population ratio of both energy levels is nearly 1, i.e., the lower energy level will contain only slightly more nuclei than the higher level. Note that this fact is the reason for the small sensitivity of NMR spectroscopy in comparison to the optical method and the development of instruments with increasingly stronger magnetic fields.

It is possible to excite the nuclei from the lower α level into the higher level β with electromagnetic radiation of the energy $E = h \cdot v$. The frequency v of radiation needed is determined by the difference in energy between the energy levels. The *consideration of resonance* is given by Eq. 4.7:

$$v = \frac{\gamma}{2\pi} \cdot B_{\mathrm{eff}} \tag{4.7}$$

The frequencies v lie in the radio wavelength range (100 MHz range), therefore, NMR spectroscopy is categorized in the range of high-frequency spectroscopy (HF spectroscopy).

The first commercial NMR spectrometers used permanent magnets or electromagnets with 60, 80, or 100 MHz, respectively. Note that the proton resonance is the usual way of describing an instrument. The spectra were obtained by the *continuous-wave (CW scan)* method using either the *frequency-sweep* (the magnetic field is constant and the frequency is altered to produce the resonance condition for every kind of nuclei step-by-step) or the *field-sweep* procedure whereby the field is altered at constant frequency. Modern instruments operate in the *pulsed Fourier transform* (FT) *mode* with frequencies of 400, 600, 750 MHz or higher based on helium-cooled superconducting magnets. This technique provides higher sensitivity, higher resolution, and enables correlation spectroscopy. The pulsed FT mode excites *all* nuclei simultaneously by a short pulse in the μs-range which includes the Lamor frequency of all nuclei. The frequencies of all nuclei which were excited during the pulse are superimposed on the emitted spectrum. The spectrum obtained in the time domain is transformed in the known frequency domain by Fourier transformation.

To understand the absorption of radiation by nuclei in a magnetic field let us consider a "classical" view of the behavior of the nuclei. Imagine nuclei of spin $I = \frac{1}{2}$ in a magnetic field. These nuclei will occupy the lower energy level. The nucleus is spinning around its axis and this axis of rotation will *precess* around the z axis of the stationary magnetic field B_0 in the same manner in which an off-perpendicular spinning top precesses under the influence of gravity. The frequency of precession is termed the Larmor frequency, which is identical to the transition frequency according to Eq. 4.6.

The addition of the magnetic moments of an assemblage of equivalent nuclei provide a net macroscopic magnetization along the z-axis, but none in the x,y plane which is outlined in Fig. 4.2a by the bold arrow.

Radiation energy is absorbed by the nuclei if the condition of resonance (Eq. 4.7) is fulfilled; i.e., the Lamor frequency is equal to the frequency of the electromagnetic radiation. The z magnetization spins around the x-axis caused by the pulse along the x-axis resulting in the precession of the nuclei around the z-axis in the x,y plane, see dotted line in Fig. 4.2b. Thus, a signal is produced in the receiver coil. The absorption of radiation energy by nuclei of spin $I = \frac{1}{2}$ can be imaged in the "classical" view by flipping of the magnetic moment so that they oppose the applied field and occupy the higher energy level.

Note that, according to Eq. 4.6, only a very small proportion of nuclei are in the lower energy state and can absorb radiation. Therefore, the possibility is given that by exciting these nuclei, the population of the higher and lower energy levels will become equal which means, the spin system is *saturated*. If this occurs, there will be no further absorption of radiation and the signal will disappear. But saturation is

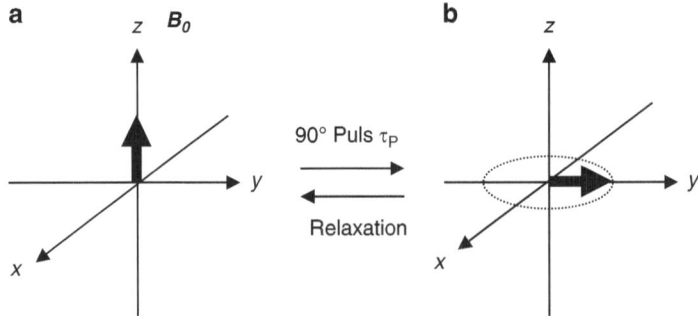

Fig. 4.2 (**a**) Representation of the macroscopic magnetization (*bold arrow*) along the z-axis in the thermodynamic equilibrium. (**b**) Effect of a 90° pulse upon the z magnetization

prevented by *relaxation processes*. Relaxation means the recreation of the thermodynamic equilibrium according to Eq. 4.6 which was disturbed by alteration of the population ratio caused by the absorption of radiation. In the pulsed FT mode the relation is observed by the free induction decay (**FID**).

The relaxation processes proceed according to a first-order rate law and are characterized by *relation times T*.

There are two major relaxation processes:

1. The *spin–lattice* (longitudinal) relaxation is characterized by the *spin–lattice relaxation time* T_1. Note that the surroundings in which the nuclei are held is called the *lattice*. The vibrational and rotational motions create the complex lattice field which has many components. But some of these components will be equal in frequency and phase to the Lamor frequency of the respective nuclei. These components of the lattice field will interact with the nuclei in the higher energy level and cause them to lose energy and to return to the lower energy state. An amount of energy equal to the energy difference between the higher and the lower energy state of the nuclei is passed to the lattice whereby the amount of vibration and rotation increases resulting in a tiny rise in the temperature of the sample.

   The spin–lattice relaxation time T_1 depends on the magnetogyric ratio of the nucleus and the mobility of the lattice. As the mobility increases, the vibrational and rotational frequencies increase and hence, there will be more components of the lattice field which can interact with excited nuclei. For example, the range of the spin–lattice relaxation time T_1 is $0.1–10$ s for ^{13}C atoms with hydrogen attached and >10 s for ^{13}C atoms with no hydrogen attached.

2. The *spin-spin* (transverse) relaxation describes the exchange of quantum states by the interaction between neighboring nuclei with identical precessional frequencies but differing magnetic quantum states. A nucleus in the lower energy state will be excited, while an excited nucleus relaxes to the lower energy level, hence, there is no net change in the population of the energy states, but the average lifetime of a nucleus in the excited state will decrease which can result in line-broadening. The *spin–spin relaxation time* T_2 describes the decay of the phase coherency with $T_1 \geq T_2$ being always valid.

Note that important structural information can be obtained by investigations of relaxation processes, but this is outside the scope of this book.

Challenge 4.1
For solutions to Challenge 4.1, see *extras.springer.com*.

1. Calculate the population ratio at 300 K for the ^1H nuclei in NMR spectrometers with the following magnetic flux density B_0 in Tesla (T):

   (a) 1.41 T ($v_0 = 60$ MHz)
   (b) 7.07 T ($v_0 = 300$ MHz)

   (continued)

2. The magnetic flux density may be 9.4 T. Calculate the respective applied frequency for

   (a) ^{1}H nuclei
   (b) ^{13}C nuclei
   (c) ^{19}F nuclei

Constants:

$h = 6.6256 \cdot 10^{-34}$ Js	$k_B = 1.3805 \cdot 10^{-23}$ J K^{-1}
$\gamma(^1\text{H}) = 26.7519 \cdot 10^7$ T^{-1} s^{-1}	$\gamma(^{13}\text{C}) = 6.7283 \cdot 10^7$ T^{-1} s^{-1}
$\gamma(^{19}\text{F}) = 25.181 \cdot 10^7$ T^{-1} s^{-1}	

NMR spectrometers are described either by their applied magnetic flux density B_0 in T (Tesla) or by the resonance frequency v_0 for ^{1}H nuclei in MHz, for example, 1.41 T ($v_0 = 60$ MHz) or 7.07 T ($v_0 = 300$ MHz). The resonance frequency of ^{13}C nuclei is 100 MHz in a 400-MHz spectrometer and so on.

4.2   Spectral Parameters

A representative NMR spectrum is shown in Fig. 4.3. In the following we will learn which parameters can be obtained by NMR spectra, for example from cumene as shown in Fig. 4.3.

4.2.1   Chemical Shift

4.2.1.1   Reason and Definition

The electrons around a nucleus will circulate in a magnetic field and create a secondary induced magnetic field which opposes the applied field as stipulated by Lenz's law. Thus, the effective magnetic field B_{eff} at the nucleus is smaller than the applied field B_0 because it is *shielded* by the surrounding electrons. The amount of the shielding constant σ (Eq. 4.5) is larger for heavy nuclei than for the proton and it is influenced by the "chemical environment" around the nucleus. Therefore, nuclear shielding is named the *chemical shift*.

   The chemical shift is a function of the nucleus and its chemical environment.

Because the resonance frequency is determined by the applied field B_0, an absolute unit doesn't exist as in optical spectroscopy. Therefore, the chemical shift is measured relative to a reference compound. For ^{1}H NMR, the reference is

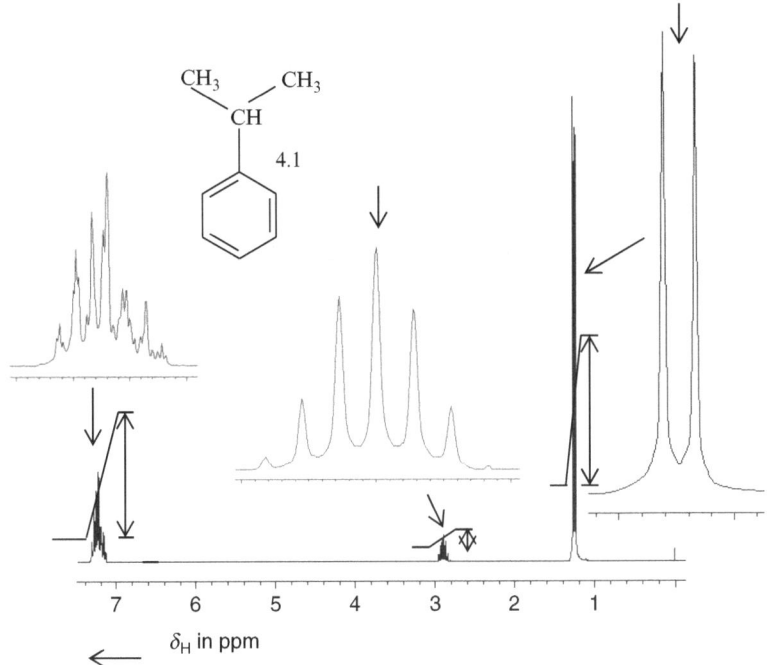

Fig. 4.3 300-MHz ^{1}H-NMR spectrum of cumene (Structure 4.1) with spread signal groups

usually tetramethylsilane (**TMS**) Si(CH$_3$)$_4$, which is promoted by the high amount of chemically equivalent protons per molecule and the fact that its resonance signal occurs at high field. For aqueous solutions, the resonance signal of the methyl group of silanes is used.

The difference (Δv) between the resonance frequency of the sample (v_i) and TMS (v_{TMS}) is the *field-dependent chemical shift* given in Hz (cycles per second):

$$\Delta v = v_i - v_{TMS} \tag{4.8}$$

Note that the chemical shift Δv of the same nuclei differs if it is measured in various applied fields.

The dimensionless *field-independent chemical shift* δ is defined as *nuclear shielding/applied magnetic field* and it is expressed in ppm:

$$\delta = \frac{v_i - v_{TMS}}{v_0} \cdot 10^6 \text{ppm}. \tag{4.9}$$

Resonance signals of the same compound are observed at equal δ values independent of the applied magnetic field. Thus, such values can be listed in tables and can be used for the recognition of structural moieties and functional groups.

Fig. 4.4 δ scale and important terms in NMR spectroscopy

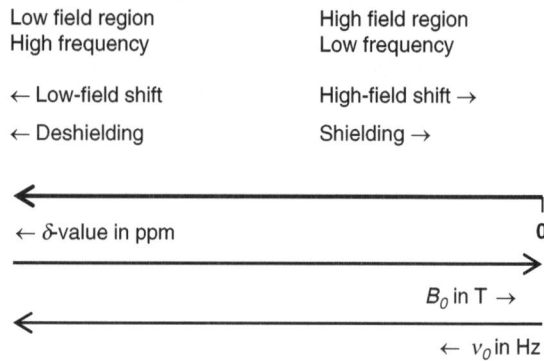

Low field region	High field region
High frequency	Low frequency
$\leftarrow$ Low-field shift	High-field shift $\rightarrow$
$\leftarrow$ Deshielding	Shielding $\rightarrow$

$\leftarrow$ δ-value in ppm   0

B_0 in T $\rightarrow$

$\leftarrow$ ν_0 in Hz

Table 4.1 Influence of the electronegativity (EN) upon the chemical shift (δ-value) for methyl halogenides CH_3–X

X =	F	Cl	Br
EN	4.5	3.5	2.7
$\delta(^1H)$ in ppm	4.27	3.06	2.69
$\delta(^{13}C)$ in ppm	71.6	25.6	9.6

Otherwise, the difference $\Delta\delta$ between two signals can be calculated in $\Delta\nu$ (signal distance in Hz) by conversion of Eq. 4.9:

$$\Delta\nu \text{ in Hz} = \Delta\delta \cdot \nu_0 \text{ in Hz} \cdot 10^{-6} \tag{4.10}$$

Conventionally, the δ values increase from right to left. The right-hand side is described as the *high-field* region and the left-hand side as the *low-field* region. Some additional terms are outlined in Fig. 4.4.

Nuclei are found to resonate in a wide range to the left (or, more rarely, to the right) of the internal standard TMS. The NMR spectrum in Fig. 4.3 shows three signal groups at 1.23, 2.9, and approximately 7.25 ppm. This is phrased, for example, by saying that the protons at 2.9 ppm resonate *downfield* from those at 1.23 ppm which are more shielded and so on.

4.2.1.2   Factors That Influence the Chemical Shift

Charge Density

The charge density is determined by inductive and mesomeric effects.

Inductive effects
Electronegative substituents (−I effect) diminish the electron density and, hence, the shielding around the nucleus. Therefore, the resonance takes place already at the lower field, i.e., at larger δ-value (see Table 4.1).

Consideration of the inductive effects can be helpful for the assignment of resonance signals.

Challenge 4.2

The ^{1}H NMR signals of the methyl groups of N-methyl *para*-anisole (Structure 4.2) measured at 2.78 and 3.73 ppm. Assign the signals to the respective methyl groups.

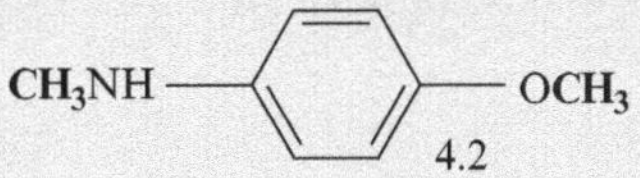

$$CH_3NH \underset{4.2}{\overset{}{\bigcirc}} OCH_3$$

Solution to Challenge 4.2

The EN of O is greater than that of N, thus, the H atoms of the CH_3O group are stronger deshielded and the resonance takes place at lower field, i.e., at greater δ-values.

Assignment: $-NH(CH_3)$: $\delta = 2.78$ ppm $-OCH_3$: $\delta = 3.73$ ppm

Mesomeric effects

Mesomerism causes electron density to vary and is, hence, responsible for different shielding at the nuclei resulting in different resonance frequencies. Resonance structures are helpful to evaluate the electron density.

Challenge 4.3

The resonance signals of the methine protons in 3-pent-2-one (Structure 4.3) are $\delta_1 = 6.18$ ppm and $\delta_2 = 6.88$ ppm. Assign the resonance signals to the respective methine groups.

$$CH_3 - \overset{1}{CH} = \overset{2}{CH} - C(CH_3) = O$$
$$4.3$$

Solution to Challenge 4.3

As the resonance structure shows,

$$CH_3 - \overset{1}{CH} = \overset{2}{CH} - C(CH_3) = O \longleftrightarrow CH_3 - \overset{+}{\overset{1}{CH}} - \overset{2}{CH} = C(CH_3) - O^-$$
$$4.3\ \mathbf{I} \qquad\qquad\qquad 4.3\ \mathbf{II}$$

(continued)

the electron density of the carbon atom (1) is smaller than that of C(2). Therefore, the hydrogen atom (1) should be stronger deshielded with the consequence that its resonance signal is low-field shifted in comparison to hydrogen (2).

Assignment: H(**1**): $\delta = 6.88$ ppm H(**2**): $\delta = 6.18$ ppm

Mostly, the resonance signal is received by concurrent influence of inductive effects in the σ skeleton and conjugation which is demonstrated by the π system of ethylene compounds with π-donor and π-acceptor groups:

π donor π acceptor

These general rules are valid:

- The β nuclei are better shielded by π-donor substituents; hence, the resonance signals lie at higher field (smaller δ value). The influence on the α nuclei is the converse.
- π-acceptor substituents cause low-field shift on both positions.

The influence of π-donor and π-acceptor substituents on the chemical shift of the resonance signals of the ethylene protons of the general Structure 4.4 is shown by the examples listed in Table 4.2.

Table 4.2 Influence of π-donor and π-acceptor substituents on the chemical shift of the resonance signals of the ethylene protons of the general Structure 4.4 (all δ values are given in ppm)

X	Electronic effect	$\delta(H_\alpha)$	$\delta(H_\beta)$	$\delta(C_\alpha)$	$\delta(C_\beta)$
H	Reference	5.28	5.28	123.3	123.3
C_2H_5	Weak π and σ donor	5.79	trans 4.87 cis 4.95	140.5	113.5
CH=CH$_2$	Weak σ donor π donor	6.31	trans 5.08 cis 5.18	137.8	117.5
Phenyl	Weak σ donor π donor	6.70	trans 5.73 cis 5.22	136.9	113.7
Cl	σ acceptor Weak π donor	6.33	trans 5.38 cis 5.43	124.9	116.0
OC_2H_5	σ donor π donor	6.45	trans 3.98 cis 4.17	151.8	86.3
C(=O)H	σ acceptor π acceptor	6.36	trans 6.62 cis 6.50	138.0	139.4
COCH$_3$	σ acceptor π acceptor	6.40	trans 5.87 cis 6.14	129.0	138.3

$$cis\ H \diagdown \qquad \diagup X$$

cis H ⟍   ⟋ X
         C = C
trans H ⟋ β   α ⟍ H
            4.4

Ring current effects (magnetic anisotropy)

The electromagnetic fields caused by electron movement are responsible for the magnetic anisotropy of the chemical bonds. The anisotropy as a result of electron movement in a cyclically conjugated π system will be explained employing the example of benzene. The delocalized π system of the aromatic ring induces a magnetic field that is directed against the external field. The direction of the magnetic field lines in respect of the external field is determined by the position as is shown in Fig. 4.5. The hydrogen atoms of benzene lie in the area in which the additional field caused by the aromatic π-electron system has the same direction as the external magnetic field. Thus, the resonance takes place at deeper field (higher δ value) because some of the required magnetic field B_{eff} (Eq. 4.7) is delivered by the ring current effect.

Conversely, hydrogen atoms situated in the area above, below, or inside the ring system experience shielding (smaller δ value) because both fields are in the opposite direction.

The anisotropy is schematically visualized by shielding cones, see Fig. 4.6.

The + denotes shielding areas and $-$ denotes deshielding areas. Note that shielding lowers the chemical shift δ and deshielding increases δ.

The ring current effect is demonstrated by the chemical shift of the methyl signals in CH_3–R:

R	Alkyl–	–CH=CH$_2$	Phenyl–	–C≡CH
δ ppm	< 1	1.6	2.3	1.8

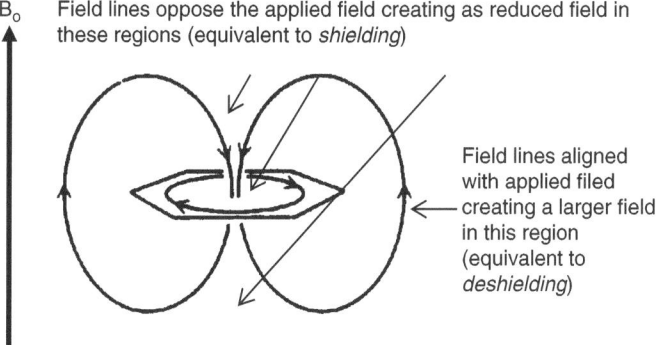

B_0 Field lines oppose the applied field creating as reduced field in these regions (equivalent to *shielding*)

Field lines aligned with applied filed creating a larger field in this region (equivalent to *deshielding*)

Fig. 4.5   Ring current effect at a cyclically conjugated π-electron system (for example, benzene)

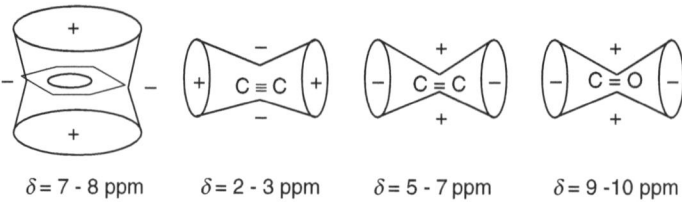

$\delta = 7 - 8$ ppm $\delta = 2 - 3$ ppm $\delta = 5 - 7$ ppm $\delta = 9 - 10$ ppm

Fig. 4.6 Schematic diagram of shielding cones for π systems

Because ^{13}C atoms cannot be influenced by the ring current effect, the sp^2-hybridized carbon atoms in alkenes and aromatic rings provide the same range of ^{13}C chemical shift values ($\delta = 100-160$ ppm). However, because of the anisotropy effect the chemical shift of the hydrogen atoms are different in both chemical classes, for example, ethylene ($\delta = 5.28$ ppm), benzene ($\delta = 7.26$ ppm).

Note that σ electrons also cause anisotropy effects that are considerably lower than that caused by π electrons. For example, hydrogen atoms in the *equatorial* position of rigid *cyclohexane* lie in the *negative* range of the anisotropy cone of the neighbored C–C σ bond and thus, they are observed at deeper field in comparison to the axially arranged hydrogen atoms.

4.2.1.3 Hydrogen bonding

Protons that are involved in hydrogen bonding are typically observed over a large range of chemical shift values. Such groups are **–OH** or **–NH**. The stronger the hydrogen bonding, the more the proton is deshielded and the higher its chemical shift will be. Especially strong hydrogen bonding is present at chelate bridges so that the chemical shift values lie beyond the usual range for ^{1}H signals ($\delta = 10-16$ ppm). Furthermore, the chemical shifts of hydrogen-bonded protons are susceptible to factors such as solvation, acidity, concentration, and temperature. Thus, it can often be difficult to predict the chemical shift values.

However, experimentally **–OH** and **–NH** protons can be identified by carrying out a D_2O exchange experiment. Because such protons are exchangeable, the signal will disappear after adding a few drops of D_2O according to Eq. 4.11:

$$R\text{-}OH + D_2O \rightleftharpoons ROD + HOD \qquad (4.11)$$

Note that also **CH** acid protons are exchangeable.

4.2.1.4 Dynamic Effects

The chemical shift of protons of nitrogen-containing groups can be influenced by dynamic effects that are attributed to restricted rotation around the carbon–nitrogen "single" bond. For example, the methyl groups of dimethylformamide

$H(C=O)N(CH_3)_2$ (Structure 4.5) show the following two ^1H NMR signals at room temperature $\delta_1 = 2.76$ ppm and $\delta_2 = 2.95$ ppm, whereas only one signal is observed at 170°C ($\delta = 2.86$ ppm).

The resonance structure II shows partial double bonding for the CN bond with the consequence of restricted rotation around the carbon–nitrogen bond. Thus, the two methyl groups are not equivalent because one methyl group is influenced by the anisotropy of the neighboring carbonyl group within the time scale of ^1H NMR spectroscopy. However, a single signal is observed at 170°C because of rapid rotation around the CN bond.

As shown, the chemical shift of magnetic nuclei is determined by *inductive*, *mesomeric*, and *anisotropic* effects. Furthermore, *hydrogen bonding* and *dynamic effects* have to be considered. An overview of the ^1H and ^{13}C NMR signals is presented in Table 4.3. Note that the range of chemical shift values for the ^1H NMR signals ($\delta \approx 0$–12 ppm) is much smaller than that for the ^{13}C NMR ones ($\delta \approx 0$–200 ppm). Thus, only small structural variations will be better observed in the ^{13}C NMR spectra.

4.2.1.5 Assignment of the Resonance Signals (Chemical Shift Values)

The assignment of experimentally observed chemical shift values to the respective atoms is an important subject with respect to structure determination.

In the simplest case, the signal can be assigned on the basis of the general relations between structure and chemical shift discussed above.

Challenge 4.4
The signal groups observed in the ^1H NMR spectrum of cumene (Structure 4.1) in Fig. 4.3 are to be assigned. Note that the chemical shift values are taken in the center of the respective signal groups (marked by vertical arrows).

Table 4.3 Typical range of chemical shift values of ^{1}H and ^{13}C NMR signals of organic chemical classes (δ in ppm)

Chemical class	^{1}H NMR signal	^{13}C NMR signal
Alkanes	1–4	10–60
CH$_3$-X X=C	≈1.0	≈10–15
N	2.4	≈30–40
O	3.4	≈60
	δ(CH$_3$) $<$ δ(CH$_2$) $<$ δ(CH)	δ(CH$_3$) $<$ δ(CH$_2$) $<$ δ(CH)
Cycloalkanes	≈2.0	25–27
Cyclopropane	≈0.2	−2.8
Alkenes	4–7.5	100–160
Alkynes	2–3	70–80
Aromatic/heteroaromatic	6.5–9	100–160
compounds	Substituent effect:	Substituent effect:
	δ(*ortho*) $<$ δ(*para*),δ(*meta*)	δ(*ypso*) $<$ δ(*ortho*)
		$<$ δ(*para*) $<$ δ(*meta*)
Carbonyl compounds	9–11	190–200
R-C(=O)-H Aldehydes		190–200
R-C(=O)-C Ketones		160– 80
R-C(=O)-X (X = O, N)		>210
R-C(=S)-C Thioketones		
-OH (exchangeable with D$_2$O)	Dependence of concentration and temperature	
	1–3	
	5–6	
	>12	
Alkyl-OH in CDCl$_3$	4–8	
Alkyl-OH in DMSO	8–12	
Chelate bridges	10–16	
Aryl-OH in CDCl$_3$		
Aryl-OH in DMSO		
Chelate bridges		
-NH (exchangeable with D$_2$O)	Dependence of concentration and temperature	
	0.6–3	
Aliphatic, cyclic	3–5	
Aromatic	5–8.5	
Amide, pyrrole, indole		

Solution to Challenge 4.4

The 300-MHz ^{1}H NMR spectrum provides three signal groups which means there are three different groups of hydrogen atoms. These signal groups can be unequivocally assigned according to the general rules given in Table 4.3: δ_1 = 1.25 ppm to **CH$_3$**-alkyl, δ_2 = 2.89 ppm to **CH**-alkyl, and δ_3 ≈ 7.2–7.3 ppm to *aromatic* hydrogen atoms.

Furthermore, there are tools such as graphics for the assignment to ^{1}H and ^{13}C chemical shift ranges (Figs. 6.3 and 6.8, respectively) or for special structural moieties (Figs. 6.4–6.6, and Table 6.18).

In addition, the expected resonance signals can be calculated by means of increment values. Examples are listed in Tables 6.19–6.22 for ^{1}H NMR signals and in Tables 6.30–6.33 for ^{13}C NMR signals. Note that nowadays computer programs can be used to calculate chemical shift values.

Challenge 4.5

Justify the assignment of the chemical shift values of cumene provided in Challenge 4.4 by calculation.

Solution to Challenge 4.5

CH$_3$	Range for CH$_3$–C–Aryl from Fig. 6.4:	$\delta = 1.2$–1.7 ppm
	Experimental:	$\delta = 1.25$ ppm
CH	Calculation according to Eq. 6.2:	
	$\delta(CH) = 1.50 + \Sigma S_i$ with $S_i(Aryl) = 1.30$:	$\delta = 2.80$ ppm
	Experimental:	$\delta = 2.85$ ppm

Aromatic H atoms

Calculation according to Table 6.21:

$\delta(CH) = 7.27 + \Sigma S_i$	
$S_i[HC(CH_3)_2]$	$\delta(CH)$
ortho H: -0.13	7.14 ppm
meta H: -0.08	7.19 ppm
para H: -0.18	7.09 ppm

The calculated chemical shift values lie in the range of the experimental values ($\delta \approx 7.2$–7.3 ppm). However, the assignment of the signals cannot be carried out because there is a spectrum of higher order (see below).

The assignment to the structure of *cumene* can be justified by the very good agreement of the experimental and the calculated chemical shift values.

Challenge 4.6

The following resonance signals are observed in the decoupled ^{13}C NMR spectrum (see Sect. 4.4.5.2) at 25.2 MHz for isoleucine (Structure 4.6):

δ in ppm (D$_2$O)	12.5	15.9	25.7	39.7	60.9	175.2

$$
\begin{array}{ccccccc}
 & & & \overset{5}{CH_3} & NH_2 & & \\
 & & & | & | & & \\
\overset{1}{CH_3} &\!\!\!\!-\!\!\!\!& \overset{2}{CH_2} &\!\!\!\!-\!\!\!\!& \overset{3}{CH} &\!\!\!-\!\!\!& \overset{4}{CH} &\!\!\!-\!\!\!& \overset{6}{COOH}
\end{array}
$$

4.6

Calculate and assign the resonance signals to the respective carbon atoms. To which hydrogen atom does the signal $\delta = 3.8$ ppm observed in the 100-MHz ^{1}H NMR spectrum belong?

Solution to Challenge 4.6

Calculation of the ^{13}C NMR signals according to Table 6.30:

$\delta(^{13}C_i) = -2.3 + 9.1 \cdot n_\alpha + 9.4 \cdot n_\beta - 2.5 \cdot n_\gamma + 0.3 \cdot n_\delta + \Sigma S_{ij}$

$\delta(C-1) = -2.3 + 9.1 \cdot 1 + 9.4 \cdot 1 - 2.5 \cdot 2 + 0(\delta NH_2) + 1.0(\delta COOH) = 12.2$ ppm

$\delta(C-2) = -2.3 + 9.1 \cdot 2 + 9.4 \cdot 2 - 5.1(\gamma NH_2) - 2.3(\gamma COOH) - 2.5 \ (S_{2°/3°}) = 24.8$ ppm

$\delta(C-3) = -2.3 + 9.1 \cdot 3 + 9.4 \cdot 1 + 11.3(\beta NH_2) + 2.7(\beta COOH) - 8.5 \ (S_{3°/3°}) = 39.9$ ppm

$\delta(C-4) = -2.3 + 9.1 \cdot 1 + 9.4 \cdot 2 - 2.5 \cdot 1 + 28.3(\alpha NH_2) + 20.8 (\alpha COOH) - 8.5(S_{3°/3°}) = 63.7$ ppm

$\delta(C-5) = -2.3 + 9.1 \cdot 1 + 9.4 \cdot 2 - 2.5 \cdot 1 - 5.1(\gamma NH_2) - 2.3(\gamma COOH) - 1.1 (S_{1°/3°}) = 14.6$ ppm

The range of the signal of the carbonyl ^{13}C atom is provided by the graphical scheme shown in Fig. 6.9.

Assignment of the experimental ^{13}C NMR signals

Number of the C atom	1	2	3	4	5	6
Calculated value, δ in ppm	12.2	24.8	39.9	63.7	14.6	No calculated
Experimental value, δ in ppm	12.5	25.7	39.7	60.9	15.9	175.2

Assignment of the ^{1}H NMR signal at $\delta = 3.8$ ppm.
The methine hydrogen atom 4-H can be calculated according to Table 6.19:
$\delta(CH) = 1.5 + 1.0\ (COOH) + 1.0\ (NH_2) = 3.5$ ppm.
Thus, the experimental ^{1}H resonance signal at $\delta = 3.8$ ppm can be assigned to the hydrogen atom 4-H. Further ^{1}H signals cannot be calculated. Note that the unequivocal assignment of the other ^{1}H NMR signals can be realized by the correlation spectroscopic methods (see later).

Challenge 4.7
Calculate the ^{1}H and ^{13}C resonance signals of nicotine acid methyl ester (Structure 4.7) and present the ^{13}C NMR signal as a line pattern.

Solution to Challenge 4.7
Calculation of the ^{1}H NMR signal according to Table 6.22:
  $\delta(H\text{-}2) = 8.59 + 0.62(Z_{32}) = 9.21$ ppm
  $\delta(H\text{-}4) = 7.75 + 0.60(Z_{34}) = 8.35$ ppm
  $\delta(H\text{-}5) = 7.38 + 0.23(Z_{35}) = 7.61$ ppm
  $\delta(H\text{-}6) = 8.59 + 0.34(Z_{36}) = 8.93$ ppm
  $\delta(H\text{-}7) \approx 3.8$ ppm, estimated according to the overview graphics given in Fig. 6.4.
  Calculation of the heteroaromatic ^{13}C NMR signal according to Table 6.33:
  $\delta(C\text{-}2) = 149.8 - 0.6(S_{32}) = 149.2$ ppm
  $\delta(C\text{-}3) = 123.7 + 1.0(S_{33}) = 124.7$ ppm
  $\delta(C\text{-}4) = 135.9 - 0.3(S_{34}) = 135.6$ ppm
  $\delta(C\text{-}5) = 123.7 - 1.8(S_{35}) = 121.9$ ppm
  $\delta(C\text{-}6) = 149.8 + 1.8(S_{36}) = 151.8$ ppm
  Calculation of the aliphatic ^{13}C NMR signal according to Table 6.30:
  $\delta(C\text{-}7) = -2.3 + 51.1(\alpha\ OCOCH_3) - 2.3(\gamma\ phenyl) = 46.5$ ppm
  $\delta(C\text{-}8) \approx 166$ ppm, estimated from the overview graphics given in Fig. 6.8.

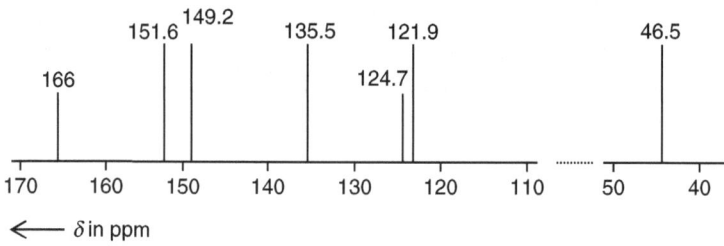

Fig. 4.7 Line pattern of the ^{13}C NMR signals for Structure 4.7

The line pattern of the ^{13}C NMR signals is shown in Fig. 4.7. Note that the intensity of quaternary ^{13}C is smaller than that of CH_x signals (see Sect. 4.2.2).

Challenge 4.8

For solutions to Challenge 4.8, see extras.springer.com.

1. The chemical shift of the hydrogen atoms in the CH group is influenced by the hybridization of the C atom. The following values are observed: ethane: $C(sp^3)H = 0.86$ ppm, ethene: $C(sp^2)H = 5.28$ ppm, ethyne: $C(sp)H = 1.8$ ppm. Explain why the ethyne signal shows a divergent behavior?

2. The ^{1}H resonance signals of the vinyl group in the two isomer compounds $R–CH=CH_2$ with $R = -O-C(=O)-CH_3$ (**A**) and $R = -C(=O)-OCH_3$ (**B**) are (δ in ppm):

Compound I	^{1}H NMR	δ in ppm	6.40	6.16	5.87
	^{13}C NMR	δ in ppm	131.3	128.1	
Compound II	^{1}H NMR	δ in ppm	7.27	4.85	4.55
	^{13}C NMR	δ in ppm	141.5	96.8	

Assign the signals to the respective compounds.

3. Explain why the hydrogen atoms in *ortho* position in pyridine compounds show chemical shift values $\delta > 8$ ppm.

4. The green compound 15,16-dimethyl dihydropyrene (Structure 4.8A) shows ^{1}H NMR signals at $\delta_1 = -2.8$ ppm and in the range $\delta_2 \approx 8.5–9$ ppm. The exposition of the compound to visible light provides a colorless compound which shows signals at $\delta_1 = 2.3$ ppm and in the range $\delta_2 \approx 7.5–8$ ppm. Structure 4.8A is formed again if the light source is removed. This process is repeatable and it is named photochromism. Can photochromism in this particular case be based on the valence tautomerism according to Eq. 4.12?

(continued)

$$\text{(4.12)}$$

    4.8A   4.8B

5. The resonance signals of the aldehyde protons R–COH lie in the range $\delta = 9\text{–}11$ ppm. Which influences are responsible for signals at such a deep field?

6. The ^{1}H NMR spectrum of acetyl acetone shows a signal at an extremely deep field ($\delta \approx 15.5$ ppm). To which hydrogen atom does this signal belong? Why does the signal lie in such a deep field?

7. The ^{13}C NMR spectrum shows two resonance signals for the methyl groups in dimethylnitrosamine $(CH_3)_2$ N–N$=$O at room temperature. Why are the methyl groups not equivalent?

8. The chemical shift values for the aryl protons in acetophenone C_6H_5–C $(=O)CH_3$ are observed at $\delta(meta) = 7.40$ ppm, $\delta(para) = 7.45$ ppm, and $\delta(ortho) = 7.91$ ppm. However, the order $\delta(ortho) < \delta(para) < \delta(meta)$ is valid in carbonyl-free aromatic compounds like anisole C_6H_5–O–CH_3. Why do α-carbonyl-substituted aromatic compounds show another order?

9. The hydrogen atoms at the C-1 atom of the two anomers of D-glucose (Structure 4.9) are observed at $\delta_1 = 5.25$ ppm and $\delta_2 = 4.65$ ppm (in D_2O). Assign the signals to the respective anomers.

        4.9 A   4.9 B

    α-D glucose β-D glucose

10. Calculate the chemical shift values for the ^{1}H and ^{13}C NMR signals in the *aliphatic* and *olefinic* part of the following molecules:

(continued)

(a) 2,3,4-Trimethylpentane	(b) 2-Hydroxypropanic acid	(c) 2-Methylbutanic acid
(d) 2-Methylbutan-1-ol	(e) 1-Pentylamine	(f) Leucine
(g) E-Croton aldehyde	(h) Butanic acid vinylester	(i) 2-Aminobutanic acid
(j) C_6H_5-CH(OCH_3)-CH_2Cl	(k) $CH_3C(Cl)=C(CH_3)OCH_3$	(l) CH_3-CH_2-CH_2-NO_2

11. Calculate the chemical shift values for the ^1H and ^{13}C NMR signals in the *aromatic* part of the following molecules:

(a) 2-Chloraniline	(b) Benzoic methyl ester	(c) Benzyl acetate
(d) 1,2-Dinitrobenzene	(e) 2-Nitro 4-hydroxybenzoic acid	(f) *o*-Cresole
(g) Nicotinic acid amide	(h) 3-Methyl 2-chlorpyridine	(i) α-Picoline

12. The observed chemical shift values of the lactone (Structure 4.10) are:

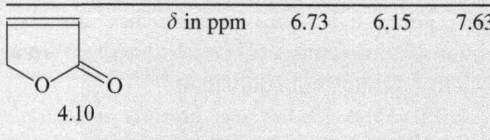

δ in ppm   6.73   6.15   7.63

4.10

Assign the ^1H NMR signals to the respective hydrogen atoms.

13. The *E/Z* isomers $CH_3HC=C(CH_3)COOCH_3$ show resonance signals at $\delta_1 = 6.73$ ppm and $\delta_2 = 5.98$ ppm, respectively. Assign the chemical shift values to hydrogen of the respective *E/Z* isomeric compound.

14. Assign the experimental ^{13}C NMR signals for the following compounds (δ in ppm):

(a) *p*-Nitrophenol	115.9	126.4	141.7	161.5	
(b) 2-Aminopyridine	108.1	113.3	137.5	147.5	158.9
(c) 3-Hydroxybutanone-2	19.4	24.9	73.1	211.2	
(d) *p*-Aminobenzoic acid methylester	49.8	115.7	120.2	130.0	151.0	167.0
(e) 2-Methylbutene-1	12.5	22.5	31.1	109.1	147.0

4.2.2 Intensity of the Resonance Signals

The analysis of the area of peaks allows determination of how many protons give rise to a particular peak. This is known as integration, i.e., a mathematical process that calculates the area under the peak. Besides *digital values* the peak intensity can also be presented by the *height* of the signals.

In ^{1}H NMR spectroscopy the observed intensity I (integrated value or height of the peak) is proportional to the number of protons N that belongs to the respective signal and the amount of substances n,

$$I \sim n \cdot N. \tag{4.13}$$

For a single pure compound is valid:

The intensity of the peaks relates to the number of the hydrogen atoms of the peaks.

Let us return to the ^{1}H NMR spectrum of cumene in Fig. 4.3. The relative peak intensities of the three signal groups measured as peak height (5:1:6) corresponds to the number of hydrogen atoms. Thus, the relation of the intensities provides further evidence for the structure of the molecule, as is shown in Table 4.4.

The proportionality of intensity I and amount of substances n according to Eq. 4.13 allows the *quantitative* determination of a mixture of substances. Let us consider a two-component system with substances A and B. Two cases must be considered:

1. A *separate* signal is present for each component

$$\frac{n_A}{n_B} = \frac{I_A \cdot N_B}{I_B \cdot N_A} \tag{4.14}$$

in which

n_A, n_B are the amount of substances A and B, respectively,

I_A, I_B are the observed intensity of components A and B, respectively,

N_A, N_B are the number of hydrogen atoms which belong to the respective signal A and B, respectively.

The signals should be chosen in a way that there is no overlapping of the signals and the number of hydrogen atoms N is as large as possible because it determines the sensitivity.

Table 4.4 Assignment of the ^{1}H NMR signals of cumene (Fig. 4.3) with the help of the intensity relations

δ in ppm	≈ 7.15	2.89	1.25
Assignment	Aromatic ring	Aryl–CH	CH_3–alkyl
Relation of the peak height	5	1	6
Information	C_6H_5–	–$CH(CH_3)_2$	

2. Overlapping signals

The intensity must by measured in *two* positions 1 and 2. The relative amount of substances A and B is calculated by Eq. 4.15

$$\frac{n_A}{n_B} = \frac{I^{(2)} \cdot N_B^{(1)} - I^{(1)} \cdot N_B^{(2)}}{I^{(1)} \cdot N_A^{(2)} - I^{(2)} \cdot N_A^{(1)}} \tag{4.15}$$

in which $I^{(1)}$, $I^{(2)}$ are the intensity observed in positions 1 and 2, respectively, $N_A^{(1)}$, $N_A^{(2)}$ are the number of hydrogen atoms of component A in the signal group 1 and 2, respectively,
$N_B^{(1)}$, $N_B^{(2)}$ are the number of hydrogen atoms of component B in the signal group 1 and 2, respectively.

In the case of ^{13}C NMR spectroscopy the integrals of the signals depend on the relaxation rate of the nucleus and its scalar and dipolar coupling constants. Therefore, there is no proportionality between intensity and number of atoms.

Because the hydrogen-free quaternary carbon atoms possess the largest relaxation time T_1, the relay time of the pulses is smaller than the time necessary for the complete regeneration of the initial population according to Eq. 4.6. Thus, the intensity of quaternary carbon atoms possesses the smallest relative intensities because it is determined by the population.

Because routine ^{13}C NMR spectra are measured by quenching of the ^{13}C/^{1}H coupling which is realized by additional irradiation of the resonance field of all hydrogen atoms (broadband decoupling!) the spin energy is transferred from hydrogen to carbon atoms resulting in enhanced intensity of the carbon resonances. This effect is known as *nuclear Overhauser enhancement* (**NOE**). Not all ^{13}C signals will be enhanced to the same extent. Generally, the larger the number of attached protons the larger the enhancement will be. Note that the intensity of hydrogen-free carbon atoms cannot be enhanced by NOE.

Furthermore, molecular size and flexibility influence the signal intensity. Thus, the intensity of CH_3 groups is somewhat smaller than that of CH_2 and CH groups, because of the higher flexibility of such groups with respect to rotation.

Quaternary carbon atoms can be assigned by their *small* intensity.

4.2.3 Half-Width of NMR Signals

The existence of relaxation implies that an NMR line must have a width in accordance with the Heisenberg uncertainty principle which states that the product of uncertainty of the frequency range at the half-height $\Delta v_{1/2}$ and the uncertainty of the time Δt is a constant

$$T_2^* \cdot \Delta v_{1/2} \cdot \Delta t = 1. \tag{4.16}$$

T_2^* is the relaxation time that is responsible for the linewidth presented as linewidth at half-height. The experimental relaxation time T_2^* is determined by the spin-spin relaxation time T_2 and a field inhomogeneity component.

All processes which reduce T_2^* cause broadening of the linewidth. Thus, nuclei with the nucleus spin $I \geq 1$ have an electrical nucleus quadruple moment which accelerates relaxation and, hence, gives rise to broader lines. For example, ^{14}N nuclei with $I = 1$ possess a quadruple moment, therefore, the N–**H** group can give rise to broad signals in the ^1H NMR spectrum. Furthermore, the linewidth is influenced by the dynamics of exchange processes of the H atoms via hydrogen bridges. Hydrogen atoms attached to N can undergo quick, middle, and slow exchange processes.

X-H signals of acidic protons with X = O, N, C can give rise to variously different broad signals in dependence of solvent, temperature, and concentration. Mostly, such groups can be recognized by broad signals. Compounds with stronger hydrogen bridges and slower exchange processes (phenols, R–OH bridges to the solvent DMSO) cause sharp signals. Note that such hydrogen atoms can easily be recognized by the D_2O experiment (see above) which is important if X–H signals are overlaid by signals of CH_x groups.

4.2.4 Indirect Nuclear Spin Coupling

Look again at the ^1H NMR spectrum of cumene in Fig. 4.3. The methyl peaks at $\delta = 1.25$ ppm are split into two peaks (*doublet*), the methine peak at $\delta = 2.89$ ppm is split into seven lines (*septet*), and the aromatic protons show many lines with various spacings (*multiplet*). Note that the spacings between the peaks of the doublet are equal to the spacings of the septet. This spacing is measured in Hertz (Hz) and is called the *indirect* or *scalar coupling constant J*. The splitting occurs because there is a small interaction (*coupling*) between neighbored protons. This coupling is named *indirect* nuclear spin coupling because it is caused by the intervening bond electrons. Note that the other type of magnetic interactions between nuclei with a nonzero spin is the *direct* coupling caused by spatial dipole-dipole interaction. Whereas direct coupling dominates in solids and very viscous liquids, in mobile isotropic liquids the random motion of molecules completely averages the dipolar coupling, hence, no direct couplings can be observed.

The scalar coupling transfers magnetization between the coupled nuclei over the bonds. Thereby, additional magnetic fields are generated at the coupled nuclei whose orientations are aligned *with* or *opposed* to the applied field and are responsible for the splitting of the resonance signals. The fact that the additional field generated at the neighboring nuclei can be aligned with or opposed to the applied magnetic field gives rise to a *sign* of the coupling constant J. The nuclear moment causes the magnetic polarization of the electrons. The lower energy state corresponds to the *antiparallel* orientation of the nuclear and electron spin,

Fig. 4.8 Fermi contact
mechanism for spin-spin
coupling of two hydrogen
atoms via three chemical
bonds

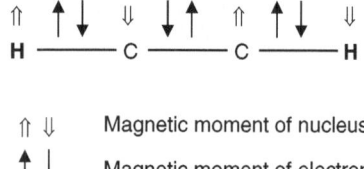

respectively. The transfer of the magnetic polarization follows according to Pauli's and Hund's rules: The magnetic polarization is *antiparallel* for orbitals at *various* centers (chemical bond) and it is *parallel* for orbitals at the *same* center (hybrid orbital).

By definition, J is positive if the lower energy state corresponds to the antiparallel orientation of the coupled nuclei.

The following rule is usually valid, but by no means always:

J is positive for coupling over one (^1J) and three (^3J) bonds, respectively and J is negative for coupling over two bonds (^2J).

The simple Fermi contact mechanism for spin-spin coupling of two hydrogen atoms via two chemical bonds is outlined in Fig. 4.8 for the example of the coupling of two protons via three chemical bonds (*vicinal* coupling). According to the definition, the coupling constant is positive.

To understand the pattern of the fine structure of NMR spectra, let us look at the simple compound chloroacetaldehyde (Structure 4.11).

$$\overset{1}{H} \!-\!\! CO \!-\!\! \overset{2}{C}H_2Cl$$

4.11

Chloroacetaldehyde possesses two different kinds of protons which give rise to chemical shifts at about 9 ppm (H-1) and 4 ppm (H-2) with the intensity ratio 1:2. There are two orientations of the H-1 nuclei: parallel and antiparallel to the applied field B_0. Both orientations produce additional magnetic fields at the H-2 nuclei. According to the definition of the sign of the coupling constant across three bonds, the parallel aligned spin produces an additional field which is antiparallel in relation to the applied field. Therefore, a slightly higher field is needed to bring them in resonance, resulting in an upfield shift. However, the additional field produced by the antiparallel aligned spin is aligned with the applied field B_0, resulting in a downfield shift. Hence, the methylene peak is split into two lines (*doublet*). Because of the same probability of the spin orientation of the H-1 nuclei, the intensity ratio of the doublet is 1:1, see Fig. 4.9a.

Similarly, the effect of the methylene protons (H-2) on the methine proton (H-1) is such that there are four possible spin combinations for the two methylene protons

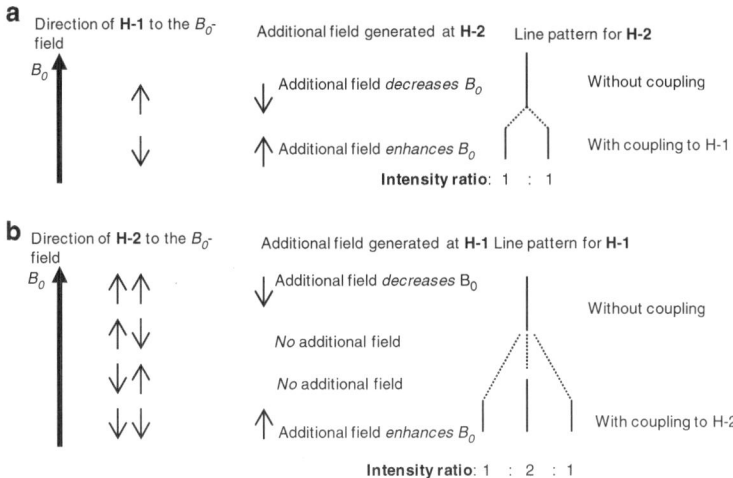

Fig. 4.9 Development of the pattern of the fine structure of the H-2 (**a**) and H-1(**b**) NMR signals of chloroacetaldehyde (Structure 4.11)

(see Fig. 4.9b). Only the combination ↑↑ and ↓↓ produce additional fields at the H-1 nuclei, resulting in an upfield and downfield shift, respectively. Neither combination of spins opposed to each other has an effect on the methine peak because no additional field is produced at H-1. Because there are two different combinations, the methine peak is observed as a *triplet*, with the ratio of areas 1:2:1. The spacing in between the doublet is equal to the spacings of the triplet because the coupling is mutual, i.e., $J_{\text{H-1,H-2}} = J_{\text{H-2,H-1}}$. Note that the spacing of the lines measured in Hertz corresponds to the coupling constant J.

Unfortunately, these simple splitting patterns can be observed only in *first-order* spectra where the chemical shift between interacting groups Δv is much larger than their coupling constant J. In general, the criterion for a first-order spectrum is

$$\frac{\Delta v}{J} > 7 \tag{4.17}$$

but sometimes you can also find $\Delta v/J > 10$.

The ^1H chemical shift values for chloracetaldehyde are $\delta_1 = 9.62$ ppm and $\delta_2 = 4.08$ ppm and the coupling constant is $J = 7$ Hz. The difference of the chemical shift Δv in a 300-Hz NMR spectrometer is

$$\Delta v = \Delta \delta \cdot v_0 \cdot 10^{-6} = 5.54 \cdot 300 \text{ MHz} \cdot 10^{-6} = 1,662 \text{ Hz}. \tag{4.18}$$

The spacing of the lines is $\Delta \delta = 0.0233$ ppm which corresponds to $\Delta v = J = 7$ Hz. Thus, the difference of the chemical shift Δv is much larger than the coupling

Table 4.5 Multiplicity and intensity ratio of multiplets caused by spin-spin coupling of ^{1}H atoms

Multiplicity	Intensity ratio
Singlet (s)	1
Doublet (d)	1:1
Triplet (t)	1:2:1
Quartet (q)	1:3:3:1
Quintet	1:4:6:4:1
Sextet	1:5:10:10:5:1
Septet	1:6:15:20:15:6:1

constant J which means, the criterion for a first-order spectrum (see Eq. 4.17) is fulfilled.

The spin-spin coupling provides detailed insight into the connectivity of atoms in a molecule. The multiplicity M of a multiplet is given by the simple relation

$$M = 2 \cdot I \cdot N + 1 \tag{4.19}$$

in which N is the number of equivalent *neighboring* atoms and I is the nuclear spin quantum number.

For ^{1}H nuclei with $I = \frac{1}{2}$, the simple $N + 1$ rule is valid:

$$M = N + 1. \tag{4.20}$$

The intensity ratios of the multiplet follow Pascal's triangle as described in Table 4.5.

Coupling combined with the chemical shift and the integrated intensities provides for protons not only information about the chemical environment of the nuclei, but also the number of *neighboring* NMR active nuclei within the molecule. In more complex spectra with multiple peaks at similar chemical shifts or in spectra of nuclei other than hydrogen, coupling is often the only way to distinguish different nuclei. Thus, the presence of fluorine can be straightforwardly recognized by multiplets of the ^{1}H-decoupled ^{13}C NMR spectra. For example, the ^{13}C signal of the CF_3 group is observed as a quartet.

Challenge 4.9

1. Which multiplet is observed for the ^{13}C signal of $CDCl_3$?
2. Figure 4.10 presents the 250-MHz ^{1}H NMR spectrum of an unknown compound having the sum formula C_4H_8O. Which structure can be derived from this information?

(continued)

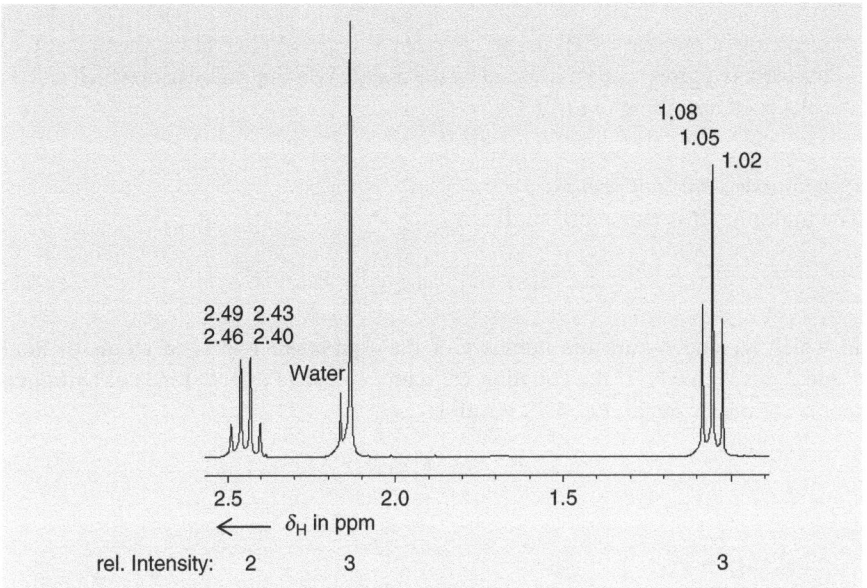

Fig. 4.10 250-MHz ^{1}H NMR spectrum of an unknown compound with the sum formula C_4H_8O

Solution to Challenge 4.9

To 1.

Because the nuclear spin quantum number for ^{2}D is $I = 1$, coupling to deuterium splits the ^{13}C NMR signal into a 1:1:1 triplet according to Eq. 4.19.

To 2.

The signal spacing of the quartet ($M = 4$, intensity ratio = 1:3:3:1) at $\delta_1 = 2.45$ ppm is $\Delta\delta = 0.03$ ppm which corresponds to $\Delta\nu = J = 7.5$ Hz for a 250-MHz NMR spectrometer. The same coupling constant is obtained in the case of the triplet at $\delta_2 = 1.05$ ppm. Therefore, both multiplets are coupled with each other and provide a spin system. According to the $N + 1$ rule, the protons of the signal group at $\delta_1 = 2.45$ ppm couple with three equivalent hydrogen atoms (CH_3 group) and the protons at $\delta_2 = 1.05$ ppm couple with the two H atoms of a CH_2 group. According to the overview of the chemical shift of ^{1}H atoms in Fig. 6.3, this group must be attached at a carbonyl group which is justified by the evaluation of the chemical shift: $\delta_2(^1H) = 1.25 + 1.2 = 2.45$ ppm (see Table 6.19, $S_1 = -CO-alkyl$). Thus, coupling combined with the chemical shift and the integration for protons provide the pattern **$CH_3-CH_2-C(=O)-$**.

According to the relative intensity (3 H) and the chemical shift (see Fig. 6.4), the singlet at $\delta_3 = 2.15$ ppm must be assigned to a **$CH_3-C(=O)-$** pattern. The structure of this compound is justified: butan-one-2 $CH_3-CH_2-C(=O)-CH_3$. Coupling over more than three σ bonds in acyclic compounds does not take

(continued)

place; therefore, the α CH_3 group provides a singlet. Note that the presence of the carbonyl group can be justified by the double bonding equivalent which is 1 calculated according to Eq. 1.19.

Coupling two kinds of protons
The multiplet M is calculated by Eq. 4.21

$$M = (N_1 + 1)(N_2 + 1) \tag{4.21}$$

in which N_1 and N_2 are the numbers of the equivalent hydrogen atoms of kind 1 and 2, respectively. If the coupling constants J_1 and J_2 to both kinds of hydrogen atoms are nearly equal, Eq. 4.22 is valid

$$M = N_1 + N_2 + 1. \tag{4.22}$$

Challenge 4.10
Figure 4.11 presents the 250-MHZ ^{1}H NMR spectrum of maleic acid N-propylamide (Structure 4.12). Develop the line pattern for the n- and *iso*-propyl group and assign the signals to the respective hydrogen atoms.

Solution to Challenge 4.10
Line pattern of the n- and the *iso*-propyl group

(continued)

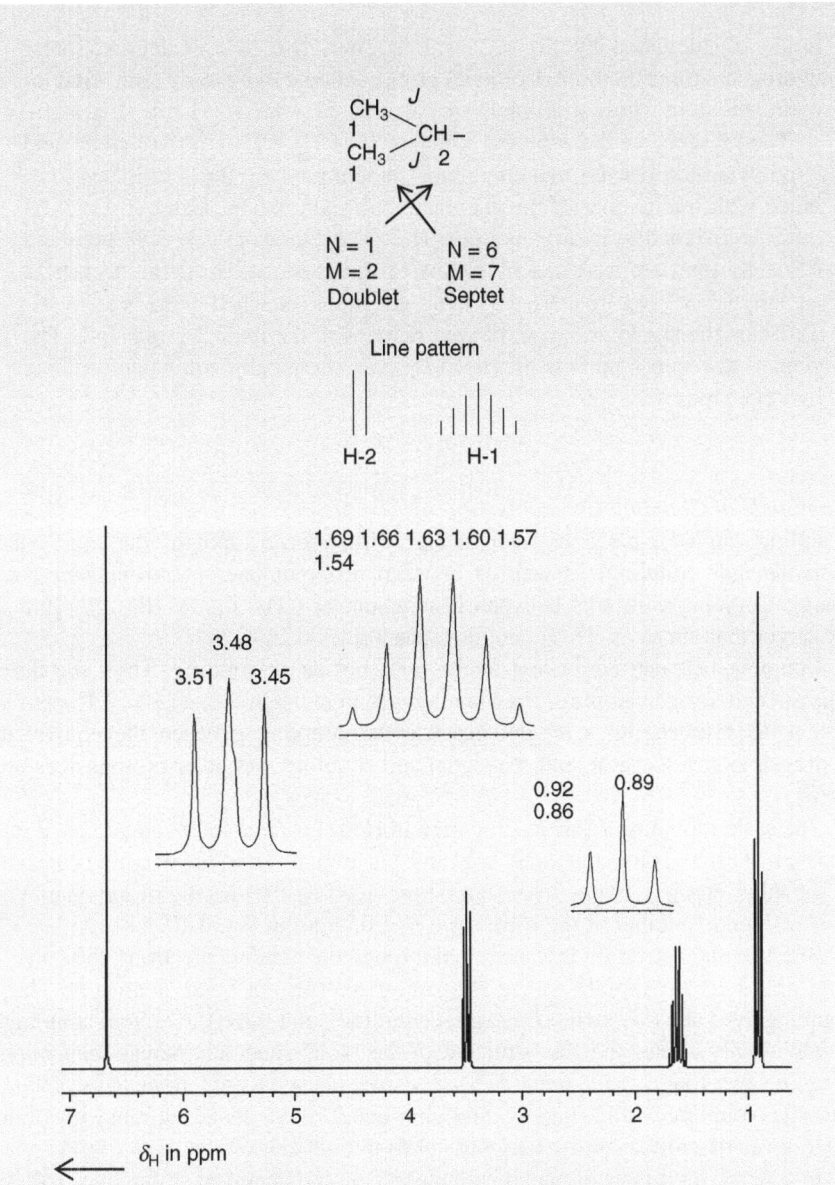

Fig. 4.11 250-MHz ^{1}H NMR spectrum of maleic acid *N*-propylamide (structure 4.12)

The hydrogen atoms H-2 of the *n*-propyl group couple with the neighboring atoms H-1 and H-3. Because the coupling constants J_1 and J_2 over each of the three σ bonds should be nearly equal, the multiplet for the methylene

(*continued*)

group is calculated by Eq. 4.22 which gives rise to a sextet. All other hydrogen atoms in both isomeric groups couple with only one kind of hydrogen atom whose multiplets are calculated by the $N + 1$ rule (Eq. 4.20).

The presence of the n-propyl group, $CH_3–CH_2–CH_2–$ in the maleic acid N-propylamide can be unambiguously determined by the multiplets combined with the number of the chemical shift values. The sextet at $\delta = 1.62$ ppm is assigned to hydrogen atoms H-2. The triplets at $\delta = 3.48$ ppm and $\delta = 0.89$ ppm are assigned by means of the chemical shift; the deep field signals at $\delta = 3.48$ are caused by hydrogen atoms attached to the N atom (H-3). Thus, the signal at $\delta = 0.89$ ppm belongs to the methyl group H-1. The singlet at $\delta = 6.7$ ppm is produced by two chemically equivalent olefinic hydrogen atoms.

Magnitude of Coupling Constants

Coupling can take place between nuclei with nonzero spin of the same kind (*homonuclear* coupling) as well as between different ones (*heteronuclear* coupling). Coupling constants between heavier nuclei ($^1H/^{19}F$, $^1H/^{31}P$, $^{19}F/^{31}P$, ...) are larger than those of $^1H/^1H$ coupling, see Tables 6.23–6.29.

Coupling between equivalent nuclei does not cause splitting. Thus, the three equivalent hydrogen atoms of the α-methyl group in butan-2-one ($\delta = 2.15$ ppm in Fig. 4.10) give rise to a singlet because the coupling between the equivalent hydrogen atoms does not split the signal and coupling into other protons does not occur.

The scalar coupling J (always reported in Hertz) is *field-independent*, i.e., J is a constant at each applied magnetic field and J is mutual. Thus, the coupling constant of the ethyl group in butan-2-one can be provided by the spacing of the quartet at $\delta = 2.45$ ppm and that of the triplet at $\delta = 1.05$ ppm in Fig. 4.10.

Because magnetization is transmitted through the bonding electrons, the magnitude of J falls off rapidly as the number of intervening bonds increases. Thus, coupling over one (^1J), two (^2J, *geminal* coupling), and three (^3J, *vicinal* coupling) bonds usually causes the fine structure of the NMR spectra, whereas *long-range coupling* over four (^4J) and five (^5J) bonds can be observed only in special structural patterns (coupling within rings or through π bonds). Note that long-range coupling often gives rise to line broadening but not to line separation.

In general, multiplets of nuclei with closely spaced chemical shift values will be overlaid resulting in a complex fine structure. The development of the line pattern is helpful to recognize the lines that belong to the respective multiplets.

In the following section we will be concerned with proton-proton couplings. The absolute magnitude of the proton-proton coupling constant is 0–20 Hz.

Geminal proton-proton couplings ($^2J_{H-H}$)

According to the simple Fermi's contact mechanisms explained above, the geminal coupling constant possesses a *negative* sign. The broad range of the absolute magnitude of the two-bond $^1H/^1H$ couplings from 0 to 20 Hz can only be understood if both magnitude and sign are taken into account.

The coupling constant of "reference compound" CH_4 is $J = -12.4$ Hz, and for acyclic and unstrained ring systems the geminal coupling constant of the CH_2 group is typically around -12 Hz. The following effects influence the geminal coupling:

Influence of a neighbored π bond
A neighbored π bond (carbonyl groups and related functions) decreases the electron density of the C–H bond; hence, the $^2J_{H\text{-}H}$ coupling becomes more *negative*. Because of the negative sign of the geminal coupling the absolute value of the gem coupling constant is larger in magnitude and is increased in the range from $|14|$ to $|15|$ Hz whereas the effect is somewhat larger at C=O than C=C groups. In rigid systems like rings the effect is again somewhat larger with the consequence that gem coupling is in the range from -16 to -18 Hz, whereas the gem coupling is around -20 Hz if two neighbored π bonds are attached at the CH_2 group.

Influence of electronegative atoms with lone electron pairs
Directly attached atoms with lone electron pairs are π-donor substituents because they increase the electron density. Therefore, the absolute magnitude of the gem coupling is decreased and hence, becomes less negative. Thus, for example, the range for the –OCH_2 group is around -10 Hz.

Ring strain effects
Ring strain has a positive effect on geminal coupling. Thus, the absolute value of the gem coupling constant in cyclopropane is decreased from -12 to -4.3 Hz. An additional positive effect causes electronegative atoms like nitrogen or oxygen. The gem coupling becomes more positive and hence, the absolute magnitude is smaller at enlargement of the H–C–H bond angle of the geminal H atoms. Thus, the gem coupling in ethylene is $+2.5$ Hz and most terminal alkenes have small coupling constants in this range. Substituents with π-acceptor or π-donor effects influence the gem coupling which is demonstrated for ^2J for the structure $RHC=CH_2$:

$$R = CH_3(+2.1 \text{ Hz}), \ F(-3.5 \text{ Hz}), \ OCH_3(-2.0 \text{ Hz}), \ COCH_3(+1.3 \text{ Hz}),$$
$$Si(CH_3)_3(+3.8 \text{ Hz})$$

Vicinal proton-proton couplings ($^3J_{H\text{-}H}$)
In general, the vic coupling constant is positive. It is influenced by C–C distance (electron density), the H–C–C angle, substituents, and the dihedral torsion angle φ which is shown in the Newman projection:

The relation between the dihedral torsion angle φ and the vic coupling constant is given by the Karplus graph, see Fig. 4.12.

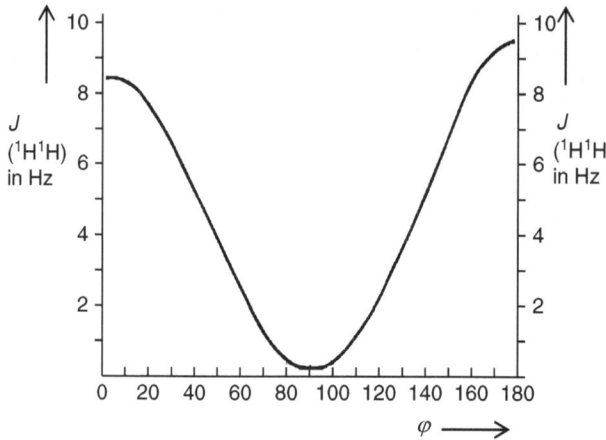

Fig. 4.12 Karplus graph: The relation between the dihedral torsion angle φ and the vic coupling constant ^3J

According to the Karplus relation, the magnitude of the vic coupling constant is smallest when the torsion angle is close to 90° and largest at angles of 0° and 180°. This relationship between local geometry and coupling constant is particularly valuable for the evaluation of torsion angles. Thus, the spatial positions of the hydrogen atoms in alicylic compounds like cyclohexane, steroids, and sugars can be determined by the vic coupling constant. Because of $\varphi = 180°$, the vic coupling of diaxial (a,a) H atoms in a cyclohexane moiety (Structure 4.13) is larger than those for the a,e and e,e positions with $\varphi \approx 60°$ (see Table 6.24). In that way epimers can be distinguished by the vic coupling constant.

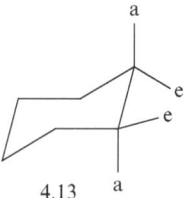

4.13   a

Furthermore, vic coupling is important to distinguish between E/Z isomers. The coupling constant ^3J for the E isomer ($\varphi = 180°$) in the range from 12 to 18 Hz is larger than that for the respective Z isomer ($\varphi = 0°$) which ranges typically from 6 to 12 Hz (see Table 6.24). Note that both hydrogen atoms in symmetrically substituted compounds of the structure R–CH=CH–R are chemically equal with the consequence that coupling does not cause splitting of the signal in the usual ^{1}H NMR spectrum. Thus, the coupling constant cannot be determined with the routine spectrum. Instead special experimental techniques are required to determine the coupling constant for such compounds.

Long-range proton-proton couplings ($^4J_{H-H}$, $^5J_{H-H}$)

Because the indirect spin-spin coupling is transmitted through the bonding electrons, the magnitude of J falls off rapidly as the number of intervening bonds increases. Therefore, couplings between nuclei that are more than three bonds apart for protons in flexible molecules are usually too small to cause observable splits and mostly give rise to broadening of the NMR lines. *Long-range* couplings can only be observed in cyclic and aromatic compounds, in compounds with the allylic structure $H_2C=CH–CH$-R, or in compounds having the so-called **W**-structure as shown in Structure 4.14 for which ^4J is ≈ 2.5 Hz.

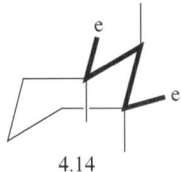

4.14

Values of long-range coupling constants for aromatic and heteroaromatic compounds are listed in Table 6.25.

4.2.5   Equivalence in NMR Spectroscopy

4.2.5.1   Chemical Equivalence

All protons found in *identical chemical environments* are chemically equivalent and will resonate at the same chemical shift. The number of peaks therefore corresponds to the number of different types of protons in the compound.

Nuclei can be chemically equivalent by *symmetry* within a molecule or by *rapid rotation* around single bonds. Thus, both protons of the symmetrically substituted ethylene derivative R–CH=CH–R are chemically equivalent because of symmetry, and the three protons of the methyl group in acetaldehyde, CH_3–C (=O)–H are equivalent because of rapid rotation of the methyl group around the C–C bond.

There are the following types of nuclei in regard to the presence of symmetry elements:

Homotopic protons
Homotopic protons are present if the molecule possesses at least one *n*-fold proper axis C_n. Thus, the methylene protons in dichloromethane (Structure 4.15) are homotopic because of the existence of a twofold proper axis C_2.

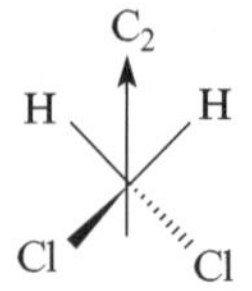

4.15

Note that homotopic nuclei are always *isochronous*, i.e., chemically equivalent in achiral as well as in a chiral environment and will give *one* signal in ¹H NMR.

Enantiotopic protons
Enantiotopic protons are present if the molecule possesses a symmetry plane σ or an inversion center *i* but does not possess a proper axis. Enantiotopic protons are chemically equivalent only in achiral environments. In chiral solvents or in the presence of chiral reagents the protons are no longer equivalent and the corresponding peak can split up.

Chlorfluormethane (Structure 4.16) possesses a symmetry plane within the paper plane σ and the cyclobutane derivate (Structure 4.18) has an inversion center *i*. However, both molecules are enantiotopic because they do not have a proper axis.

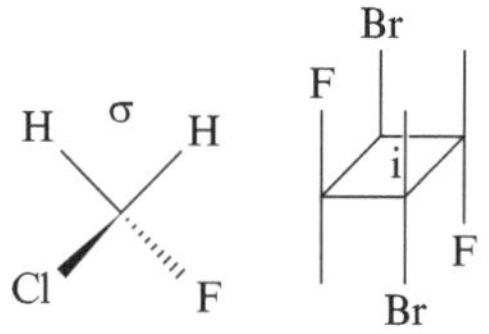

4.16   4.17

Diastereotopic protons
Diastereotopic protons are present if the molecule does not have a symmetry element. Such molecules are chiral. The methylene protons in 2-chloro butanic acid (Structure 4.18) are diastereotopic because the molecule possesses a chiral carbon atom which is termed by an asterisk, *.

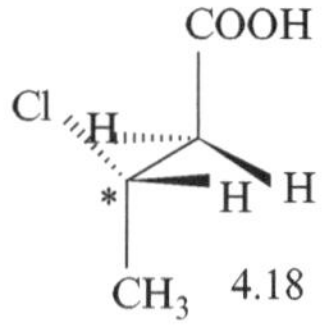

Note that diastereotopic protons are not equivalent and will usually give different signals in ¹H NMR.

Diastereotopic nuclei are existent in molecules that have a *prochiral* structure. Thus, a chiral center is not necessarily necessary as is shown by the example of amino glutaric acid (Structure 4.19).

4.19

Both methylene groups at C-1 and C-2 are enantiotopic because they can be transferred into each other via a symmetry operation involving a symmetry plane. However, the protons of each methylene group are diastereotopic because they cannot be transferred into each other by any symmetry operation. The chemical shift of both methylene groups coincides in achiral environments. Thus, the C-1 and C-3 nuclei show only *one* signal in the ^{13}C NMR spectrum. However, each proton of the methylene group has its own chemical shift. Coupling of the chemically nonequivalent protons gives rise to a multiplet.

To what extent the existence of a chiral center causes a splitting of signals depends on the spacing between the respective nuclei and the chiral center. Although not all nuclei of the chiral molecule 25-hydroxy provitamin D$_3$ (Structure 4.20) are equivalent, both methyl groups at C-24 and C-25 lead to *one* chemical shift in the ^{1}H NMR spectrum because the spacing to the next chiral center at C-18 is too large ^{1}H NMR. However, *two* ^{13}C signals are observed in ^{13}C NMR.

4.20

Chemical equivalence by dynamic processes
Chemically nonequivalent protons can become equivalent if the chemical environment causes them to be identical by rapid intra or intermolecular processes. Because dynamic processes depend on temperature, catalyst, and concentration, the equivalence on the basis of dynamic processes is determined by experimental conditions. According to a rule of thumb, dynamic processes faster than 10^{-1} to 10^{-3} Hz can coalesce to *one* signal, i.e., both nuclei are chemically equivalent in dependence of the applied magnetic field whereas slower processes provide two different signals. Some examples of rapid processes:

– Rotation around partial double bonds
   Dimethylformamide (Structure 4.5) is an example for chemical nonequivalence that is attributed to restricted rotation around the carbon nitrogen bond resulting

in two different signals for the methyl groups at room temperature; however, only *one* signal is observed at higher temperatures because of faster rotation.
- Keto-enol tautomerism
  Keto-enol tautomerism refers to a chemical equilibrium between a keto form (a ketone or an aldehyde) and an enol (an alcohol) which is demonstrated by acetylacetone (Structure 4.21).

4.21

Two signals are observed for the methyl groups in the ^{1}H NMR spectrum and, in addition to the methylene signal, the methine signal of the enol isomer is seen at deep field. However, the methyl groups become chemically equivalent at higher temperatures.
- Ring inversion in cyclohexane
  The transformation of two chair conformations into each other causes an effect where axial protons become equatorial and vice versa. The process occurs so rapidly at room temperature that only *one* signal is observed for the diastereotopic methylene groups. Axial and equatorial protons can only be distinguished in rigid ring systems like steroids.
- Rotation around a C–C single bond
  Generally, the three protons of the methyl groups are always chemically equivalent, also in chiral molecules because of rapid rotation of the methyl group at room temperature. Note that the lifetime of rotamers is about 10^{-6} s at room temperature.

Random chemical equivalence
Nuclei can be *randomly* isochronous although they have a different chemical environment and rapid exchange processes are absent. Randomly equivalent nuclei complicate the interpretation of NMR spectra; however, such nuclei can be recognized applying correlation spectroscopy (see below).

4.2.5.2 Magnetic Equivalence

Chemically equivalent nuclei must be checked to see whether they are also magnetically equivalent or not.

Magnetically equivalent nuclei are isochronous nuclei that have the same coupling constant J to all nonisochronous nuclei; in other words, chemically equivalent nuclei are not magnetically equivalent if the coupling is different. Coupling between magnetically equivalent nuclei does not affect NMR spectra and can be ignored.

Two magnetically inequivalent nuclei are given an AA′ designation where A and A′ refer to protons that are symmetry equivalent but not magnetically equivalent.

Magnetic equivalence and inequivalence are most easily understood based on specific examples like a *para*-disubstituted aromatic (Structure 4.22) and a 1,2,3-trisubstituted aromatic compound (Structure 4.23). Note that the protons are termed by capitals of the alphabet which is explained in the next section.

Compound 4.22 possesses two types of nuclei, those that are attached in an *ortho* position to substituent R_1, termed by A and both nuclei B that are in an *ortho* position to substituent R_2. In order to check if both chemically equivalent protons are also magnetically equivalent let us consider the coupling of the chemically equivalent protons of type A. Proton A couples over three bonds to proton B, but proton A′ couples over five bonds to the same proton. Hence, the coupling constants $J_{A\text{-}B}$ and $J_{A'\text{-}B}$ are different with the consequence that the chemically equivalent protons A are magnetically nonequivalent. Therefore, they are marked by A and A′. The same result is obtained by checking the chemically equivalent protons of type B. For example, the coupling of B to proton A is different from the coupling of B′ to A. Therefore, the protons BB′ are magnetically nonequivalent.

Compound 4.23 also possesses two types of nuclei, the proton termed A and the chemically equivalent protons B. But in this case, the coupling of both protons B to proton A is equal. Therefore, the two chemically equivalent protons B are also magnetically equivalent.

4.2.6 Nomenclature of Spin Systems

Spin systems are nuclei that are connected by scalar coupling. Generally, spin systems are termed by *capitals of the alphabet*. The following rules are valid:

- Chemically equivalent nuclei are termed by the same letter. The number of nuclei is given by subscript indices.
  Example: CH_3 $\Rightarrow A_3$

– Different letters are chosen for various chemically equivalent nuclei. Which letter is used, depends on the order of the NMR spectrum, i.e., from the ratio $\Delta v/J$.

If the ratio is large ($\Delta v/J > 7$) the letters **M, N** and **X, Y** are used to designate the chemical shift for spin systems with middle and weak coupling, respectively. This case gives rise to *first-order* spectra.

Examples: $-CH=CH_2$ $\Rightarrow$ **AMX**

                 CH_3-CH_2- $\Rightarrow$ **A_3X_2**

If the ratio is small ($\Delta v/J < 7$) letters are used to designate the chemical shifts are close, **B, C**.

Example: $R-CH=CH-R'$ $\Rightarrow$ **AB**

– Two chemically equivalent nuclei that are magnetically nonequivalent are designed **AA'**.

Example: Aromatic protons in Structure 4.22:

                 $\Rightarrow$ **AA'XX'** or

                 AA'BB' (if the chemical shifts are close)

Because the chemical shift differences in, for example, *para*-nitrophenol (Structure 4.22 with $R_1 = NO_2$, $R_2 = OH$) are large, the spin system of the aromatic protons is **AA'XX'**. Instead **AA'BB'** must be used for *para*-chlorotoluene because the difference is small.

The chemical shifts of all aromatic protons are equal in compounds of Structure 4.22 with $R_1 = R_2$, but their coupling is different resulting in an AA'A"A''' spin system. Because this spin system does not cause splitting, a singlet is observed. Therefore, the notation for this spin system is **A_4**.

Let us now consider the spin system of the ethylene moiety **$X-CH_2-CH_2-Y$**. Both methylene protons are chemically equivalent because a mirror plane is existent in the staggered conformation which is shown in Structure 4.24.

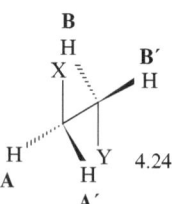

4.24

The other possible conformations do indeed no longer possess a mirror plane, but the chemical equivalence is maintained because of rapid rotation. However, the different vicinal couplings are retained so that the chemically equivalent protons of each methylene group are magnetically nonequivalent. Thus, an AA'BB' spin system is existent which can become at most the pattern of a simple A_2X_2 spin system if the vic coupling constants are nearly equal ($^3J_{AB} \approx {}^3J_{A'B'}$) as is shown in the ^1H NMR spectrum of *bis*-(chlorethyl) ether, $(Cl-CH_2-CH_2)_2-O$ in Fig. 4.13.

Fig. 4.13 400-MHz ^{1}H NMR
spectrum of *bis*-(chlorethyl)
ether, (Cl–CH$_2$–CH$_2$)$_2$–O

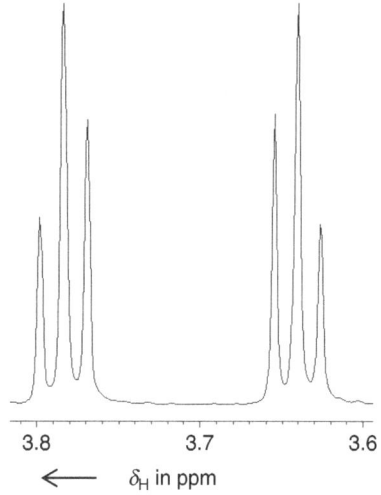

3.8   3.7   3.6

⟵ δ_H in ppm

Challenge 4.11

For solutions to Challenge 4.11, see *extras.springer.com*.

1. The range of ^{1}H chemical shifts of carbonic acids R–COOH and aldehydes R–COH partially overlaps. How can the signals be unambiguously assigned?
2. In general, the coupling between the proton attached at N and the neighbored methylene group in amines R–**CH$_2$–NH**–R′ is experimentally not observed. However, the couplings of the methylene group of amides R–C(=O)–**NH–CH$_2$**–R′ are observed as a doublet (further absent is the coupling to R′) whereas the signal of the **NH** group is broad without any fine structure. Explain these facts.
3. Determine the mol ratio of the tautomeric keto and enol forms of acetylic acetone CH$_3$–C(O)–CH$_2$–C(O)–CH$_3$ from the ^{1}H NMR spectrum in Fig. 4.14.
4. The synthesis of butyraldoxime, CH$_3$–CH$_2$–CH$_2$–CH=N–OH provides a syn/anti mixture. Determine the mol ratio of the *syn/anti* mixture from the 250-MHz ^{1}H NMR spectrum in Fig. 4.15. Note that $\delta_{syn} > \delta_{anti}$ is valid for the methine proton in oximes R–**CH**=N–OH.

   Develop the line pattern for the signal groups and assign the chemical shifts. Determine the coupling constants and assign these to the respective signal groups.

(continued)

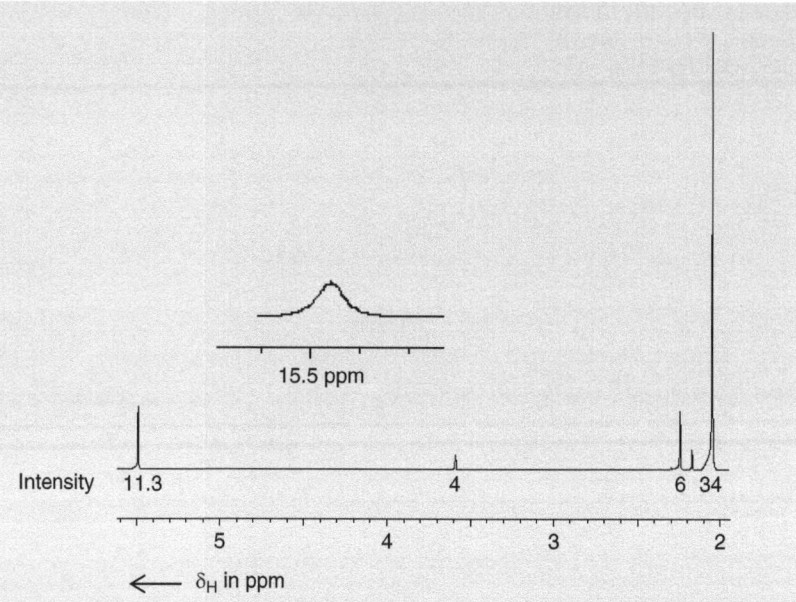

Fig. 4.14 400-MHz ^{1}H NMR spectrum of acetylic acetone CH_3–$C(O)$–CH_2–$C(O)$–CH_3

5. Determine the mol ratio of the isomers hexene-1, CH_3–CH_2–CH_2– CH_2–CH=CH_2 and 4-methylpentene-1, $(CH_3)_2CH$–CH_2–CH=CH_2 from the 400-MHz ^{1}H NMR spectrum shown in Fig. 4.16. Note that the respective intensities are given in the NMR spectrum.

6. The 400-MHz ^{1}H NMR spectrum of morpholine (Structure 4.25) is shown in Fig. 4.17. Designate the spin system and assign the signals.

4.25

7. Draw the line pattern for the ^{1}H NMR spectrum for the following compounds and designate the respective spin system.

| (a) Ethyl *iso*-propyl ether | (b). $(CH_3)_3P$=O | (c) Methyl *tert.*-butyl ketone |
| (d) CF_3CH_2Cl | (e) $ClCH$=$C(F)P(O)(OR)_3$ | |

8. Draw the line pattern for the 250-MHz ^{1}H NMR spectrum of the vinyl group R–CH=CH_2 with the following data: $\delta_1 = 5.22$ ppm, $\delta_2 = 5.73$ ppm, $\delta_3 = 6.70$ ppm, $J(H_1/H_2) = 16$ Hz, $J(H_1/H_3) = 10$ Hz, $J(H_2/H_3) = 2.5$ Hz.

(continued)

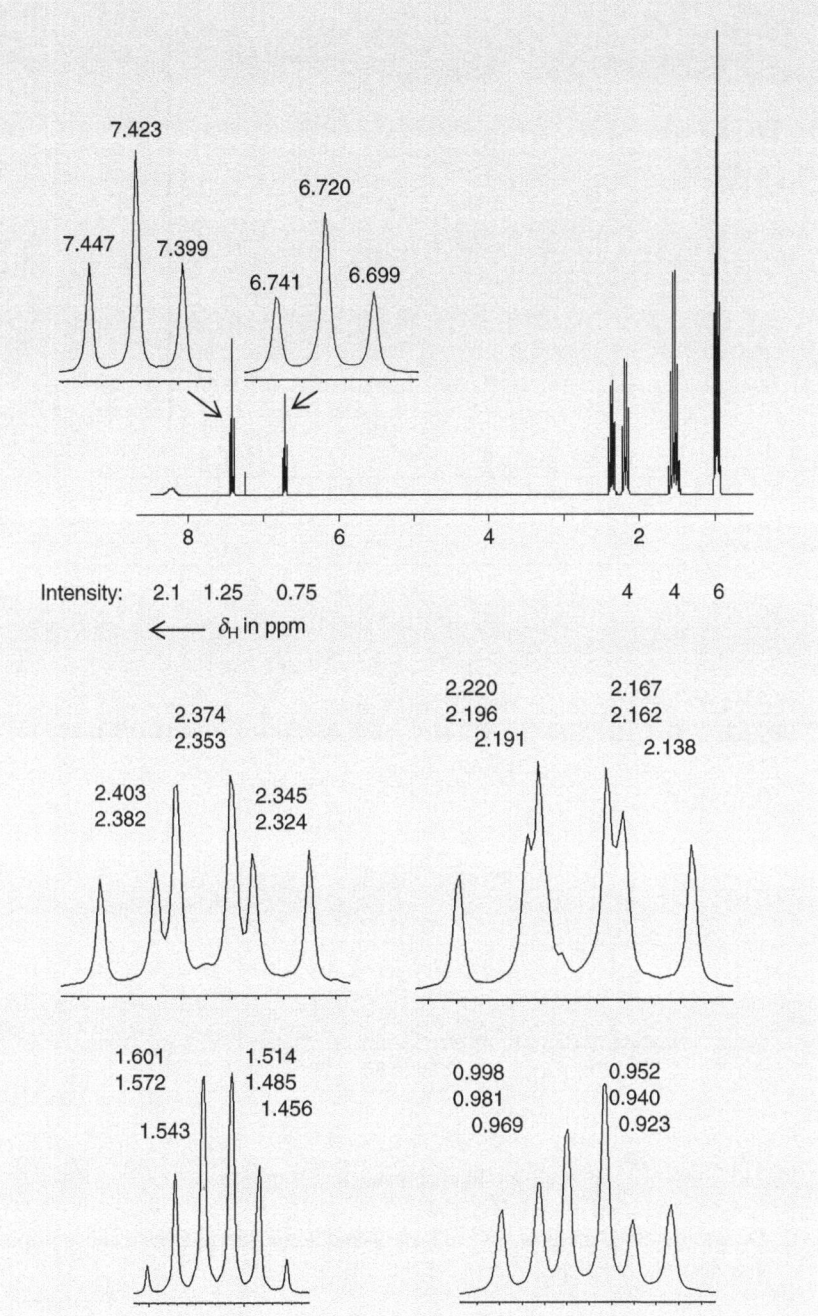

Fig. 4.15 250-MHz spectrum of the *syn/anti* mixture of butyraldoxime, $CH_3-CH_2-CH_2-CH=N-OH$

(continued)

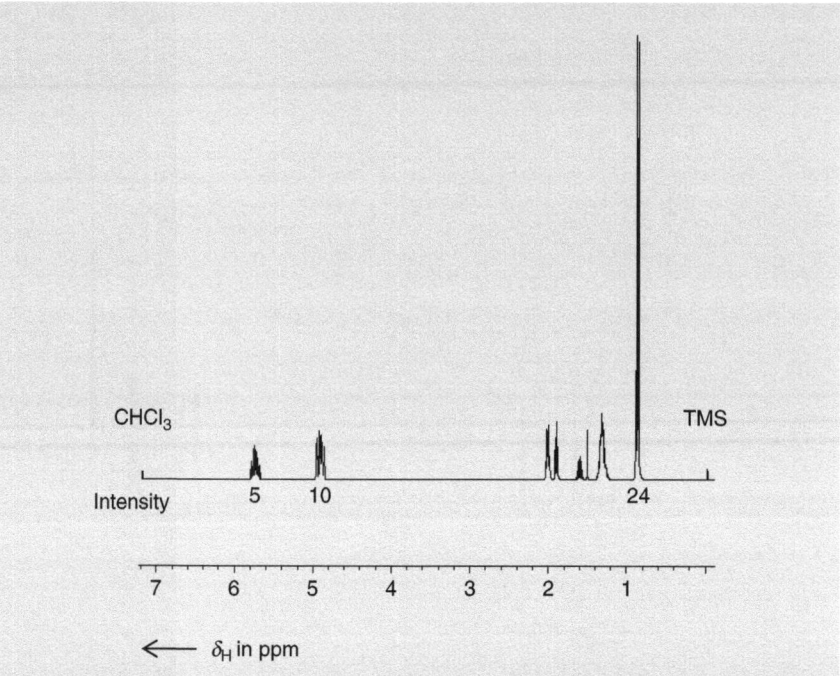

Fig. 4.16 400-MHz ^{1}H NMR spectrum of a mixture of hexene-1, CH_3–CH_2–CH_2–CH_2–CH=CH_2 and 4-methylpentene-1, $(CH_3)_2CH$–CH_2–CH=CH_2

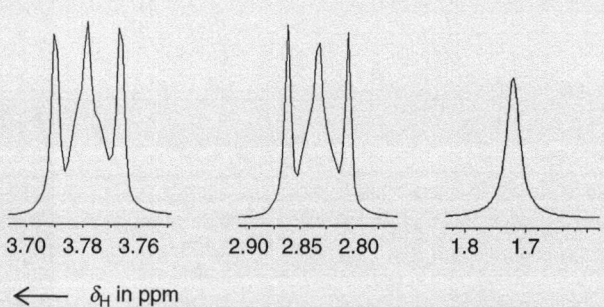

Fig. 4.17 400-MHz ^{1}H NMR spectrum of morpholine (Structure 4.25)

9. Decide whether the protons are homotopic, enantiotopic, or diastereotopic and designate the spin system of the following compounds:

(continued)

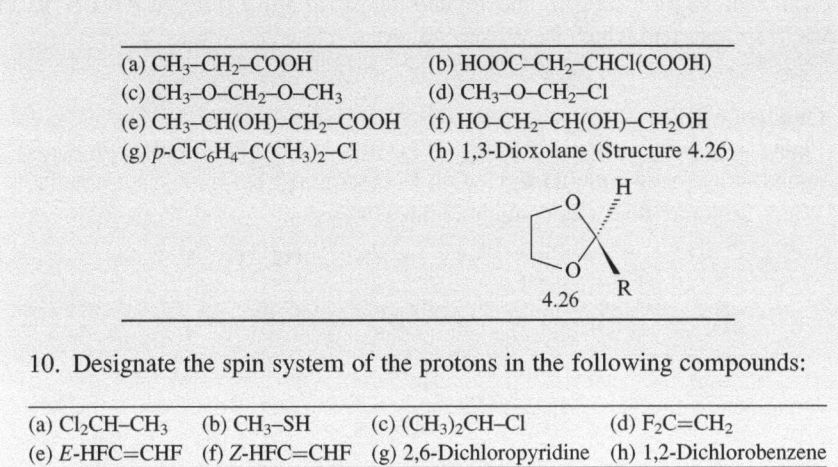

(a) CH_3–CH_2–COOH	(b) HOOC–CH_2–CHCl(COOH)
(c) CH_3–O–CH_2–O–CH_3	(d) CH_3–O–CH_2–Cl
(e) CH_3–CH(OH)–CH_2–COOH	(f) HO–CH_2–CH(OH)–CH_2OH
(g) p-ClC$_6$H$_4$–C(CH_3)$_2$–Cl	(h) 1,3-Dioxolane (Structure 4.26)

4.26

10. Designate the spin system of the protons in the following compounds:

(a) Cl$_2$CH–CH$_3$	(b) CH_3–SH	(c) (CH$_3$)$_2$CH–Cl	(d) F$_2$C=CH$_2$
(e) E-HFC=CHF	(f) Z-HFC=CHF	(g) 2,6-Dichloropyridine	(h) 1,2-Dichlorobenzene

4.3   Analysis of ^{1}H NMR Spectra

4.3.1   Simple First-Order Spectra

First-order NMR spectra are present if the difference of the chemical shift Δv is substantially larger than the coupling constant J (see Eq. 4.17).

The multiplet of the signals is determined by the simple $N + 1$ rule (Eq. 4.18). Thus, the number of coupled neighboring protons can be obtained by the number of lines of the respective multiplet.

Because the difference of the chemical shift Δv is determined by the spectrometer frequency whereas the coupling constant J is field-independent, increasing the spectrometer frequency gives rise to an enhancement of the ratio $\Delta v/J$. Thus, many ^{1}H NMR spectra become suitable for a first-order analysis at least in a rough approximation by increasing the spectrometer frequency.

First-order NMR spectra with two different nuclei are designated according to the rules for spin systems as $\mathbf{A_nX_m}$. The chemical shifts of the nuclei A (δ_A) and X (δ_X) are taken from the center of the respective multiplet and the coupling constant is given by the spacing of each of the neighbored lines in both multiplets. Note that J is always given in Hz; therefore, the difference of the lines in $\Delta \delta$ must be converted into Δv according to Eq. 4.10.

In an AA'BB' spin system there are two pairs of magnetically nonequivalent protons which gives a second-order spin system. If the spectrometer field is increased the spin system will move to an AA'XX' system. However, this spin system is also of second order.

Spin systems with magnetically nonequivalent nuclei will always be second-order spectra regardless of the magnetic field applied. The strongest simplification one can achieve by increasing the applied field is to move from an AA′BB′ to an AA′XX′ spin system which is still second order.

Challenge 4.12

Figure 4.18 presents the 250-MHz ^{1}H NMR spectrum of an unknown compound whose sum formula $C_4H_{10}O$ was determined by mass spectrometry. Which structure does this compound have?

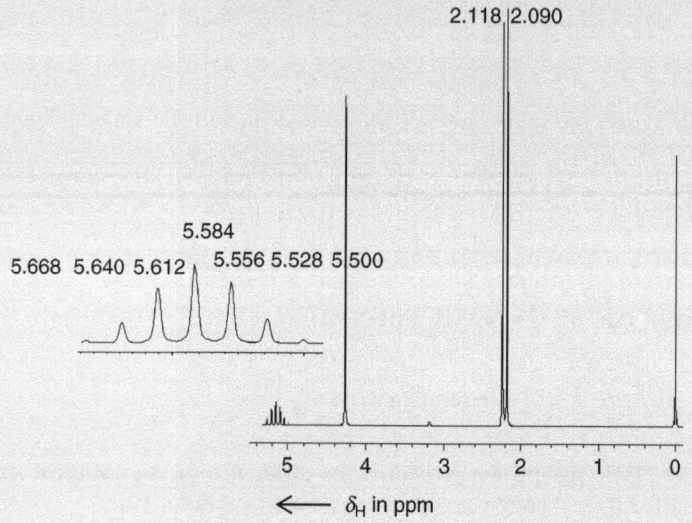

Fig. 4.18 250-MHz ^{1}H NMR spectrum of an unknown compound with the sum formula $C_4H_{10}O$

Solution to Challenge 4.12

The ^{1}H NMR spectrum shows three signal groups: $\delta_1 = 5.584$ ppm (septet), $\delta_2 = 4.13$ ppm (singlet), and $\delta_3 = 2.104$ ppm (doublet). The protons at $\delta_3 = 2.104$ ppm couple with the single proton according to the $N + 1$ rule and the proton at $\delta_1 = 5.584$ ppm couples with six chemically equivalent protons. The spacing of each multiplet is $\Delta\delta = 0.028$ ppm. Thus, the coupling constant is $J = \Delta\delta \cdot \Delta\nu_0 \cdot 10^{-6}$ Hz $= 0.028 \cdot 250 \cdot 10^6 \cdot 10^{-6}$ Hz $= 7$ Hz. This coupling constant has a typical value for a vicinal coupling over three σ bonds (see Table 6.24). Thus, the signals at δ_1 and δ_3 show evidence of the *iso*-propyl group, $(CH_3)_2CH$–. Because the double binding equivalent is zero, a carbonyl group cannot be present. Hence, the *iso*-propyl group is attached to an oxygen atom $(CH_3)_2CH$–O–R. The singlet at $\delta_2 = 4.13$ ppm is assigned to a methyl group which is not attached to other CH_x groups. Thus, the unknown compound is *iso*-propyl-methyl ether $(CH_3)_2CH$–O–CH_3.

The assignment of signal groups is achieved by the influence of the electronegative oxygen on the chemical shift in connection with the multiplet pattern.

(continued)

The CH$_x$ groups which are directly attached to oxygen give rise to signals in the deep field, i.e., >4 ppm. The two signal groups are distinguished by the multiplet pattern. Thus, the following assignment is valid: $\delta_1 = 5.584$ ppm $\Rightarrow$ **CH**, $\delta_2 = 4.13$ ppm $\Rightarrow$ **CH$_3$-(O)**, $\delta_3 = 2.104$ ppm $\Rightarrow$ **(CH$_3$)$_2$(CH)**.

In general, NMR spectra of the type **AMX** can be handled with simple first-order rules at least as a rough approach. The chemical shifts δ_A, δ_M, and δ_X are taken from the respective center of the signal group, and the multiplets provide the coupling constants J_{AX}, J_{MX}, and J_{AM}.

An **AMX** spin system is present if the smallest difference of the chemical shift $\Delta\nu_{min}$ is larger than the 2.5-fold of the largest coupling constant J_{max}, Eq. 4.23:

$$\Delta\nu_{min} > 2.5 \times J_{max} \qquad (4.23)$$

An example for an AMX system is the 400-MHz ^{1}H NMR spectrum of 2,3-dichlorophenol (Structure 4.27) which is shown in Fig. 4.19 with the line pattern. The criterion for an AMX spin system is fulfilled. The smallest difference of the chemical shifts $\Delta\nu_{min} = \Delta(\delta_A - \delta_M) \cdot 400$ MHz $\cdot 10^{-6} = 44$ Hz is greater than the 2.5-fold of the greatest coupling constant $2.5 \cdot 8$ Hz $= 20$ Hz.

Because both *ortho* coupling constants belonging to proton A are nearly equal the double doublet (dd) for nuclei A is "fused" to a triplet, $J_{AM} = J_{AX} = 8$ Hz. Each of the two other protons gives rise to a double doublet (dd). The *meta* coupling constant is $J_{MX} = 2$ Hz.

The assignment of the chemical shifts can easily be achieved by consideration of the substituents and the chemical shift values in combination with the coupling pattern. Thus, the triplet at $\delta = 7.15$ ppm unambiguously belongs to H$_X$ because it provides only an *ortho* coupling. The signal at $\delta = 6.93$ ppm should be assigned to H$_A$ because of the highly negative increment (Table 6.21, $S = -0.53$) of the *ortho*-substituted OH group. Note that an unambiguous assignment of the signals can be realized by correlation spectroscopy (see Sect. 4.4.8).

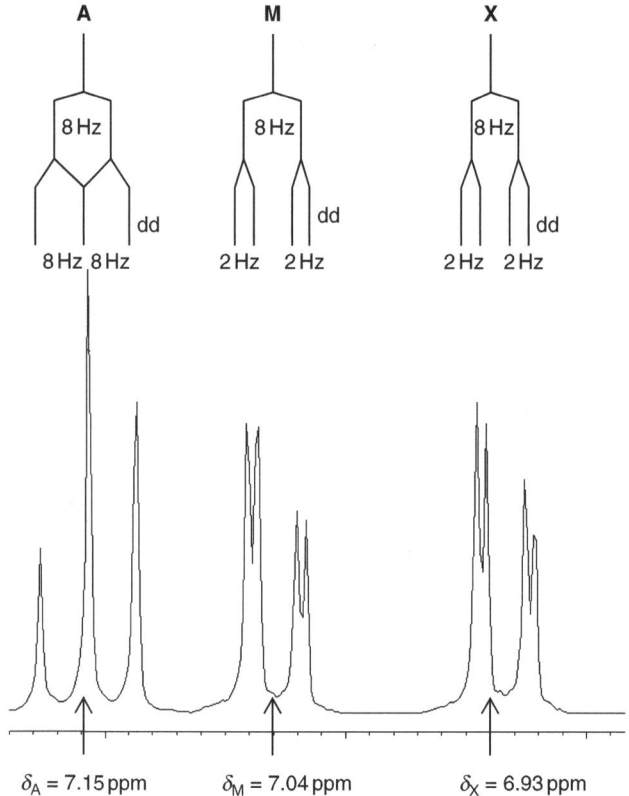

Fig. 4.19   400-MHz ^{1}H NMR spectrum of 2,3-dichlorophenol (Structure 4.27) and the line pattern of the AMX spin system

4.3.2   Second-Order Spectra

First-order spectra assume that the coupling constant is small in comparison with the differences of the chemical shifts. If the shift separation decreases or the coupling strength increases, the multiplet intensity patterns are first distorted, and then become more complex. The first-order spectrum is transferred into one of second order. Second-order NMR spectra are caused by *strong* coupling, i.e., if the magnitude of the coupling constant and the difference of the chemical shifts are similar. The rule of thumb $\Delta v / J < 7$ is valid.

Second-order NMR spectra can no longer be analyzed by simple first-order rules, i.e., in general, chemical shift δ and coupling constant J cannot be directly obtained from the spectrum. The only way to accurately analyze these systems is by means of quantum mechanics calculations.

However, the parameters δ and J can be calculated by formula based on quantum mechanical results for some spin systems and certain spin systems can be recognized by their multiplet pattern which is the subject of this section.

Recognition of Second-Order Spectra
The general difference between first-order and second-order spectra is explained employing the example of a three-spin system, i.e., the line patterns of the first-order AX_2 spin system is compared with that of the second-order one AB_2.

First-order spin system AX_2 Example:	Second-order spin system AB_2 Example:

Line pattern	

Comparison	
Equal spacing of the multiplet lines	Unequal spacing of the multiplet lines
Multiplet: $M = N + 1$	*More* lines than for a first-order system
Intensity: According to Pascal's triangle	The inner lines are more intense than the outer lines and there is an evident lack of symmetry (*roofing* intensity pattern)
Parameters δ, J: Directly obtained from the spectrum	Only via quantum mechanics calculation

Two lines with the same intensities in a first-order spectrum will be altered in the second-order spectrum: The line being closer to the corresponding coupling partner becomes more and the other one less intense resulting in a roofing intensity pattern. This effect is very useful because the connecting lines point in the direction of the coupling partners.

Two-spin system **AB**
Example:

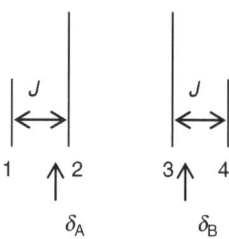

Line pattern:

δ_A δ_B

Calculation of the parameters chemical shift in Hz (v_A, v_B) and coupling constant J in Hz:

$$v_{A,B} = v_Z \pm 0.5 \cdot \sqrt{(1-4) \cdot (2-3)} \tag{4.24}$$

$$|J| = |1 - 2| = |3 - 4| \tag{4.25}$$

The parameters of the AB spin system can be directly obtained from the NMR spectrum. Whereas the line intensity is nearly equal in a first-order spectrum and the chemical shift lies in the middle of both line pairs, the intensity of the inner lines increases and the intensity of the outer ones decreases at the same time. The difference of the line intensities is the greater the stronger the coupling is. In the same manner, the chemical shifts move towards the more intense inner lines.

Three-spin systems
AB₂ spin system
The line pattern and examples are given above. The parameters of the AB₂ spin system can be directly obtained from the spectrum. Note that the numbers refer to the number of lines in the line pattern given above.

$$n_A = 3 \tag{4.26}$$

$$n_B = 0.5 \times (5 + 7) \tag{4.27}$$

$$|J| = \frac{1}{3} \cdot |(1-4) + (6-8)| \tag{4.28}$$

ABX spin system
An ABX spin system is present if the difference of the chemical shifts of two nuclei is small but that of the third nucleus is substantially greater.
    Example:

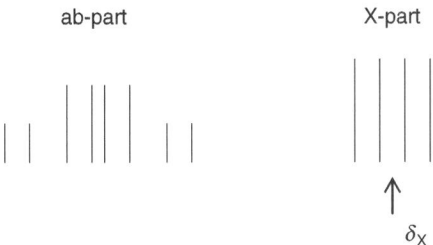

An ABX spin system is represented by the following line pattern:

The chemical shift of the X proton δ_X lies in the center of the X-part which consists of four equally intense lines (sometimes two additional lines are observed). The chemical shifts of protons A and B as well as the coupling constants cannot be directly obtained by the NMR spectrum, but the recognition of the spin system according to the line pattern is important for structure determination.

ABC spin system
An ABC spin system is present if the chemical shifts of all three protons are substantially equal. Sometimes the difference of chemical shifts can be smaller than the coupling constants. Parameters cannot be directly obtained from the spectrum.
    Theoretically, an ABC spin system consists of 15 lines but, in general, not all lines are resolved. A typical ABC system is shown in Fig. 4.20 employing the example of 2-methyl 6-chlorobenzonitrile. The 15 lines are marked by dots and the roofing intensity, typical for second-order spectra, is marked by lines.

Four-spin systems
The spin systems AA'XX' and AA'BB' are important for structure determination because these spin systems, in general, can easily be recognized by the symmetry of two half-patterns, as is shown by the examples in Figs. 4.21 and 4.22. Exact chemical shifts and coupling constants can only be obtained by simulation of the

Fig. 4.20 250-MHz ^{1}H NMR
spectrum of 2-methyl
6-chlorobenzonitrile. The 15
possible lines are dotted

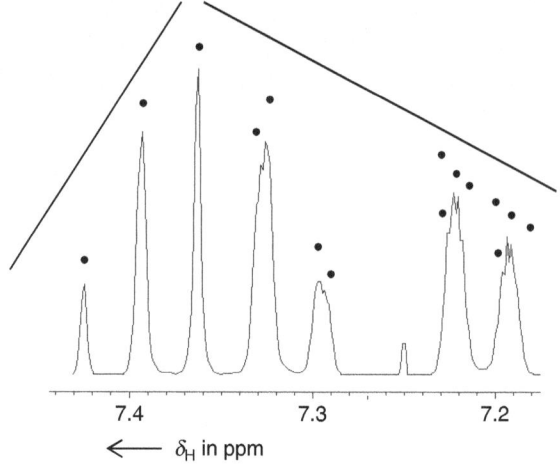

Fig. 4.21 250-MHz ^{1}H NMR
spectrum of *para*-chlor-
acetophenone (Structure
4.31)

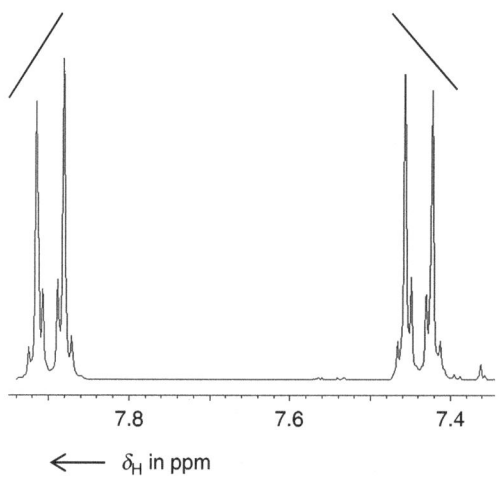

spectra but the chemical shifts of both nuclei will lie within the respective half-
patterns.

Fig. 4.22 250-MHz ^{1}H NMR
spectrum of *para*-chlor-
benzonitrile (Structure 4.32)

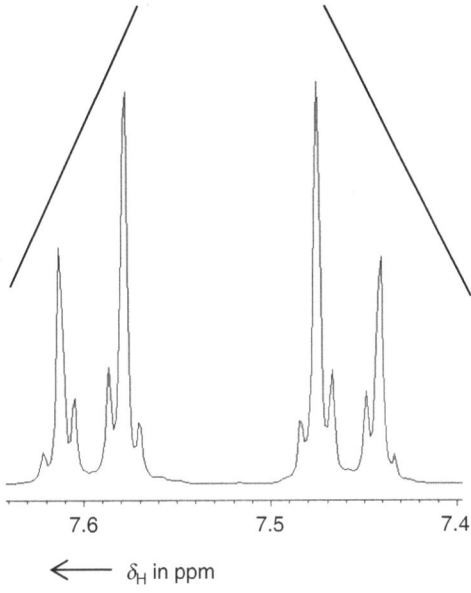

δ_H in ppm

The difference between the centers of the two half-spectra of *para*-chlor-
acetophenone (Structure 4.31) is about 108 Hz. This value is greater than seven
times the highest coupling constant J_{ortho} which is about 8 Hz. Therefore, the nuclei
must be designated by the letter A and X. However, the two pairs of chemically
equivalent protons are magnetically nonequivalent. Thus, a second-order spin
system AA'XX' is present which has a complex pattern.

Because the difference of the two half-spectra of *para*-chlorobenzonitrile (Structure 4.32) ($\approx$32 Hz) is only about four times greater than the coupling constant
J_{ortho}, the letters A and B must be employed to designate the spin system. Note that
the roofing pattern is substantially steeper in the AA'BB' spin system than in an
AA'XX' one.

The recognition of the AA'XX' and AA'BB' system, respectively in the range
from 6.5 to 8.5 ppm gives rise to important hints for structure patterns like
Structures 4.33–4.35. Thus, other isomers can be excluded.

R$_1$

A H — [ring] — H A′

(X) B H — [ring] — H B′(X′)

4.33

R$_2$

Para-disubstituted aromatic compounds R$_1$ ≠ R$_2$

R

A H — [ring] — H A′

(X) B H — N — H B′(X′)

4.34

4-Substituted pyridine compounds

H A

(X) B H — [ring] — R

(X′) B′ H — [ring] — R

4.35   H A′

Ortho-disubstituted aromatic compounds

$$R_1 = R_2$$

Para-disubstituted aromatic compounds with R$_1$ ≠ R$_2$ give rise to the same spin system as *ortho*-disubstituted aromatic compounds with equal substituents. However, both types can be easily distinguished by the number of aromatic signals in the ^{13}C NMR spectrum; Structure 4.33 gives four signals, whereas Structure 4.35 gives only three.

The AA′BB′ spin system is also present in anthracene (Structure 4.36) and 9,10-disubstitued derivatives with equal substituents. The spin systems A$_2$ + AA′BB′ is easily recognized in the 250-MHz ^{1}H NMR spectrum of anthracene (Fig. 4.23).

H$_A$ H$_A$ H$_A$

H$_B$ — [ring] — H$_B$

9

10

H$_{B'}$ — [ring] — H$_{B'}$

H$_{A'}$ H$_A$ H$_{A'}$

4.36

As explained above, the AA′X(B)X′(B′) spin system is also present in the structure patterns –CH$_2$–CH$_2$– especially in rigid structures like rings.

Fig. 4.23 250-MHz ^{1}H NMR spectrum of anthracene

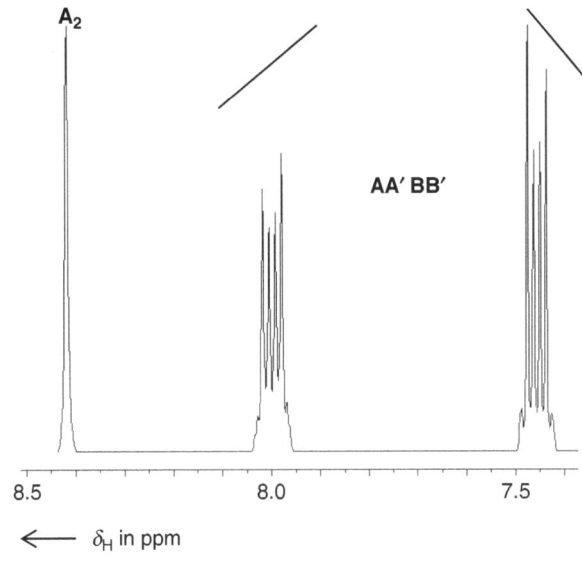

Challenge 4.13

For solutions to Challenge 4.13, see *extras.springer.com*.

1. Figure 4.24 shows a section of the 250-MHz ^{1}H NMR spectrum of 3-ethyl-2-methyl glutaric acid (Structure 4.37). Determine the chemical shifts and the coupling constant. Justify that the correct designation for the spin system is AB. To which protons does the signal group belong? Why are these protons not equivalent?

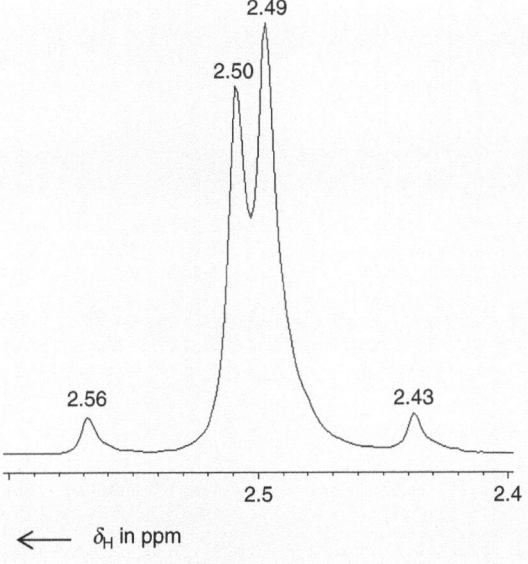

Fig. 4.24 Section of the 250-MHz ^{1}H NMR spectrum of 3-ethyl 2-methyl glutaric acid

(continued)

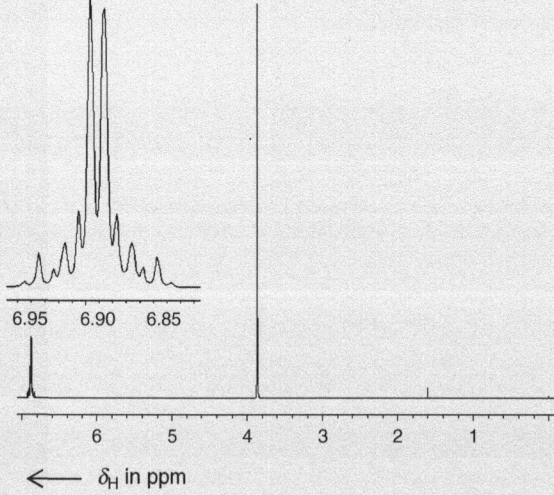

4.37

2. Designate the spin system of the protons in thiophene (Structure 4.38).

4.38

3. An A_2X_2 spin system is a first-order NMR spectrum. Why does an AA'XX' spin system produce a second-order spectrum and why can such a spectrum not be transferred into one of first order?
   The protons of a compound provide an AB_2 spin system in a 60-MHz NMR spectrometer. How can the AB_2 spin system possibly be transferred into an AX_2 system?

4. The 250-MHz ^1H NMR spectrum of an unknown compound with the sum formula $C_8H_{10}O_2$ is presented in Fig. 4.25. Determine the structure of the compound and designate the spin system. Calculate the chemical shift values.

Fig. 4.25 250-MHz ^1H NMR spectrum of an unknown compound with the sum formula $C_8H_{10}O_2$

(continued)

5. Figures 4.26 and 4.27 show the 250-MHZ ^{1}H NMR spectra of 2,6-
   dibromopyridine (Structure 4.39) and 2,6-dichloropyridine (Structure
   4.40), respectively. Determine the chemical shifts and the coupling
   constants for both compounds and designate the respective spin system.
   Justify that first-order and second-order spin systems, respectively, are
   present. Would a first-order spectrum be expected for Structure 4.39 in a
   600-MHz spectrometer?

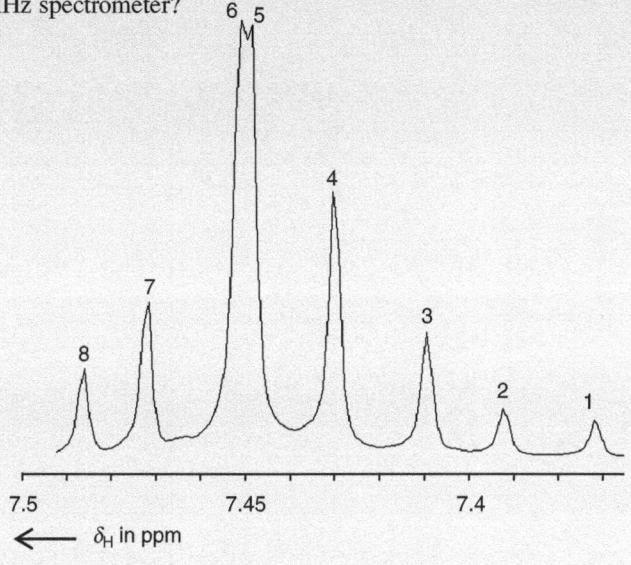

Fig. 4.26 250-MHZ ^{1}H NMR spectra of 2,6-dibromopyridine (Structure 4.39)

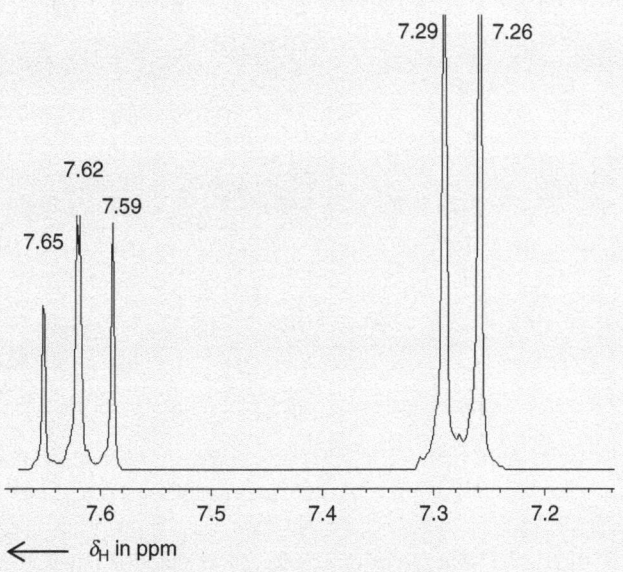

Fig. 4.27 250-MHZ ^{1}H NMR spectra of 2,6-dichloropyridine (Structure 4.40)

(continued)

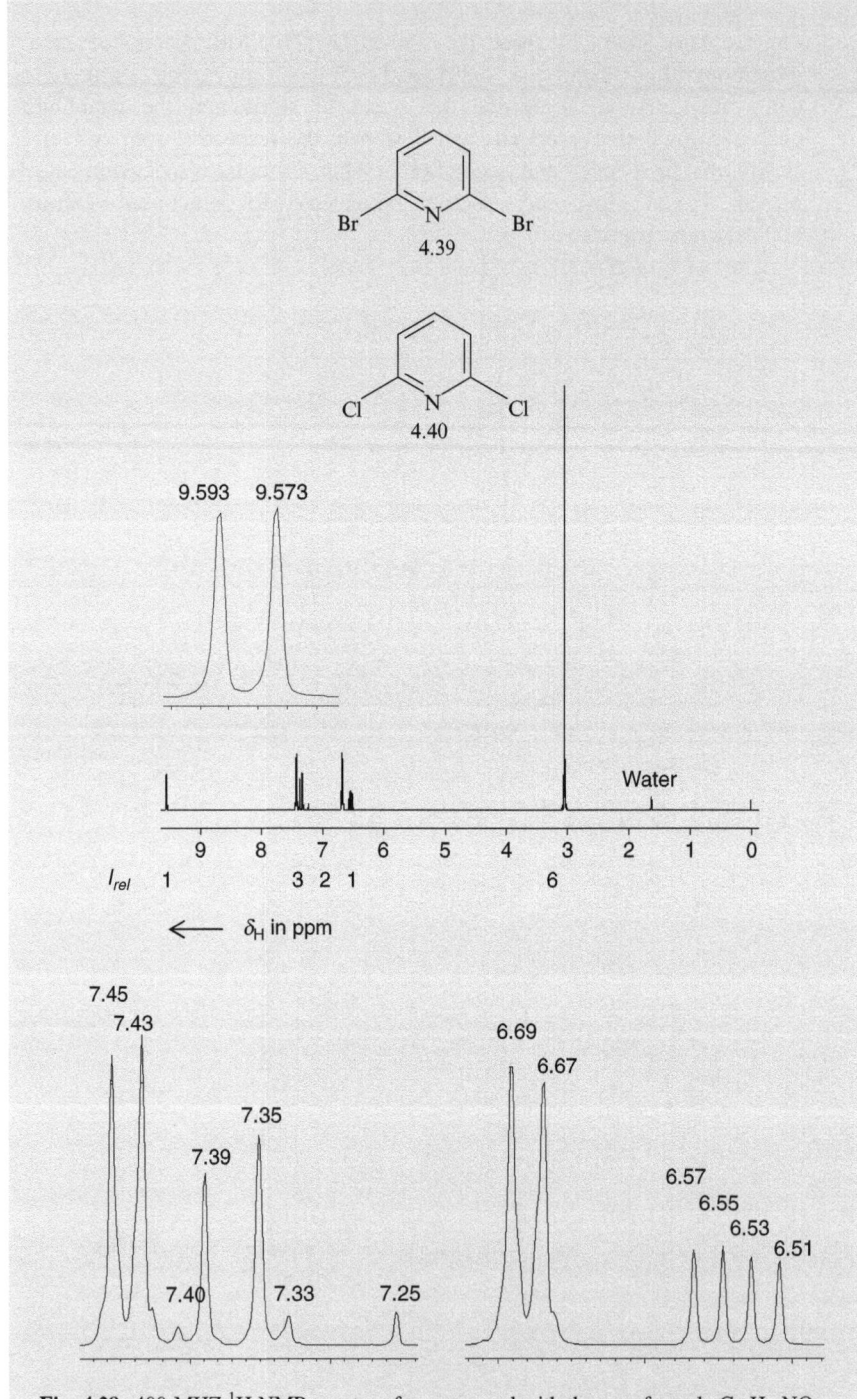

Fig. 4.28 400-MHZ ^{1}H NMR spectra of a compound with the sum formula $C_{11}H_{13}NO$

(continued)

Fig. 4.29 Section from the
300-MHz ^{1}H NMR spectrum
of a steroidal reaction product
(Structure 4.41)

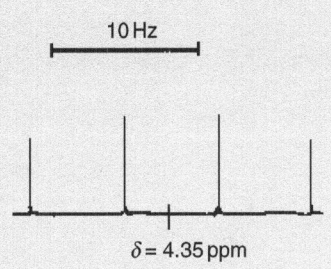

$\delta = 4.35\,\text{ppm}$

6. An unknown compound with the sum formula $C_{11}H_{13}NO$ provides the 400-MHz ^{1}H NMR spectrum which is shown in Fig. 4.28. The protons of the compound cannot be exchanged by D_2O. Determine the structure, chemical shifts, coupling constants (as far as possible), and designate the spin systems.

7. The ^{1}H NMR spectrum of an allylic metal compound of the type CH_2CHCH_2MeX provides a quintet at $\delta = 6.2$ ppm and a doublet at $\delta = 2.5$ ppm. Decide whether the compound has a Me–C moiety or an ionic structure with an allylic cation or allylic anion.

8. The introduction of a CH_3COO group into a $^1\Delta$ steroid provides a product whose molecular mass is in accordance with Structure 4.41. Solve the following structural problems: Is the CH_3COO group attached to C-2 and not to C-1? Is the α- or β-epimeric structure present? Figure 4.29 shows the respective section from the 300-MHz ^{1}H NMR spectrum. The integration corresponds to one proton.

R

H   H

H

?

CH_3COO

O

4.41

9. Figure 4.30 presents the 400-MHz ^{1}H NMR spectrum of an aniline derivative with the substituents NH_2, 2 CH_3, and NO_2 (Structure 4.42). Which structural information can be achieved by the ^{1}H NMR?

(continued)

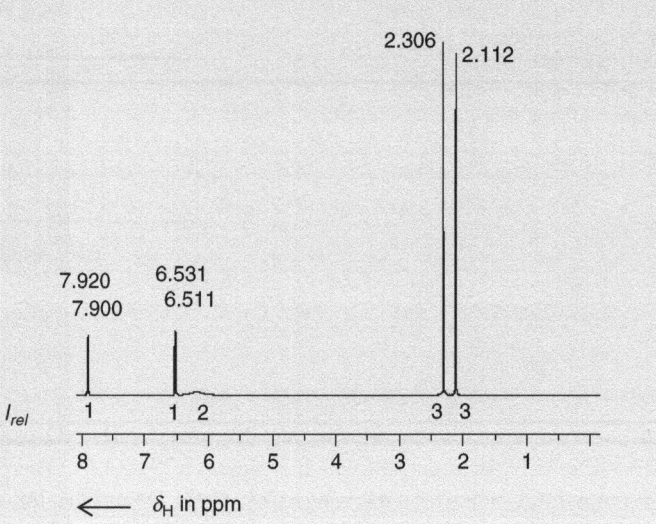

Fig. 4.30 400-MHz ^{1}H NMR spectrum of an aniline derivative with the substituents NH_2, 2 CH_3, NO_2

10. Figure 4.31 shows the ^{1}H NMR spectrum of a dimethoxy benzonitrile isomer (Structure 4.43). Which structure is correct? Assign the chemical shifts and coupling constants and designate the spin systems.

11. A liquid compound that was isolated from bran was transferred into another liquid compound by catalytic reaction and the molecular peak is $m/z = 98.0369$ amu. The intensity of a [M + 2] peak is present but it is

(continued)

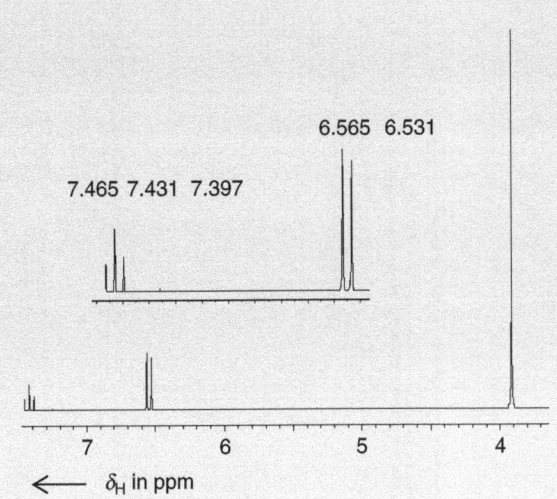

Fig. 4.31 250-MHz ^{1}H NMR spectrum of a benzonitrile derivative (Structure 4.43)

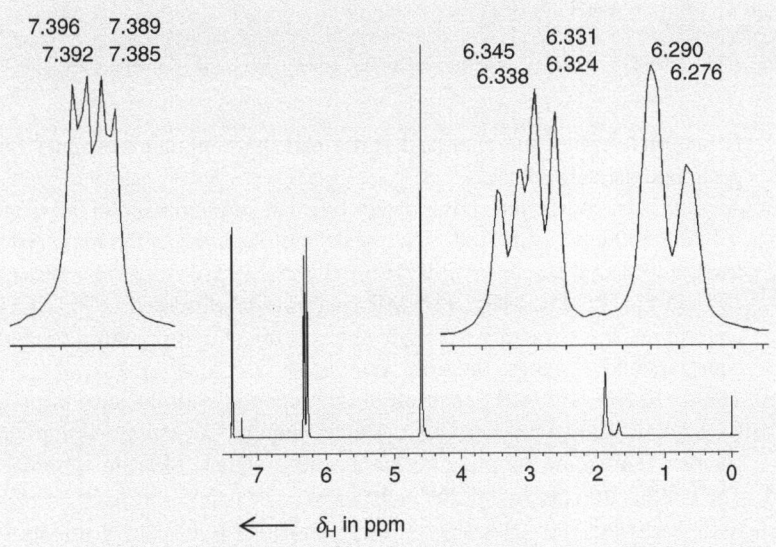

Fig. 4.32 250-MHz ^{1}H NMR spectrum of a reaction product with the molecular peak $m/z = 98.0369$ amu

very small. The IR spectrum is dominated by an intense broad band at 3,365 cm^{-1} and a very strong band at 1,055 cm^{-1}. Furthermore, mass and IR spectra show information pointing towards an aromatic compound. The 250-MHz ^{1}H NMR spectrum is shown in Fig. 4.32. Which structure

(continued)

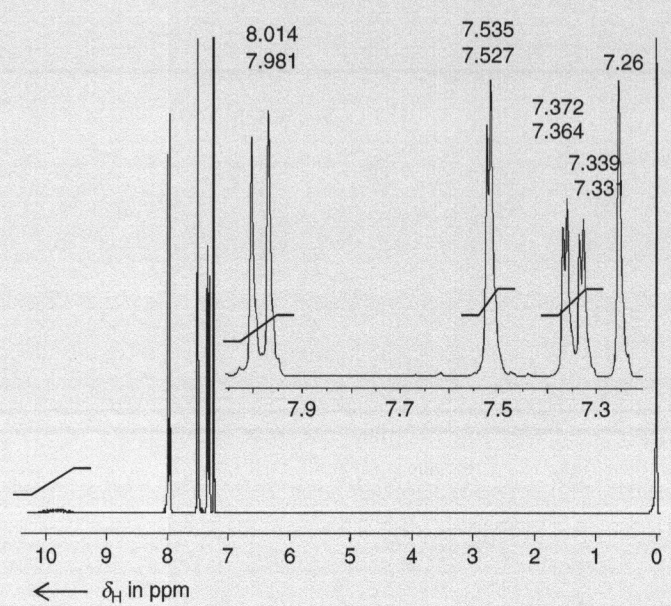

Fig. 4.33 250-MHz ^{1}H NMR spectrum of an unknown compound (Solvent: CDCl$_3$)

is present? Assign the chemical shifts and the coupling constants and designate the spin systems.

12. An unknown compound shows a very broad absorption band in the range 3,400–2,500 cm^{-1}. The following peaks are observed in the mass spectrum; m/z in amu (I_{rel} in %): 190 (36), 191 (2.8), 192 (22), 193 (1), 194 (6), 195 (< 1). The 250-MHz ^{1}H NMR spectrum is presented in Fig. 4.33. Determine the structure. Assign the chemical shifts and coupling constants and designate the spin systems.

13. The 250-MHz ^{1}H NMR spectrum of a compound with the sum formula C$_6$H$_6$OS is shown in Fig. 4.34. Determine the structure. Assign the chemical shifts and coupling constants and designate the spin systems.

14. The degradation product of the main alkaloid of pepper extract gives rise to a compound whose mass spectrum provides the following information: Molecular peak $m/z = 150$ amu, C$_8$ compound, base peak at 149 amu. The 250-MHz ^{1}H NMR spectrum is presented in Fig. 4.35. Assign the chemical shifts and coupling constants and designate the spin systems.

15. The structure of a phosphoric compound is to be determined. The structure CHCl=C(F)P(=O)(OC$_2$H$_5$)$_2$ is suggested by the IR and mass spectroscopy. The ^{19}F NMR spectrum provides a doublet doublet (dd) with the coupling constants $J_1 = 82$ Hz and $J_2 = 24.5$ Hz. Can the geminal position of F and P be confirmed? Which E/Z isomer is given?

(continued)

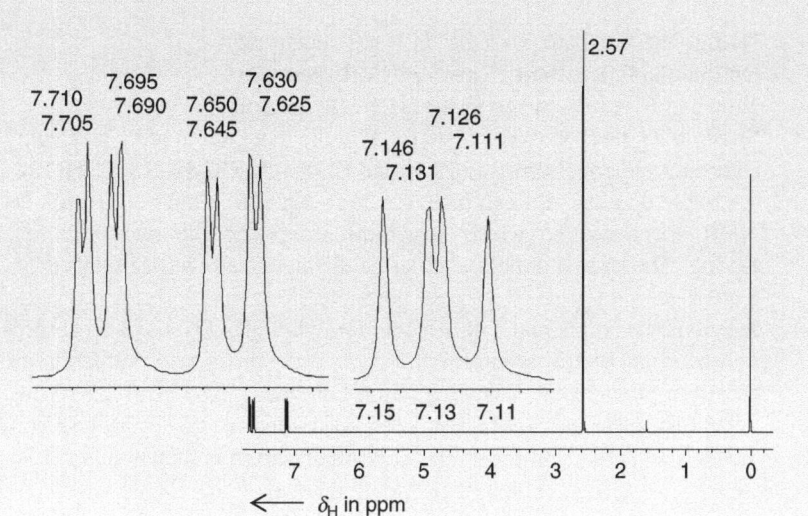

Fig. 4.34 250-MHz ^{1}H NMR spectrum of a compound with the sum formula C_6H_6OS

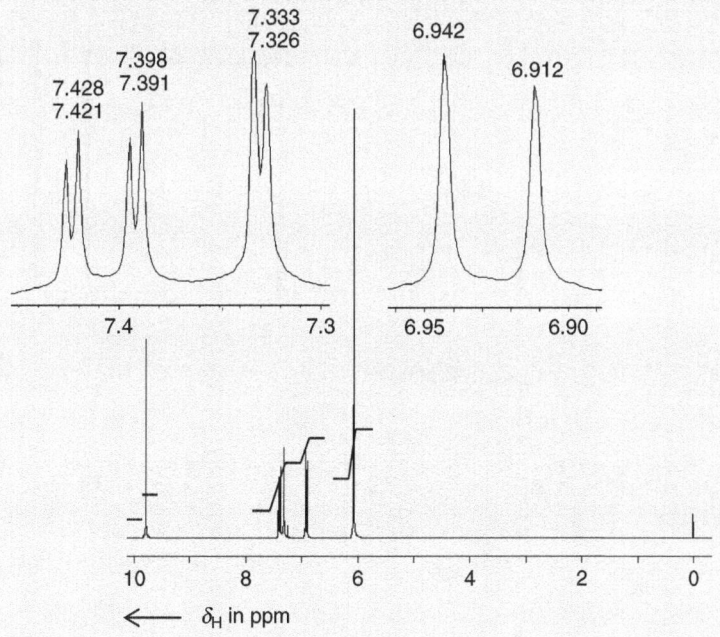

Fig. 4.35 250-MHz ^{1}H NMR spectrum of an unknown C_8 compound

(continued)

Outline the line pattern of the ^1H NMR spectrum.

Outline the line pattern of the ^{31}P NMR spectrum.

How can P be recognized in the ^1H NMR spectrum?

Summarize the spectroscopic method for the recognition of the following structural moieties: Presence of P, and Cl, P=O, -OC$_2$H$_5$, C=C, HC=C.

16. A double doublet (dd) can have a pattern equal to a quartet (q) in a ^1H NMR spectrum. To which structural groups do the two multiplets belong? How can a double doublet be distinguished with certainty from a quartet?

17. An unknown compound with the sum formula $C_6H_{10}O_2$ shows an absorption band in the UV spectrum at $\lambda_{max} = 205$ nm (lg α = 3.95). The IR spectrum provides the following absorption bands: 3,052 (m), 2,980 (m), 2,918 (m), 2,876 (m), 1,722 (vs), 1,652 (s), 1,446 (m), 1,377 (m), 1,185 (s), 1,042 (s), 967 (s). The 250-MHz ^1H NMR spectrum is shown in Fig. 4.36.

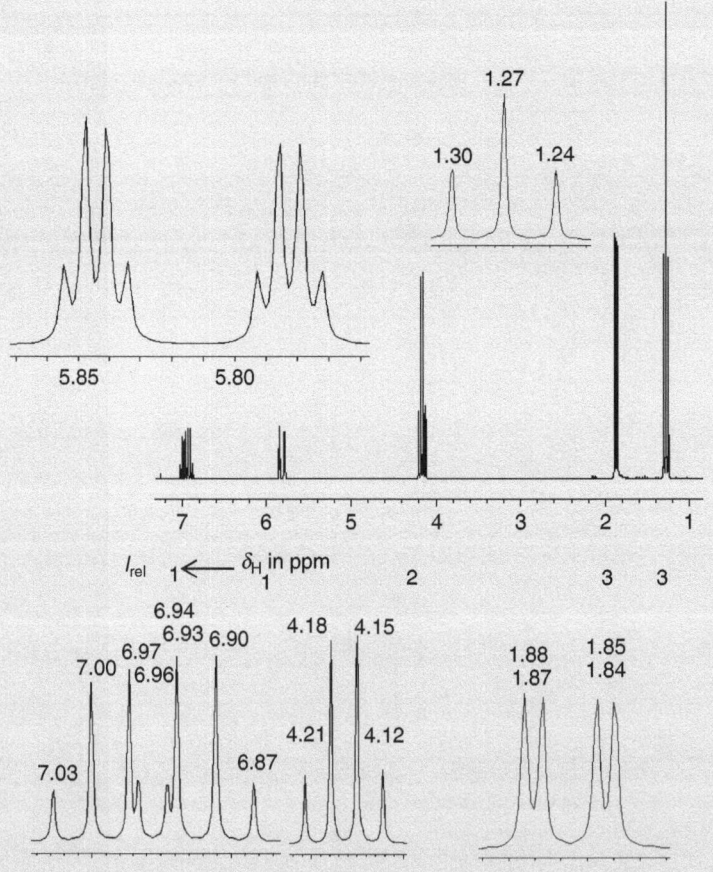

Fig. 4.36 250-MHz ^1H NMR spectrum of an unknown compound

(continued)

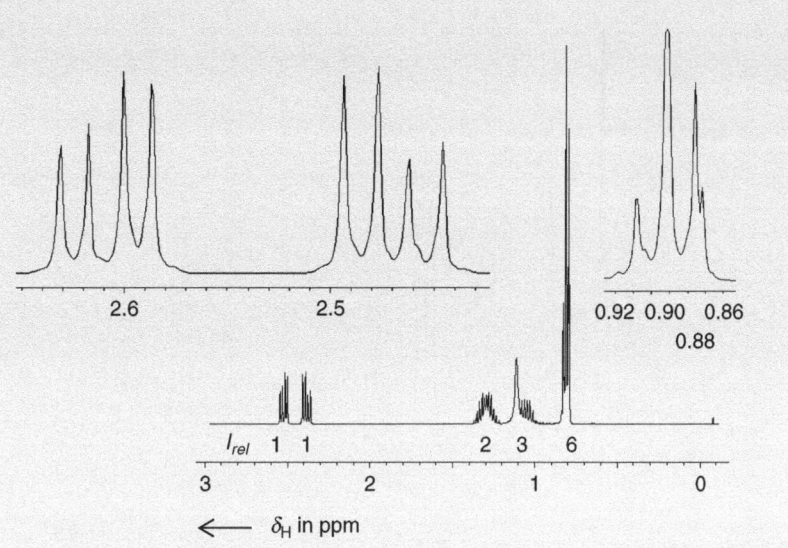

Fig. 4.37 400-MHz ^{1}H NMR spectrum of a compound with the sum formula $C_5H_{11}N$

Develop the structure of the compound. Assign the IR bands and the chemical shifts. Determine the coupling constants and designate the spin systems. Evaluate λ_{max} for the structure.

18. Figure 4.37 presents the 400-MHz ^{1}H NMR spectrum of a compound with the sum formula $C_5H_{11}N$. The structure is to be determined. Assign the chemical shifts and coupling constants.

19. An aromatic compound shows in the upper range of the mass spectrum the following peaks; m/z in amu (I_{rel} in %): 162 (100), 163 (6.8), 164 (63), 165 (4.5), 166 (11), 167 (0.5). The 250-MHz ^{1}H NMR spectrum measured in D_2O is shown in Fig. 4.38. One exchangeable proton is present. Determine the structure of the compound. Evaluate the chemical shifts. Is an unambiguous assignation of the chemical shift values possible? Determine and assign the coupling constants.

20. The structure of an unknown compound with the sum formula C_6H_6NClO is to be determined. The 400-MHz ^{1}H NMR spectrum is presented in Fig. 4.39. The relative intensity ratio is (beginning from the left) 1:1:1:3.

21. A 400-MHz ^{1}H NMR spectrum of a compound with the sum formula C_6H_7N is shown in Fig. 4.40. The structure is to be determined. Calculate the chemical shifts; assign the chemical shifts and coupling constants.

(continued)

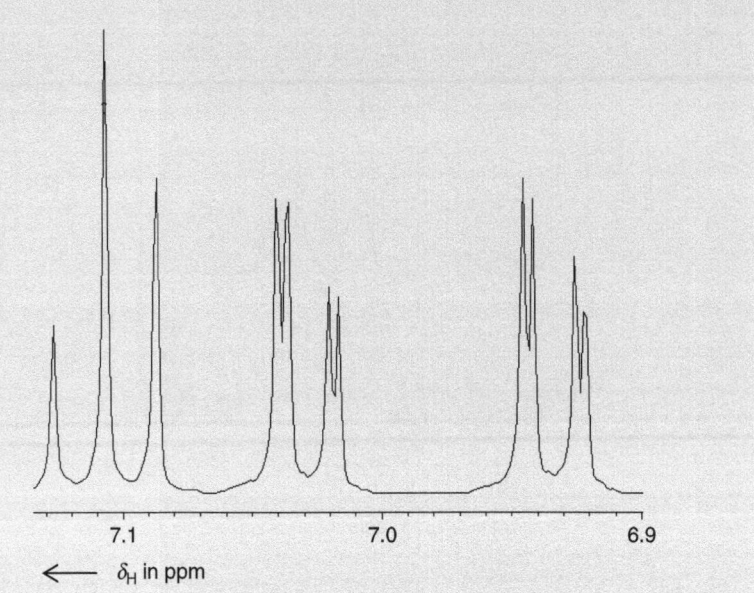

Fig. 4.38 250-MHz ^{1}H NMR spectrum of an aromatic compound

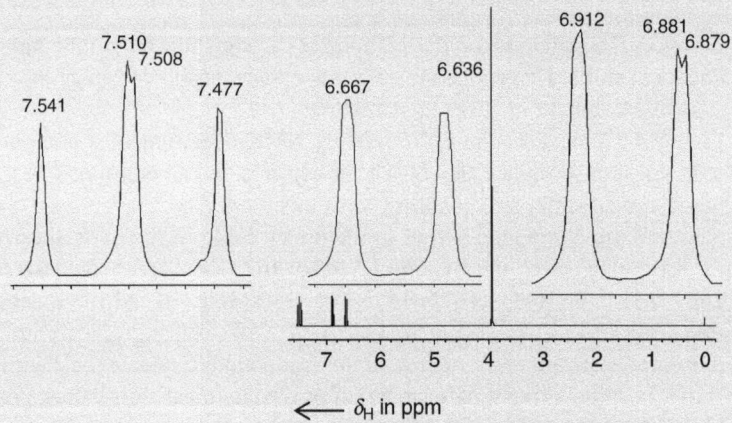

Fig. 4.39 250-MHz ^{1}H NMR spectrum of an unknown compound with the sum formula C_6H_6NClO

22. A compound with the sum formula $C_7H_4O_6N_2$ provides the 400-MHz ^{1}H NMR spectrum presented in Fig. 4.41. The IR spectrum is characterized by a very intense absorption band in the range 3,300–2,500 cm^{-1}. What is the structure of this compound? Calculate and assign the chemical shifts and coupling constants.

(continued)

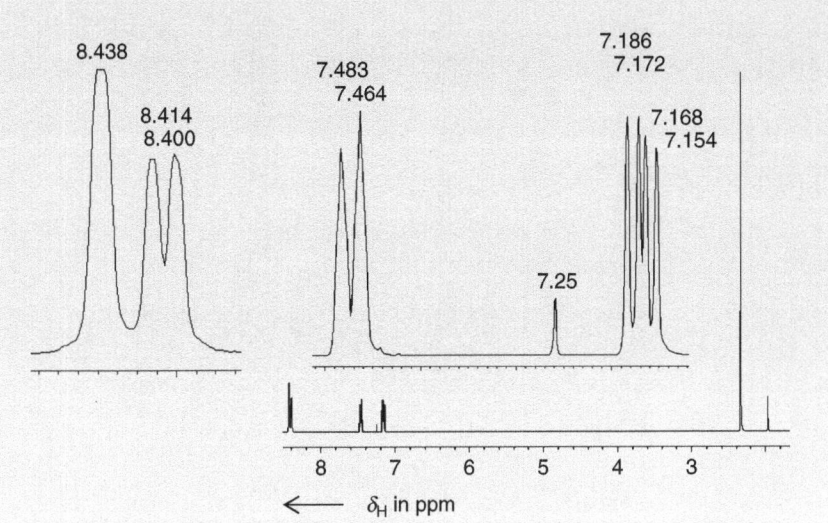

Fig. 4.40 400-MHz ^{1}H NMR spectrum of an unknown compound with the sum formula C_6H_7N

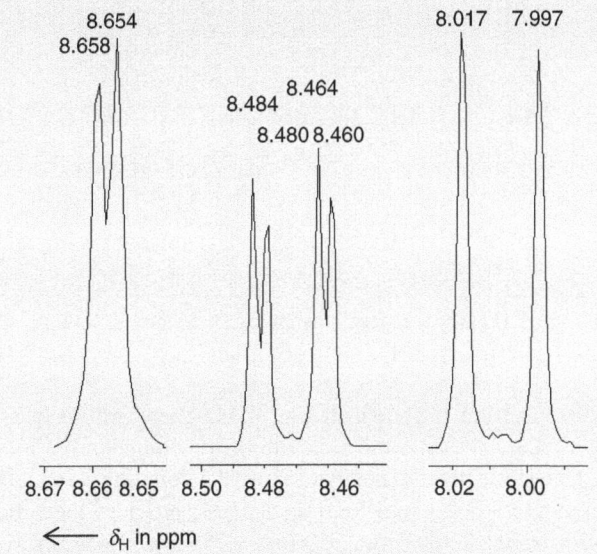

Fig. 4.41 400-MHz ^{1}H NMR spectrum of an unknown compound with the sum formula $C_7H_4O_6N_2$

23. The sum formula of a compound obtained by the mass spectrum is $C_9H_7NO_3$. Intense and very intense, respectively absorption bands are observed in the IR spectrum at 1,714, 1,522, and 1,350 cm^{-1}. The 400-MHz

(continued)

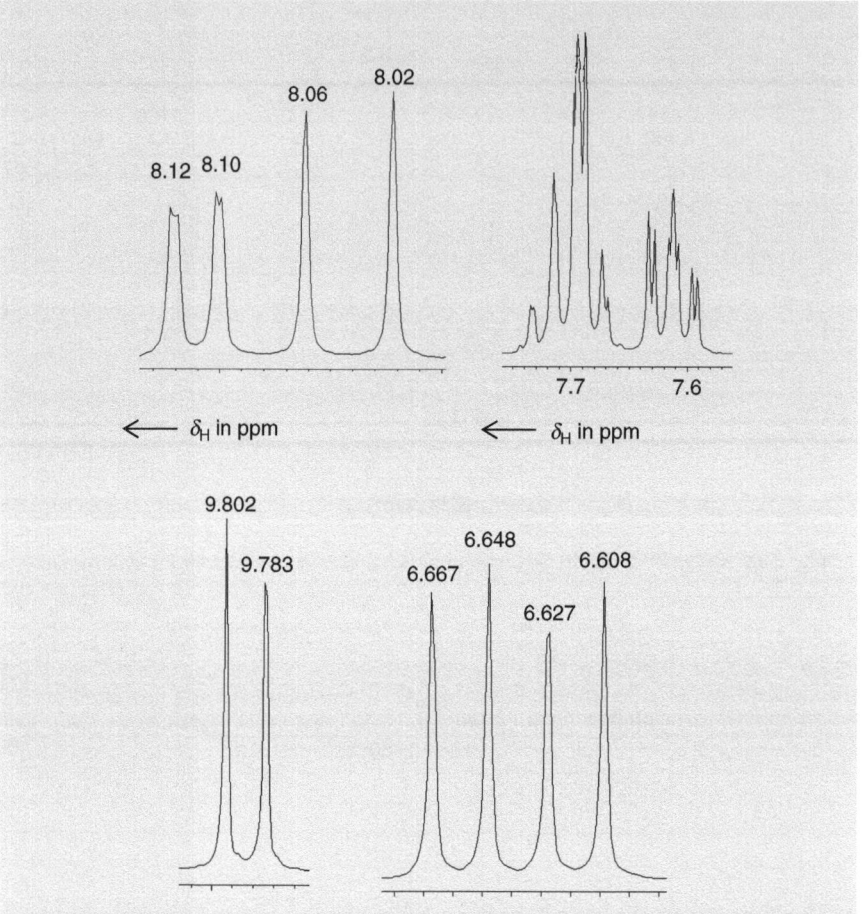

Fig. 4.42 400-MHz ^{1}H NMR spectrum of an unknown compound with the sum formula $C_9H_7O_3N$

^{1}H NMR spectrum is shown in Fig. 4.42. There are no exchangeable protons by D_2O. What is the structure of the compound? Calculate and assign the chemical shifts and determine the coupling constants.

24. A compound with the sum formula $C_4H_6O_3$ provides the following IR absorption bands: 2,969 (m), 2,933 (m), 1,789 (vs), 1,484 (m), 1,389 (m), 1,364 (m), 1,340 (m), 1,186 (s). The 400-MHz ^{1}H NMR spectrum is presented in Fig. 4.43. Determine the structure and justify your findings.

(continued)

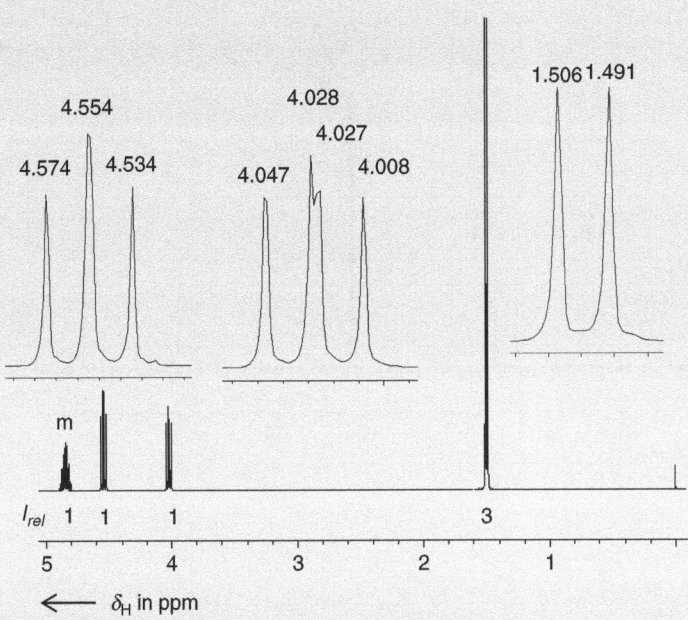

Fig. 4.43 400-MHz ^{1}H NMR spectrum of an unknown compound with the sum formula $C_4H_6O_3$

25. Two doublets are observed in the ^{1}H NMR spectrum of a compound with the sum formula C_3H_5FO at $\delta_1 = 4.75$ ppm ($J_1 = 48$ Hz), $\delta_2 = 2.7$ ppm ($J_2 = 4.3$ Hz). What is the structure of the compound? Outline the line pattern for the ^{19}F NMR spectrum.

26. Let us return to Challenge 2.12. The following structural moieties were determined by mass and IR spectroscopy: Aromatic compound with the substituents NO_2, NH_2, and CH_3. The IR spectrum gives hints of a 1,2,4-trisubstituted aromatic structure by the respective out-of-plane and γ_{CH} vibrations. Further information is obtained by the UV spectrum (see Challenge 3.11.18). Note that the information provided by the mass spectrum is not in line with the *ortho* position of the CH_3 and NO_2 groups because of the lack of an *ortho* effect, i.e., elimination of an OH radical. In continuation of the structural determination the ^{13}C NMR (Fig. 4.44a) and the 400-MHz ^{1}H NMR spectra (Fig. 4.44b) are measured. What are the locations of the protons at the aromatic moiety (substitution type)? Justify your decision. Is it possible to assign an unequivocal structure based on the NMR spectra?

(continued)

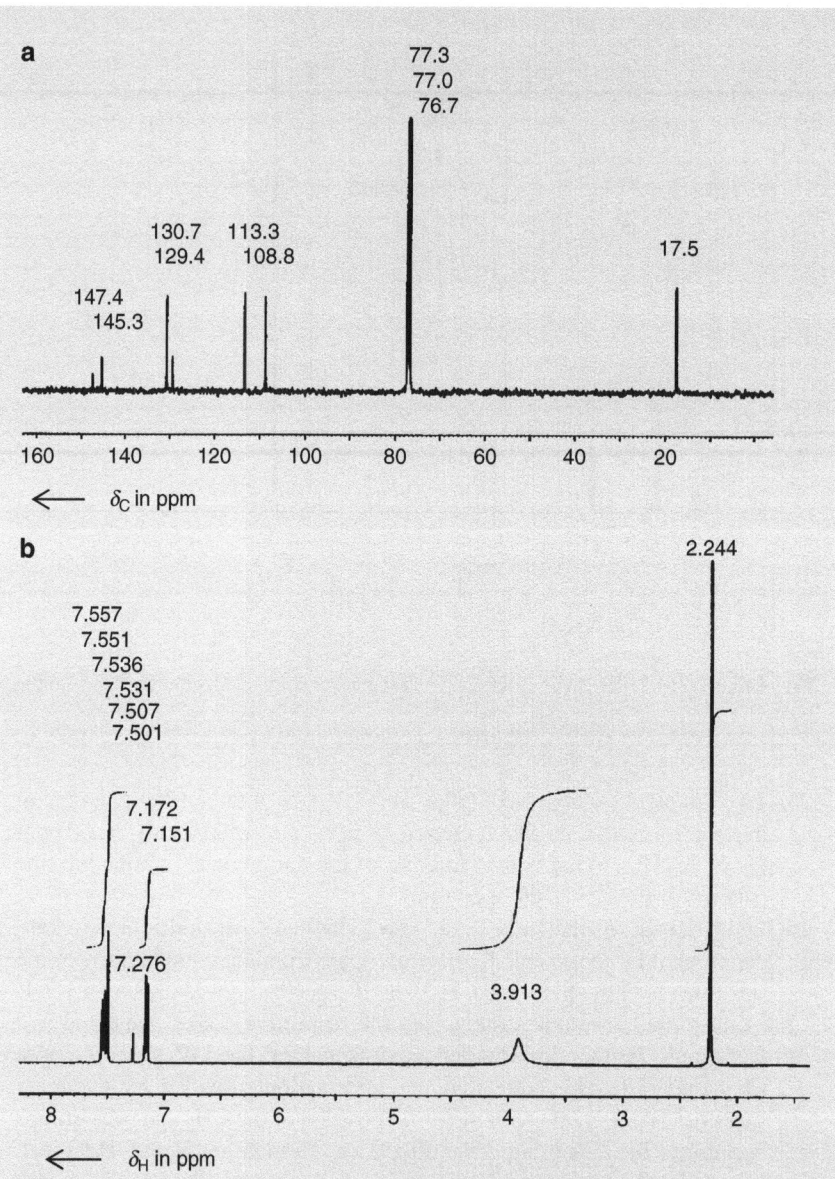

Fig. 4.44 (a) 100-MHz ^{13} C NMR spectrum of an aromatic compound with the substituents NO_2, NH_2, and CH_3 (continuation of Challenge 2.12). (b) 400-MHz ^{1} H NMR spectrum of an aromatic compound with the substituents NO_2, NH_2, and CH_3 (continuation of Challenge 2.12)

4.4 Experimental Techniques

4.4.1 General Remarks

As the challenges in the previous section reveal, the structures of a wide variety of simple to moderately complex organic molecules can be elucidated by the so-called *one*-dimensional 1D NMR spectra using the complementary information obtained by mass, IR, and UV–vis spectroscopy. However, structural determination of more complex molecules poses difficulties using only 1D NMR spectra even in the case of flanking information. Some causes for this may be: Overlaying of signals; coupling constants cannot be obtained because of multiple coupling; signal assignment is prevented because of nearly equal chemical shifts; randomly degenerated NMR signals and others.

Nowadays, many experimental techniques are available to circumvent these difficulties. However, in the scope of this book, only such techniques can be considered which are incorporated into the challenges given in Chap. 5. For the theoretical background and further experimental techniques see respective NMR books.

Some aims of the experimental techniques are:

– Simplification of complex spectra in order to assign superimposed signals, eliminate undesired couplings, or to produce first-order spectra
– Recognition of diastereotopic atoms
– Recognition of coupling spin systems
– Determination of the CH_x structure
– Assignment of chemical shifts in complex molecules, especially if they have a high similarity.

4.4.2 Increasing of Magnetic Field Strength

As already explained above, sometimes second-order ^1H NMR spectra can be transferred into first-order spectra by increasing the magnetic field strength. According to Eq. 4.10 Δv depends on the applied magnetic field strength v_0, whereas the coupling constant J is not dependent on the applied field. Thus, the ratio according to Eq. 4.17 can become greater than seven and the criterion for a first-order spectrum is fulfilled. Note that a spin system with magnetically non-equivalent protons can never be transferred in a simple first-order spectrum.

4.4.3   Lanthanide Shift Reagents (LSR)

Although lanthanide ions are paramagnetic ions they do not give rise to remarkable broadening of the linewidth of the NMR signals. However, the signals of molecules with coordination-enabled substituents are shifted more or less strongly after addition of β-diketo complex compounds of lanthanides because of the generation of coordinative bonding between the molecule and the lanthanide complex compound. The direction of the shift is determined by the lanthanide ion. Thus, Eu(III) complexes give rise to a deep field shift whereas Pr(III) complexes shift the signals to the higher field.

An example of a lanthanide shift reagent (**LSR**) is $Eu(dpm)_3$ with β-diketone $(CH_3)_2C–C(=O)–CH_2–C(=O)–C(CH_3)_2$ as the ligand. The magnitude of the shift Δv depends on the concentration of the LSR and the spacing of the respective proton to the lanthanide central ion r according to the relation $\Delta v \sim 1/r^3$.

All CH_x groups in n-hexanol can be recognized by addition of the LSR $Eu(dpm)_3$ as is shown in Fig. 4.45.

Nowadays, experiments with LSR are no longer common because of the use of high-field magnetic strength but they are still important for the recognition of *enantiomers* if chiral lanthanide shift reagents are used. Thus, for example, this technique is the best method to determine the optical purity of chiral molecules.

4.4.4   Recognition of OH Groups

In general, ^1H NMR signals can be recognized by their relatively large linewidth. Furthermore, the proton of the OH group can be exchanged by deuterium after addition of D_2O with the consequence that the respective ^1H NMR signal disappears. Note that this test is not specific for OH groups because also NH, NH_2 and CH-acid protons are exchanged by D_2O.

But the D_2O technique fails if the OH signal is superimposed by the signals of many CH_x groups which can be the case, for example, in steroids. Thus, the removal of *one* proton with D_2O will not significantly alter the intensity which is mainly determined by the very large number of CH_x groups in this range. However, reliable knowledge of the number of OH groups is an essential condition for the structural determination of steroids and related compounds.

A reliable determination of the numbers of OH groups is the transformation into urethane derivatives by the reaction with added trichloracetyl isocyanate, $CCl_3–C(=O)–N=C=O$ (**TAI**) according to Eq. 4.29:

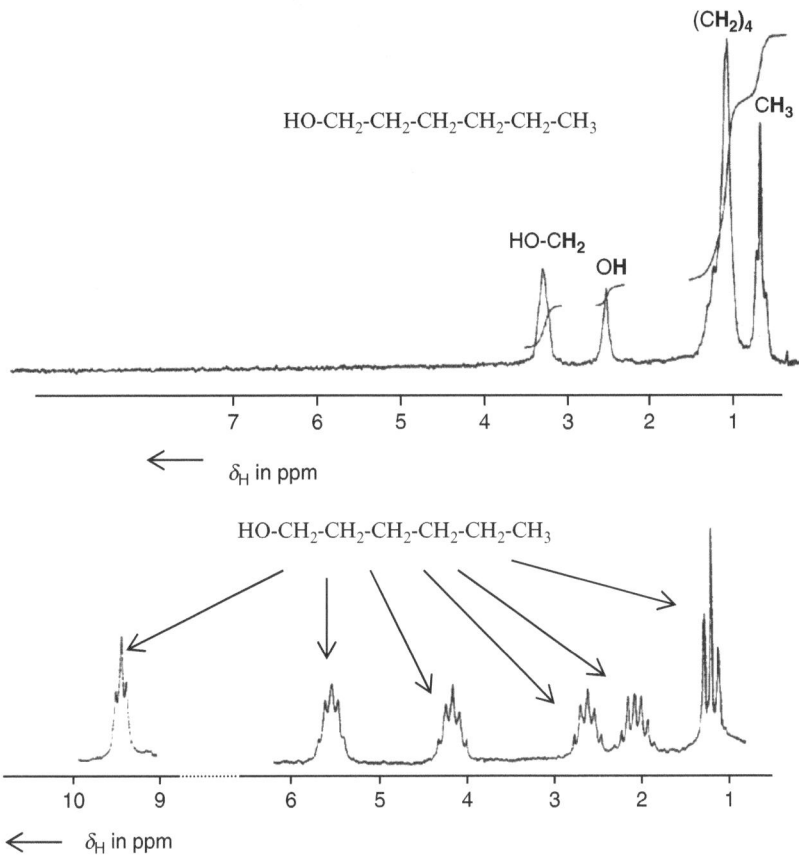

Fig. 4.45 100-MHz ^{1}H NMR spectrum of *n*-hexanol *without* LSR (**a**) and after addition of the LSR Eu(dpm)$_3$ (**b**)

$$R\text{-}OH + O=C=N\text{-}C(=O)\text{-}Cl_3 \longrightarrow O=C\overset{\displaystyle NH\text{-}CCl_3}{\underset{\displaystyle OR \quad 4.44}{\Big<}} \qquad (4.29)$$

The **NH** group of the urethane (Structure 4.44) gives rise to sharp ^{1}H NMR signals in the range 8–9 ppm.

The number of **NH** signals in the ^{1}H NMR spectrum corresponds to the number of the various bonded OH groups in the molecule.

Furthermore, the alcohol type can be recognized by the chemical shift of the **NH** signal because $\delta_{\text{tertiary}} < \delta_{\text{secondary}}$ is valid.

4.4.5   Spin Decoupling

4.4.5.1   Selective Decoupling

The application of a second (strong) field with the frequency v_2 in a spin-coupled system removes its coupling effect to neighboring protons to which it has been coupled. The second field gives rise to a quick exchange of the spin orientation of this proton. Thus, a zero-field is transferred to the neighboring proton with the consequence that a multiplet is transformed to a singlet. This technique can be used for recognizing the connectivity of protons through bonds and assignment of proton signals. Furthermore, undesired couplings which make it difficult to determine coupling constants or overlapping peaks can be simplified by removing selective couplings.

Let us return to the spin coupling of chloroacetaldehyde (Structure 4.11) whose line pattern is shown in Fig. 4.8. The doublet of the CH_2 group (H-2) is changed to a singlet by irradiation of a second field with the resonance frequency of the aldehyde proton, and the triplet of the CHO group (H-1) becomes a singlet upon application of the resonance field of the CH_2 group.

4.4.5.2   Broadband Decoupling {^{1}H}

Because of the appreciable $^{13}C/^1H$ couplings the ^{13}C NMR spectra would be very complex and, thus, they are difficult to interpret. Therefore, *all* $^{13}C/^1H$ couplings are removed in routine ^{13}C NMR spectroscopy by broadband decoupling which is, generally, denoted by the symbol {^{1}H}. Thus, the pattern of ^{13}C NMR spectra becomes very simple because only singlets are observed.

The number of singlets in the spin-decoupled ^{13}C NMR spectrum corresponds to the number of chemically nonequivalent (nonisochronous) carbon atoms, provided that no randomly isochronous C atoms are present.

The simplification of ^{13}C NMR spectra leads to the fact that the whole intensity of each ^{13}C NMR signal is concentrated in one signal. Furthermore, the intensity of the ^{13}C signals is amplified up to 200% by the NOE. Because the NOE amplification is caused by the attached protons the signal of the proton-free quaternary carbon atom cannot be amplified and leads to signals with small intensity. Thus, in general, quaternary ^{13}C atoms can be recognized by their small intensity. Note that the simplification of the ^{13}C NMR spectra by spin decoupling removes information important for structure determination like, for example, information about CH_x moieties which must be gained by other techniques.

4.4.5.3 Off-Resonance Decoupling

Applying a field which is about 100–500 Hz out of the range of the proton resonances eliminates all coupling up to the couplings of the protons bonded to a carbon atom.

In an off-resonance decoupled spectrum, only ^{1}H atoms bonded to a carbon atom will split its ^{13}C signal. The CH_x groups provide quasi-first-order spectra which are important for the recognition of the CH_x structure.

The multiplets of an off-resonance decoupled ^{13}C NMR spectrum are:

$CH_3 \Rightarrow$ quartet	$CH_2 \Rightarrow$ triplet	$CH \Rightarrow$ doublet	$C^{quaternary} \Rightarrow$ singlet

Note that nonequivalent protons of the CH_2 group are observed as the X part of an ABX spin system instead of the expected triplet.

Figure 4.46 shows the line pattern of the off-resonance decoupled ^{13}C NMR spectrum of crotonic acid (Structure 4.45) with irradiation at the deep field. Thus, the CH_x moieties are recognized which facilitate the assignment of the signals. The coupling constants increase with spacing of the respective signal to the irradiation field v_2. The coupling constants would be smaller for the quartet with irradiation at the high field side. In general, the multiplets obtained in the off-resonance ^{13}C NMR spectrum are marked in the routine spectrum by the symbols s, d, t, and q for singlet, doublet, triplet, and quartet, respectively.

$$H_3C \diagdown \overset{\displaystyle H}{\diagup} \diagdown OH$$

4.45

However, this experimental technique for the determination of the CH_x moieties fails in the case of ^{13}C NMR spectra that possess many signals in close vicinity because of the overlapping of multiplets. Nowadays, the CH_x moieties can be unambiguously determined by DEPT so that the off-resonance technique is only of historical interest.

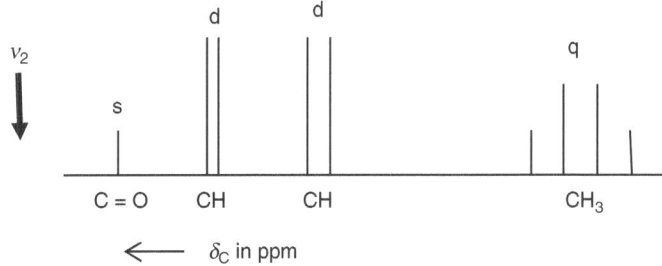

Fig. 4.46 Line pattern of the off-resonance decoupled ^{13}C NMR spectrum of crotonic acid (Structure 4.45)

4.4.6   Distortionless Enhancement by Polarization Transfer (DEPT)

DEPT (Distortionless Enhancement by Polarization Transfer) belongs to the techniques that use complex pulse series. It is a very useful method for determining the presence of primary, secondary, and tertiary carbon atoms.

The DEPT experiment differentiates between CH, CH_2, and CH_3 groups by variation of the selection angle parameter:

- The DEPT45 spectrum (45° angle) gives, regardless of number, positive signals for *all* **CH$_x$** groups.
- The DEPT90 spectrum (90° angle) shows signals only for **CH** groups, the others being suppressed.
- The DEPT135 spectrum (135° angle) gives positive signals for **CH** and **CH$_3$** and negative signals for **CH$_2$** groups.
- The signals of quaternary carbon atoms are suppressed in all three experiments. However, $C^{quaternary}$ atoms are shown in phase with the CH_2 groups in the *PENDANT* representation.

Note that the time τ of the complex impulses must be equal to the coupling constant $^1J_{C,H}$. The standard time $\tau = 3.6$ ms corresponds to a C,H coupling constant of 139 Hz. However, this requirement is not always fulfilled because the coupling constants are somewhat different with the consequence that weak signals of the CH_3 groups can be observed in the DEPT90 spectrum. This fact must be considered to avoid incorrect conclusions.

4.4.7   NOE Difference Spectroscopy

Applying an additional strong field enhances the intensity of neighbored protons by the NOE if the spacing between the protons is ≤ 4 Å (0.4 nm). Because this enhancement is very weak the *difference* between the spectra *without* and *with* the additional field is measured. This subtraction leaves only the enhanced absorption. Thus, only such protons give rise to signals whose distance to the proton in resonance is up to about 4 Å. This procedure can be applied, for example, to distinguish E/Z (*trans/cis*) isomers.

Figure 4.47 shows the range of the ^{1}H NMR spectrum which is important for the determination of the E isomer of the moiety $R(CH_3)C=CHCH_2OH$ (Structure 4.46). The signals at $\delta = 5.85$ and $\delta = 5.16$ ppm (see Fig. 4.47a) are assigned to the methine (C–H) and methylene (CH_2) protons, respectively. Resonance with methyl protons gives rise to enhancement of the methine signal at $\delta = 5.85$ ppm (see Fig. 4.47b) which means that the methyl and methine protons must be located on the same side of the double bond according to Structure 4.46.

Fig. 4.47 Section of the ^{1}H
NMR spectrum of a
compound with the
structural moiety
R–(CH$_3$)C=CHCH$_2$OH (**a**)
and the 1D NOE difference
spectrum which is obtained
upon resonance with the
methyl group (**b**)

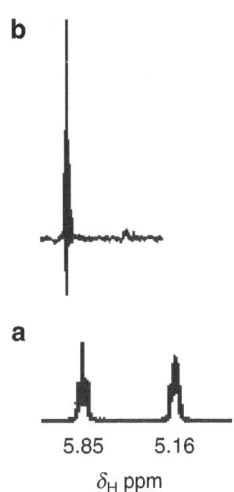

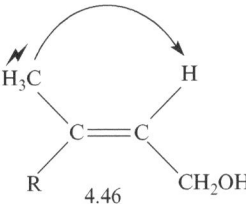

As explained above, *E/Z* isomers can be distinguished by the vicinal coupling
constant. However, if only *one* or *no* proton is attached to the ethylene moiety,
this procedure cannot be applied for the determination of the geometrical
isomers. Thus, the NOE procedure is a powerful tool for distinguishing *E/Z*
(*trans/cis*) isomers. However, nowadays, 1D NOE difference spectroscopy has
been substantially replaced by **2D NOESY** spectroscopy which is explained in
the next section.

Challenge 4.14
For solutions to Challenge 4.14, see *extras.springer.com*.

1. The aromatic protons of a 1,2,3-trisubstituted aromatic compound provide
   an AB$_2$ spin system in a 200-MHz spectrometer with the chemical shifts
   $\delta_1 = 8.19$ ppm and $\delta_2 = 8.01$ ppm. The coupling constant is $J = 8$ Hz.
   Check whether the spin system can be transferred into a first-order AX$_2$
   spin system in a 600-MHz spectrometer.
2. The ratio of the enantiomers of 1-phenyl 1-methyl ethylamine (Structure
   4.47) has to be determined. Propose an appropriate method.

(continued)

3. Which spin systems of 4-vinyl pyridine can be transferred into a first-order spin system by enhancement of the magnetic field strength? Justify your decision.

4. Explain how the chemical shift values of the methine groups ($\delta_1 = 6.09$ ppm, $\delta_2 = 6.47$ ppm) as well as that of the methyl groups ($\delta_1 = 1.28$ ppm, $\delta_2 = 1.87$ ppm) for Structure 4.48 can be unambiguously assigned.

5. The reaction between $1\alpha,2\alpha$-epoxy-25-hydroxy-provitamin D_3 (Structure 4.49) with KHF_2 at 180°C gives a product for which Structure 4.50 is suggested. The spatial structure of the steroidal ring A is shown in Fig. 4.51. The suggested structure has to be justified by the following spectroscopic information:

(continued)

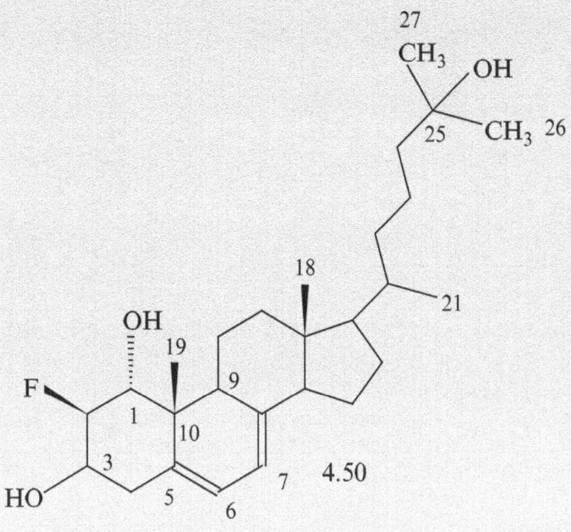

The sum formula $C_{27}H_{43}FO_3$ is justified by h-MS. The presence of an intact 5,7-diene moiety can be justified by the vibrationally structured UV spectrum with λ_{max} 282 nm. Further structural information can be obtained by the ^{1}H NMR spectra which are presented in Fig. 4.48a–e.

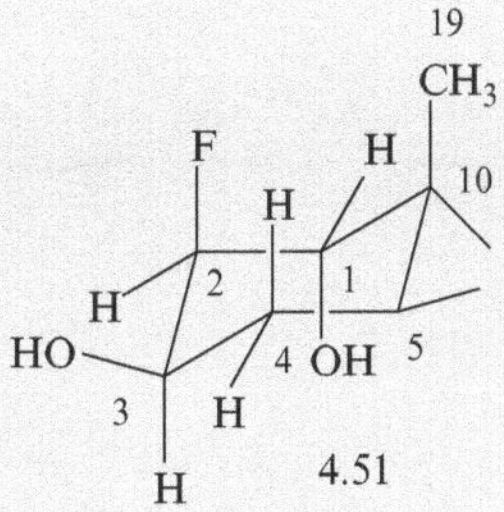

6. An amino acid is transferred by a condensation reaction into a product which has the sum formula C_5H_9NO. The ^{13}C NMR spectrum with the multiplets of the off-resonance experiment is presented in Fig. 4.49. What is the structure of the compound?

7. The compound which was isolated by a HPLC separation from a hydrolysis reaction of a peptide provides a molecular peak in the mass spectrum at $m/z = 117$ amu. The ^{13}C NMR spectrum with the multiplets of the off-resonance experiment is shown in Fig. 4.50. What is the structure of the compound?

8. An unknown compound with the sum formula $C_8H_{10}O_2$ shows a very intense IR absorption band at $3{,}400$ cm^{-1}. The ^{13}C/DEPT NMR spectra are presented in Fig. 4.51. What is the structure of the compound?

(continued)

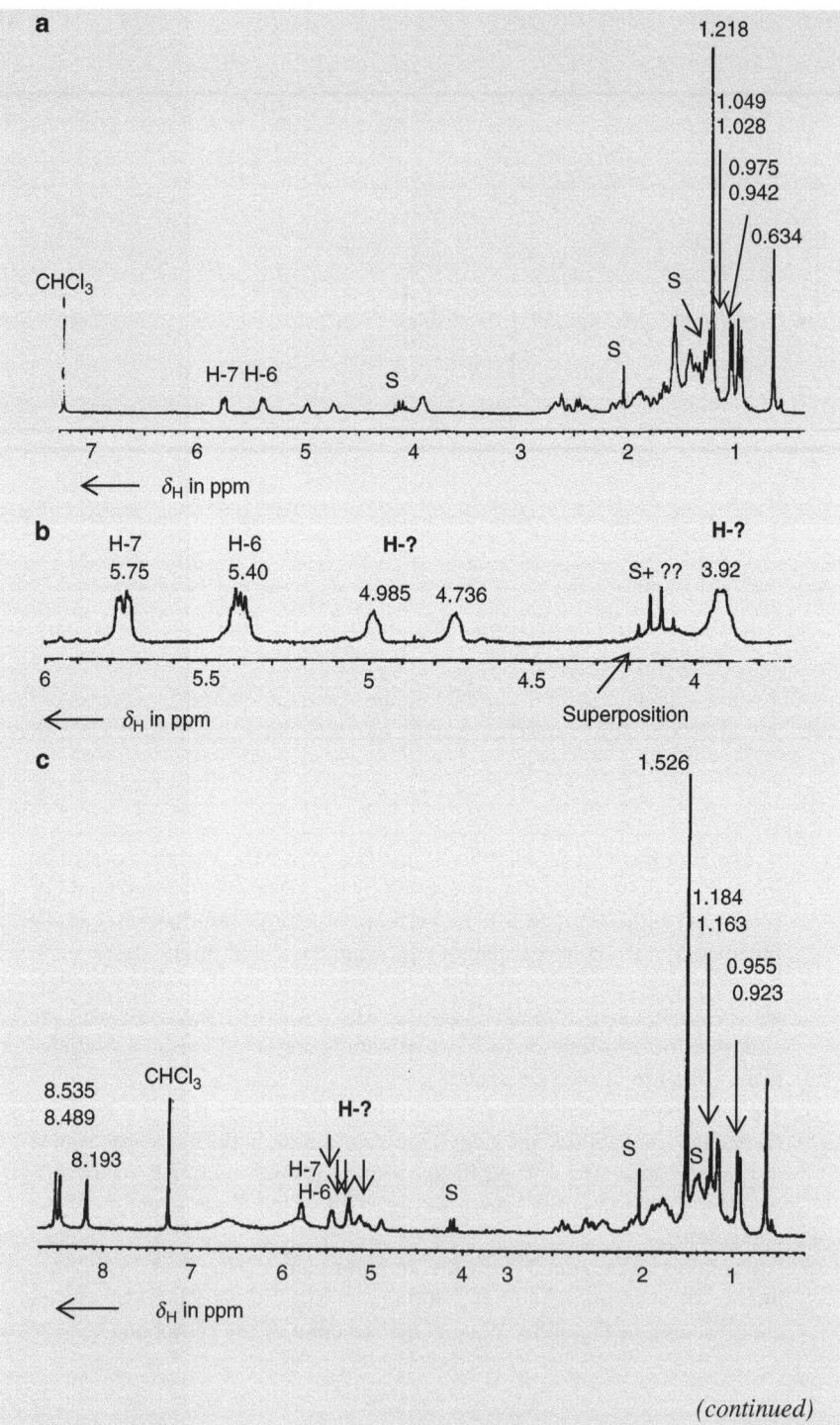

(continued)

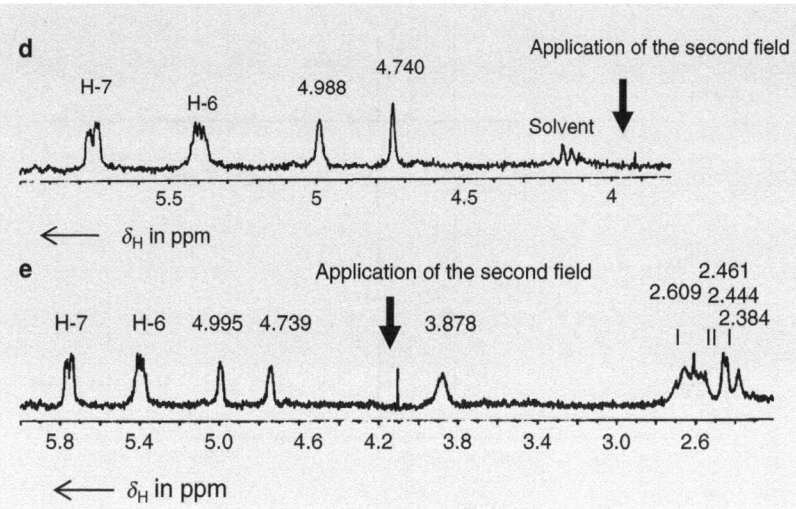

Fig. 4.48 (a) 200-MHz ^{1}H NMR spectrum of the steroidal reaction product (suggested Structure 4.50); S - solvent. (b) Section of the 200-MHz ^{1}H NMR spectrum of the steroidal reaction product (suggested Structure 4.50); S - solvent. (c) 200-MHz ^{1}H NMR spectrum of the steroidal reaction product (suggested Structure 4.50) after addition of TAI; S - solvent. (d) Selective spin decoupling of the steroidal reaction product (suggested Structure 4.50) with irradiation at $\delta = 3.91$ ppm. (e) Selective spin decoupling of the steroidal reaction product (suggested Structure 4.50) with irradiation at $\delta = 4.12$ ppm

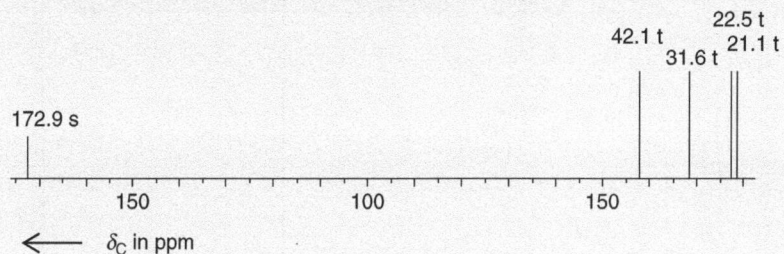

Fig. 4.49 Line pattern of the ^{13}C NMR spectrum and the multiplets of the off-resonance experiment of the condensation product of an amino acid with the sum formula C_5H_9NO

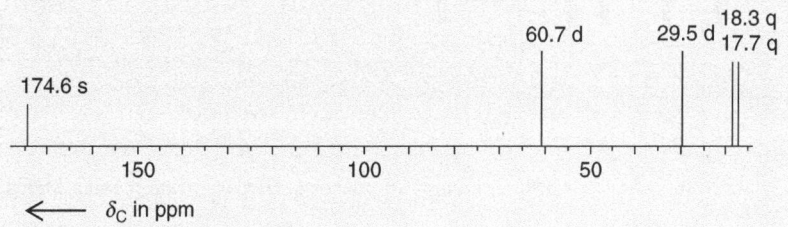

Fig. 4.50 Line pattern of the ^{13}C NMR spectrum and the multiplets of the off-resonance experiment of the chromatographically separated hydrolysis product from a peptide with a molecular peak at $m/z = 117$ amu

(continued)

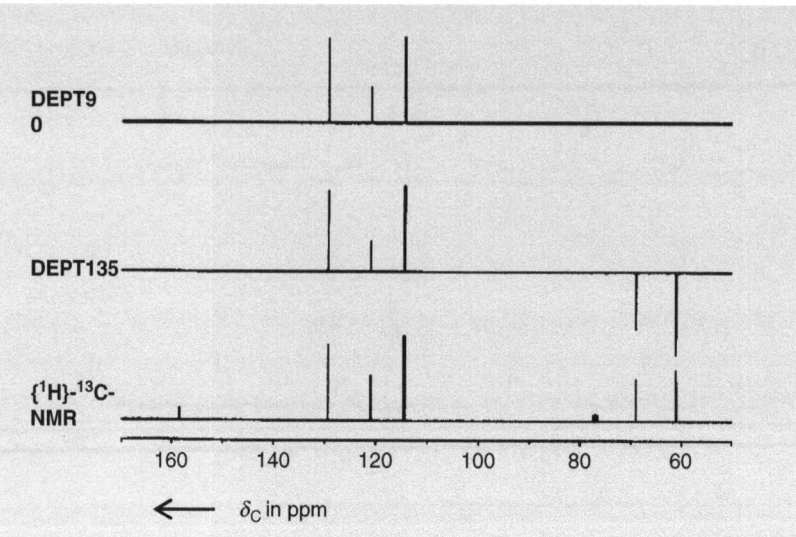

Fig. 4.51 ^{13}C/DEPT NMR spectra of a compound with the sum formula $C_8H_{10}O_2$

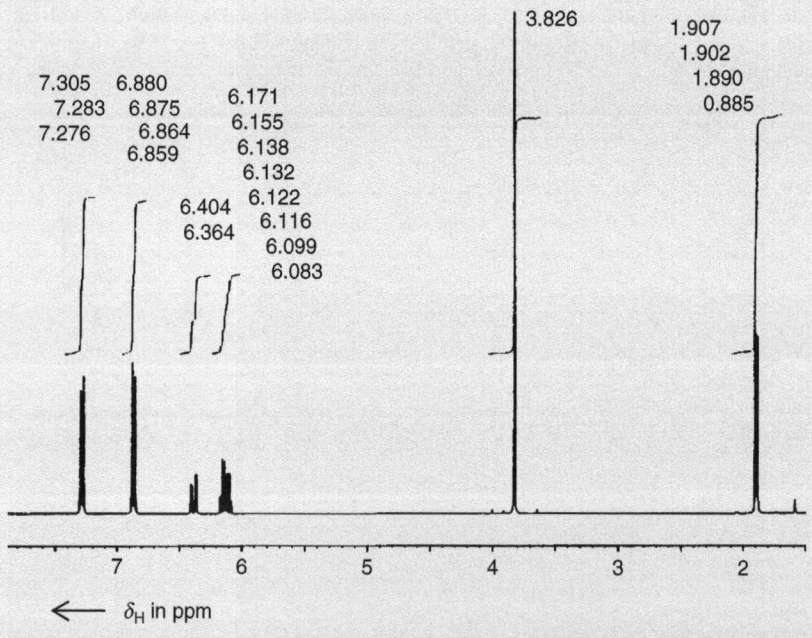

Fig. 4.52 400-MHz ^{1}H NMR spectrum of an unknown compound with the sum formula $C_{10}H_{13}O$

(continued)

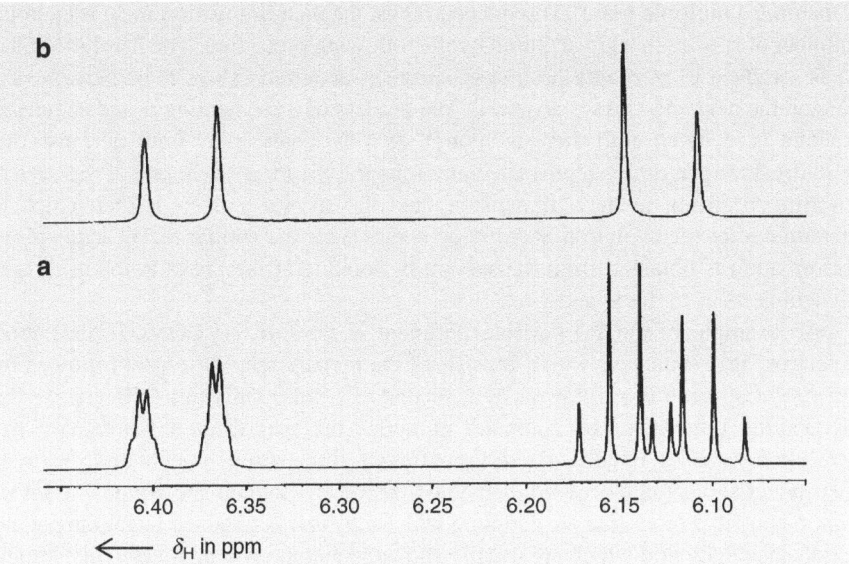

Fig. 4.53 Section of the 400-MHz ^{1}H NMR spectrum (**a**); double resonance ^{1}H NMR spectrum with irradiation at $\delta = 1.9$ ppm (**b**)

9. The sum formula $C_{10}H_{12}O$ of an unknown compound was obtained by mass spectroscopy. The conventional ^{1}H NMR spectrum and its decoupled counterpart resulting from irradiation at $\delta = 1.9$ ppm are presented in Figs. 4.52 and 4.53. What is the structure of the compound? Designate the spin systems and determine the coupling constants as far as is possible.

4.4.8   Correlation Spectroscopy

4.4.8.1   General Remarks

Note that the theoretical basis and experimental details of 2D NMR spectroscopy are far beyond the scope of this book; therefore further reading will be necessary to fully understand this topic. The following remarks are limited to the fundamental difference between 1D and 2D NMR spectra and to learn enough to interpret the various 2D spectra presented in challenges.

Whereas a conventional 1D NMR experiment only consists of the *preparation* period (perturbation of the spin system with multiple pulses) and the *acquisition* period t_2, two further periods are added in the 2D experiment: the *evolution time* t_1 and the *mixing* event in which the information from one part of the spin system is relayed to the other parts. It is the timing, frequencies, and intensities of these pulses that distinguish different NMR experiments from one another. 1D NMR deals with multiple pulses in a single dimension, whereas the basic 2D NMR involves *repeating* a multiple pulse 1D sequence. Thus, the data acquisition includes a large number of spectra that are acquired by the following procedure: The first time value t_1 is set close to zero and the first spectrum is acquired. Then, t_1 is increased by Δt and the next spectrum is acquired. The process of incrementing t_1 and acquiring spectra is repeated until there is enough data for analysis by Fourier transform. Usually, the spectrum is represented as a topographic map where one of the axes is F_1 (the spectrum in the t_1 dimension) and the second axis F_2 is that which is acquired after the evolution and mixing stages which is similar to 1D acquisition. Information is obtained from the spectra by looking at the peaks in the grids and matching them to the x- and y-axes.

An example of the 2D NMR experiment is the (^{1}H,^{1}H) COSY (*CO*rrelation *S*pectroscop*Y*) sequence, which consists of the preparation (90° pulse) followed by an evolution time (t_1) followed by a second 90° pulse (mixing) followed by the acquisition time (t_2). The computer compiles the spectra as a function of the evolution time t_1. Finally, the Fourier transform is used to convert the time-dependent signals into a two-dimensional frequency-domain spectrum. The information on the protons that are coupling with each other is obtained by looking at the peaks inside the grid which are usually shown in a *contour-type format*, like height intervals on a map.

4.4.8.2 Homonuclear 2D Spectra

There are three 2D *homonuclear* experiments which are widely used in structural determination: 2D COSY, 2D TOCSY, and 2D NOESY. Both axes correspond to protons in the NMR spectra. The signals on the diagonal across the spectrum correspond to the usual 1D spectrum. However, the interesting information from the 2D spectra is obtained by those peaks in which the magnetization evolves with one frequency during t_1 and a different frequency during t_2. These peaks are called *off-diagonal* or *cross peaks*. These peaks indicate such nuclei which have exchanged magnetization during the evolution time and, thus, coupling interactions. Note that the spectra are symmetrical with respect to the diagonal, i. e., there are always two mirror-like cross peaks which is helpful to find the respective correspondence.

2D COSY

In the COSY experiment, magnetization is transferred by *scalar* coupling. In general, protons that are more than three chemical bonds apart give no or weak *cross peaks* because the long-range coupling constants are close to zero. Therefore, mostly signals of protons that are two or three bonds apart are visible in the COSY spectrum. Long-range couplings can be observed as cross peaks if protons are connected through π-electrons. Let us consider an example shown in Fig. 4.54, the COSY for Structure 4.52.

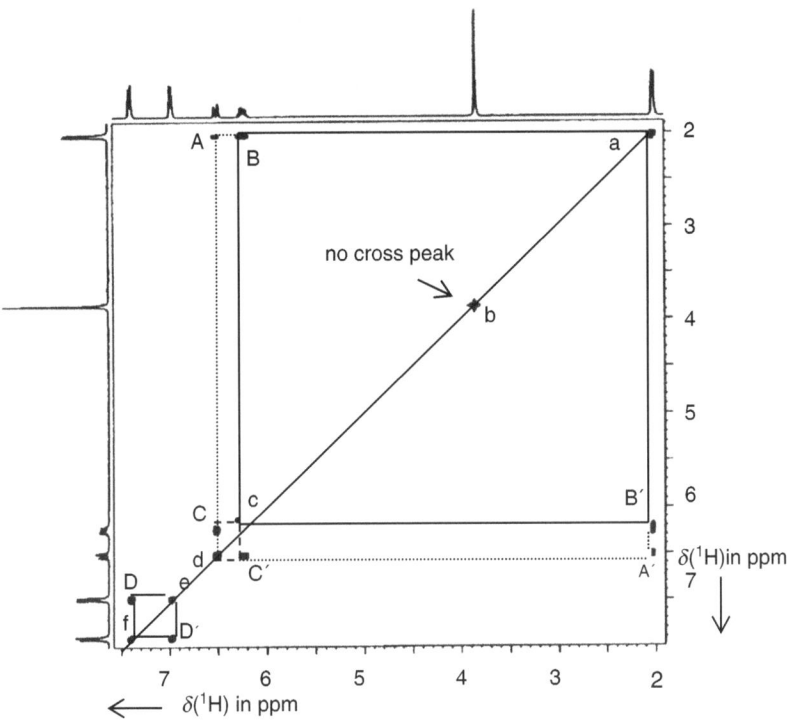

Fig. 4.54 (^{1}H,^{1}H) COSY spectrum of Structure 4.52 and the outlined correlations

$$\underset{\text{4.52}}{CH_3O-\underset{2\quad3}{\bigcirc}-\underset{4}{CH}=\underset{5}{CH}-\underset{6}{CH_3}}$$

Looking at Fig. 4.54, one can recognize six diagonal peaks (a–f) which correspond to the six chemically nonequivalent protons (1–6 in Structure 4.52). Furthermore, one can recognize cross peaks marked by capital letters that belong to four
couplings according to the four square patterns: Square BaB′c corresponds to the vic
coupling of the methyl protons (H-6) with the methine proton (H-5), square AaA′d
indicates the long-range coupling through the π-electrons between the protons H-6
with the methine proton H-4, square CcC′d is caused by the coupling of the olefinic
methine protons, and the coupling between the aromatic protons (AA′XX′ spin
system) gives rises to the cross peaks D and D′ marked by the square DeD′f.

In contrast, the signal at $\delta = 3.83$ ppm (peak b) does not show any cross peaks
which means that the protons must be located more than three bonds apart from the
next protons. This is the case for the protons of the methoxy group (H-1).

COSY spectra using cross peaks along with the diagonal spectrum are especially helpful to elucidate structural data for molecules that are not satisfactorily represented in complicated 1D NMR spectra.

2D NOESY (*Nuclear Overhauser Enhancement Spectroscopy*)

In NOESY, the NOE effect is used to establish the correlation, i.e., *dipolar* interaction of spins instead of scalar coupling is used. In general, cross peaks are indicated if the distance between two protons is smaller than 4 Å (0.4 pm). Thus, the NOESY experiment correlates all protons that are spatially close. Thus, important information can be obtained for structure determination.

Let us return to Challenge 2.12. The mass and IR spectra cannot unambiguously distinguish between Structures 2.23, 2.24, and 2.25. However, the 2D NOESY spectrum shown in Fig. 4.55 provides a unique result. Structure 2.23 is correct because the signals from both the CH_3 and NH_2 protons indicate only *one* connection to various aromatic protons but two correlations should be observed for the

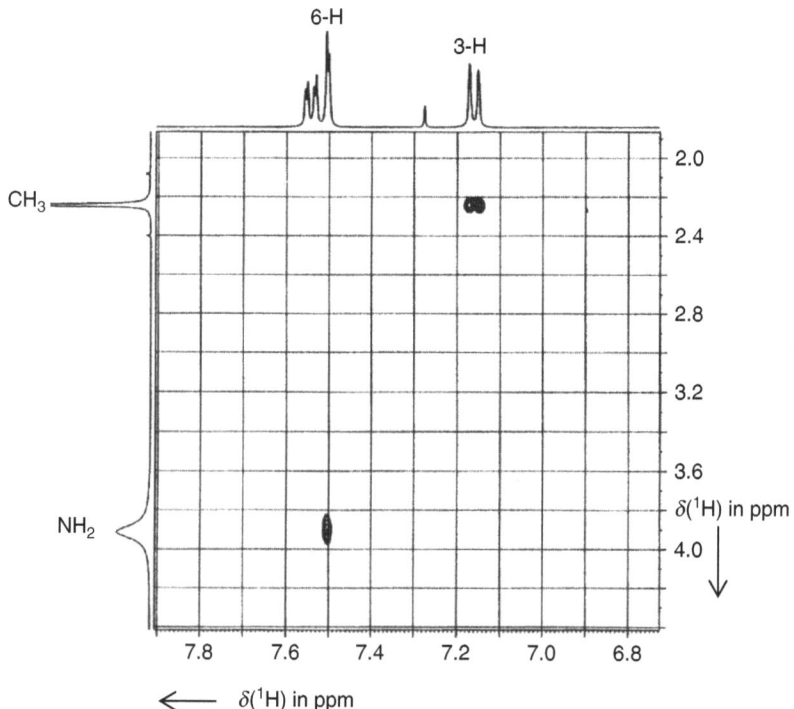

Fig. 4.55 Section from the 2D NOESY spectrum of Structure 4.23 (see Challenge 2.12)

CH_3 signal in the two other structures. The corresponding NOESY correlations are marked by arrows.

2D TOCSY (*TO*tal *C*orrelation *S*pectroscop*Y*)

In the TOCSY experiment, magnetization is dispersed over the complete spin system by a scalar multistage coupling. TOCSY correlates *all* protons of a spin system. Not only cross peaks are indicated between coupling via two and three bonds as in the COSY experiment but additional signals are visible which originate from interactions of all protons of a spin system. Especially, the 2D TOCSY experiment is very helpful for the structural determination of high-molecular compounds with spatially limited spin systems like polysaccharides or peptides. Thus, a characteristic pattern of signals result from each amino acid in proteins from which the respective amino acid can be identified.

So-called *relay spectra* are produced in the *1D TOCSY* experiment in which the mixing time is systematically increased while the proton being irradiated is kept constant. Thus, the series of one-dimensional TOCSY experiments show successively the protons in relation to the distance from the irradiated signal. Sometimes, this experiment provides better information if the resolution is poor and signal overlapping is given in the 2D TOCSY spectrum.

4.4.8.3 Heteronuclear 2D Spectra

Heteronuclear NMR spectroscopy instrumentalizes signals in which magnetization is transferred between ^1H and hetero nuclei (^{13}C or ^{15}N). The cross peaks indicate the interaction between *various* nuclei. Note that no signals are situated at the diagonal of the spectrum like in COSY spectra. The usual 1D spectra are pictured at the F_1 and F_2 axes.

In the following the 2D NMR specialties are listed that are used in the challenges of this book.

$^1J_{C,H}$-correlated NMR spectra

This is a 2D method that correlates chemical shifts of ^{13}C atoms and the *directly* bonded protons. Therefore, cross peaks enable the unambiguous assignment of the directly bonded protons to the respective carbon atoms from the known ^{13}C signals and vice versa. Unfortunately, quaternary C atoms cannot be recognized.

$^1J_{C,H}$-correlated NMR spectroscopy allows the recognition of *chemically non-equivalent* protons attached at a carbon atom like diastereotopical H atoms of the CH_2 group or the H atoms of a $-CH_2-$ structural moiety: By ^{13}C outgoing cross peaks are observed towards the two protons bonded at the same carbon atom. Furthermore, *randomly isochronic* nuclei can be recognized in addition to the DEPT spectra.

$^1J_{C,H}$-correlated NMR spectra can be produced by various techniques. The historical *HETCOR* experiment in which the ^{1}H signals lie on the F_1 axis and the F_2 axis has the ^{13}C signals produced during the time t has been replaced by the *HMQC* (*H*eteronuclear *M*ultiple *Q*uantum *C*oherence) and *HSQC* (*H*eteronuclear *S*ingle *Q*uantum *C*oherence) experiments. The F_1 and F_2 axes are exchanged in comparison to the HETCOR experiment.

The following three rules can be observed for these techniques:
The line starting from a ^{13}C NMR signal

• Does not intersect a cross peak.

   ⇒ This C atom is not attached by a hydrogen atom; this is a *quaternary* C-atom

• Intersects one cross peak.

   ⇒ A CH, CH_2, or CH_3 group is present.

• Intersects *two* cross peaks.

   ⇒ A CH_2 group with diastereotopic protons or the structural moiety $R_1R_2C=CH_2$ is present.

The $^1J_{C,H}$-correlated NMR spectra techniques enable the assignment of ^1H NMR signals from unambiguously assigned ^{13}C NMR signals and vice versa. Furthermore, diastereotopical CH_2 signals (in addition to the DEPT spectra) and randomly isochronic nuclei can be recognized. However, quaternary carbon atoms cannot be detected.

Let us again return to Challenge 2.13. The 2D NOESY spectrum presented in Fig. 4.55 provides the assignment of the aromatic protons as well as the order of the three substituents at the aromatic moiety. The assignment of the ^{1}H NMR signal can be unambiguously carried out by the coupling pattern and the coupling constants, respectively. Thus, the HSQC spectrum (Fig. 4.56) provides the assignment of the ^{13}C NMR signal of the aromatic methine carbon atoms. The numbers of the carbon atoms refer to Structure 4.53.

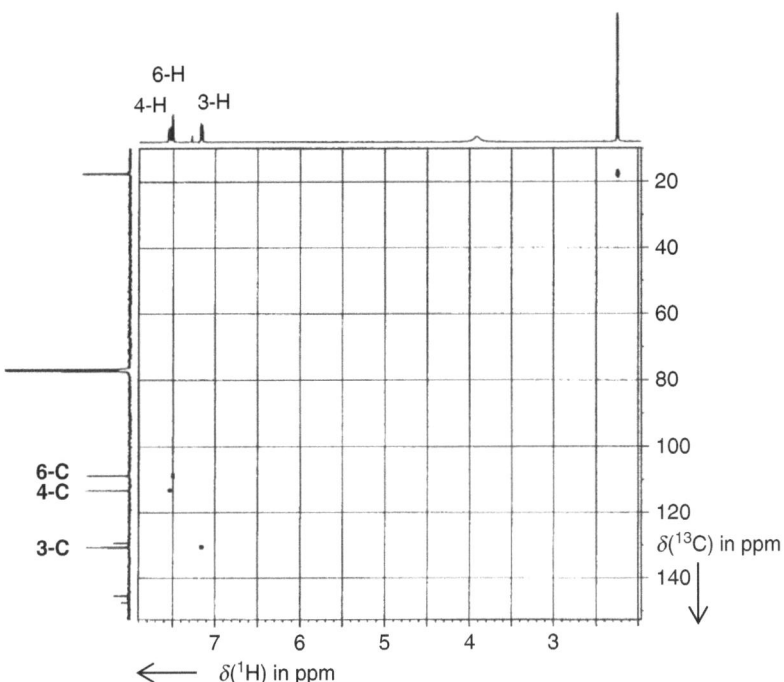

Fig. 4.56 HSQC spectrum of the unknown compound in Challenge 2.13. The numbers of the carbon atoms refer to Structure 4.53

The *HSQC-DEPT* and *HSQC-TOCSY* experiments combine the information of the corresponding two experiments. Note that some HSQC-DEPT and HSQC-TOCSY spectra are presented in the challenges.

(^{1}H,^{13}C)-correlated NMR spectra over *two* and *three* bonds

Whereas HMQC is selective for *direct* coupling HMBC (*Heteronuclear Multiple Bond Coherence*) gives longer range couplings, especially between two ($^2J_{C,H}$) and three bonds ($^3J_{C,H}$). Thus, *quaternary* carbon atoms can now be also assigned by this experiment.

Note that $^1J_{C,H}$ coupling results in satellites whenever it cannot be completely suppressed. However, such peaks can be easily recognized because they are not directed to ^{1}H signals, see the crossed-out peaks in Fig. 4.57c.

Let us complete the assignment of the *quaternary* signals via ^{13}C NMR by means of the HMBC spectra (Fig. 4.57a–c) for Structure 4.53 obtained by mass, IR, and NMR spectra.

Fig. 4.57 (**a**) Section of the
HMBC spectrum of Structure
4.53. (**b**) Section of the
HMBC spectrum of Structure
4.53. (**c**) Section of the
HMBC spectrum of Structure
4.53

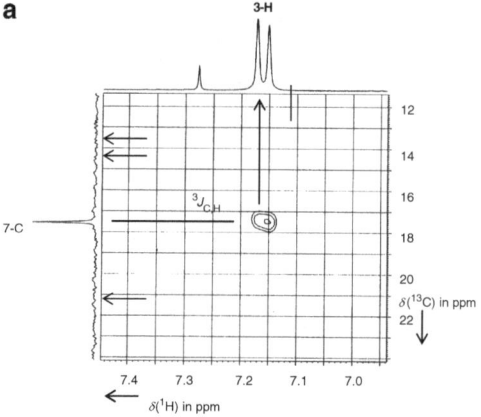

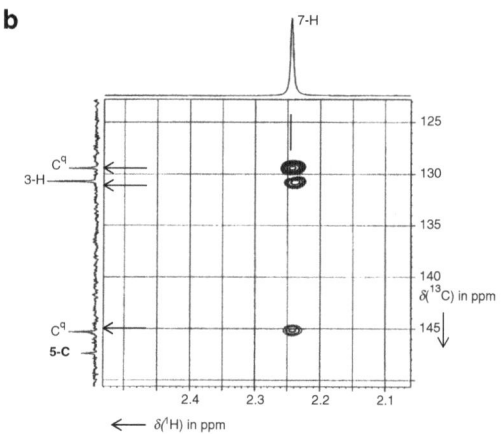

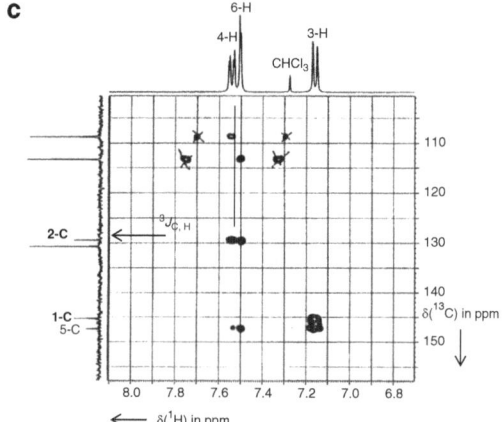

Table 4.6 Correlations obtained from the HBMC spectrum for the justification of Structure 4.53 (2-methyl-5-nitro-aniline), the unknown compound of Challenge 2.13

H-atom	^{1}H-signal δ = in ppm	Coupling	^{13}C signal δ = in ppm	C-atom
7	2.34 (s)	$^2J_{C,H}$	129.5	2
		$^3J_{C,H}$	145.3	1
		The third quaternary C-atom belongs to 5-C!		
		$^3J_{C,H}$	130.7	3
3	7.16 (d)	$^3J_{C,H}$	17.5	7
		$^3J_{C,H}$	145.3	
		$^3J_{C,H}$	147.4	5
6	7.50 (d)	$^3J_{C,H}$	113.4	4
		$^3J_{C,H}$	129.5	2
		$^3J_{C,H}$	147.4	5
4	7.54 (dd)	$^3J_{C,H}$	108.8	6
		$^3J_{C,H}$	129.5	2
		The distance to 1-C is too far		

The remaining three quaternary ^{13}C NMR signals, giving no cross peaks in the HSQC spectrum, are the signals at δ = 147.4, 145.3, and 129.5 ppm. As Fig. 4.57b indicates, the methyl protons (H-7) show cross peaks to two quaternary carbon atoms at δ = 145.3 and 129.5 ppm but no cross peak to the signal at δ = 147.4 ppm which must therefore be assigned to the quaternary C-5 atom. Note that the correlation between H-7 and C-5 involves five bonds, but such a correlation cannot be observed. Thus, the remaining quaternary signals belong to the carbon atoms 1 and 2, which cannot be distinguished by cross peaks started from H-7 because cross peaks can be caused by $^2J_{C,H}$ and $^3J_{C,H}$ correlations, respectively. The section of the HMBC spectrum shown in Fig. 4.57c shows a $^3J_{C,H}$ correlation between H-4 and the quaternary carbon atom at δ = 129.5 ppm. Because the distance to the carbon atom 1 is too great (over four bonds) the signal at δ = 129.5 ppm must be assigned to C-2. Thus, the remaining signal of a quaternary carbon atom at δ = 145.3 belongs to C-1. The assignment of 3-H obtained via coupling constants is justified by a $^3J_{C,H}$ correlation to 7-C. The most important correlations for the justification of Structure 4.53 (2-methyl 5-nitro aniline) are summarized in Table 4.6.

Challenge 4.15
For solutions to Challenge 4.15, see *extras.springer.com*.

1. The unknown compound of Challenge 1.5 possesses the sum formula $C_6H_4Cl_2N_2O_2$ with 5 DBE. The substituents NH_2 and NO_2 were obtained by the mass spectrometric fragmentation and the IR spectrum. Determine the proper structure by means of the NMR spectra presented in Figs. 4.58a–d. Assign all NMR signals.

(continued)

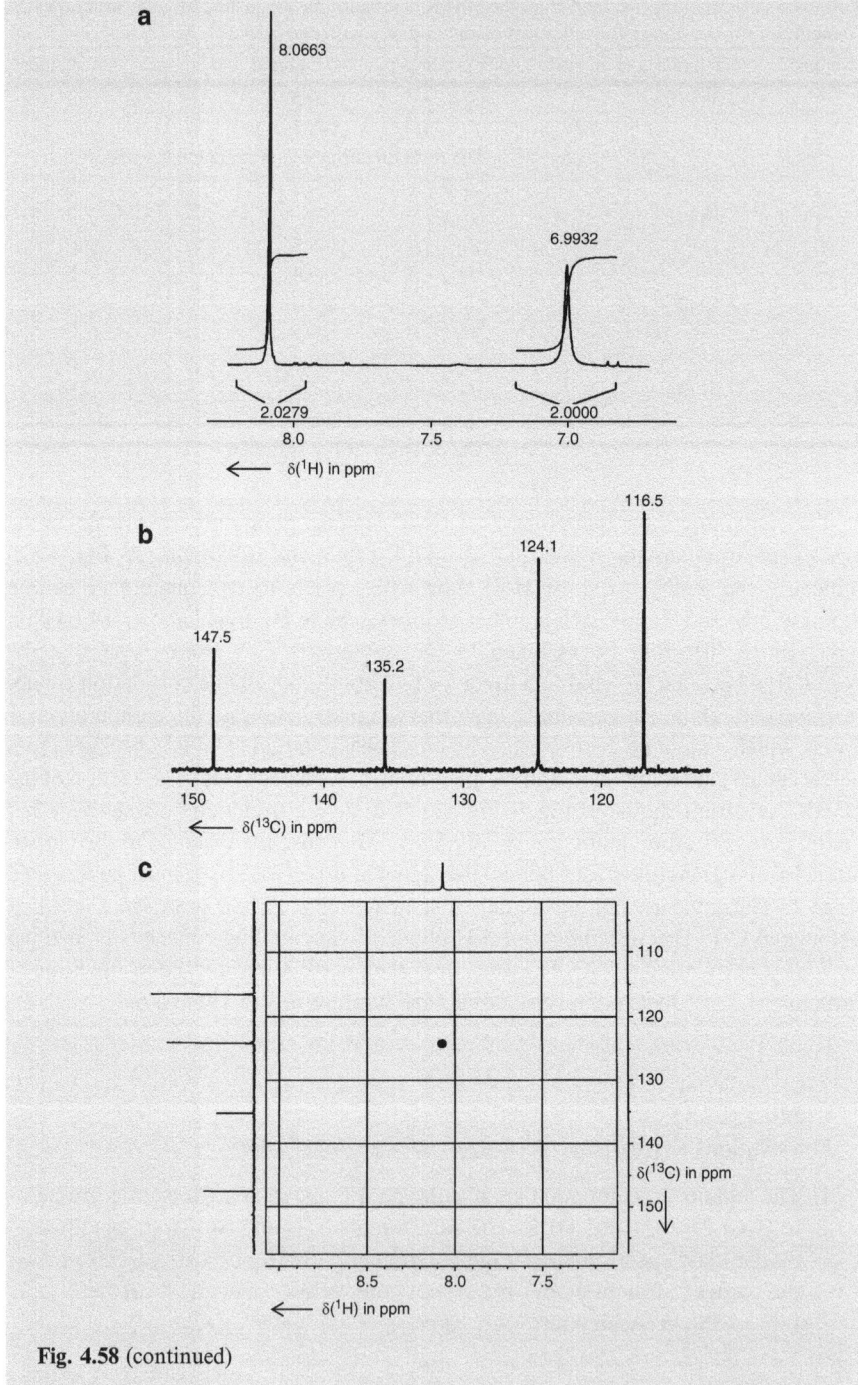

Fig. 4.58 (continued)

(continued)

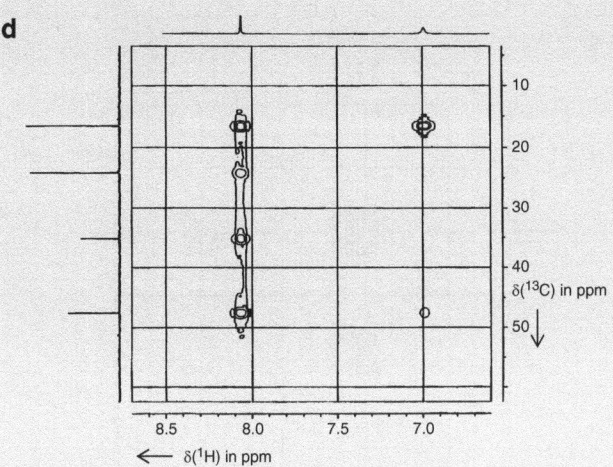

Fig. 4.58 (**a**) 400-MHz ^{1}H NMR spectrum of the unknown compound of Challenge 1.5. (**b**) ^{13}C NMR spectrum of the unknown compound of Challenge 1.5. (**c**) HSQC spectrum of the unknown compound of Challenge 1.5. (**d**) HMBC spectrum of the unknown compound of Challenge 1.5

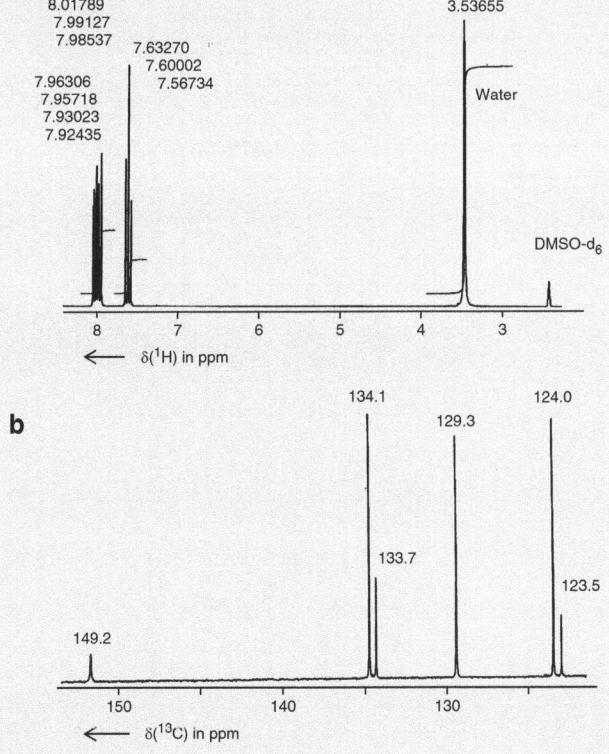

Fig. 4.59 (continued)

(continued)

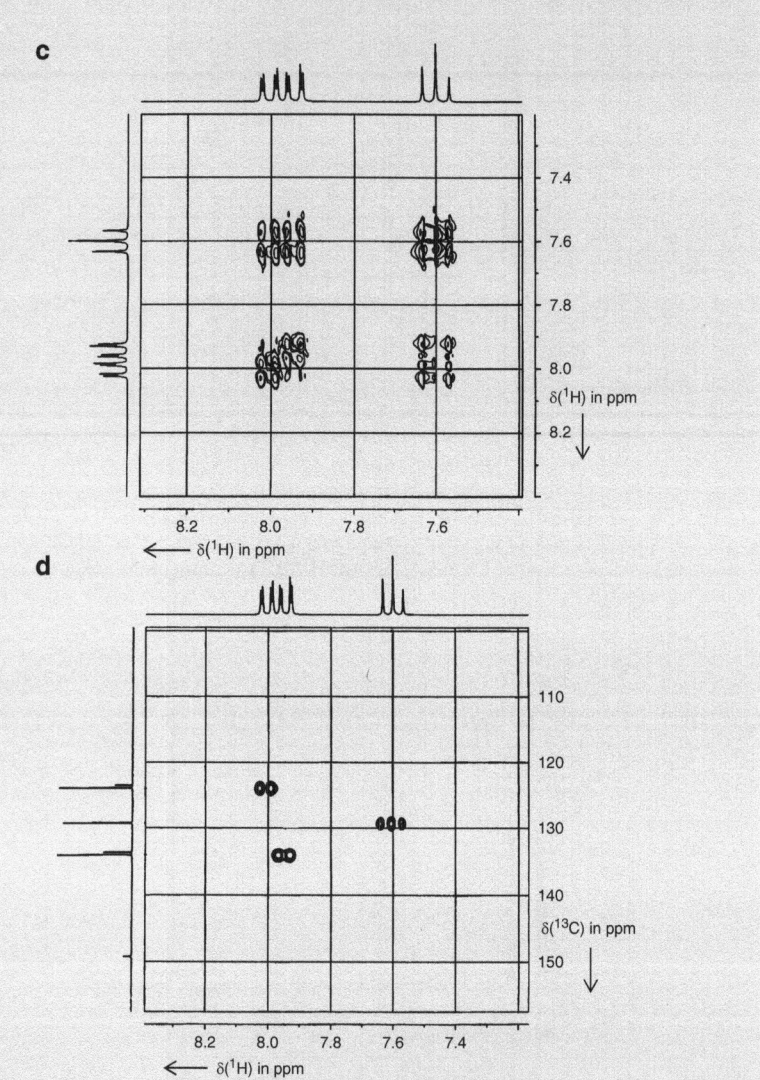

Fig. 4.59 (continued)

(continued)

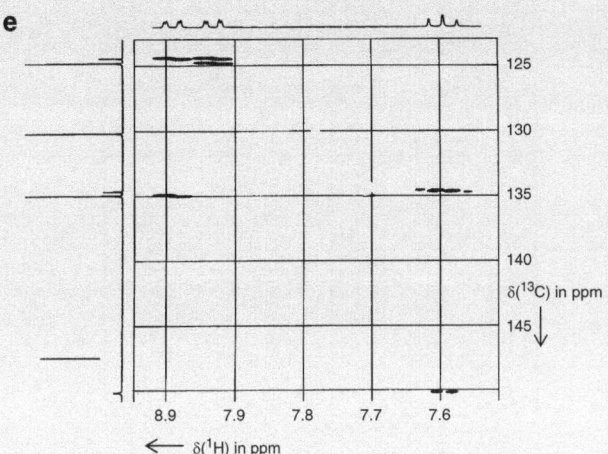

Fig. 4.59 (**a**) 250-MHz ^{1}H NMR spectrum of the unknown compound with the sum formula $C_6H_3Cl_2NO_2$. (**b**) ^{13}C NMR spectrum of the unknown compound with the sum formula $C_6H_3Cl_2NO_2$. (**c**) Section of the (^{1}H,^{1}H) COSY spectrum of the unknown compound with the sum formula $C_6H_3Cl_2NO_2$. (**d**) Section of the HSQC NMR spectrum of the unknown compound with the sum formula $C_6H_3Cl_2NO_2$. (**e**) Section of the HMBC NMR spectrum of the unknown compound with the sum formula $C_6H_3Cl_2NO_2$

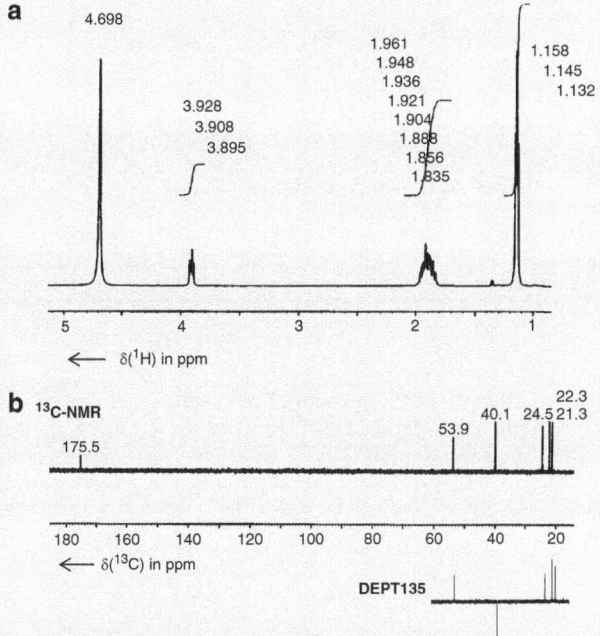

Fig. 4.60 (continued)

(continued)

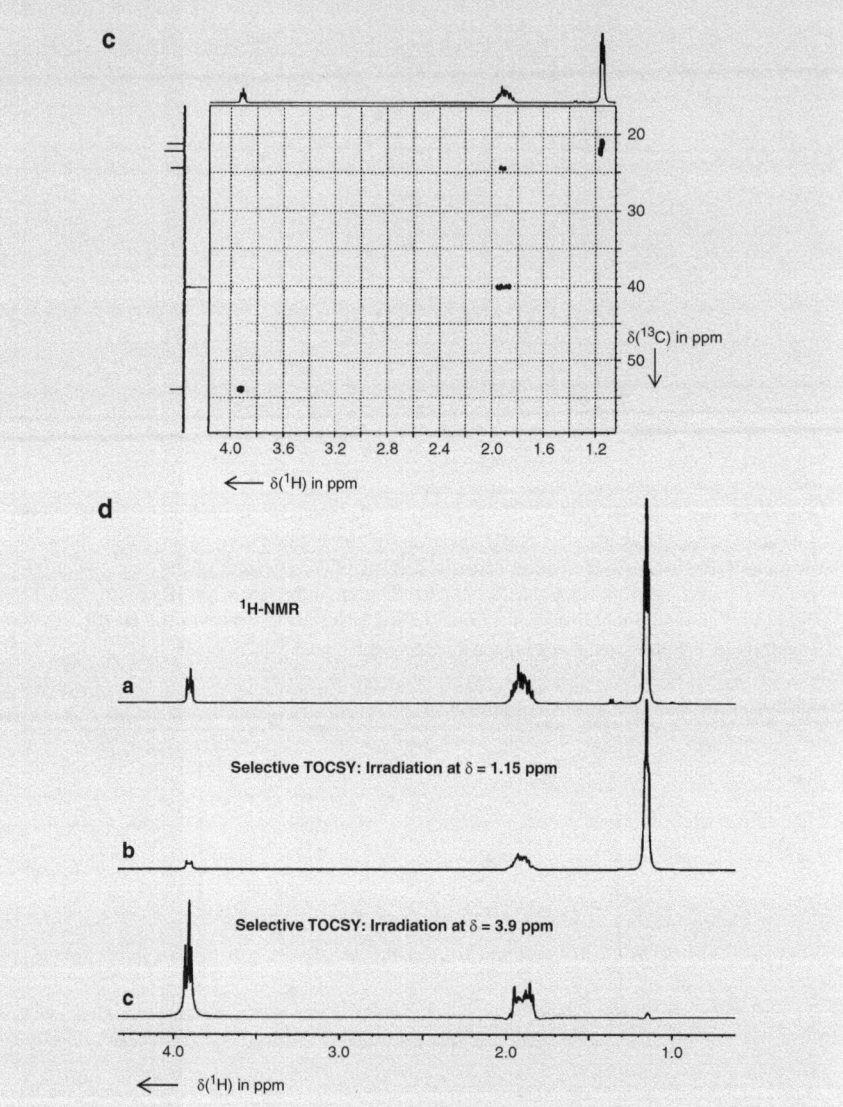

Fig. 4.60 (**a**) 400-MHz ^{1}H NMR spectrum of the unknown compound with the sum formula $C_6H_{13}NO_2$. (**b**) ^{13}C NMR and DEPT135 spectra of the unknown compound with the sum formula $C_6H_{13}NO_2$. (**c**) HSQC-DEPT spectrum of the unknown compound with the sum formula $C_6H_{13}NO_2$. (**d**) ^{1}H NMR (**a**) and selective TOCSY spectra with irradiation at 1.15 ppm (**b**) and 3.9 ppm (**c**) of the unknown compound with the sum formula $C_6H_{13}NO_2$

(continued)

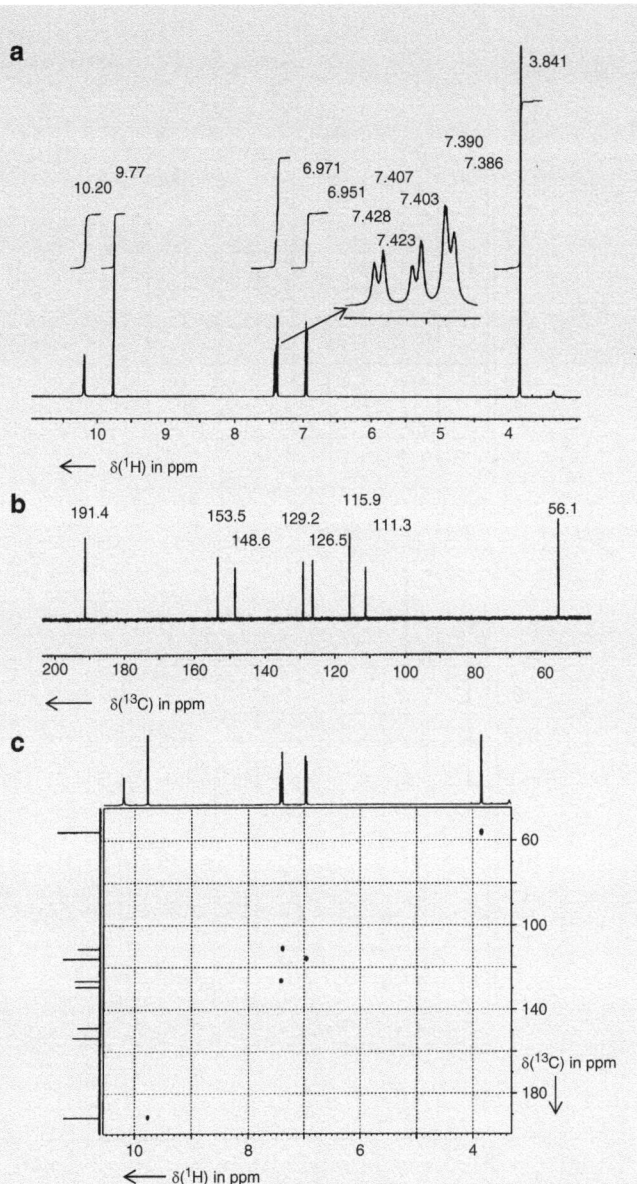

Fig. 4.61 (continued)

(continued)

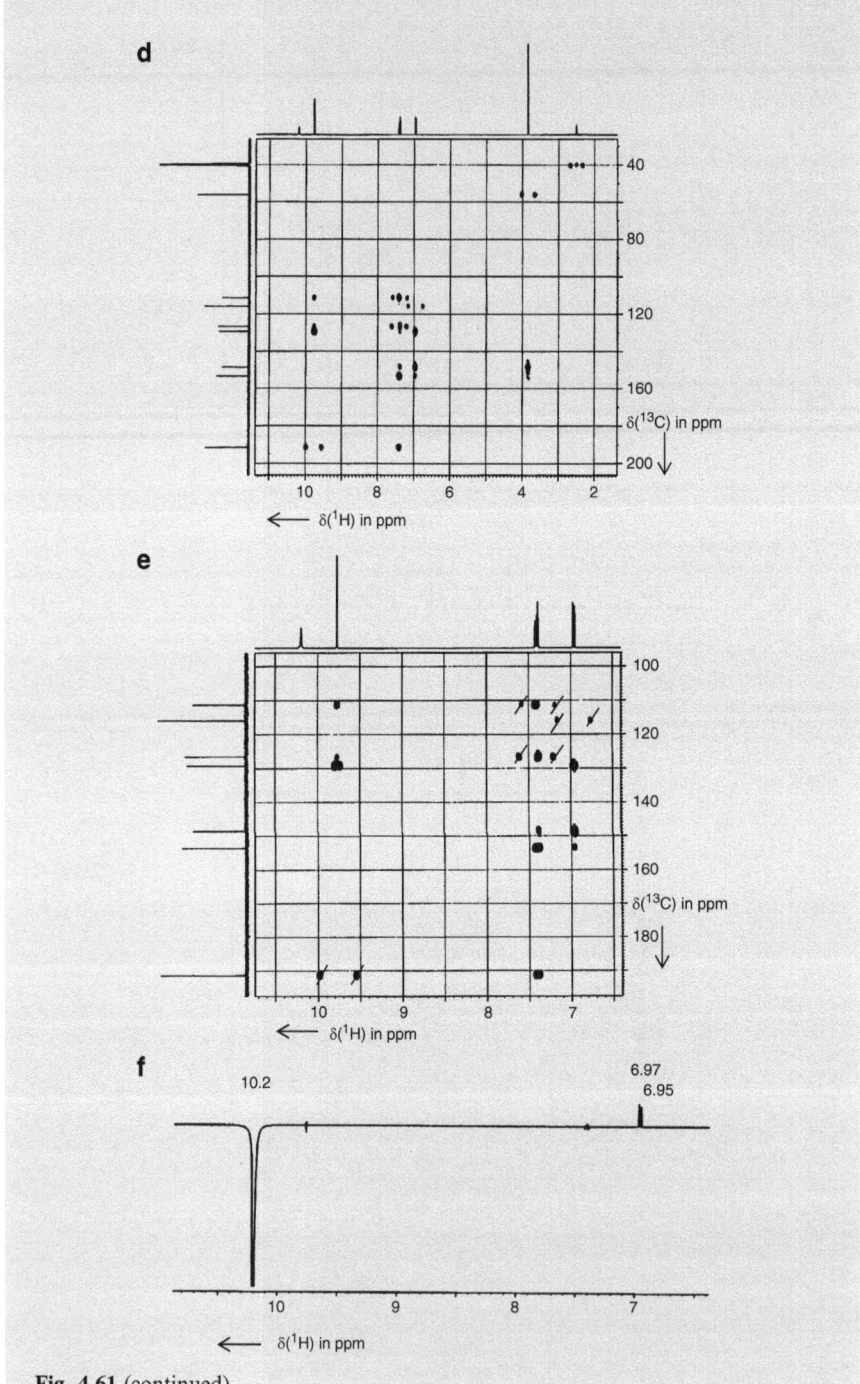

Fig. 4.61 (continued)

(continued)

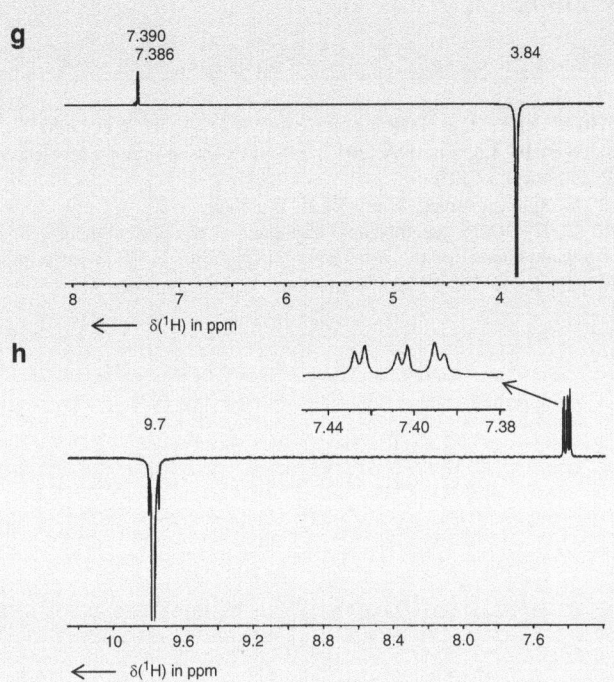

Fig. 4.61 (a) 400-MHz ^{1}H NMR spectrum of the unknown compound with the sum formula $C_8H_8O_3$ in DMSO. (b) ^{13}C NMR spectrum of the unknown compound with the sum formula $C_8H_8O_3$ in DMSO. (c) HSQC spectrum of the unknown compound with the sum formula $C_8H_8O_3$. (d) HMBC spectrum of the unknown compound with the sum formula $C_8H_8O_3$. (e) Section of the HMBC spectrum of the unknown compound with the sum formula $C_8H_8O_3$. (f) 1D NOESY spectrum of the unknown compound with the sum formula $C_8H_8O_3$ with selective irradiation at $\delta = 10.2$ ppm. (g) 1D NOESY spectrum of the unknown compound with the sum formula $C_8H_8O_3$ with selective irradiation at $\delta = 3.84$ ppm. (h) 1D NOESY spectrum of the unknown compound with the sum formula $C_8H_8O_3$ with selective irradiation at $\delta = 9.7$ ppm

2. The sum formula $C_6H_3Cl_2NO_2$ for an unknown compound was obtained by mass spectroscopy. The ^{1}H NMR, ^{13}C NMR, (^{1}H,^{1}H) COSY, HSQC, and HMBC spectra are presented in Fig. 4.59a–e. What is the structure of the compound? Assign all NMR signals in order to justify the structure.

3. Develop the structure of an unknown compound with the sum formula $C_6H_{13}NO_2$ by means of the NMR spectra shown in Fig. 4.60a–d. Assign all NMR signals in order to justify the structure. The NMR spectra were measured in D_2O.

4. The compound extracted from a natural product has the sum formula $C_8H_8O_3$ determined by mass spectrometry. Fig. 4.61a–h show the following spectra: 400-MHz ^{1}H NMR, 100-MHz ^{13}C NMR, HSQC, HMBC, 1D NOESY with various applied fields. Develop the structure for this compound. Assign all NMR signals and coupling constants in order to justify the structure.

Further Reading

1. Friebolin H (2011) Basic one- and two-dimensional NMR spectroscopy, 5th edn. Wiley-VCH, Weinheim
2. Keeler J (2010) Understanding NMR spectroscopy. Wiley-VCH, Weinheim
3. Huggins MT, Gurst JE, Lightner DA (2010) 2D NMR-based organic spectroscopy problems. Prentice Hall, Englewood Cliffs
4. Levitt MH (2008) Spin dynamics. Wiley-VCH, Weinheim
5. Jacobsen NE (2007) NMR spectroscopy explained – simplified theory, applications and examples for organic chemistry and structural biology. Wiley-VCH, Weinheim
6. Balci M (2005) Basic ^{1}H- ^{13}C-NMR spectroscopy. Elsevier, Amsterdam

Chapter 5
Molecular Structure Determination by Means of Combined Application of Methods

5.1 Overview of Determination of Heteroatoms

5.1.1 Nitrogen

Method	Information	
MS	• N rule	
	• N-specific fragment ions and fragment loss peaks	
	$\Rightarrow$ Mass correlation table (Tables 6.3 and 6.4)	
IR	Characteristic vibrations	
	$\Rightarrow$ Table 6.15	
	$v_{as} + v_s(NH_2)$, $\delta(NH_2)$	1° Amine or amide
	$v(NH)$	2° Amine or amide
	$v_{as} + v_s(NO_2)$	Nitro compounds
	$v(N{=}O)$	Nitroso compounds
	$v(N{\equiv}C)$	Nitriles (frequently very weak)
UV–VIS	N containing base chromophore	N-heterocyclic compounds
		Polymethine compounds
		Azo compounds
^{1}H NMR	• Exchangeable protons with D_2O	
	• Signals in the deep field ($\delta > 8$ ppm)	Ar–NO_2; pyridine compounds
	• Signals in the range $\delta < 7$ ppm)	Ar–NH_2

M. Reichenbächer and J. Popp, *Challenges in Molecular Structure Determination*,
DOI 10.1007/978-3-642-24390-5_5, © Springer-Verlag Berlin Heidelberg 2012

5.1.2   Chlorine, Bromine

Method	Information
MS	Isotope pattern $\Rightarrow$ Table 6.2

5.1.3   Iodine

Method	Information
MS	Unusually large mass differences because of loss of I ($\Delta m = 127$ amu) and HI ($\Delta m = 128$ amu), respectively

5.1.4   Fluorine

Method	Information	
MS	Unusual mass differences because of elimination of F ($\Delta m = 19$ amu) and HF ($\Delta m = 20$ amu), respectively	
IR	Intense ν(CF)	$1{,}370$–$1{,}120$ cm^{-1} (Alkyl-F)
		$1{,}270$–$1{,}100$ cm^{-1} (Aryl-F)
NMR	• Signals in the ^{19}F NMR spectrum, mostly as multiplets because of coupling with protons	
	• Unusually large spacing of the lines of the multiplets in the ^{1}H NMR spectrum because of ^{1}H/^{19}F coupling	
	• Multiplets in the {^{1}H} ^{13}C NMR spectrum instead of singlets because of ^{13}C/^{19}F coupling	

5.1.5   Sulfur

Method/structural group	Information	
MS	• Recognition and determination of the S number with the [M + 2] peak at absence of Cl and/or Br	
	• Specific S-containing fragment ions and fragment loss peaks $\Rightarrow$ Mass correlation table (Tables 6.3 and 6.4)	
–SH	IR	$2{,}500$–$2{,}600$ cm^{-1}
	MS	• Elimination of SH ($\Delta m = 33$ amu)

		• Specific fragment ions like CHS^+ ($m/z = 45$ amu)
–S–	IR	No specific information (Excluding of –SH by IR)
	MS	Information as for –SH
–S–S–	Raman	Intense line at 500 cm^{-1} (IR: weak)
>C=S	IR	Intense v(C=S) in the range 1,270–1,030 cm^{-1}
	^{13}C NMR	Signal at unusual deep field ($\delta > 210$ ppm)
R–SO$_2$–R′(H)	IR	$v_{as} + v_s$(SO$_2$); intense bands at 1,370 + 1,030 cm^{-1}
	MS	Elimination of SO$_2$ ($\Delta m = 64$ mu)
Thiophenene	MS	• Specific aromatic peaks
		• DBE (3) is smaller than for benzene (DBE = 4)
	^{1}H NMR	Coupling constants ($\Rightarrow$ Table 6.25)

5.1.6 Oxygen

Method	Information
MS	O-specific fragment ions and fragment loss peaks
	$\Rightarrow$ Mass correlation table (Tables 6.3 and 6.4)
^{1}H NMR	–OH Changeable H atom
	R–O–R Chemical shift values for R at deep field
^{13}C NMR	RR′(H)C=O $\delta = 200$–180 ppm
	RC(=O)X (X = heteroatom) $\delta = 180$–160 ppm
IR	Characteristic vibrations: v(OH), v(C–O), v(C=O), v(NO$_2$), ...
	$\Rightarrow$ Tables 6.15
UV–VIS	Weak $n \rightarrow \pi^*$ transitions
	Polymethine compounds: Very intense $\pi \rightarrow \pi^*$ transitions

5.1.7 Phosphorous

Method	Information
MS	P-specific fragment ions
	$\Rightarrow$ Mass correlation table (Tables 6.3 and 6.4)
	Isotope-free peaks: $m/z = 65$ amu (PO$_2$H$_2$), 97 amu, 99 amu (PO$_4$H$_4$)
NMR	• Signals in ^{31}P NMR, mostly as multiplets because of coupling with ^{1}H atoms
	• Unusually large spacing of the multiplets because of ^{1}H/^{31}P coupling
	• Multiplets in the {^{1}H} ^{13}C NMR spectrum instead of singlets because of ^{13}C/^{31}P coupling
IR	Characteristic vibrations: v(P=O), v(P–O), but coincidence with fundamental vibrations of C=S, C–N, C–O, C–F, S=O

5.1.8   Silicium

Method	Information
MS	Relatively intense [M + 1] *and* [M + 2] peaks
IR	Characteristic vibrations:
	$v(\text{Si–O}) \approx 800 \text{ cm}^{-1}$,
	$v_{as}(\text{Si–O–Si})$: 1,100–1,000 cm^{-1}, $v_s(\text{Si–O–Si})$: 600–500 cm^{-1}

5.2   Determination of the Sum Formula

Method	Number of atoms	Information
MS	C	From the [M + 1] peak
	S	From the [M + 2] peak
	Cl/Br	From isotopic pattern
	N	N rule
	O	From specific O fragment ions and loss peaks
	H	From the difference between molecular peak and the sum of all other elements
^{1}H NMR	H	From the relative intensities of the signals
^{13}C NMR	C	Nonisochronic C atoms: From the number of signals
	CH$_x$	From DEPT
		From correlation spectroscopy

The sum formula is the base for the calculation of the double binding equivalent (*DBE*), see Sect. 1.3.

5.3   Geometrical Isomers of the Structural Type RR′C=CR″R‴

H number	Structural type	Method	Information
3H	RHC=CH$_2$	IR	γ(CH) $\Rightarrow$ Table 6.15.I.2.2.3
		^{1}H NMR	AMX system (multiplet, *J*)
2H	*trans*-RHC = CHR′	IR	γ(CH) $\Rightarrow$ Table 6.15.I.2.2.1
		^{1}H NMR	$^3J(^1\text{H}/^1\text{H})$ if R $\neq$ R′
	cis-RHC=CHR′	IR	γ(CH) $\Rightarrow$ Table 6.15.I.2.2.2
		^{1}H NMR	$^3J(^1\text{H}/^1\text{H})$ if R $\neq$ R′
			NOESY
	RR′C=CH$_2$	IR	γ(CH) $\Rightarrow$ Table 6.15.I.2.2.4
		^{1}H NMR	$^2J(^1\text{H}/^1\text{H})$
			NOESY
1H	RR′C=CR″H	IR	v(C=C), no γ(CH)
		^{1}H NMR	NOESY
0H	RR′C=CR″R‴	IR	v(C=C), intense in Raman!
			No v(CH,sp^2)
		^{1}H NMR	NOESY

5.4   Molecular Structure Determination

Challenge 5.1
Deduce the structure of the unknown compound by means of the following spectra.

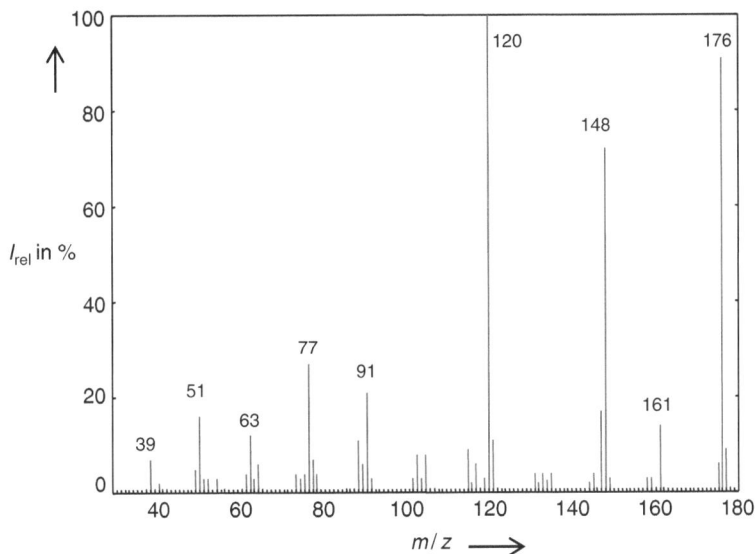

Fig. 5.1A   Mass spectrum (EI 70 eV). m/z in amu (I_{rel} in%): 176 (94.2), 177 (11.8)

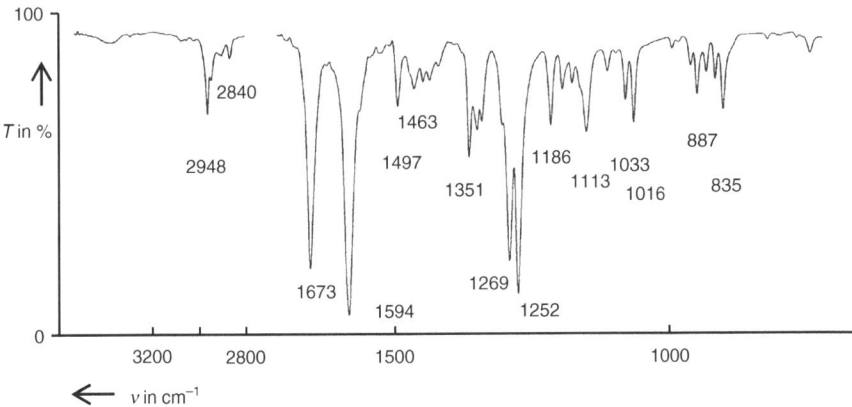

Fig. 5.1B   IR spectrum (KBr pellet)

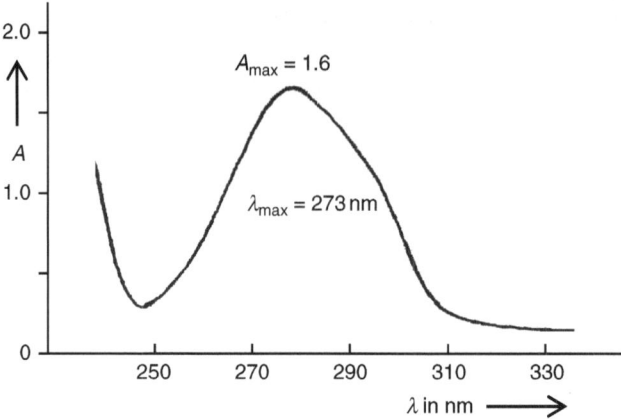

Fig. 5.1C UV spectrum. Sample: 0.45 mg in 25-mL methanol; cuvette: l = 1 cm

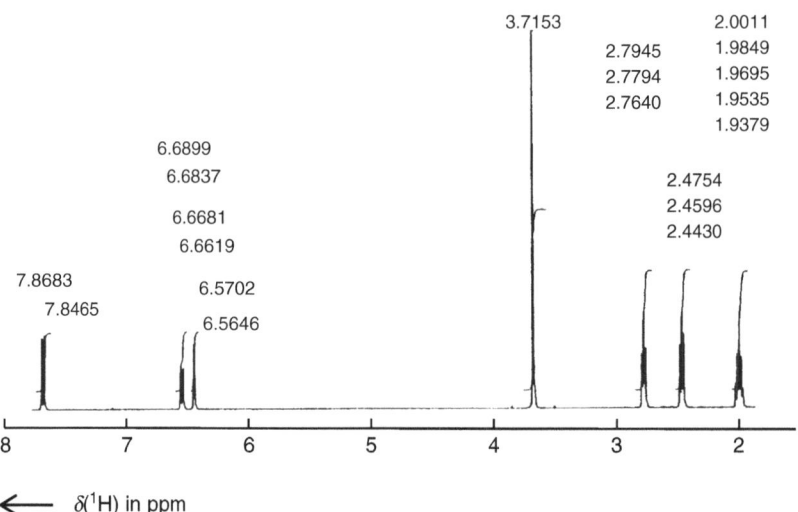

$\delta(^1H)$ in ppm

Fig. 5.1D ^{1}H NMR spectrum ($\nu_0 = 400$ MHz; solvent: CDCl$_3$)

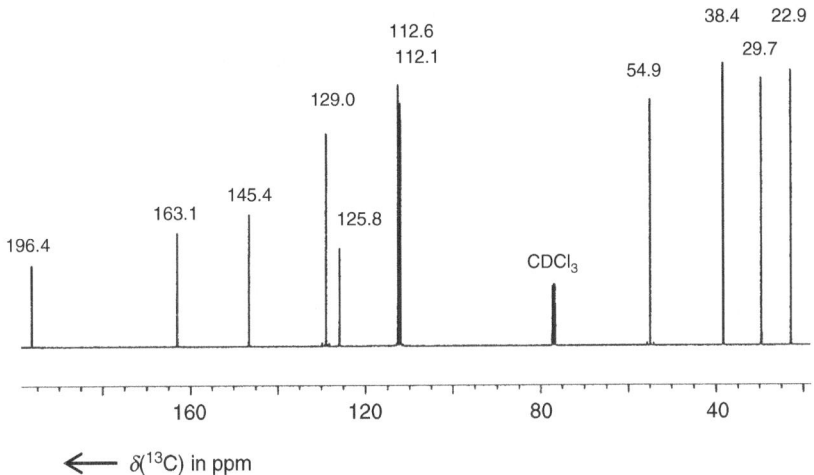

Fig. 5.1E ^{13}C NMR spectrum ($v_0 = 400$ MHz; solvent: CDCl$_3$)

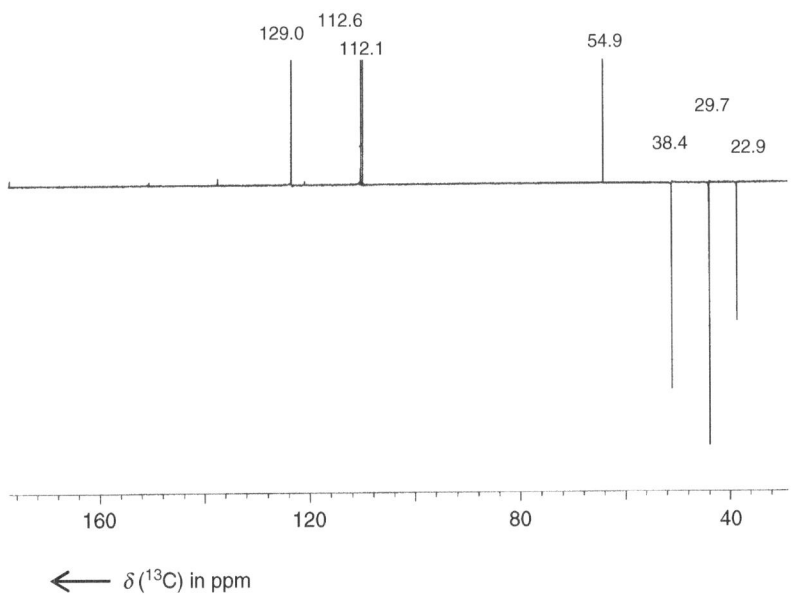

Fig. 5.1F DEPT135

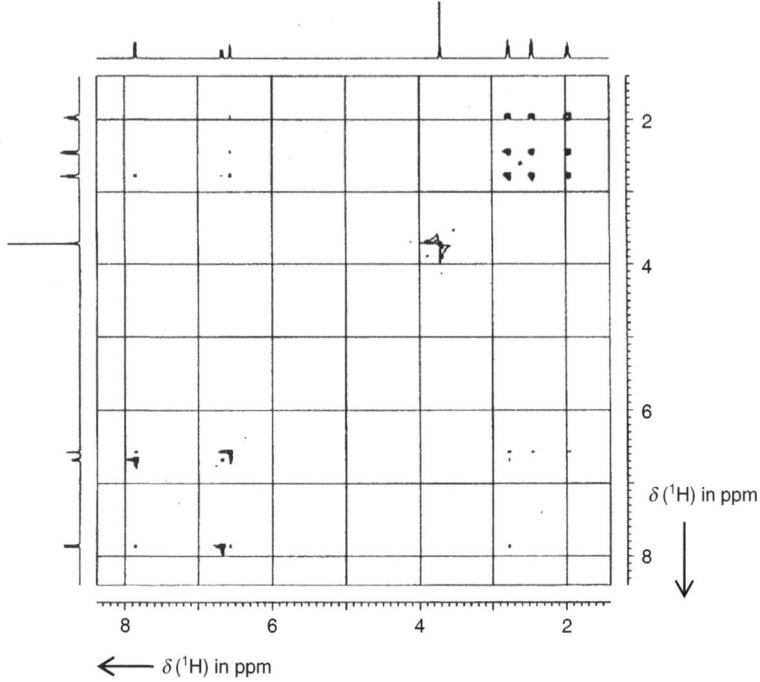

Fig. 5.1G (^{1}H,^{1}H) COSY spectrum

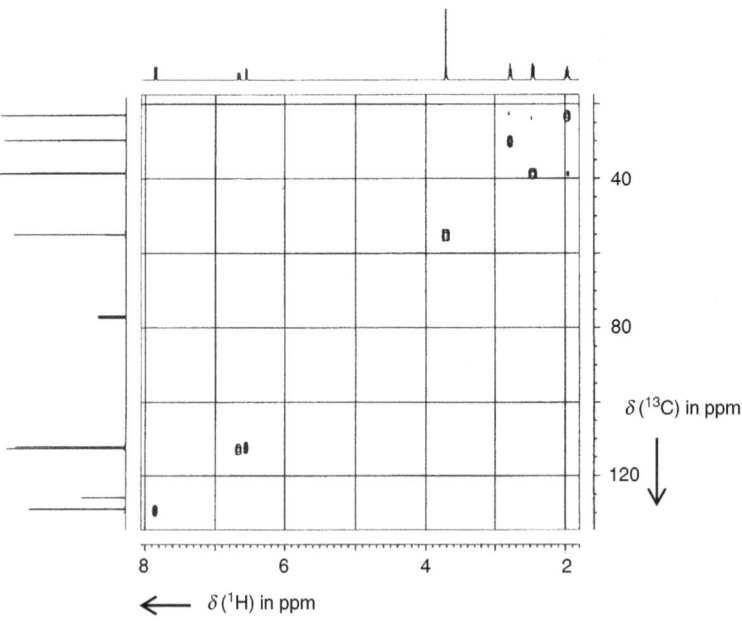

Fig. 5.1H Section of the HSQC spectrum

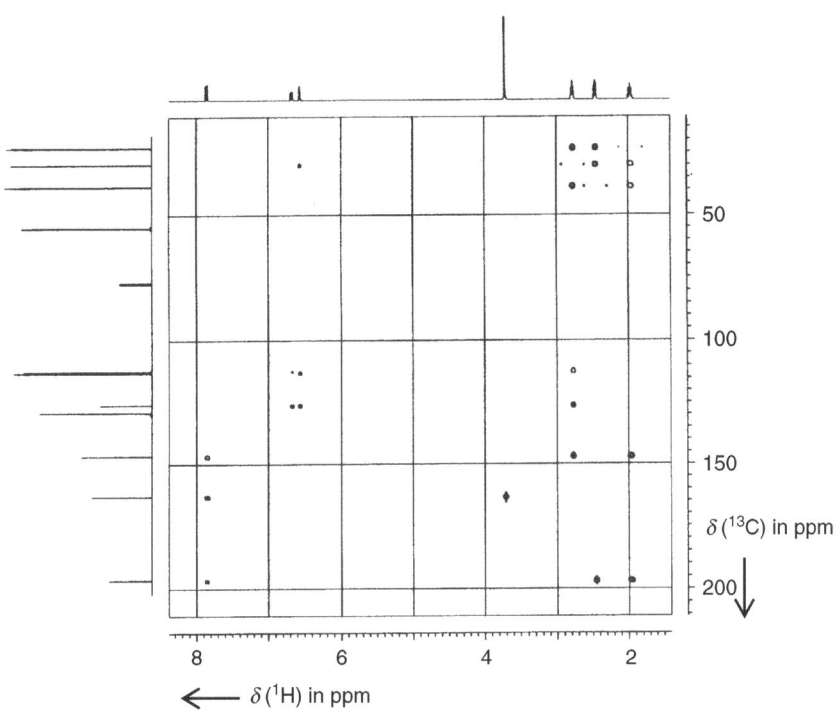

Fig. 5.1I HMBC spectrum

5.4.1 Solution to Challenge 5.1

A. Sum formula

Number of C:

MS: $M^{\bullet+}$: $m/z = 176$ amu ($I_{rel} = 94.2\%$)

$\Rightarrow$ Reasonable differences to fragment peaks

[M+1]: $m/z = 177$ amu ($I_{rel} = 11.8\%$)

$\Rightarrow n_C = 11 \pm 1$

^{13}C NMR: 11 signals $\Rightarrow \boldsymbol{n_C = 11}$

Number of H atoms:

^{1}H NMR: Integration (from left): 1:1:1:3:2:2:2 $\Rightarrow n_H = 12$

^{13}C NMR/DEPT135:

CH_x (from left): CH, CH, CH, CH_3, CH_2, CH_2 CH_2 $\Rightarrow \boldsymbol{n_H = 12}$

Search for heteroatoms:

N: MS: $M^{\bullet+}$ is even $\Rightarrow n_N = 0$ or 2 (N rule!)

IR: no vibrations for N-containing groups

$\Rightarrow$ no N

Cl, Br: no isotopic patterns $\Rightarrow$ no Cl, no Br

F: MS: no differences $\Delta m = 19$ and $\Delta m = 20$

^{13}C NMR: no multiplets of the broad-band decoupled ^{13}C NMR spectrum

$\Rightarrow$ no F

S: MS: no [M+2] peak with $I_{rel} \geq 4.4\%$
$$\Rightarrow \text{no S}$$
O: IR: $v(C{=}O) = 1{,}673$ cm^{-1}
MS: $m/z = 148 \rightarrow m/z = 120 \Rightarrow \Delta m = 28$ amu ($-C{=}O$)
^{1}H NMR: $\delta = 3.72$ ppm (s, 3H) $\Rightarrow$ OCH$_3$
$$\Rightarrow \text{O is present (least 2 O)}$$
Summation of the detected atoms:
132 amu (11C) + 12 amu (12H) + 32 amu (2 O) = 176 amu ($\Rightarrow$ M$^{\bullet+}$)
Sum formula: **C$_{11}$H$_{12}$O$_2$**

B. Numbers and assignment of double binding equivalents (DBE)

C$_{11}$H$_{12}$O$_2$ (-2 O) $\rightarrow$ C$_{11}$H$_{12}$ **$\Rightarrow$ 6 DBE**
Assignment: 4 DBE $\Rightarrow$ aromatic structure (from MS, NMR, and IR)
          1 DBE $\Rightarrow$ C=O (from MS, IR, and ^{13}C NMR)
          Difference:
          1 DBE $\Rightarrow$ no C=C (IR, ^{13}C NMR)
          $\Rightarrow$ 1 ring
Structural moieties: **Aromatic** structure **+ 1 ring**

C. UV spectrum

Evaluation of the molecular absorptivity at $\lambda = 273$ nm (α_{273}) according to Lambert-Beer's law:

$$\alpha_{273} = \frac{A \cdot M \text{in gmol}^{-1} \cdot V_{sol} \text{in mL}}{m \text{ in mg} \cdot l \text{ in cm}} = \frac{1.6 \cdot 176 \cdot 25}{0.45 \cdot 1} = 15650 \text{ L mol}^{-1}\text{cm}^{-1}$$

The longest wavelength absorption band can be assigned to the *K* band of the *aromatic* chromophore with a *conjugated* substituent.

D. IR spectrum

Table 5.1 Assignment of the IR absorption bands

v in cm^{-1}	Intensity	Assignment	Structural moiety
3,100–3,000	w	$v(CH,sp^2)$	Aromatics/alkene
2,948	m	$v(CH,sp^3)$	Alkyl
2,840	m	$v(OCH_3)$	OCH$_3$
1,673	vs	$v(C{=}O)$, conjugated	α-Aryl-ketone
1,594	s	$v(C{=}C)$, aryl	Aromatic structure, no alkene
1,497	m		
1,467	m	$\delta_{as}(CH_3) + \delta(CH_2)$	Alkyl
1,446	m		
1,351	m	$\delta_s (CH_3)$	CH$_3$
887	m	$\gamma(CH, 1H, \text{isolated})$?	1,2,4-tri-substituted benzene?
835	m	$\gamma(CH, 2H, \text{neighbored})$?	

Structural moieties: Aryl (1,2,4-tri-substituted?), OCH$_3$, Alkyl,
α-Aryl-ketone

E. ^{1}H NMR spectrum

Table 5.2 Assignment of ^{1}H NMR signals

δ in ppm	Z^{rel}(H)	Multiplet	J in Hz	Assignment
7.86	1	d	8.7	Aryl with *ortho*-H
6.68	1	dd	8.7	Aryl with *ortho*-H
			2.5	Aryl with *meta*-H
6.56	1	d	2.5	Aryl with *meta*-H
3.72	3	s		O–CH$_3$
2.78	2	"t"	6.1	C–CH$_2$–CH$_2$
2.46	2	"t"	6.5	C–CH$_2$–CH$_2$
1.97	2	"quintet"	6.3	H$_2$C–CH$_2$–CH$_2$

Structural moieties: Aryl (1,2,4-trisubstituted)

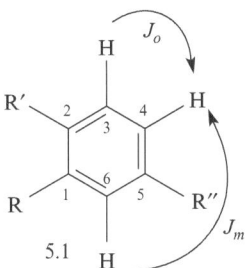

–OCH$_3$, –CH$_2$–CH$_2$–CH$_2$–

F. ^{13}C NMR/DEPT135 spectra

Table 5.3 Assignment of the CH$_x$ groups

δ in ppm	CH$_x$	Assignment	Comment
196.4	C^q	C=O	Ketone
163.1	C^q	Aryl	C attached by a substituent
145.4	C^q	Aryl	C attached by a substituent
129.0	CH	Aryl	
125.8	C^q	Aryl	C attached by a substituent
112.6	CH	Aryl	
112.1	CH	Aryl	
54.9	CH$_3$–O		Unambiguous assignment
38.4	C–CH$_2$–C		
29.7	C–CH$_2$–C		
22.9	C–CH$_2$–C		

Sum C_xH_y: $C_{11}H_{12}$ which is in accordance to the sum formula, therefore, there are no isochronic C atoms.

Structural moieties:
Aryl attached by three substituents (no symmetric pattern)
CH_3–O, C=O (Ketone), 3 CH_2

G. Structural suggestions

The numbers of DBE requires the presence of an additional ring. Furthermore, there is no end group for the –H_2C–CH_2–CH_2– moiety. Therefore, the moiety –H_2C–CH_2–CH_2– must be part of an *alicyclic ring* that is attached at the benzene ring. According to the IR spectrum, the carbonyl group must be ordered in conjugation to the benzene ring and is the second substituent and the third substituent is given by the OCH_3 group. The 1,2,4-position of the aromatic protons is unambiguously provided by the coupling constants. Thus, there are two structural suggestions (Structures 5.2 and 5.3) for the unknown compound.

The proton H-1 (marked in bold) should be observed mostly at deep field because of its *ortho* position to the carbonyl group. The doublet at $\delta = 7.86$ ppm shows an *ortho* coupling with $J = 8.7$ Hz which is only possible in Structure 5.3. Note that H-1 in Structure 5.2 gives rise also to a doublet but with a *meta* coupling in the range 2–3 Hz.

Structure 5.3 is also favored by the UV spectrum. The evaluation of the λ_{max} value according to Table 6.17 gives rise to the following results for Structure 5.3: 246 + 3 + 25 = 274 nm (experimental: 273 nm) whereas $\lambda_{max} = 256$ nm is calculated for Structure 5.2. However, the true structure recovers the assignment of all NMR signals.

H. Evaluation of the NMR signals by the software SpecTool®

Note that the numeration of the atoms refers to that which is used in Structure 5.3. All values are given in ppm.

As assumed above, the chemical shift for H-1 (7.73 ppm) is also calculated being the signal at the lowest field. However, the NMR signals for only the CH_3O and $C=O$ groups can be unambiguously assigned whereas all other signals must be assigned by the NMR correlation methods.

I. ($^1H,^1H$) COSY

The seven diagonal signals are denoted by capitals A to G and the cross peaks are denoted by lower cases. Two correlations are marked in the COSY spectrum; the other ones are given by the respective squares in Table 5.4. Note that the numeration of the atoms refers to that which is used in Structure 5.3.

All protons can be unambiguously assigned by means of the coupling constants and the connections which can be obtained by the COSY spectrum. Because the protons are assigned, the corresponding carbon atoms can also be assigned by the HSQC spectrum.

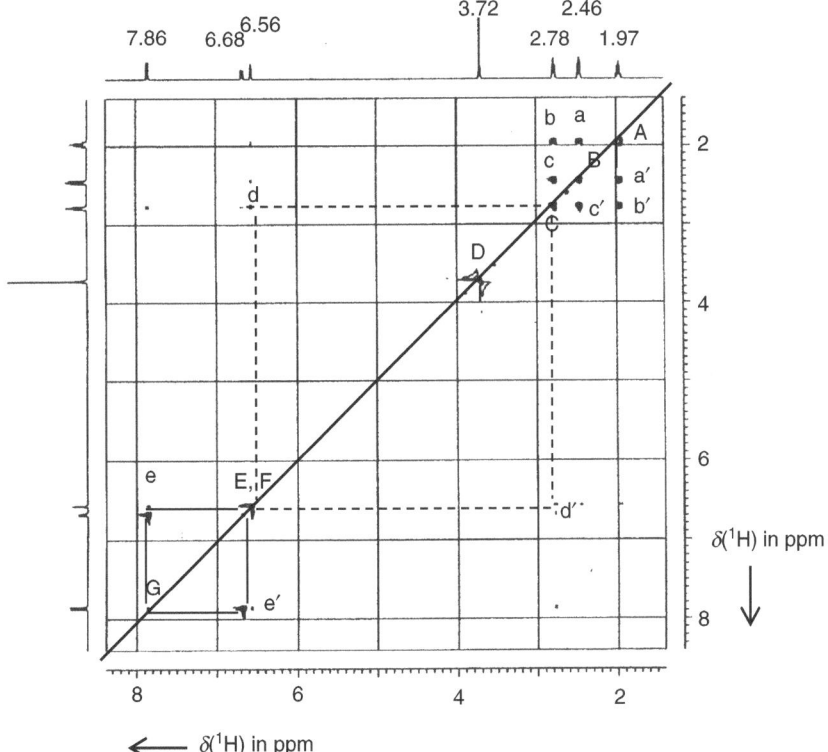

Fig. 5.1J   Correlations in the ($^1H,^1H$) COSY spectrum

For reasons of clarity only two correlations are marked, the other ones are given in Table 5.4.

Table 5.4 Assignment of the ^{1}H NMR signals

δ in ppm	Cross peak to signal	Assignment of the H atom number	Comment
1.97	1. AaBa' 2. AbCb'	7	Unambiguous assignment because of coupling to *two* CH$_2$ groups
2.78	Cd(E,F)d'	6	Unambiguous assignment because of long-range coupling to an aromatic proton
2.46	BcCc'	8	Coupling to H-7
3.72	No cross peak!	11	Unambiguous assignment because of no coupling
6.56		4	Assignment occurs about the J_{meta}
6.68		2	Assignment occurs about the *two* couplings: $J_{ortho} + J_{meta}$
7.86	Ge(E,F)e'	1	Assignment occurs about the J_{ortho}

J. Correlations in the HSQC spectrum ($^1J_{C,H}$ coupling)

The cross peaks enable the unambiguous assignment of the directly bonded protons to the respective carbon atoms from the known ^{13}C signals and vice versa. The corresponding correlations are marked in the HSQC spectrum in Fig. 5.1K beginning with the known chemical shift values of the protons. The assignment of the ^{13}C NMR signals is listed in Table 5.5. Note that the ^{13}C NMR signal at 125.8 ppm does not show any cross peak which means that this signal belongs to a quaternary carbon atom.

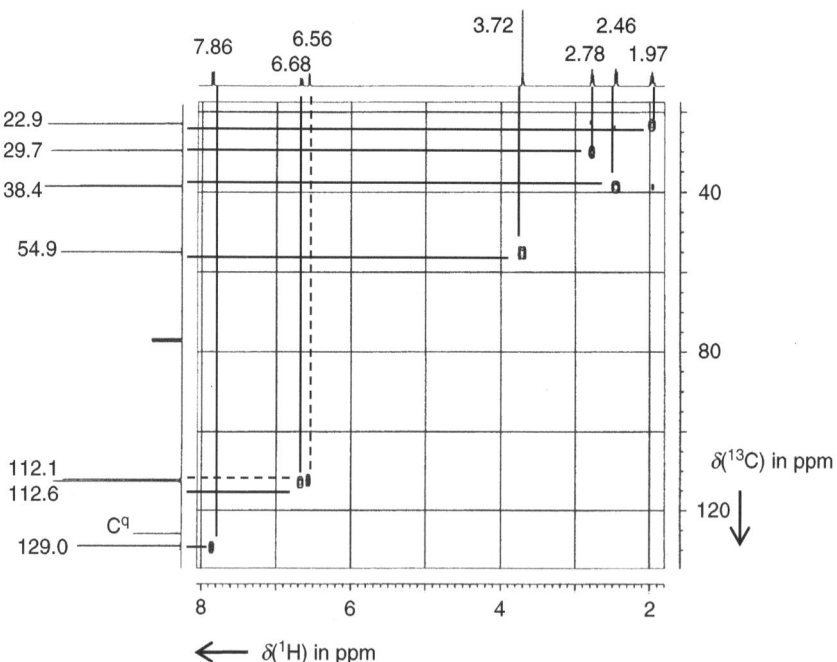

Fig. 5.1K Assignment of the ^{13}C–H$_x$ NMR signals by means of the HSQC spectrum beginning with the assigned proton signals

Table 5.5 ^{13}C–H$_x$ NMR signals beginning with the assigned ^{1}H NMR ones

^{1}H signals		^{13}C signals	
δ in ppm	No. of the H atom	δ in ppm	No. of the C atom
7.86	1	129.0	**1**
6.68	2	112.6	**2**
6.56	4	112.1	**4**
3.72	11	54.9	**11**
2.78	6	29.7	**6**
2.46	8	38.4	**8**
1.97	7	22.9	**7**

K. Correlations in the HMBC spectrum ($^2J_{C.H}$ and $^3J_{C.H}$ couplings)

The four quaternary carbon atoms at 125.8, 145.4, 163.1, and 196.4 ppm can be assigned by the HMBC spectrum by the cross peaks which show couplings over two and three bonds. The correlations are marked in the HMBC spectrum (Fig. 5.1L) and are listed in Table 5.6. The corresponding correlations for the assignment of the *quaternary* carbon atoms are summarized in Table 5.7 and are graphically illustrated in Fig. 5.1M.

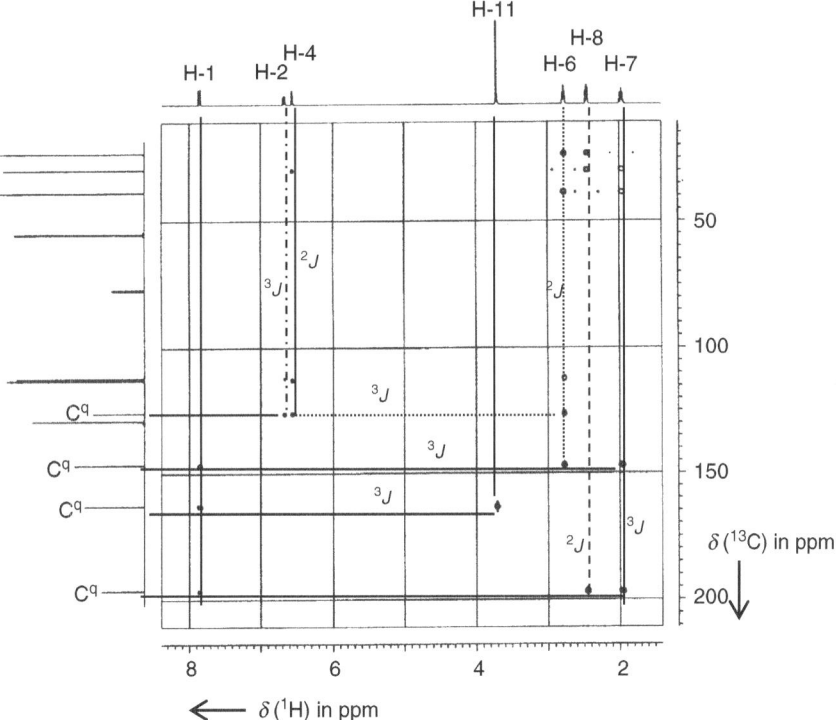

Fig. 5.1L Assignment of the quaternary ^{13}C^q NMR signals by means of the HMBC spectrum beginning with the assigned proton signals. The numeration of the protons refers to Structure 5.1

Table 5.6 ^{1}H, ^{13}C correlations over two and three bonds

^{1}H		Correlation to C atom number			
δ in ppm	H atom number	Over $^2J_{C,H}$	δ in ppm	Over $^3J_{C,H}$	δ in ppm
1.97	7	6	29.7	5	145.4
		8	38.4		196.4
2.46	8	7	22.9	7	22.9
		9	196.4		
2.78	6	7	22.9	8	38.4
				10	125.8
3.72	11			3	145.4
6.56	4			2	112.6
				10	125.8
6.68	2			4	112.1
				10	125.8
7.86	1			5	145.4
				3	163.1
				9	196.4

Table 5.7 ^{1}H, ^{13}C correlations to quaternary carbon atoms C^q

Start (δ in ppm)	Coupling	^{13}C signal (δ in ppm)	Assignment to the C^q atom number
H-11 (3.72)	$^3J_{C,H}$	163.1	3
H-7 (1.97)	$^3J_{C,H}$	196.4	9
H-8 (2.46)	$^2J_{C,H}$	196.4	9
H-6 (2.78)	$^3J_{C,H}$	125.8	10
	$^2J_{C,H}$	146.5	5
H-2 (6.68)	$^3J_{C,H}$	125.8	10
H-7 (1.97)	$^3J_{C,H}$	146.5	5

The assignment of the quaternary carbon atom C^q-3 is unequivocal because of the cross peak to H-11. Also unequivocally can the ^{13}C NMR signal at 196.4 ppm be assigned to the carbonyl carbon C^q-9. The signal at 125.8 ppm must be unequivocally assigned to C^q-10 because the distance to C^q-5 is too large. Thus, the remaining ^{13}C NMR signal at 146.5 ppm belongs to the quaternary carbon atom C^q-5 which is confirmed by the cross peak to H-7. Thus, *all* NMR signals could be unambiguously assigned.

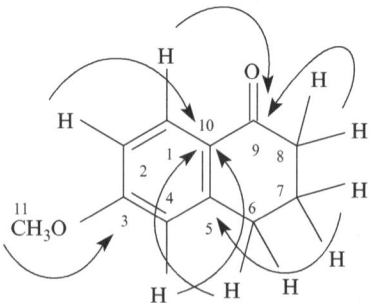

Fig. 5.1M Important correlations to quaternary carbon atoms

L. Mass spectrometric fragmentations

The mass spectrometric fragmentations (Fig. 5.1N) are in accordance with the deduced structure, *3-methoxy-tetralone* (Structure 5.3). However, the isomeric Structure 5.2 should give rise to a similar mass spectrum.

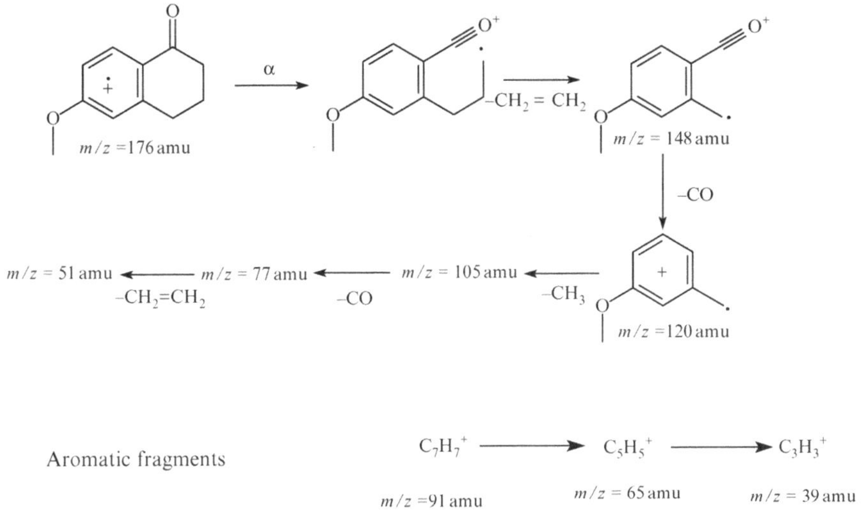

Fig. 5.1N Mass spectrometric fragmentations

Whereas isomeric structures do not give rise to significantly different fragmentation patterns for the aromatic moiety, structures possessing a β-ordered carbonyl group (Structure 5.4) should provide quite different mass spectra. Thus, the elimination of ketene gives the base peak at $m/z = 134$ amu:

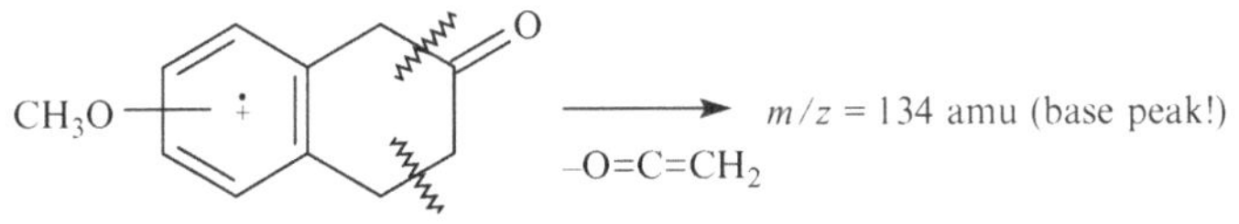

Challenge 5.2

Deduce the structure of the unknown compound by means of the given spectra and assign the spectroscopic data as is presented in Challenge 5.1.
   For solutions to Challenge 5.2, see *extras.springer.com*.

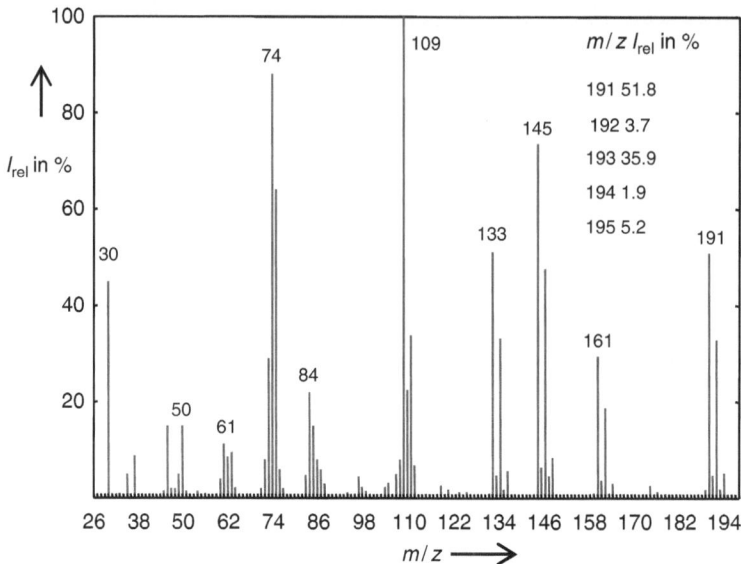

Fig. 5.2A Mass spectrum (EI, 70 eV)

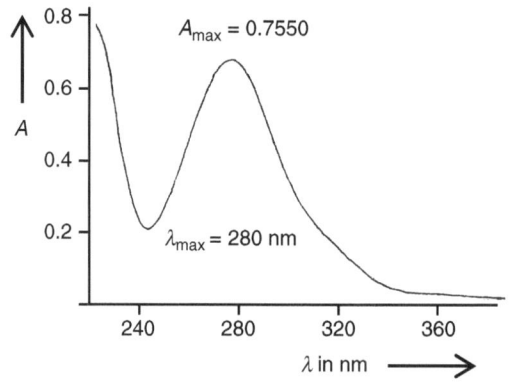

Fig. 5.2B UV spectrum

Stock solution:	0.776 mg in 5 mL
Analytical sample:	1-mL stock solution diluted to 10 mL
Solvent:	Methanol
Cuvette:	$l = 1$ cm

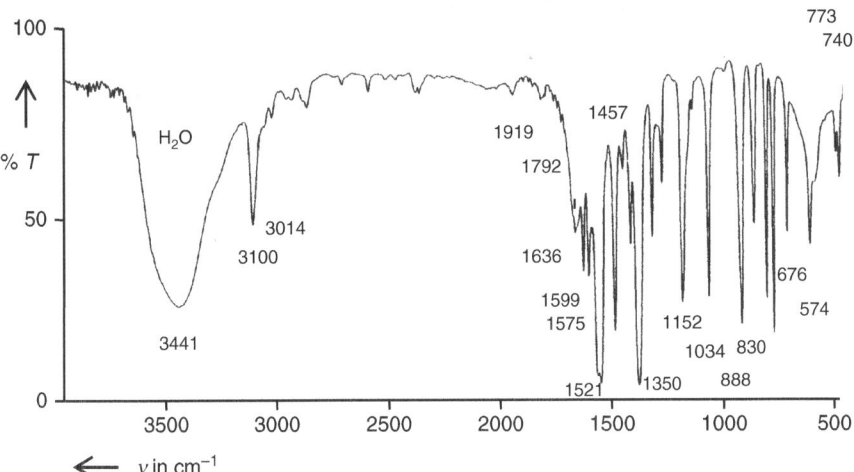

Fig. 5.2C IR spectrum (KBr pellet)

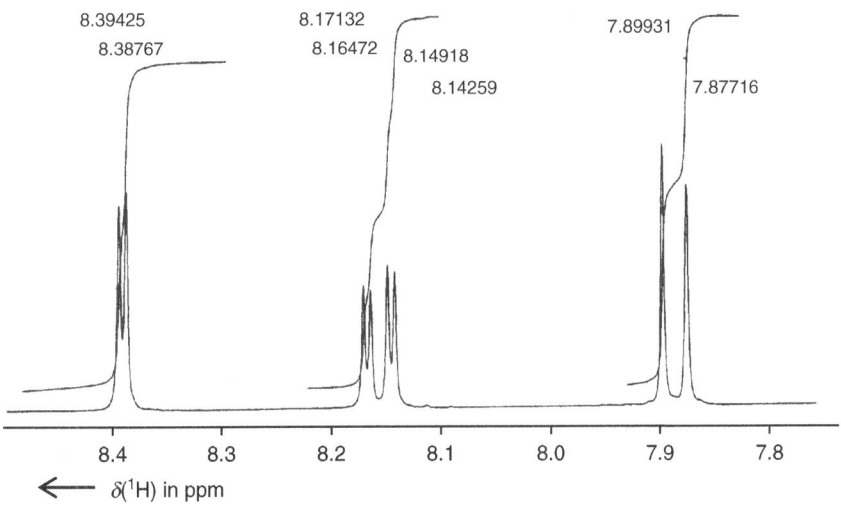

Fig. 5.2D ^{1}H NMR spectrum ($\nu_0 = 400$ MHz; solvent: DMSO)

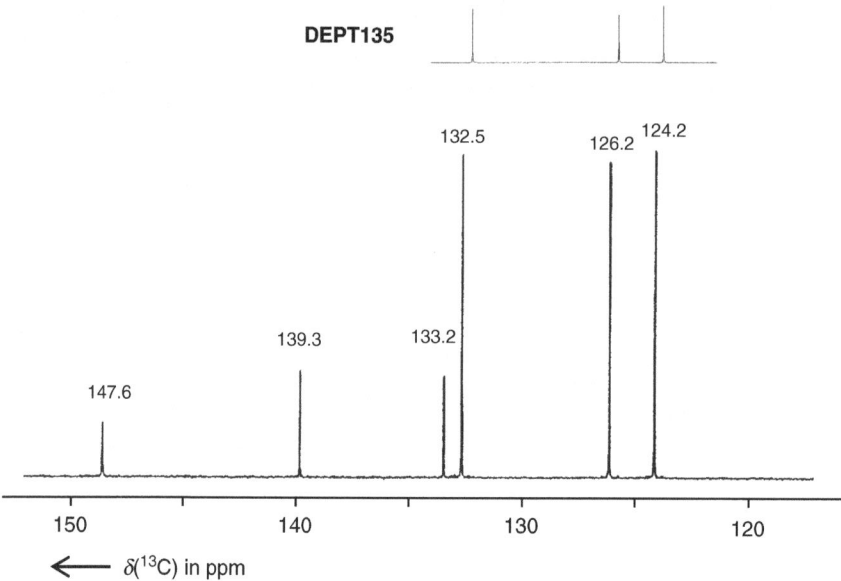

Fig. 5.2E ^{13}C NMR and DEPT135 spectra ($v_0 = 100$ MHz; solvent: DMSO)

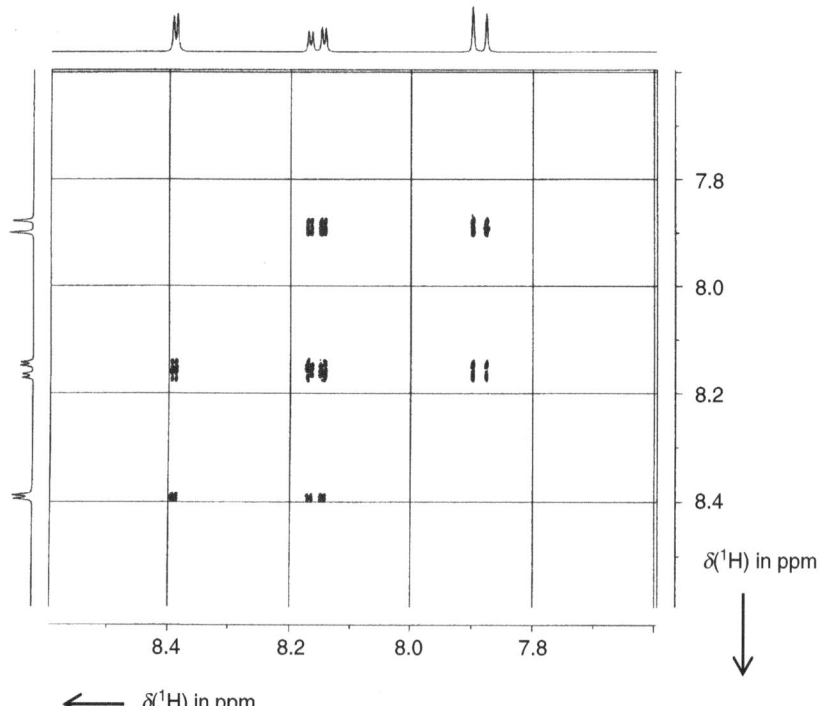

Fig. 5.2F (^{1}H,^{1}H) COSY spectrum

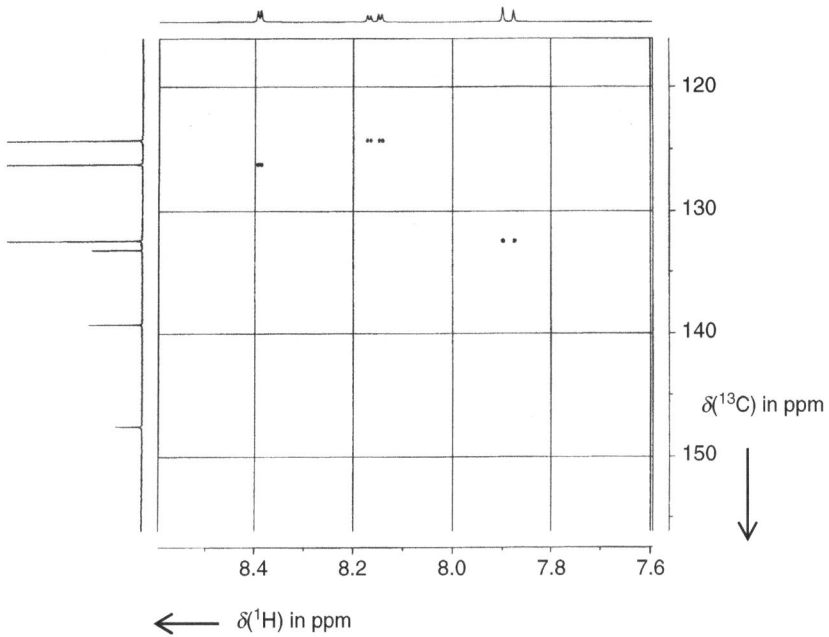

Fig. 5.2G HSQC spectrum

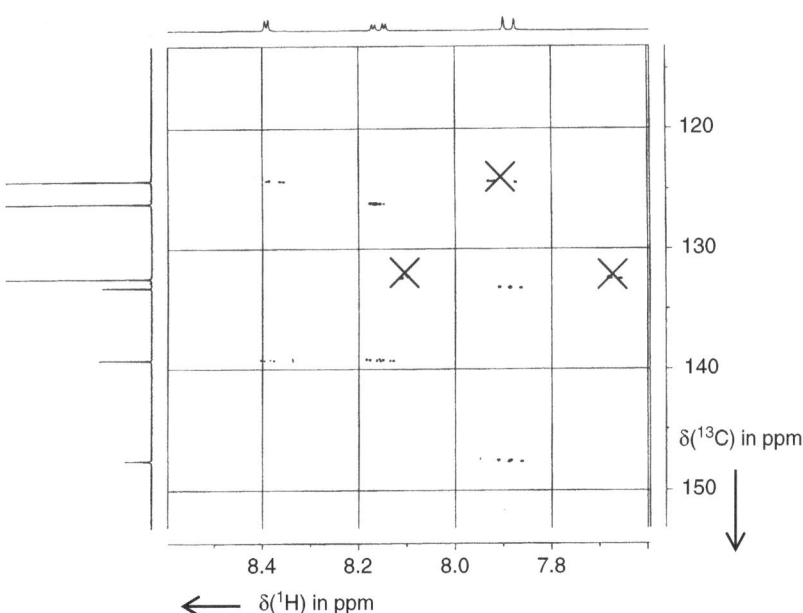

Fig. 5.2H HMBC spectrum

Challenge 5.3
Deduce the structure of the unknown compound by means of the given spectra and assign the spectroscopic data as is presented in Challenge 5.1.
For solutions to Challenge 5.3, see *extras.springer.com*.

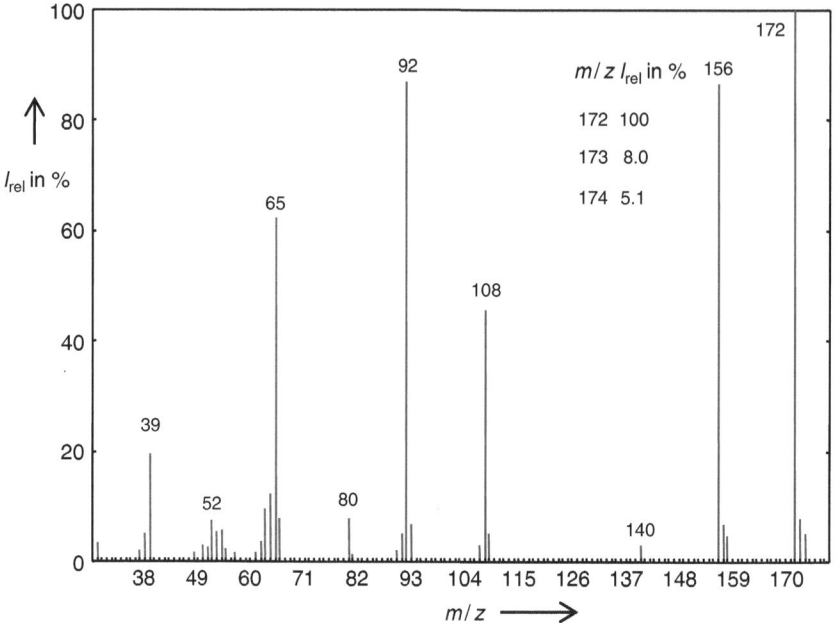

Fig. 5.3A Mass spectrum (EI, 70 eV)

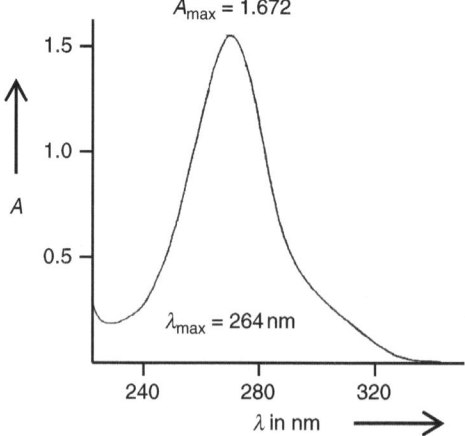

Fig. 5.3B UV spectrum

Stock solution:	0.756 mg in 5 mL
Analytical sample:	1-mL stock solution diluted to 10 mL
Solvent:	Methanol
Cuvette:	l = 1 cm

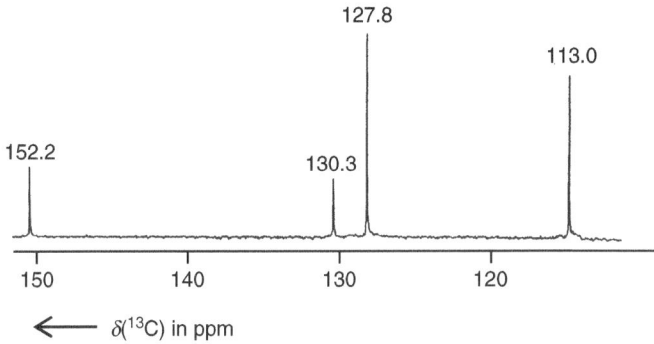

Fig. 5.3C ^{13}C NMR spectrum (v_0 = 100 MHz; solvent: DMSO)

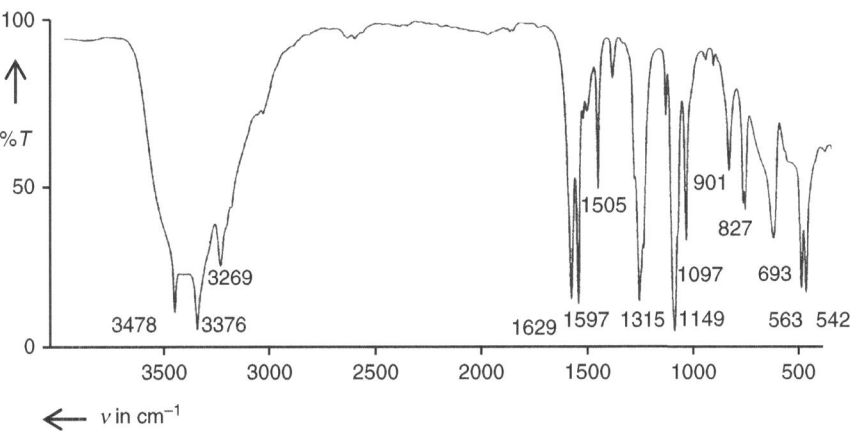

Fig. 5.3D IR spectrum (KBr pellet)

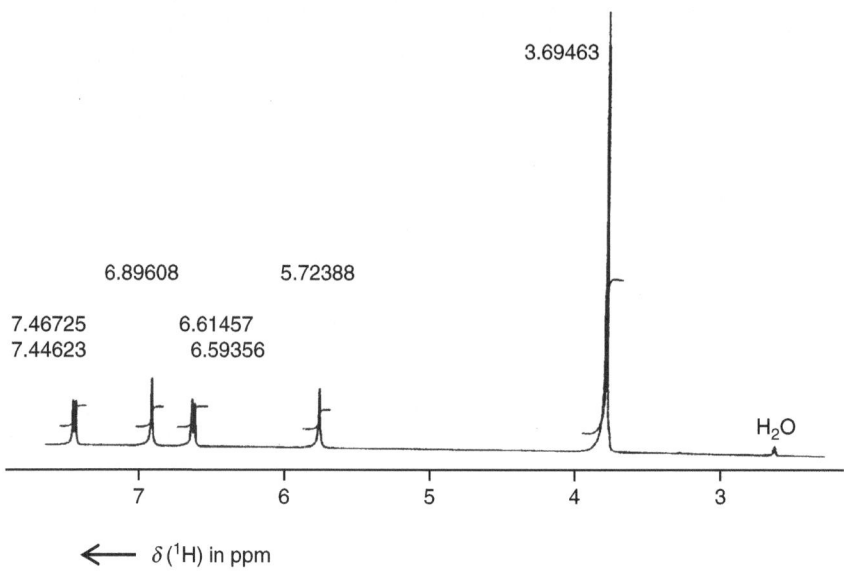

Fig. 5.3E ^{1}H NMR spectrum ($\nu_0 = 400$ MHz; solvent: DMSO)

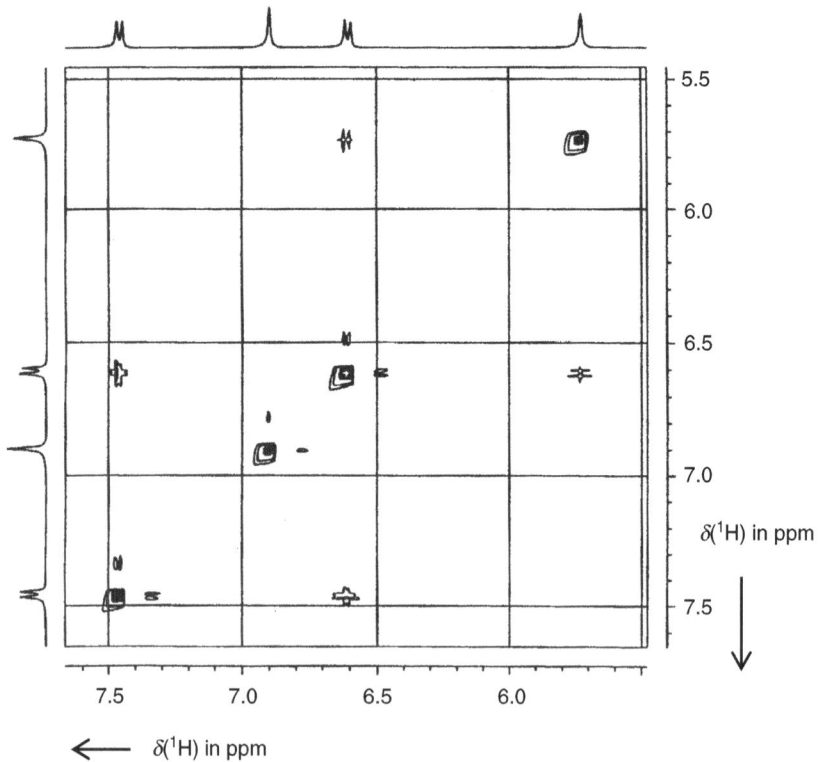

Fig. 5.3F   NOESY spectrum

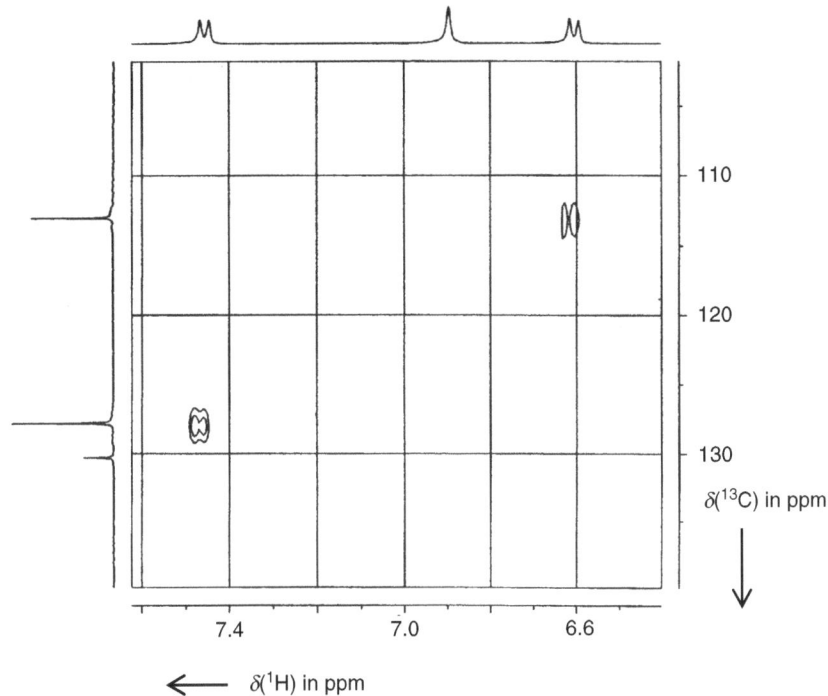

Fig. 5.3G HSQC spectrum

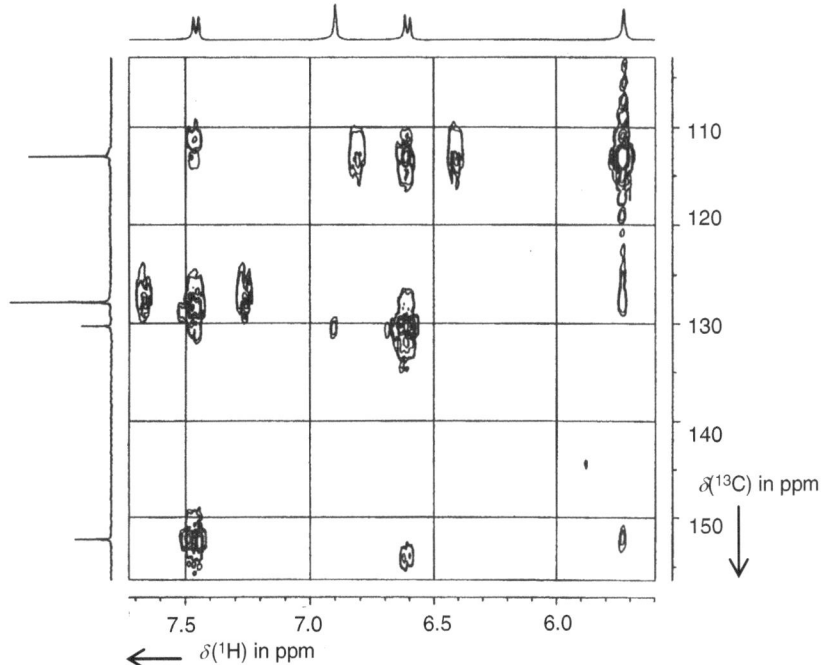

Fig. 5.3H HMBC spectrum

Challenge 5.4

Deduce the structure of the unknown compound by means of the given spectra and assign the spectroscopic data as is presented in Challenge 5.1.

For solutions to Challenge 5.4, see *extras.springer.com*.

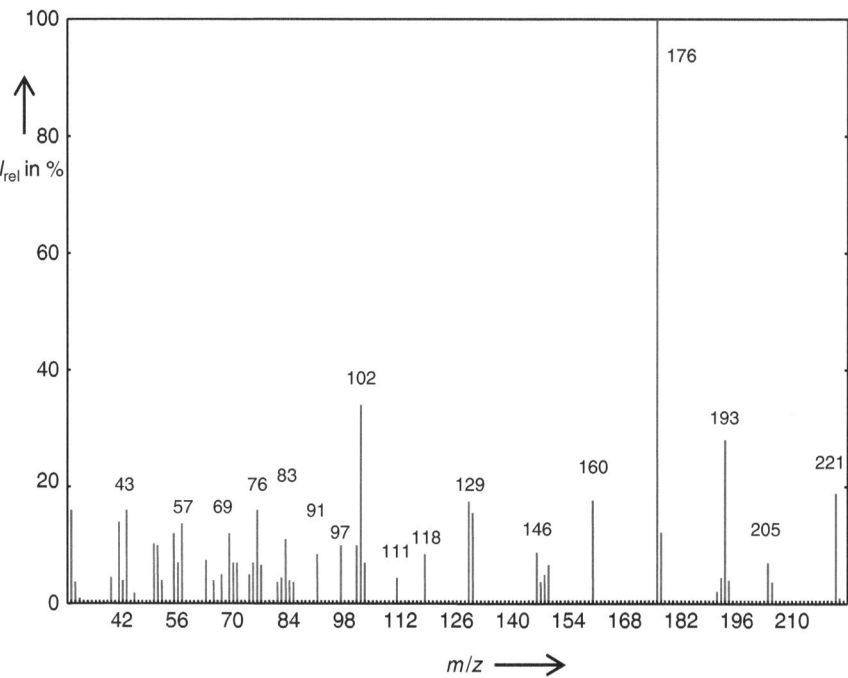

Fig. 5.4A   Mass spectrum (EI, 70 eV)

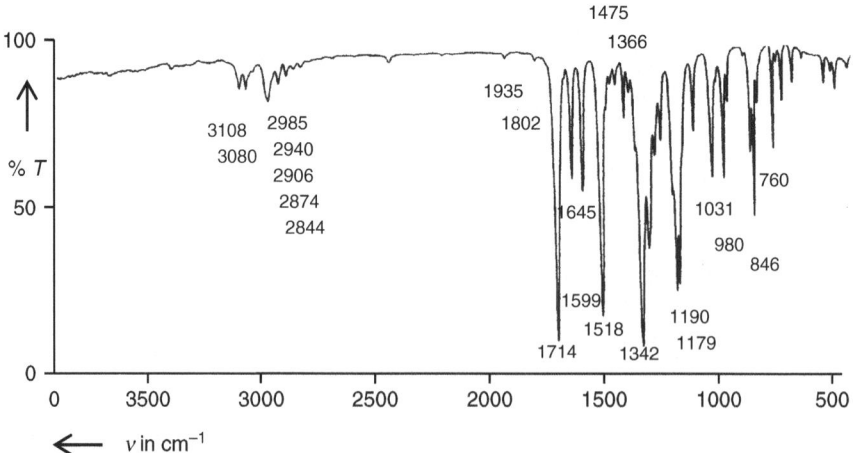

Fig. 5.4B   IR spectrum (KBr pellet)

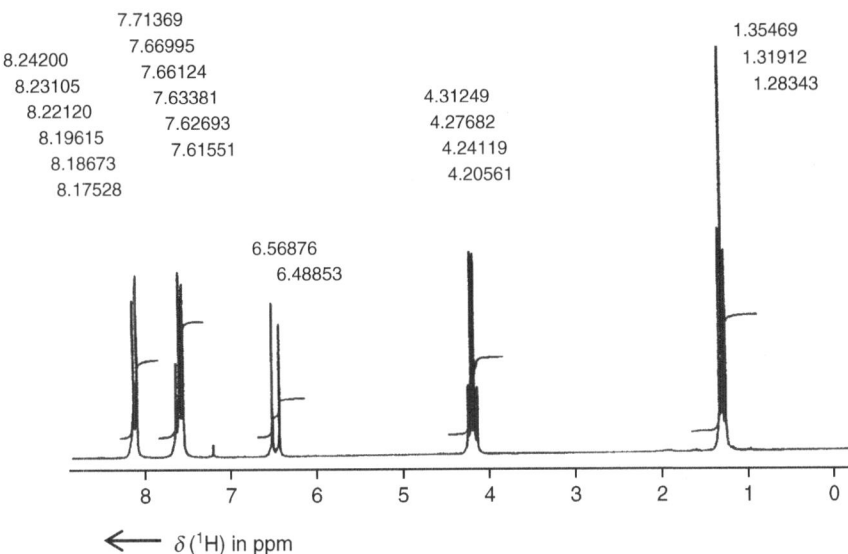

Fig. 5.4C ^{1}H NMR spectrum (v_0 = 200 MHz; solvent: CDCl$_3$)

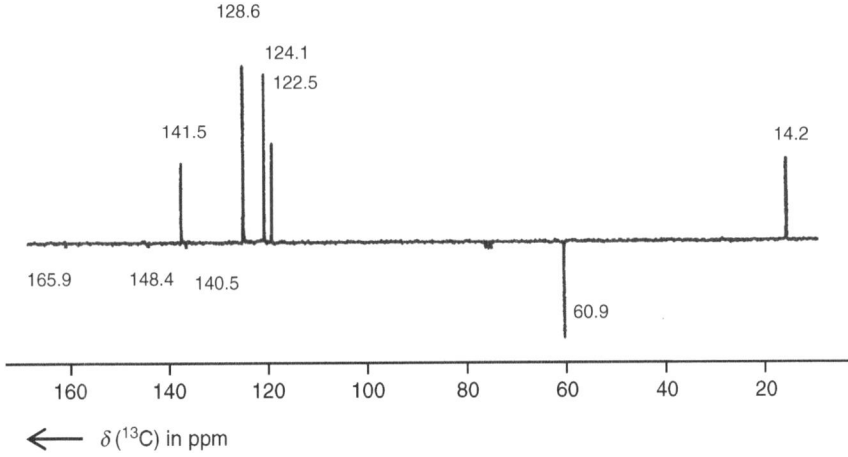

Fig. 5.4D ^{13}C NMR pendant spectrum (v_0 = 25 MHz; solvent: CDCl$_3$)

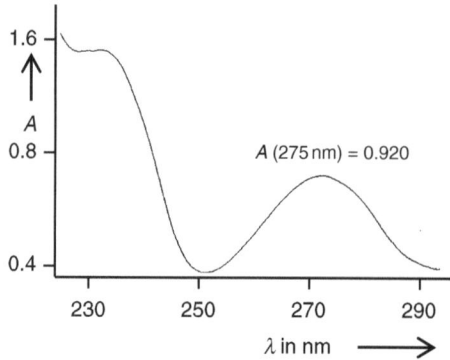

Fig. 5.4E UV spectrum

Stock solution:	0.358 mg in 5 mL
Analytical sample:	1-mL stock solution diluted to 10 mL
Solvent:	Methanol
Cuvette:	$l = 1$ cm

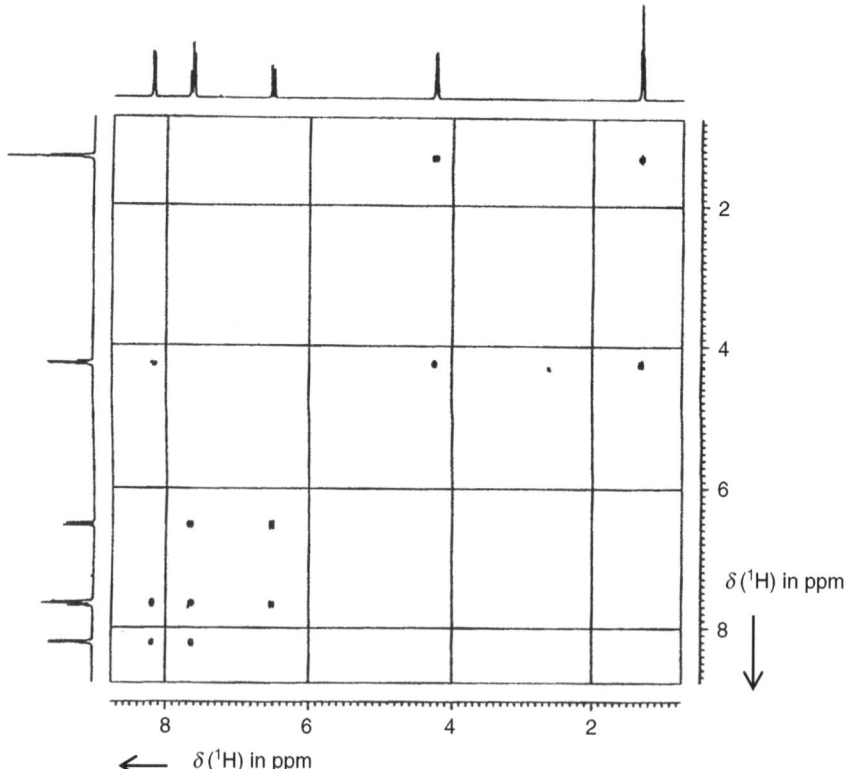

Fig. 5.4F (^{1}H,^{1}H) COSY spectrum

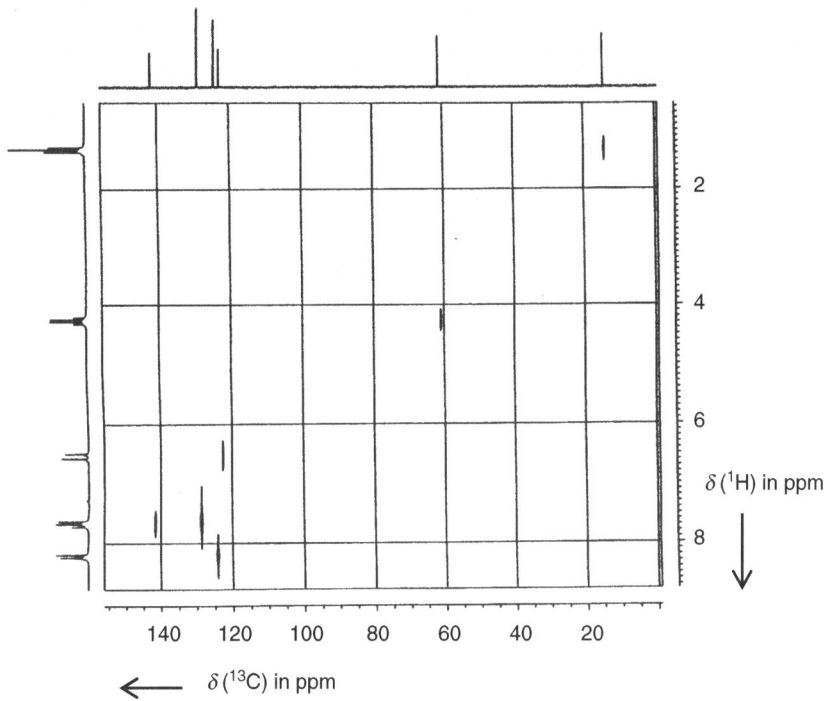

Fig. 5.4G HSQC spectrum

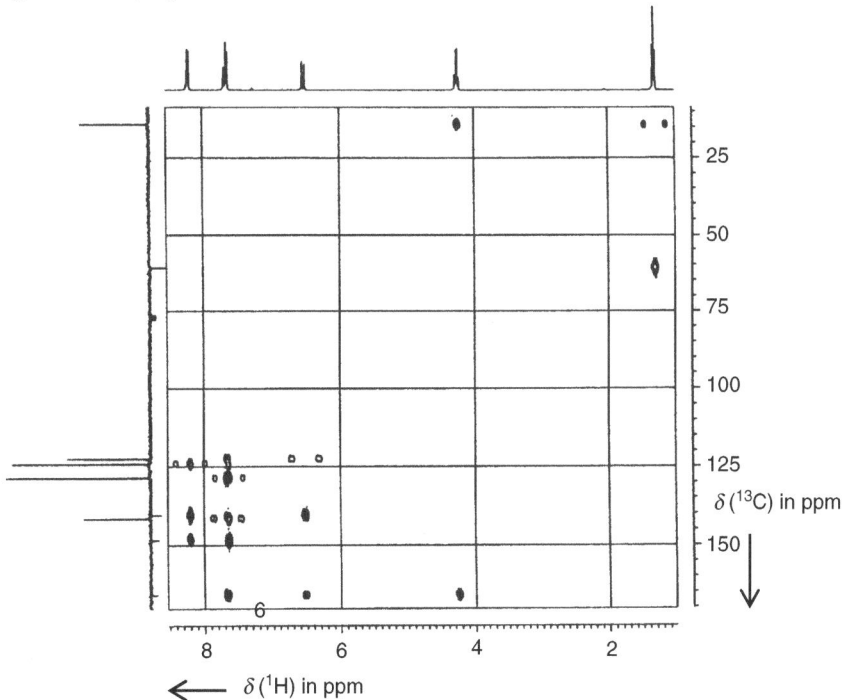

Fig. 5.4H HMBC spectrum

Challenge 5.5
Deduce the structure of the unknown compound by means of the given spectra and assign the spectroscopic data as is presented in Challenge 5.1.
   For solutions to Challenge 5.5, see *extras.springer.com*.

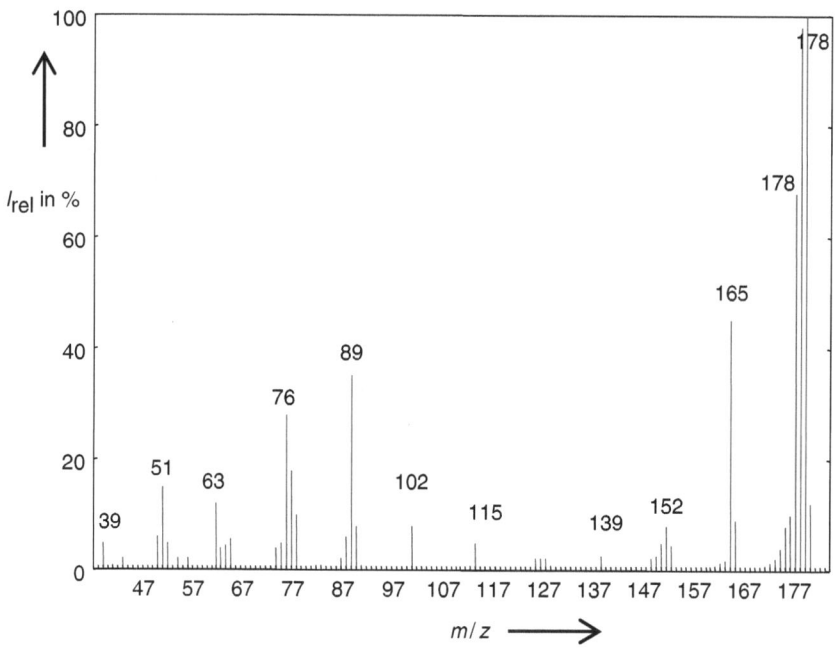

Fig. 5.5A Mass spectrum (EI, 70 eV)

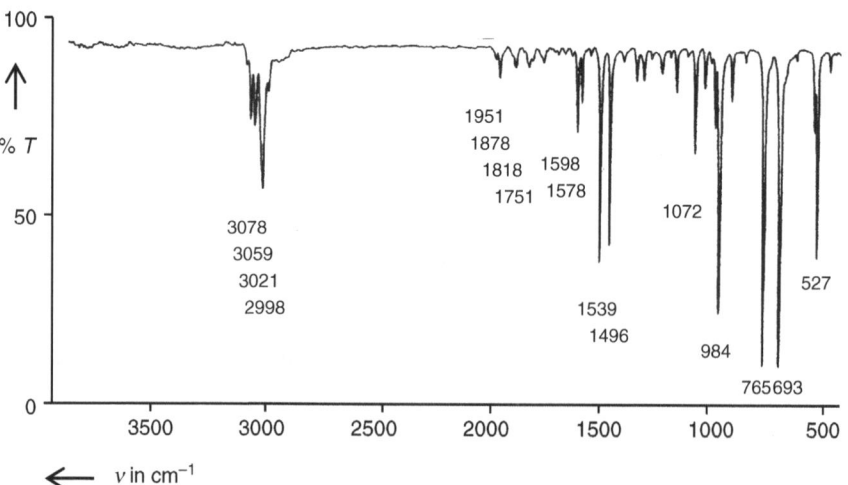

Fig. 5.5B IR spectrum (KBr pellet)

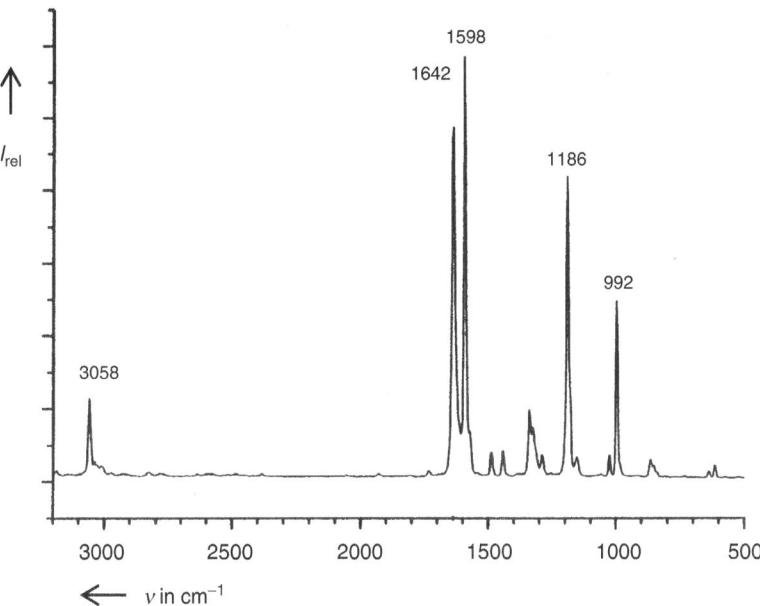

Fig. 5.5C Raman spectrum

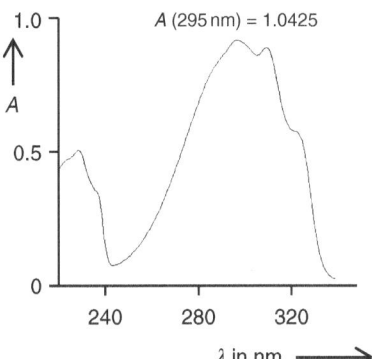

Fig. 5.5D UV spectrum

Stock solution:	0.339 mg in 5 mL
Analytical sample:	1-mL stock solution diluted to 10 mL
Solvent:	Methanol
Cuvette:	l = 1 cm

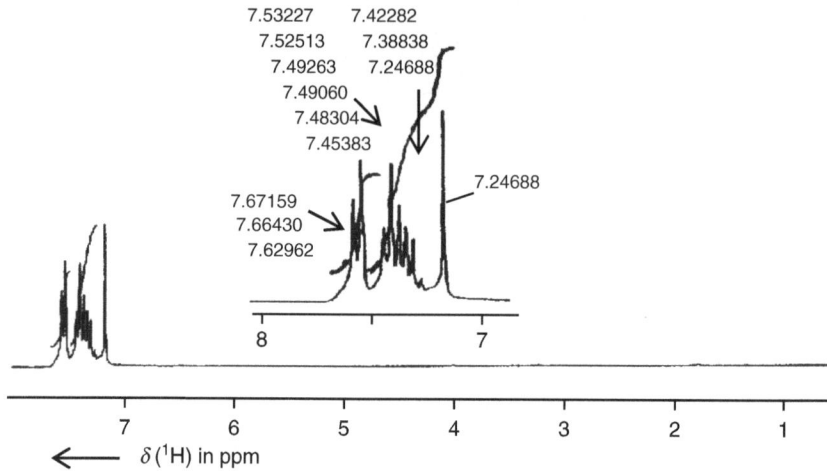

Fig. 5.5E ^{1}H NMR spectrum ($\nu_0 = 200$ MHz; solvent: CDCl$_3$)

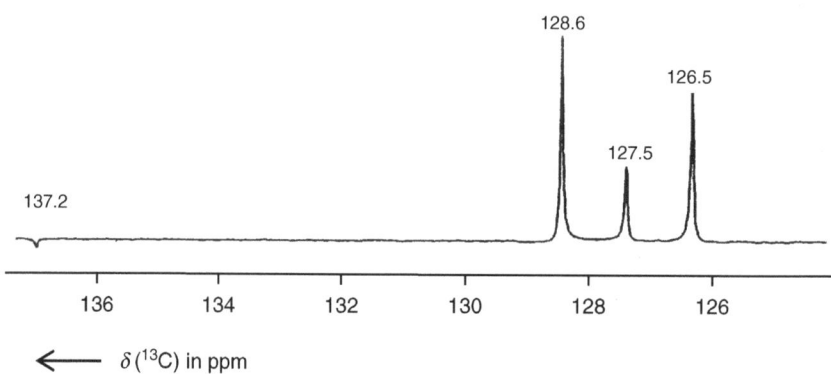

Fig. 5.5F ^{13}C NMR pendant spectrum ($\nu_0 = 25$ MHz; solvent: CDCl$_3$)

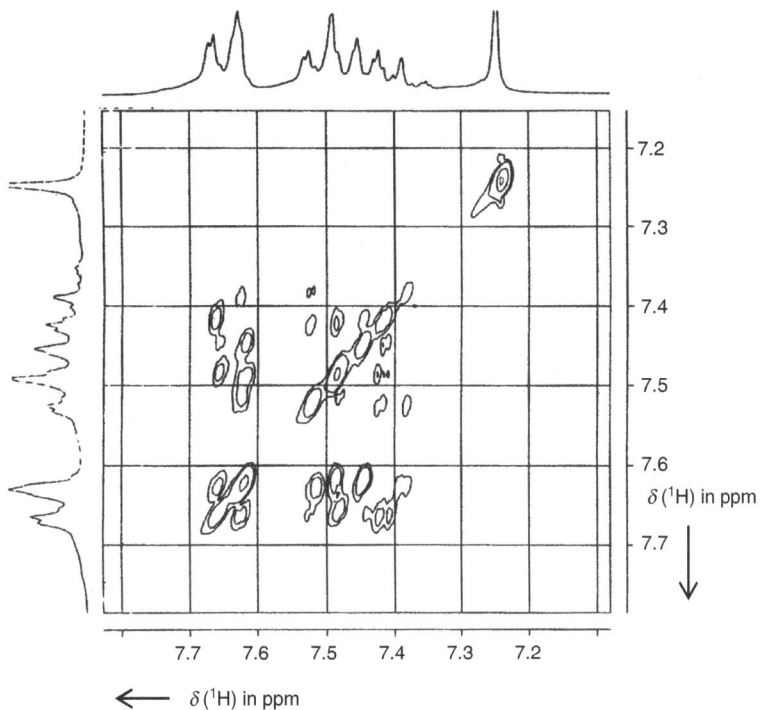

Fig. 5.5G (^{1}H,^{1}H) COSY spectrum

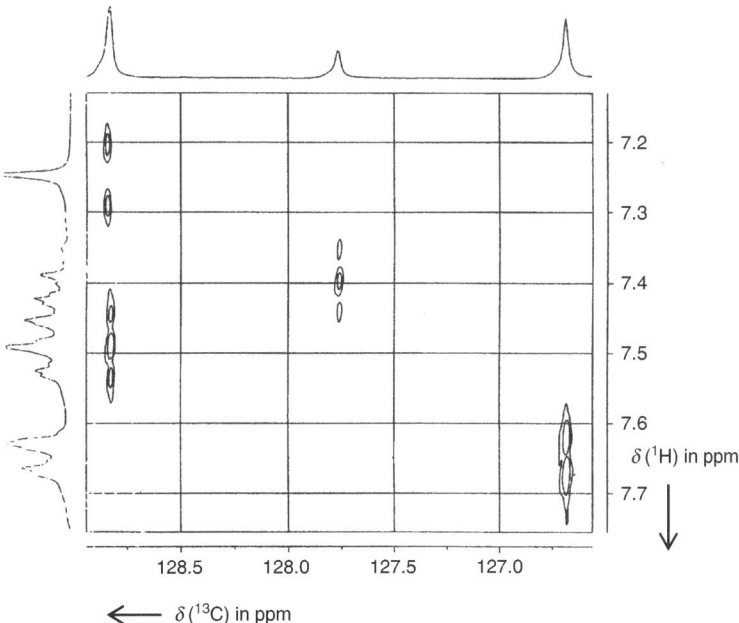

Fig. 5.5H HSQC spectrum

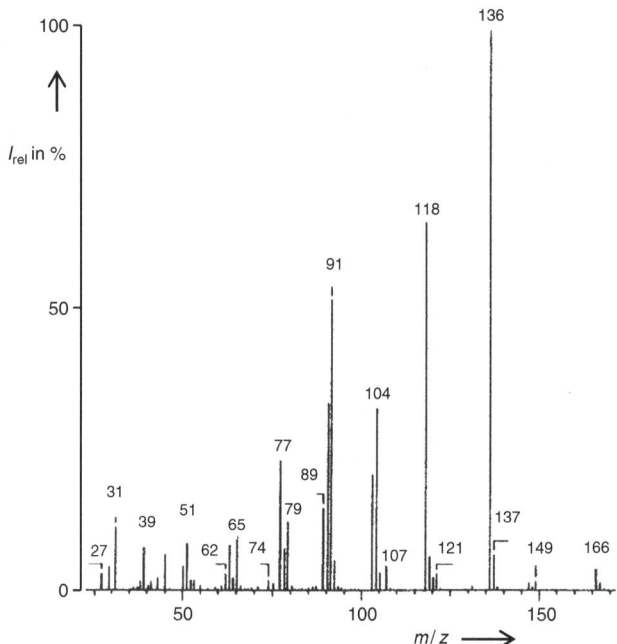

Fig. 5.6A Mass spectrum (EI, 70 eV)

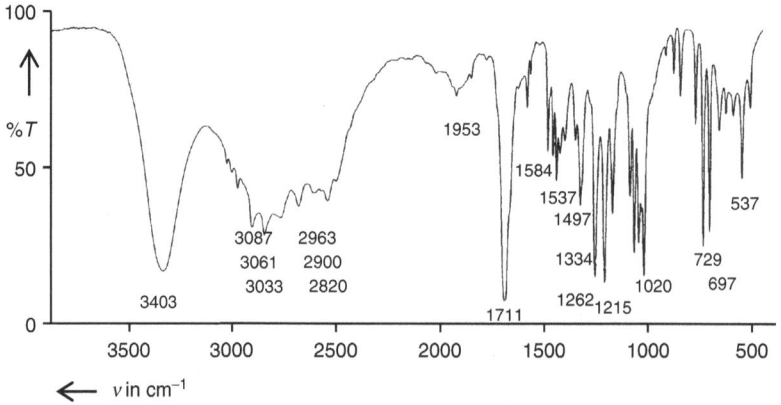

Fig. 5.6B IR spectrum (KBr pellet)

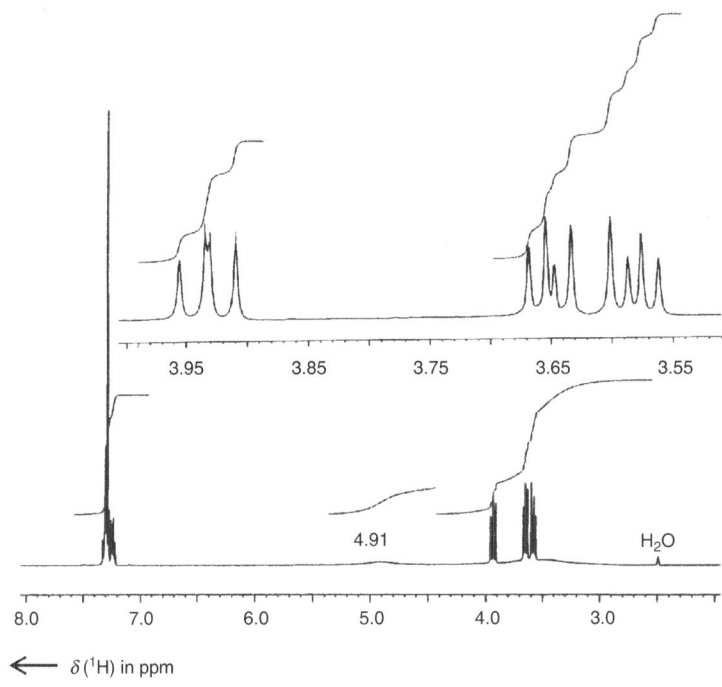

Fig. 5.6C ^{1}H NMR spectrum ($\nu_0 = 400$ MHz; solvent: DMSO)

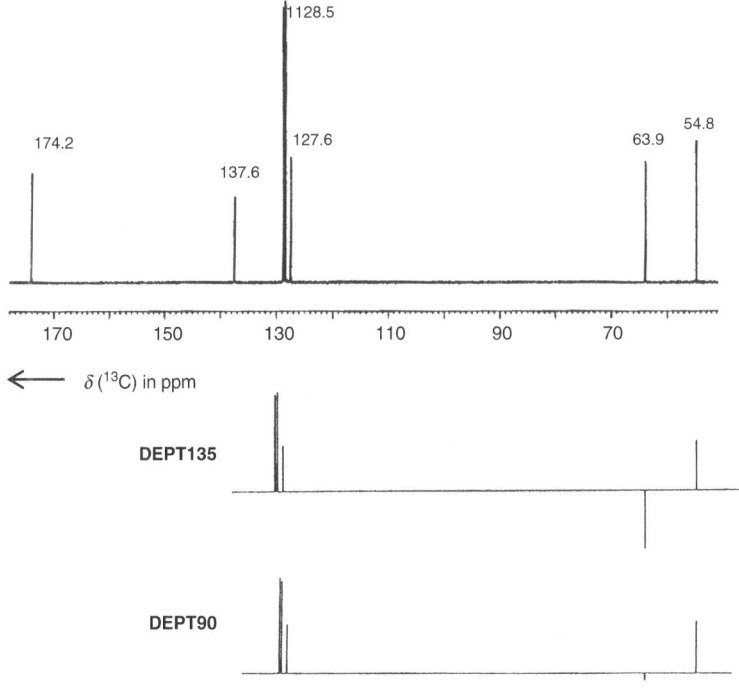

Fig. 5.6D ^{13}C NMR and DEPT spectra ($\nu_0 = 100$ MHz; solvent: DMSO)

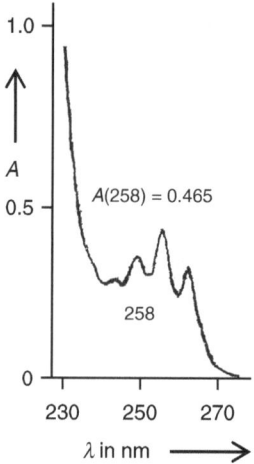

Fig. 5.6E UV spectrum

Analytical sample:	2.043 mg in 5 mL
Solvent:	Methanol
Cuvette:	l = 1 cm

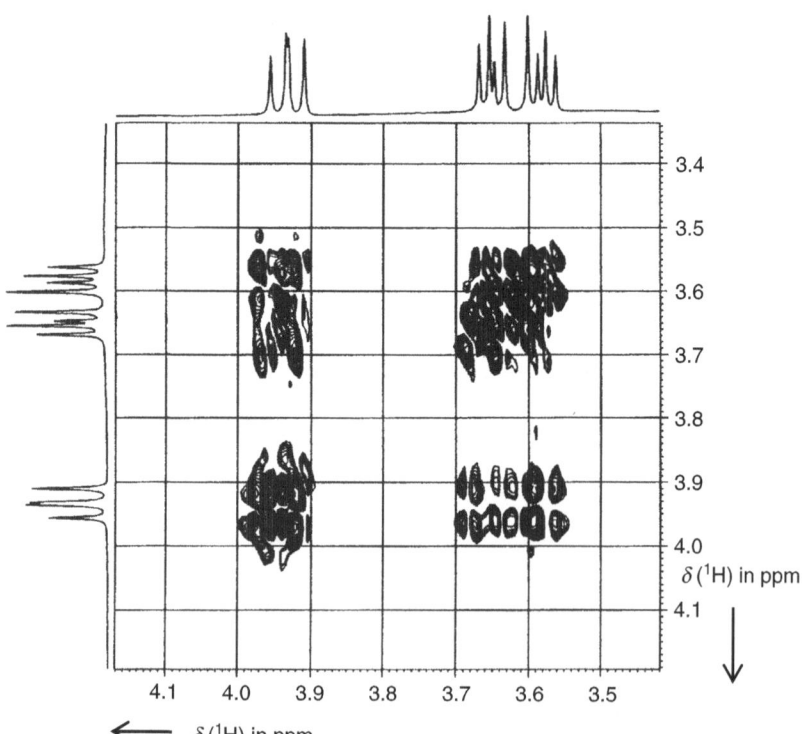

Fig. 5.6F (^{1}H,^{1}H) COSY spectrum

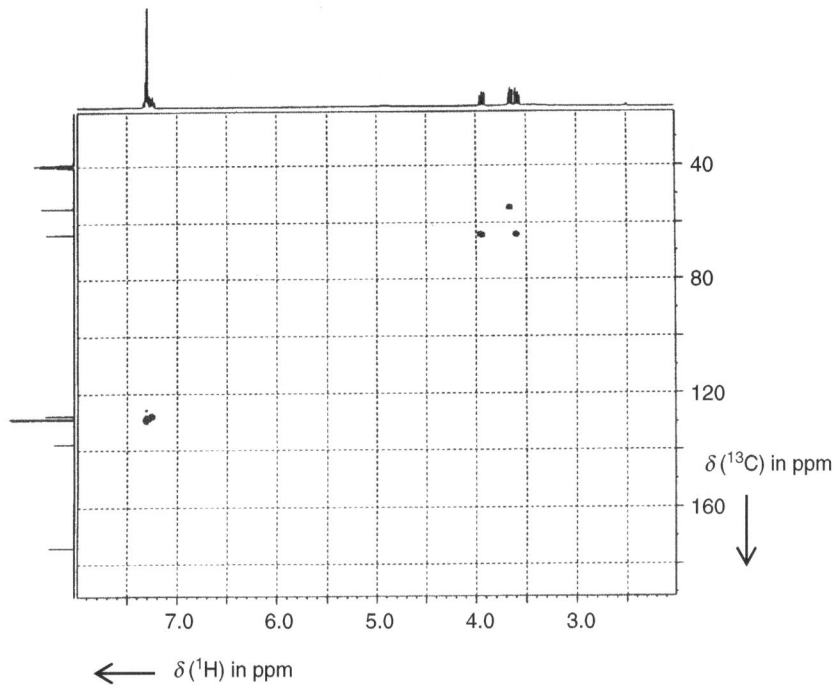

Fig. 5.6G   HMQC spectrum

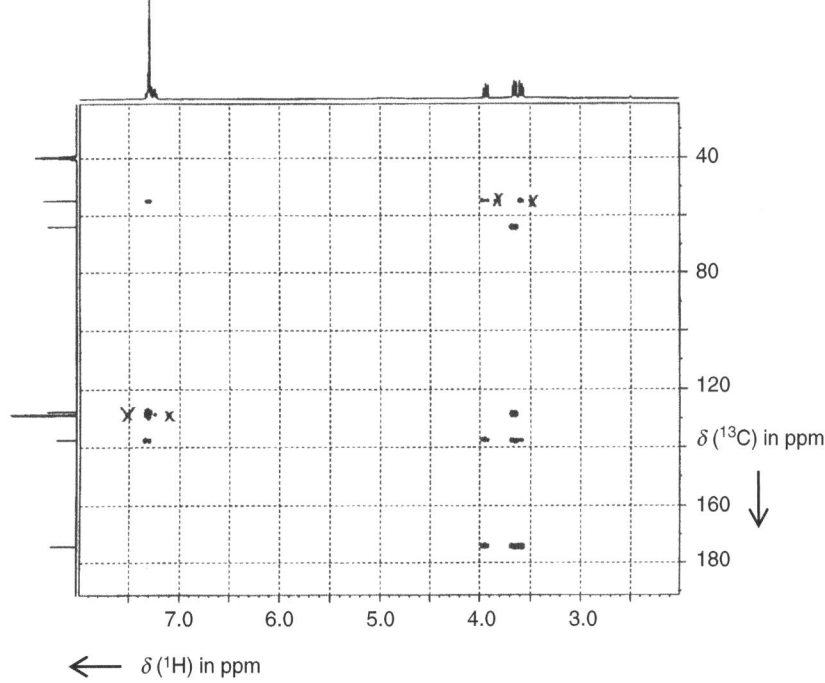

Fig. 5.6H   HMBC spectrum

Challenge 5.7

Deduce the structure of the unknown compound by means of the given spectra and assign the spectroscopic data as is presented in Challenge 5.1.

   For solutions to Challenge 5.7, see *extras.springer.com*.

UV: Analytical sample: 8.3 mg in 5 mL
   Cuvette: $l = 1$ cm
   Structureless absorption band: $\lambda_{max} = 285$ nm; $A(285$ nm$) = 0.653$

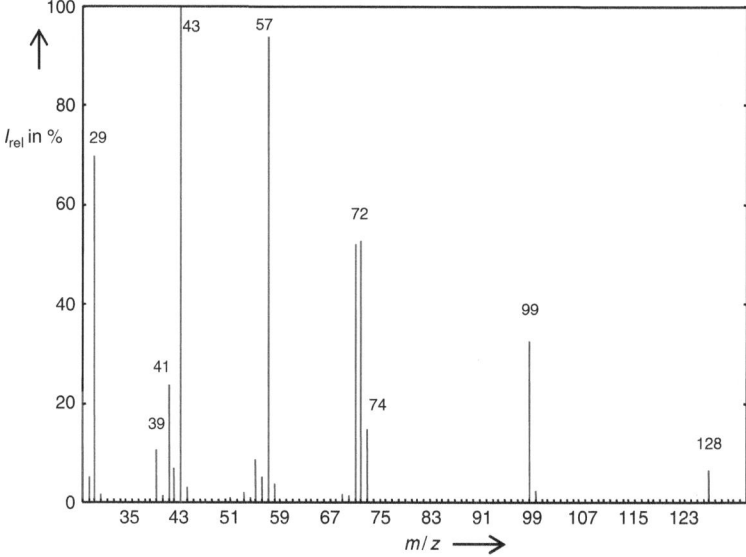

Fig. 5.7A Mass spectrum (EI, 70 eV)

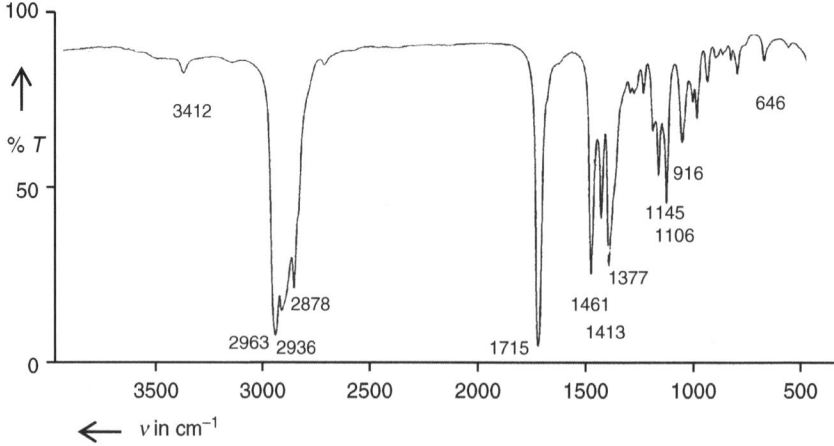

Fig. 5.7B IR spectrum (capillary thin film)

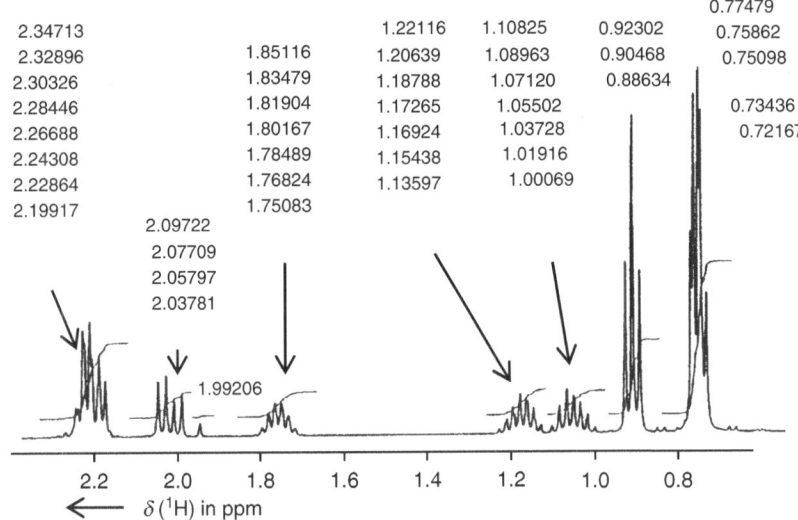

Fig. 5.7C ^{1}H NMR spectrum ($\nu_0 = 400$ MHz; solvent: CDCl$_3$)

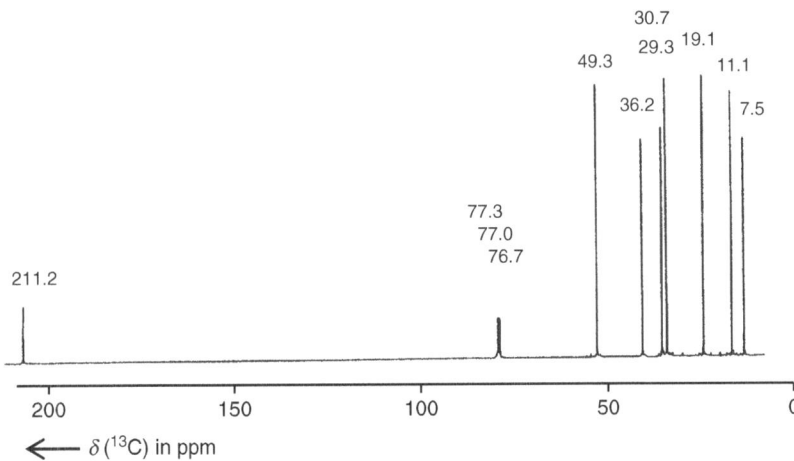

Fig. 5.7D ^{13}C NMR spectrum ($\nu_0 = 100$ MHz; solvent: CDCl$_3$)

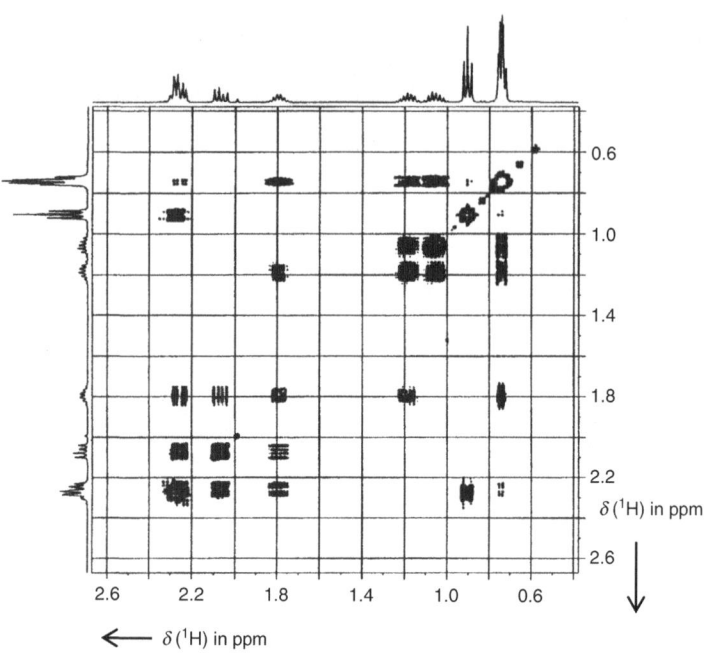

Fig. 5.7E   (¹H,¹H) COSY spectrum

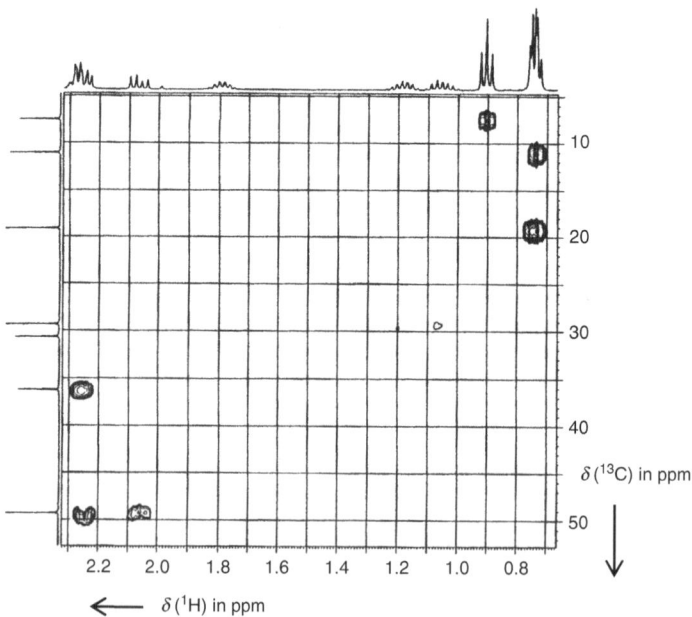

Fig. 5.7F   HSQC spectrum

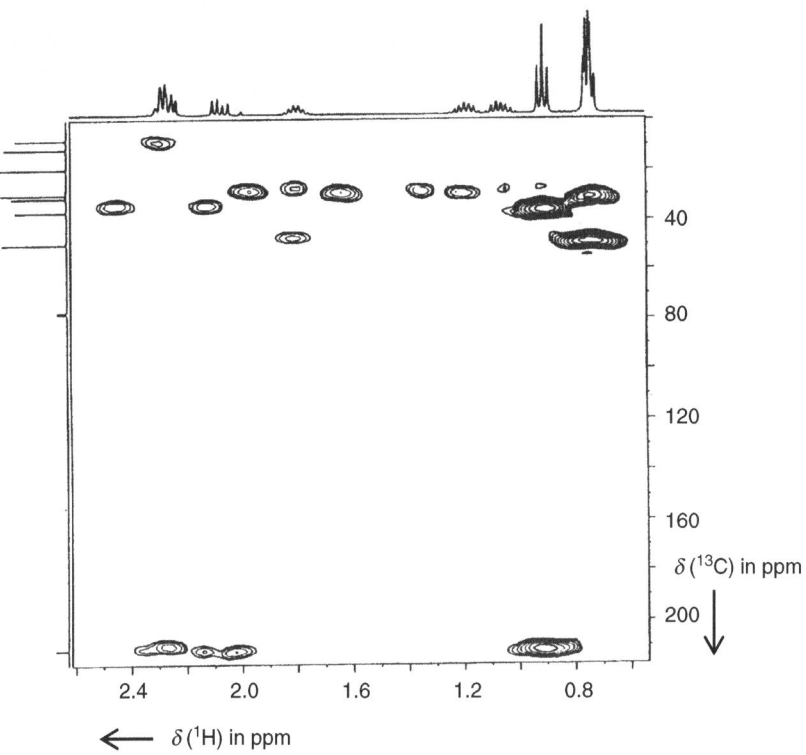

Fig. 5.7G HMBC spectrum

Challenge 5.8

Deduce the structure of the unknown compound by means of the given spectra and assign the spectroscopic data as is presented in Challenge 5.1.

For solutions to Challenge 5.8, see *extras.springer.com*.

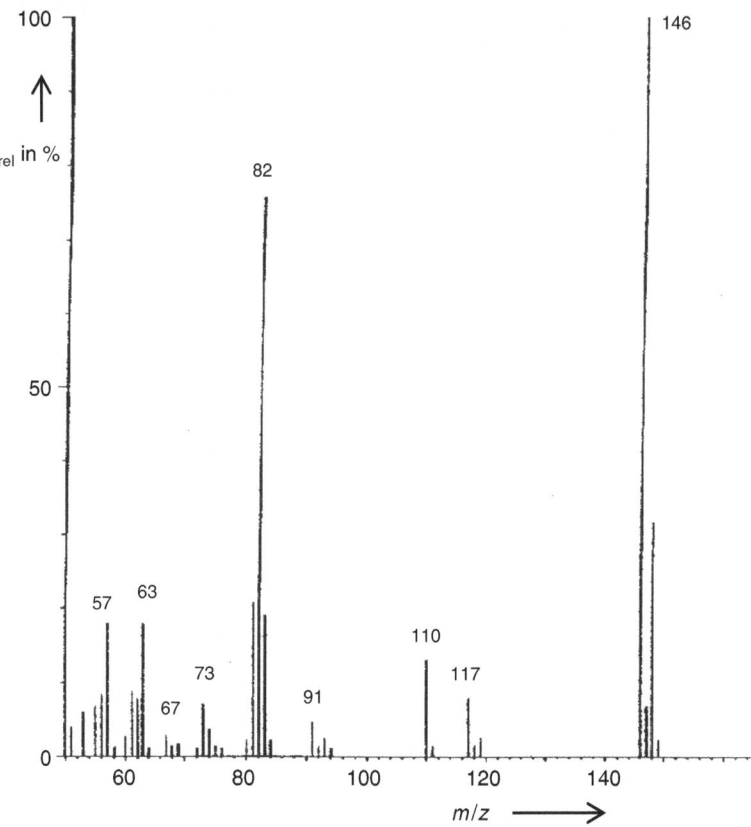

Fig. 5.8A　Mass spectrum (EI, 70 eV)

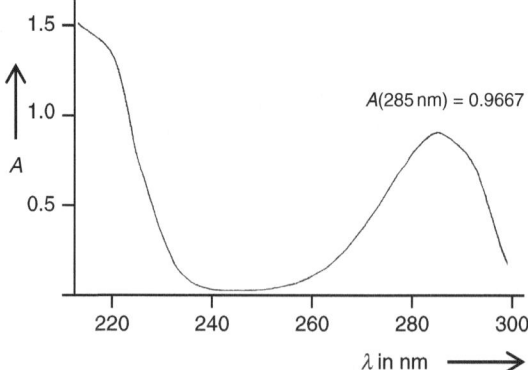

Fig. 5.8B　UV spectrum

Stock solution:	15 μL in 5 mL, $\rho = 1.35$ g mL^{-1}
Analytical sample:	1-mL stock solution diluted to 10 mL
Solvent:	Methanol
Cuvette:	$l = 1$ cm

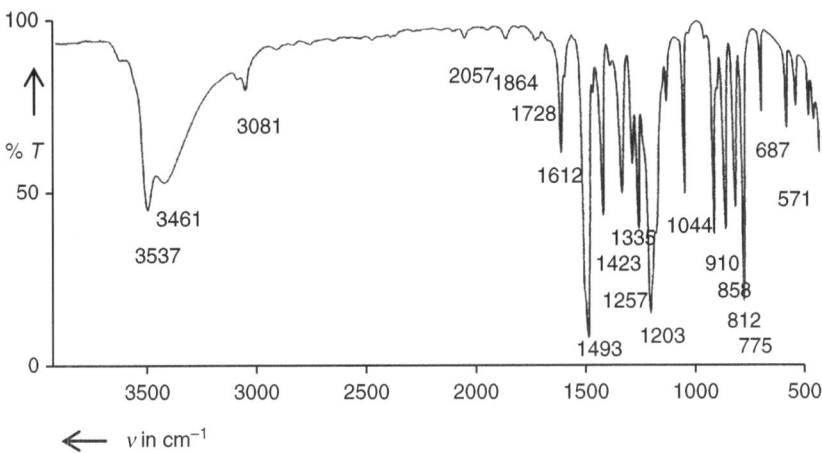

Fig. 5.8C   IR spectrum (capillary thin film)

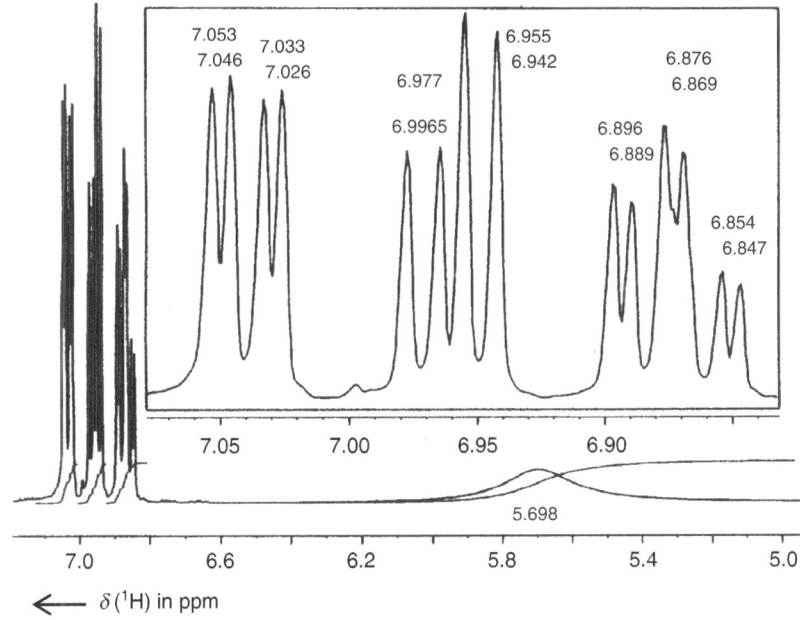

Fig. 5.8D ^{1}H NMR spectrum ($v_0 = 400$ MHz; solvent: CD$_3$CN)

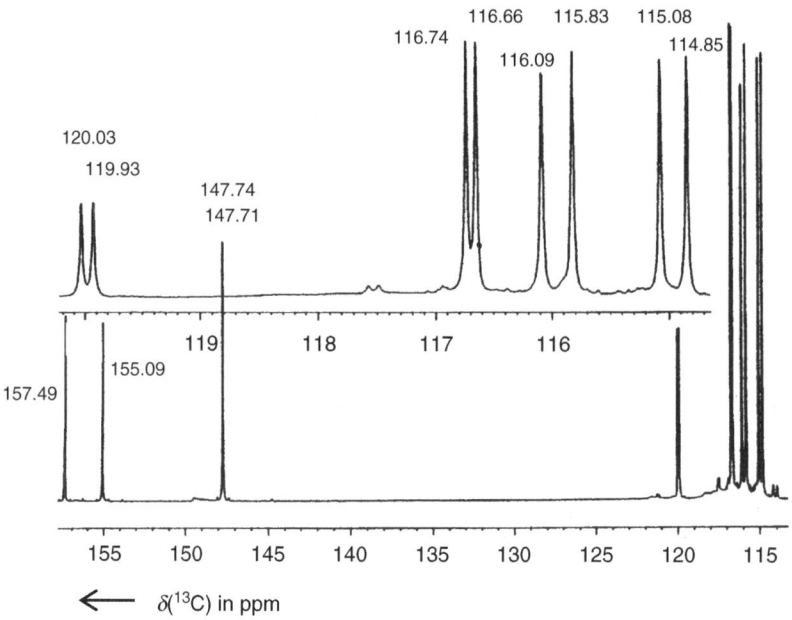

Fig. 5.8E ^{13}C NMR spectrum ($\nu_0 = 100$ MHz; solvent: CDCl$_3$)

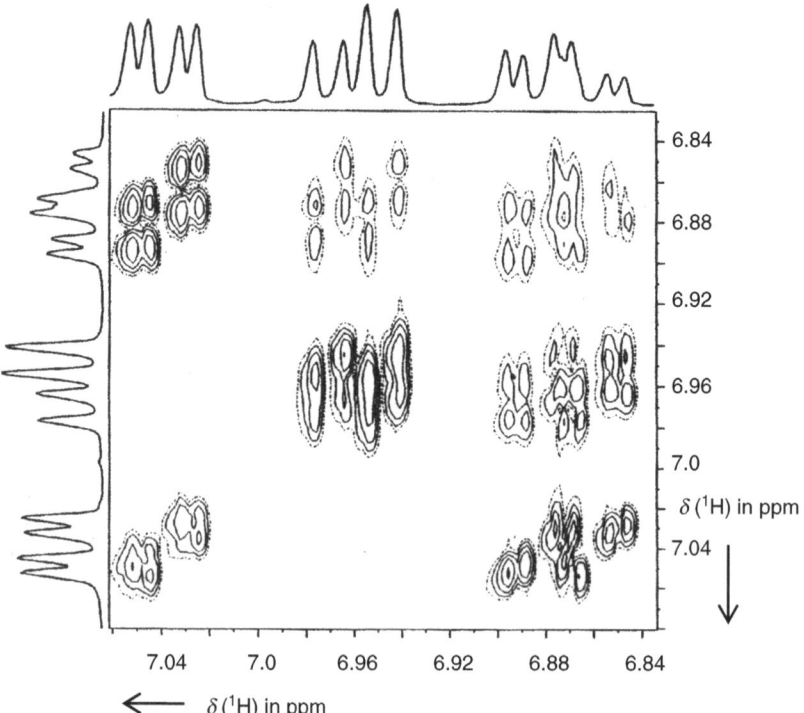

Fig. 5.8F (^{1}H,^{1}H) COSY spectrum

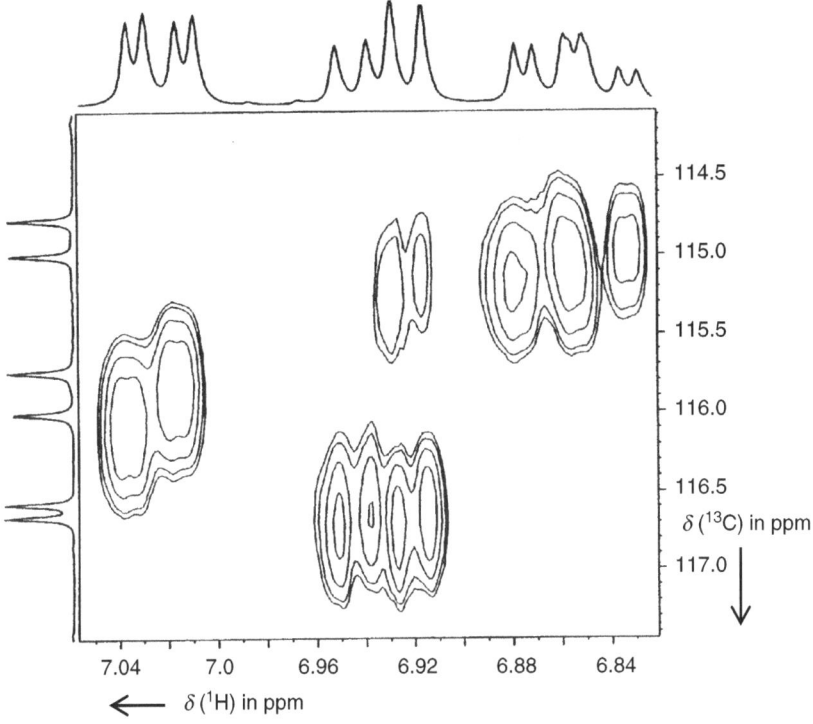

Fig. 5.8G HSQC spectrum

Challenge 5.9
Deduce the structure of the unknown compound by means of the given spectra and assign the spectroscopic data as is presented in Challenge 5.1.
  For solutions to Challenge 5.9, see *extras.springer.com*.

UV: Stock solution: 0.751 mg in 5 mL
      Analytical sample: 1-mL stock solution diluted to 10 mL
      Solvent: Methanol
      Cuvette: $l = 1$ cm
      Structureless absorption band: $\lambda_{max} = 272$ nm; $A(272$ nm$) = 0.7398$

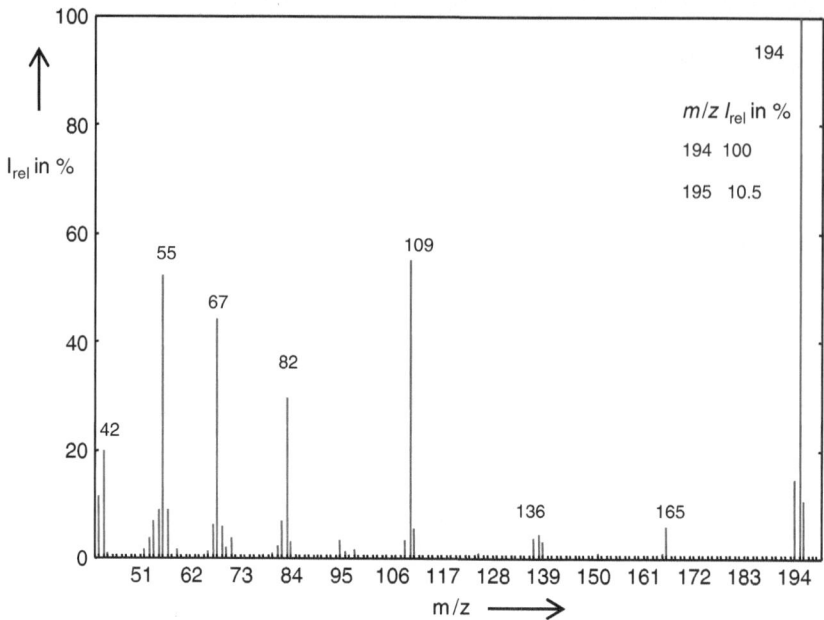

Fig. 5.9A Mass spectrum (EI, 70 eV)

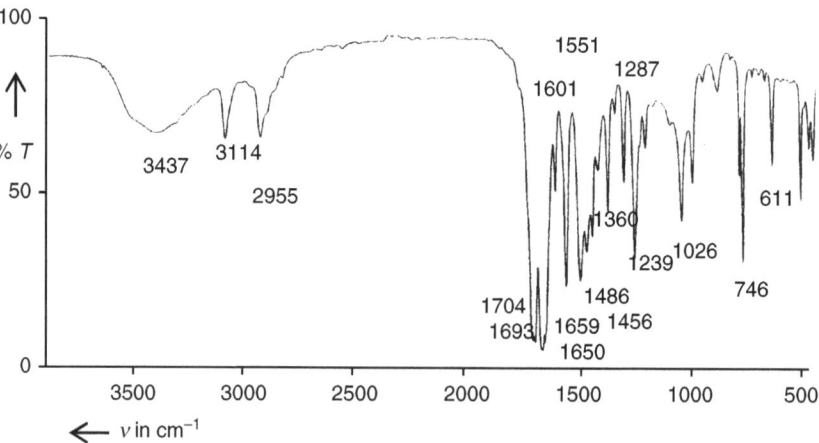

Fig. 5.9B IR spectrum (KBr pellet)

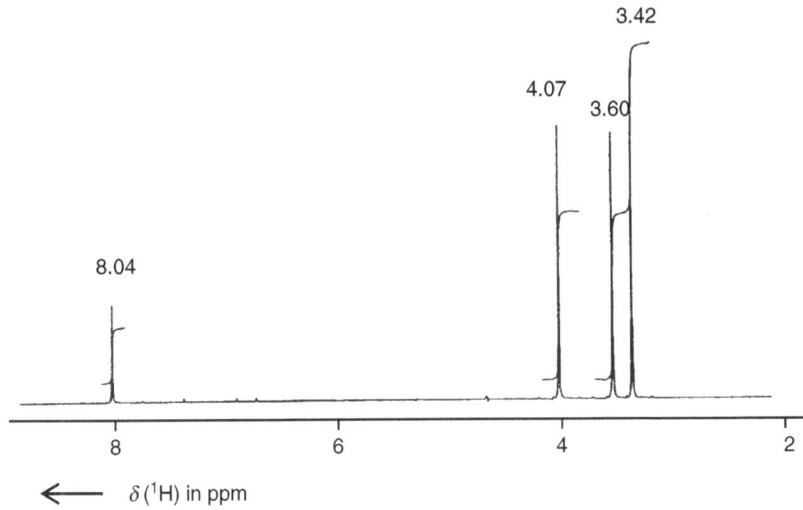

Fig. 5.9C ^{1}H NMR spectrum ($v_0 = 400$ MHz; solvent: D_2O)

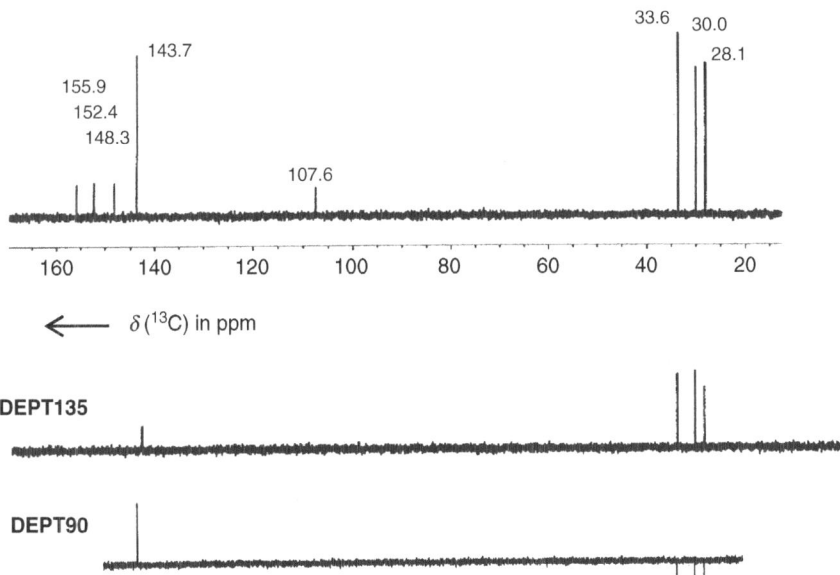

Fig. 5.9D ^{13}C NMR and DEPT spectra ($v_0 = 100$ MHz; solvent: D_2O)

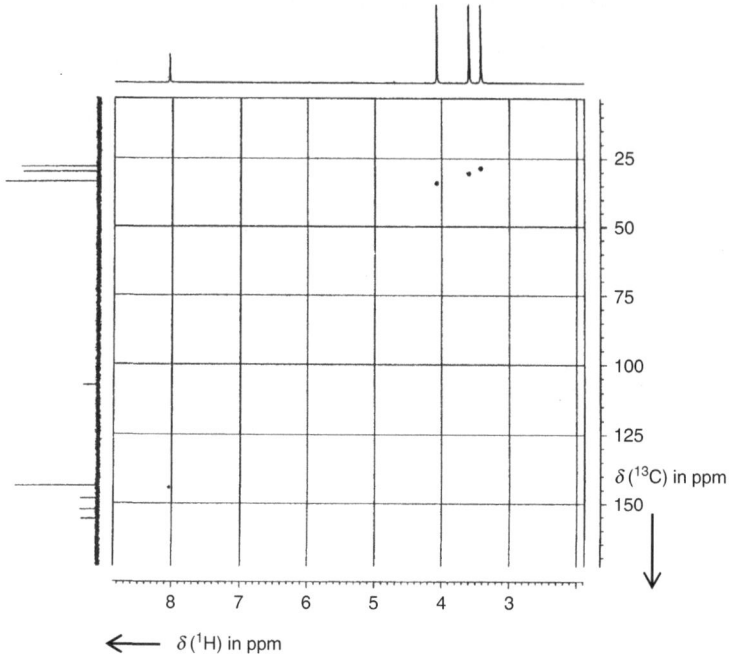

Fig. 5.9E HMQC spectrum

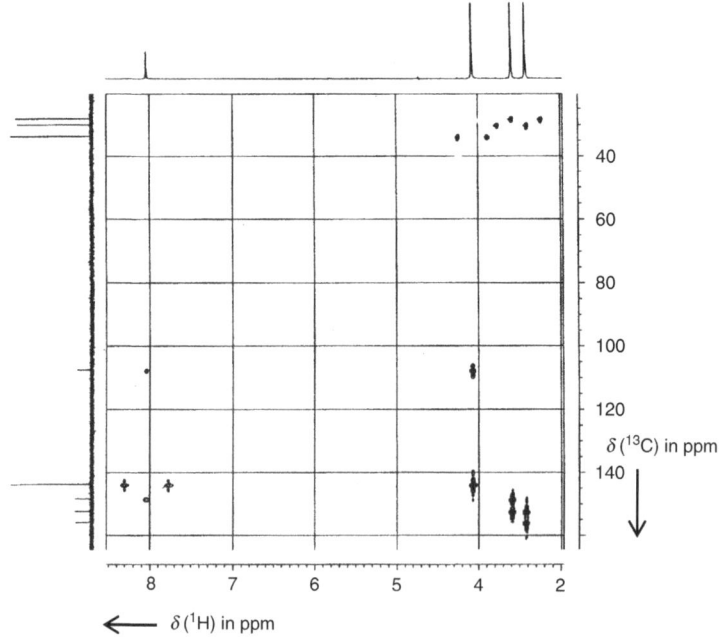

Fig. 5.9F HMBC spectrum

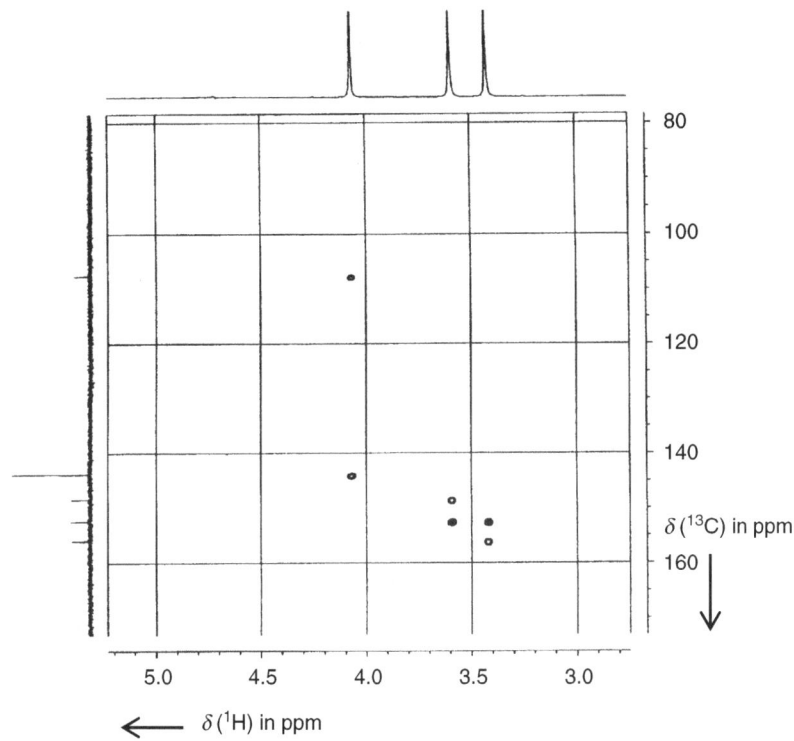

Fig. 5.9G Section of the HMBC spectrum

Challenge 5.10

Deduce the structure of the unknown compound by means of the given
spectra and assign the spectroscopic data as is presented in Challenge 5.1.

For solutions to Challenge 5.10, see *extras.springer.com.*

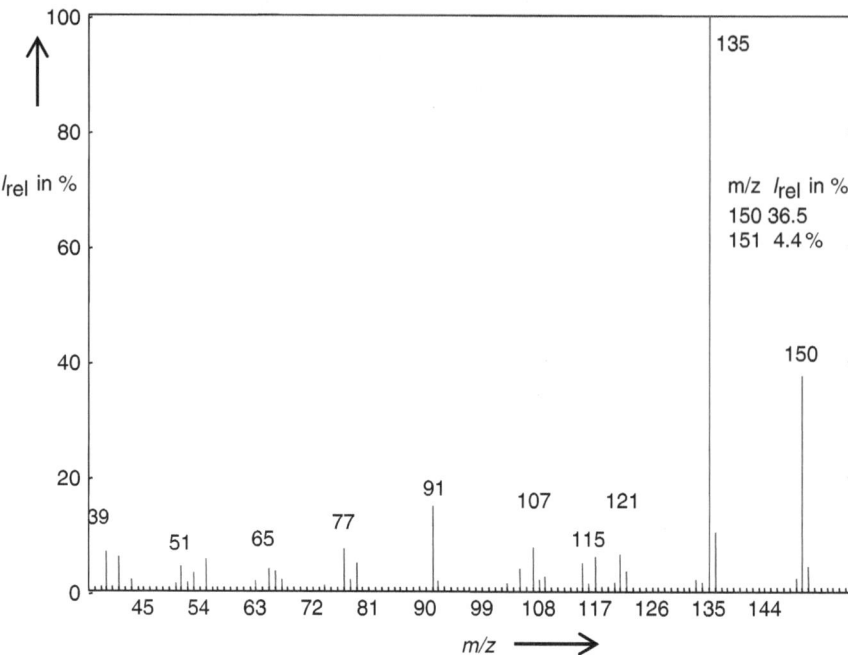

Fig. 5.10A Mass spectrum (EI, 70 eV)

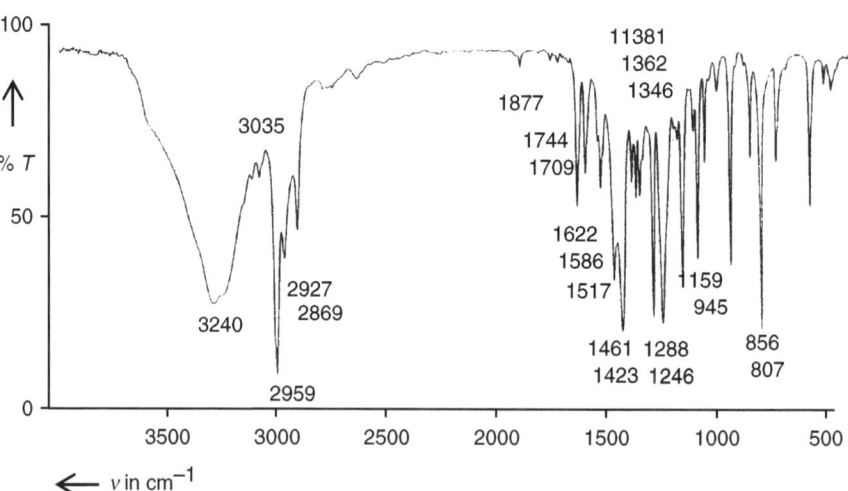

Fig. 5.10B IR spectrum (KBr pellet)

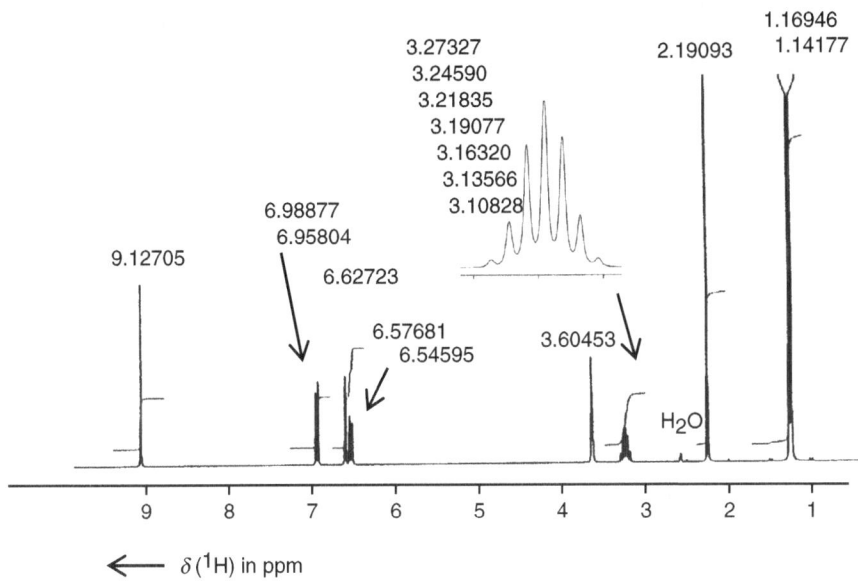

Fig. 5.10C ^{1}H NMR spectrum ($\nu_0 = 250$ MHz; solvent: DMSO)

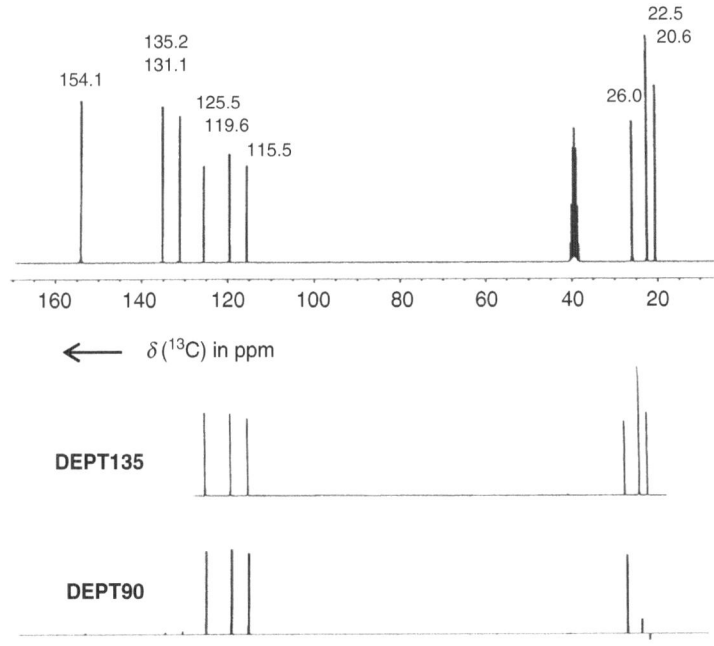

Fig. 5.10D ^{13}C NMR and DEPT spectra (solvent: DMSO)

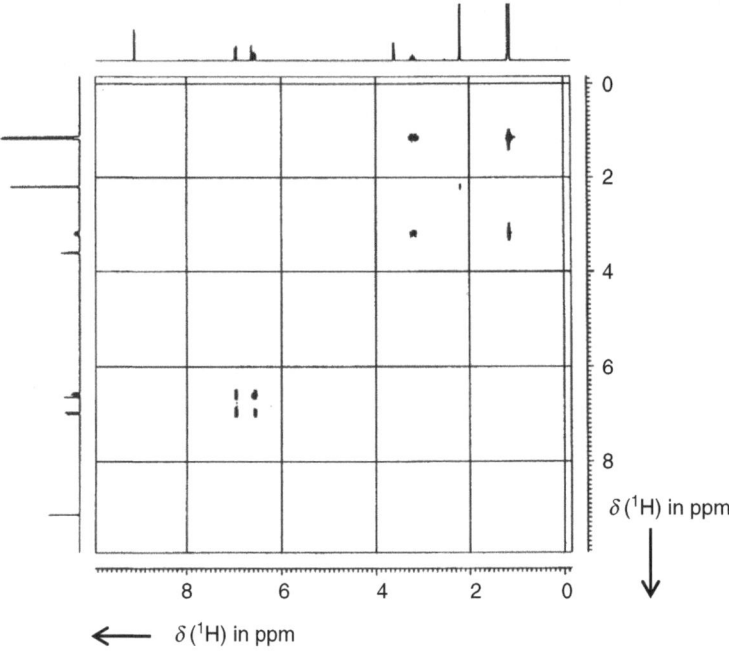

Fig. 5.10E (¹H,¹H) COSY spectrum

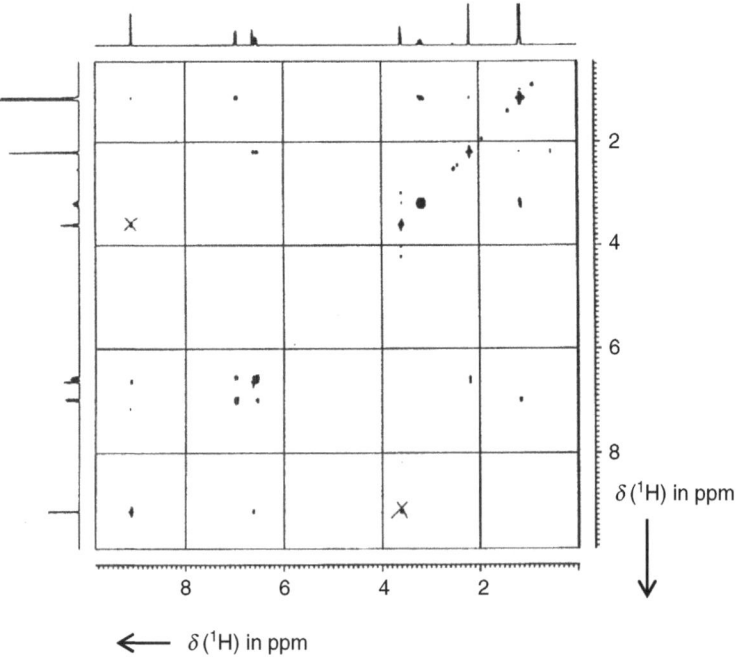

Fig. 5.10F NOESY spectrum

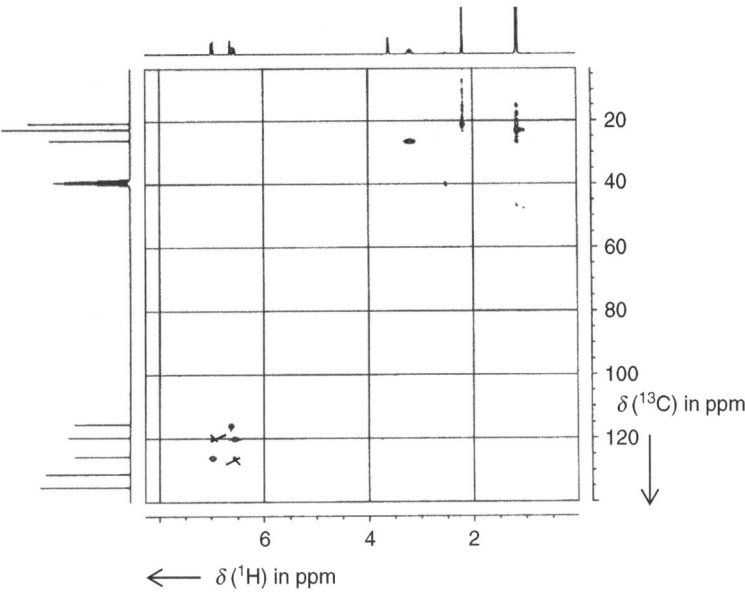

Fig. 5.10G HMQC spectrum

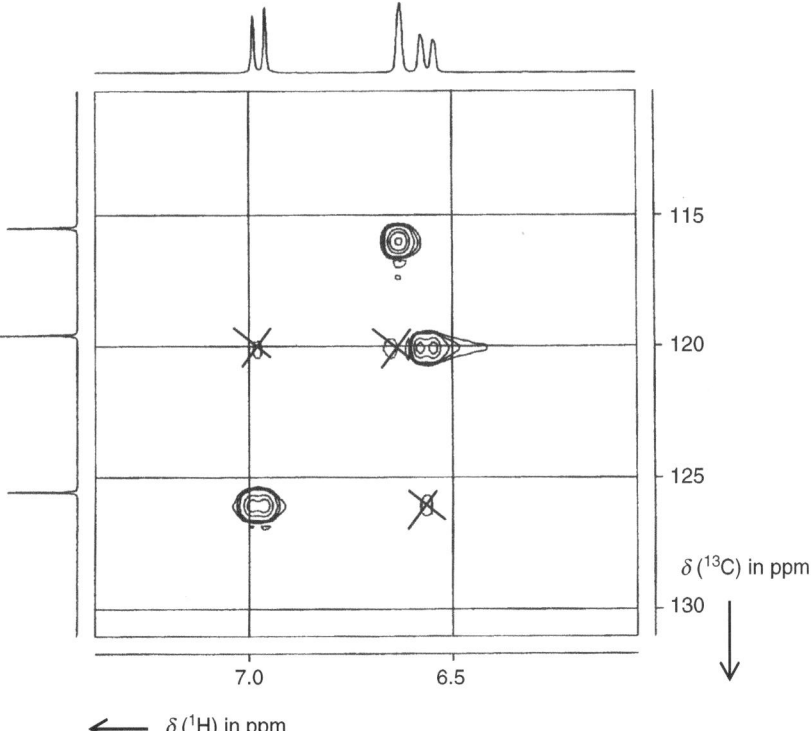

Fig. 5.10H Section of the HMQC spectrum

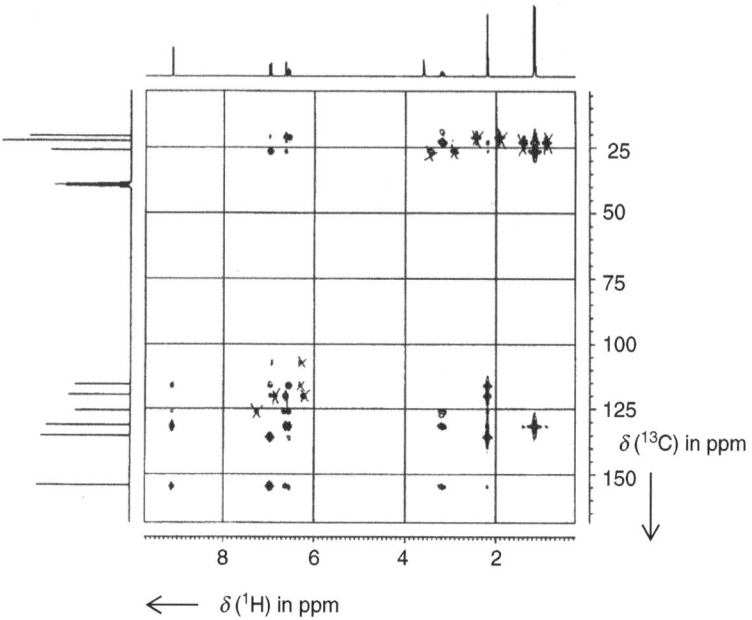

Fig. 5.10I HMBC spectrum

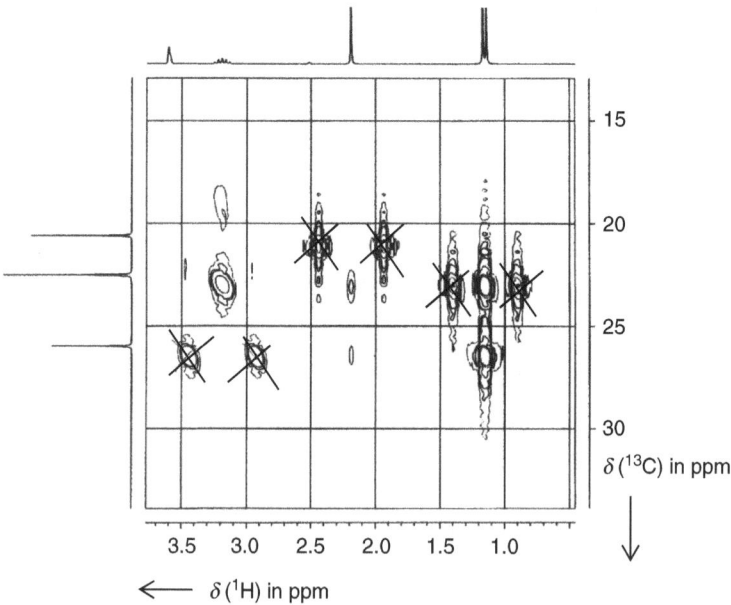

Fig. 5.10J Section of the HMBC spectrum

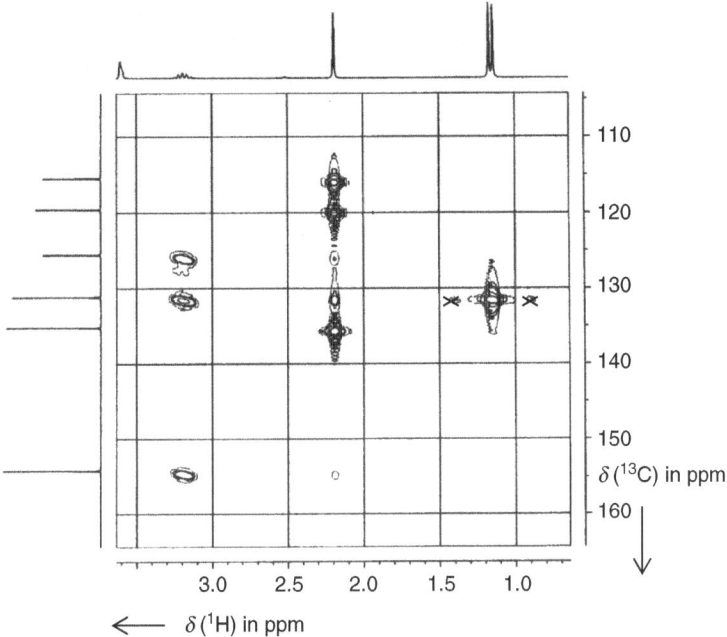

Fig. 5.10K   Section of the HMBC spectrum

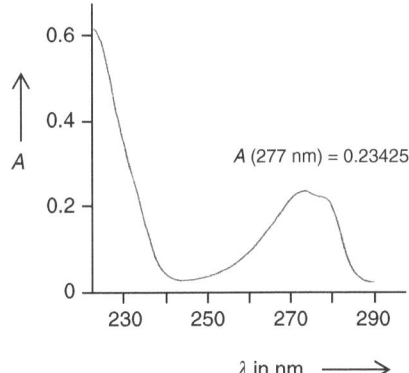

Fig. 5.10L   UV spectrum

Stock solution:	0.769 mg in 5 mL
Analytical sample:	1-mL stock solution diluted to 10 mL
Solvent:	Methanol
Cuvette:	$l = 1$ cm

Challenge 5.11
Deduce the structure of the unknown compound by means of the given spectra and assign the spectroscopic data as is presented in Challenge 5.1.
    For solutions to Challenge 5.11, see *extras.springer.com*.

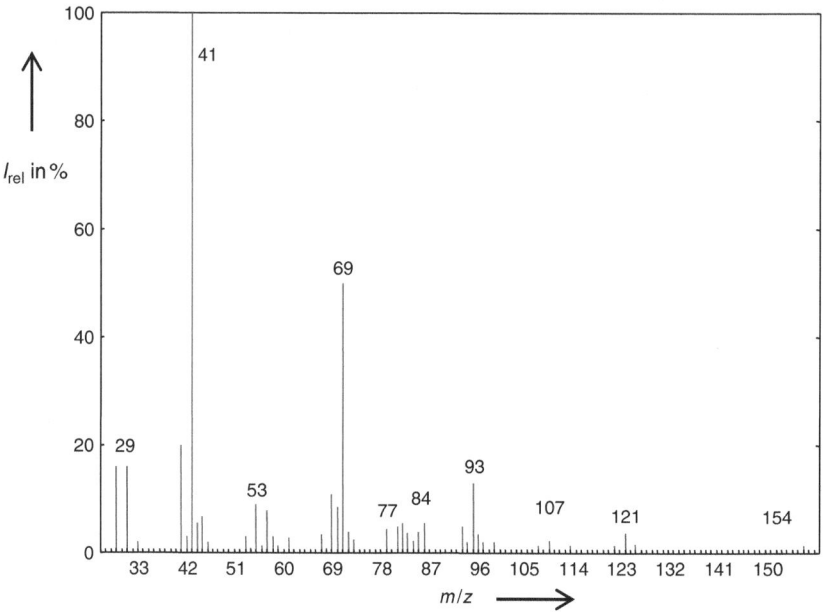

Fig. 5.11A Mass spectrum (EI, 70 eV)

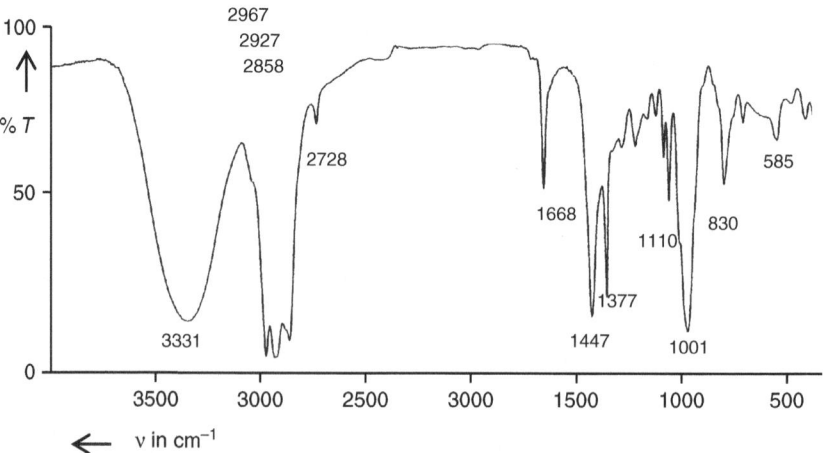

Fig. 5.11B IR spectrum (capillary thin film)

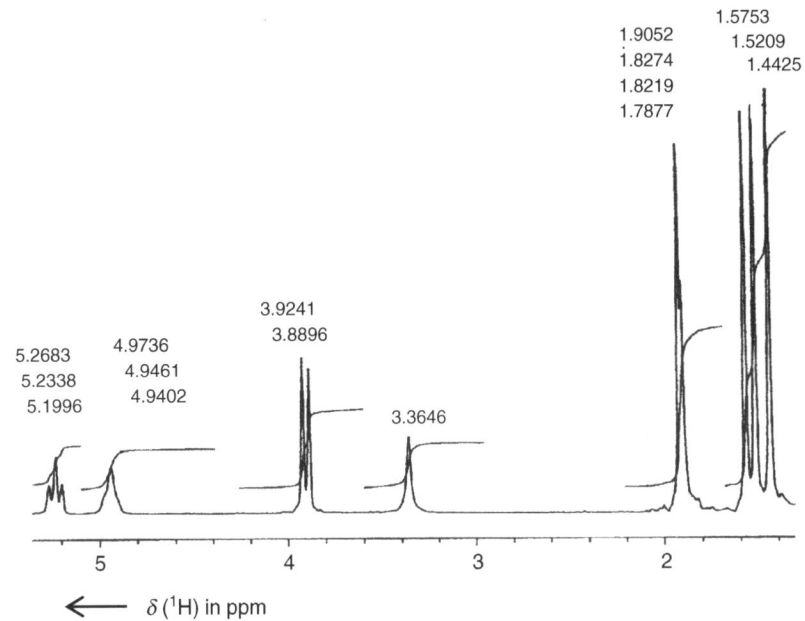

Fig. 5.11C ^{1}H NMR spectrum ($\nu_0 = 200$ MHz; solvent: CDCl$_3$)

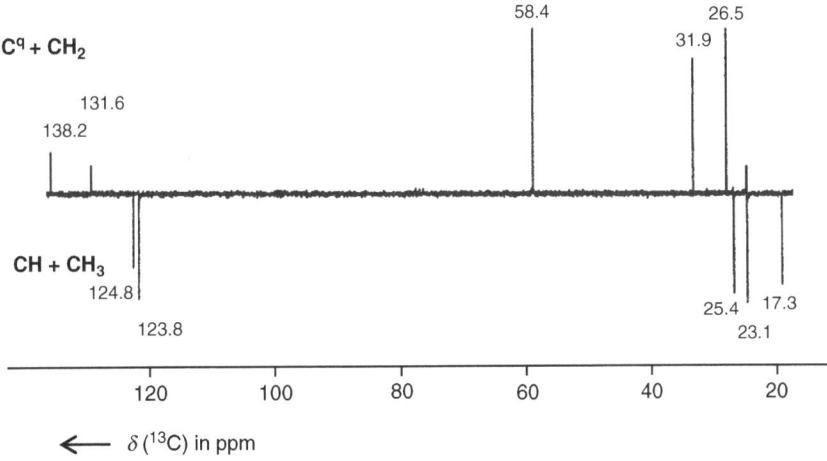

Fig. 5.11D ^{13}C NMR (pendant) and DEPT spectra (solvent: CDCl$_3$)

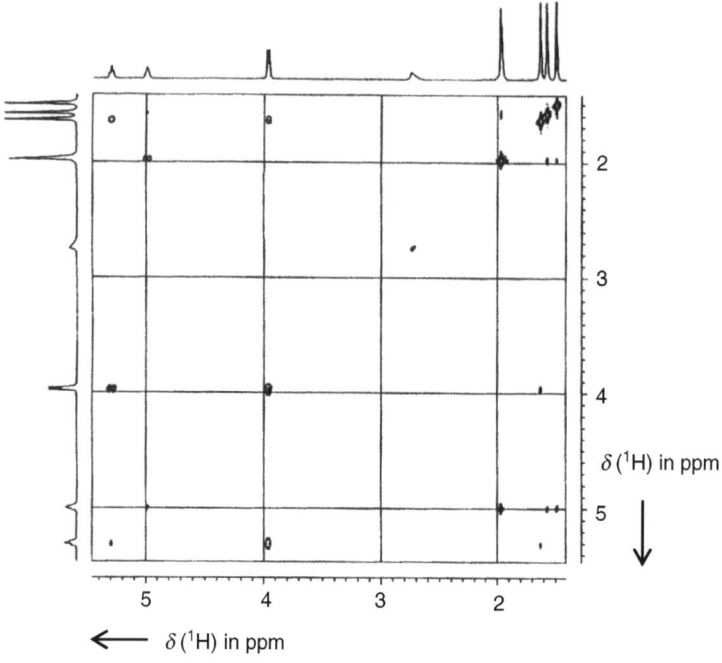

Fig. 5.11E (^{1}H,^{1}H) COSY spectrum (solvent: CDCl$_3$)

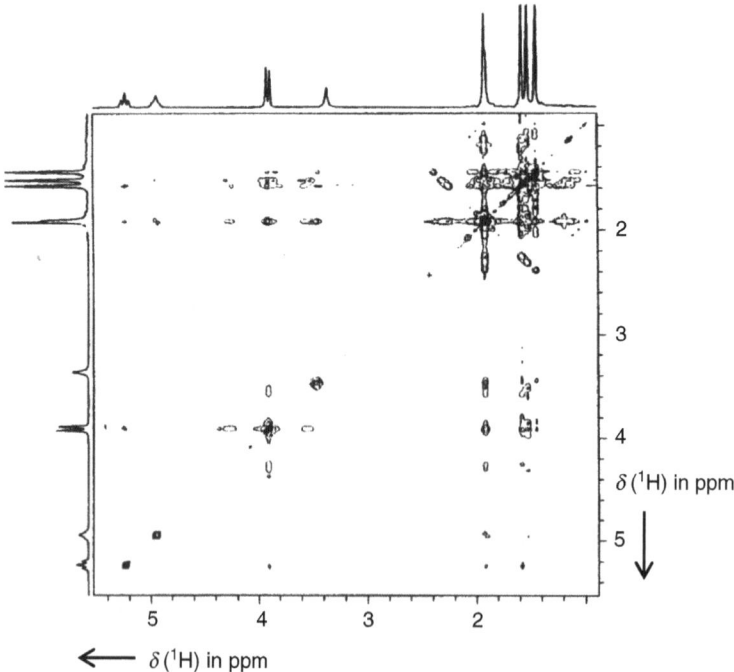

Fig. 5.11F NOESY spectrum (solvent: CDCl$_3$)

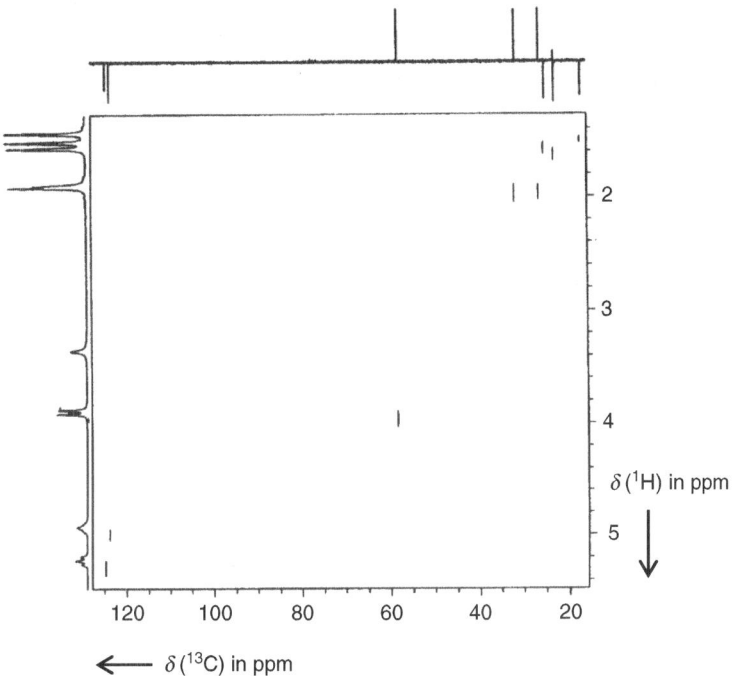

Fig. 5.11G HSQC spectrum (solvent: CDCl$_3$)

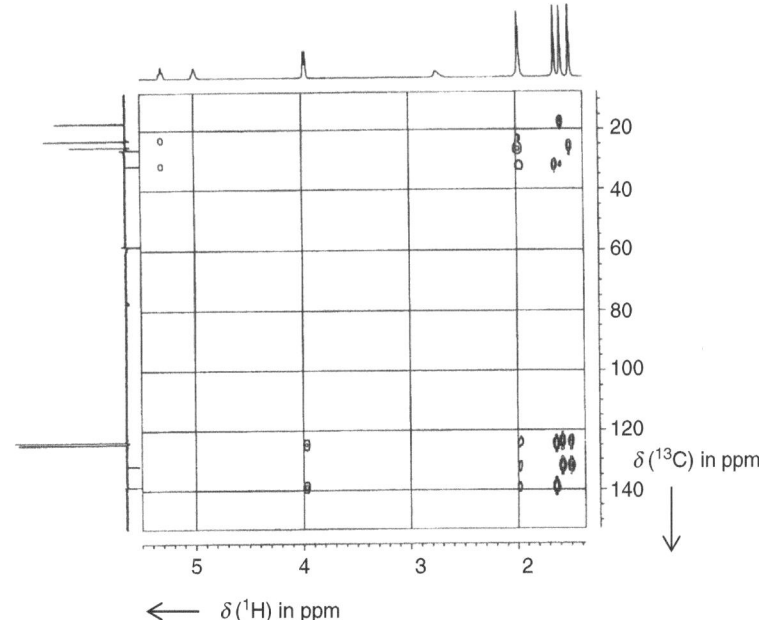

Fig. 5.11H HBMC spectrum (solvent: CDCl$_3$)

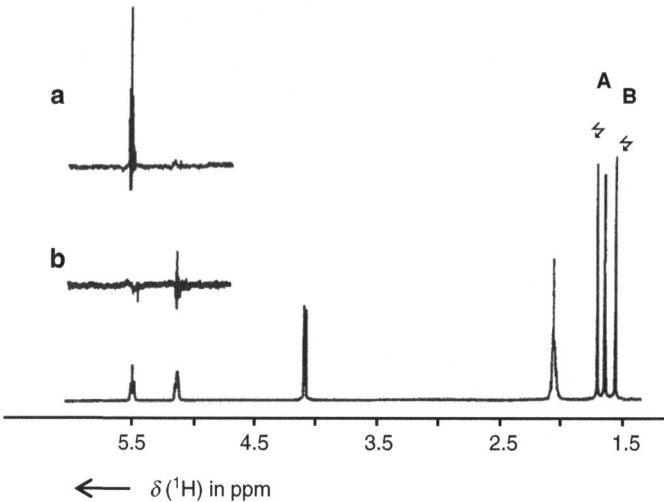

Fig. 5.11I 1D NOE spectrum. (**a**) Irradiation at signal A. (**b**) Irradiation at signal B

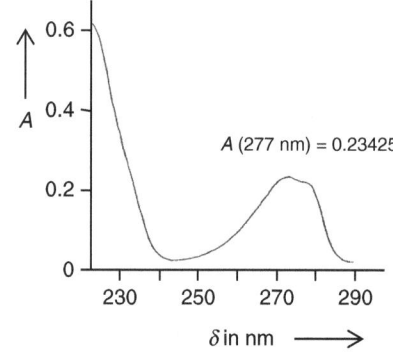

Fig. 5.11J UV spectrum

Analytical sample:	15 μL in 5 mL ACN, $\rho = 0.88$ g mL^{-1}
Cuvette:	$l = 1$ cm

Challenge 5.12

Deduce the structure of the unknown compound by means of the given spectra and assign the spectroscopic data as is presented in Challenge 5.1.

For solutions to Challenge 5.12, see *extras.springer.com*.

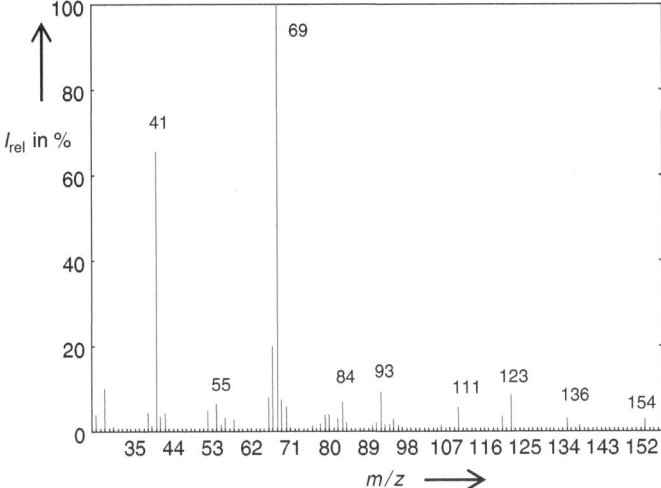

Fig. 5.12A Mass spectrum (EI, 70 eV)

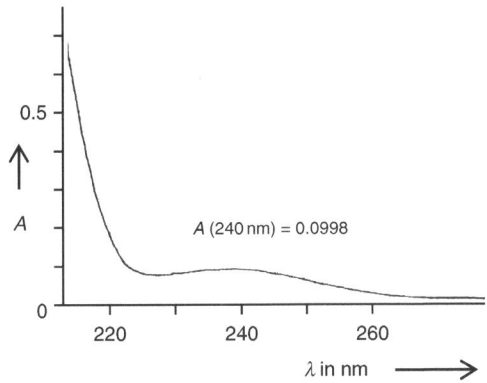

Fig. 5.12B UV spectrum. Analytical sample: 5 μL in 5 mL ACN, $\rho = 0.89$ g mL^{-1}

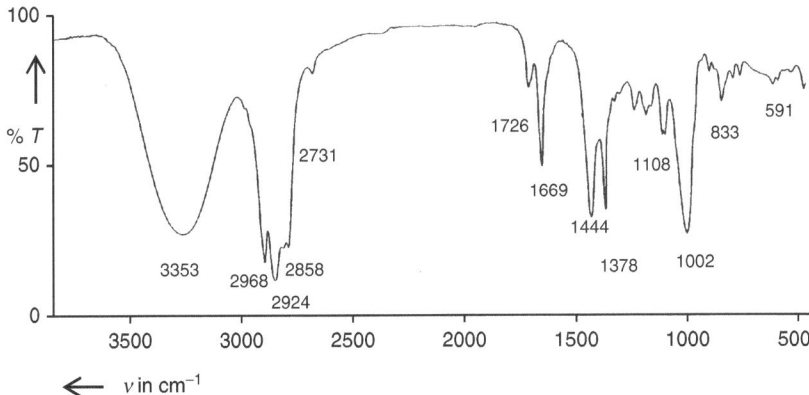

Fig. 5.12C IR spectrum (capillary thin film)

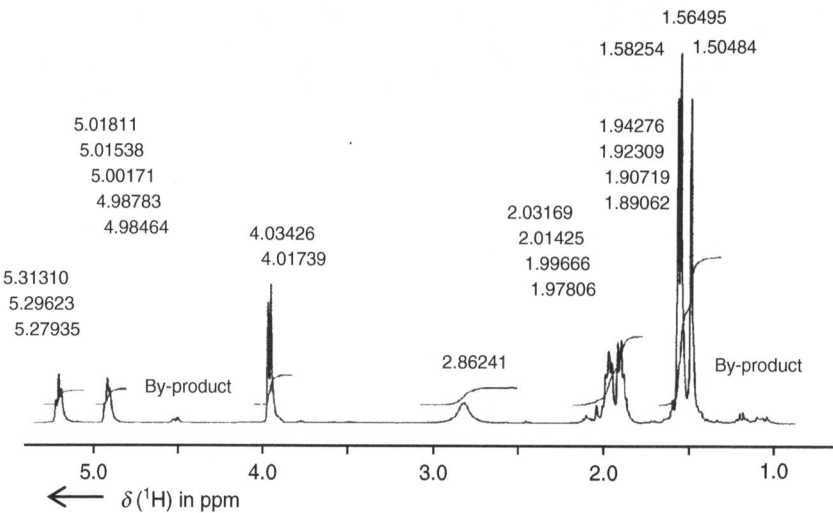

Fig. 5.12D ^{1}H NMR spectrum ($v_0 = 400$ MHz; solvent: CDCl$_3$)

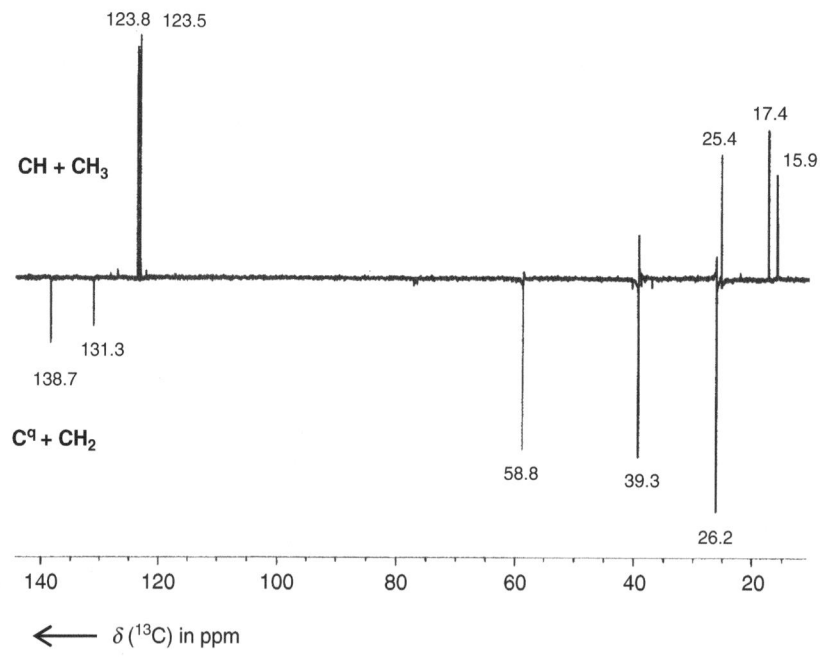

Fig. 5.12E ^{13}C NMR (pendant) spectrum (solvent: CDCl$_3$)

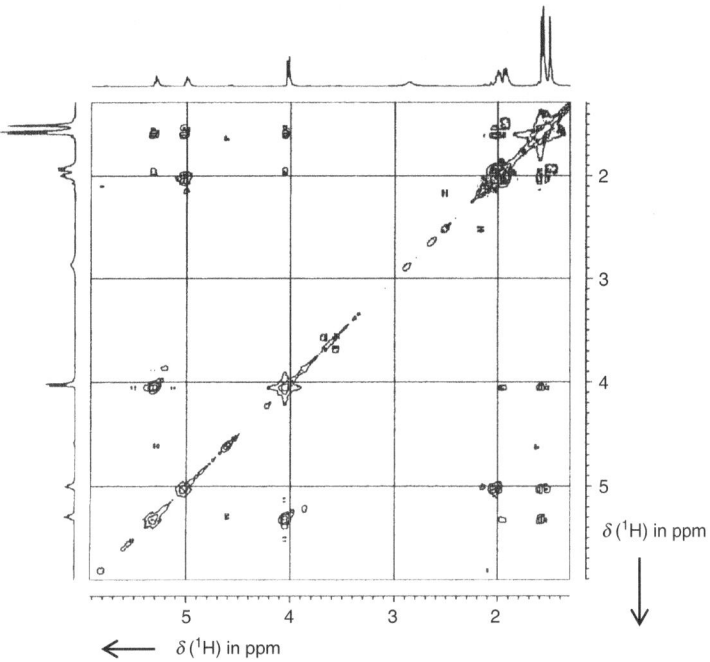

Fig. 5.12F (^{1}H,^{1}H) COSY spectrum

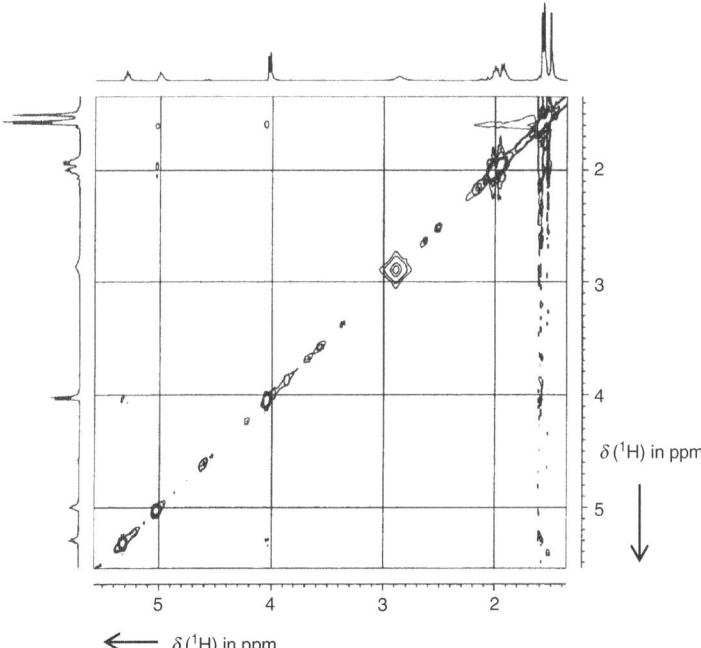

Fig. 5.12G NOESY spectrum

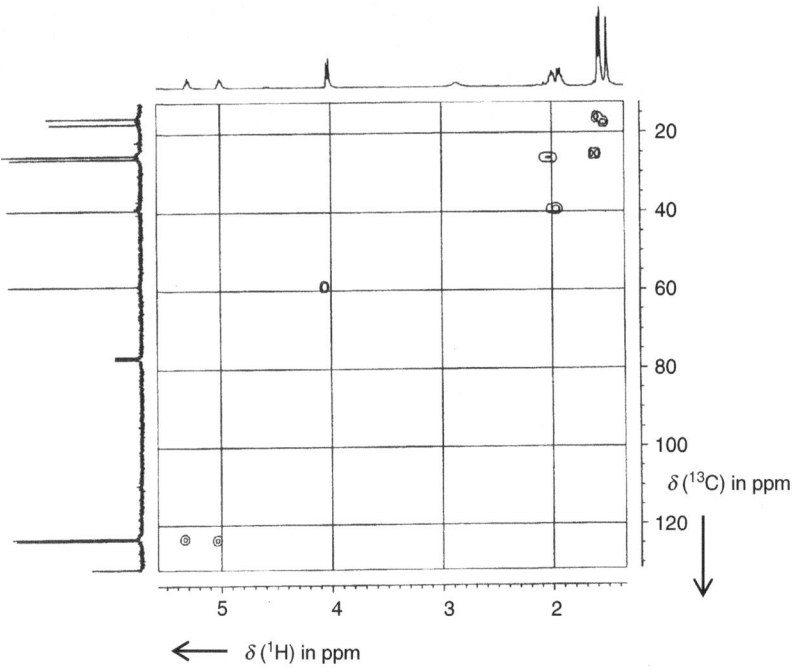

Fig. 5.12H HSQC spectrum

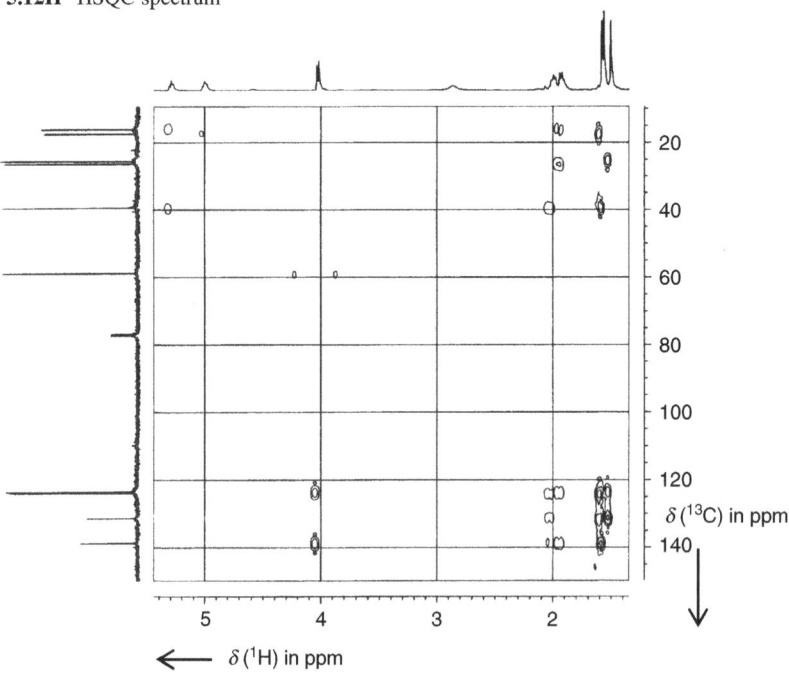

Fig. 5.12I HBMC spectrum

Challenge 5.13
Deduce the structure of the unknown compound by means of the given spectra and assign the spectroscopic data as is presented in Challenge 5.1.
  For solutions to Challenge 5.13, see *extras.springer.com.*

UV spectrum:
    Analytical sample: 1.5 μL in 10 mL methanol, $\rho = 0.87$ g mL^{-1}
    Cuvette: $l = 1$ cm
    Structureless absorption band: $\lambda_{max} = 240$ nm; $A(240$ nm$) = 0.095$

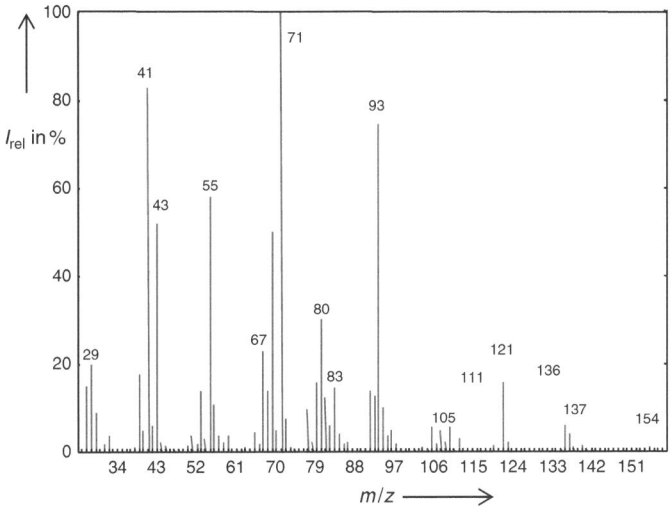

Fig. 5.13A Mass spectrum (EI, 70 eV)

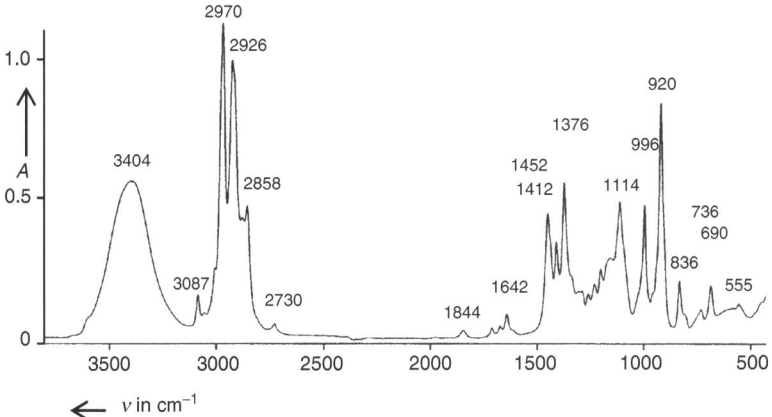

Fig. 5.13B IR spectrum (capillary thin film)

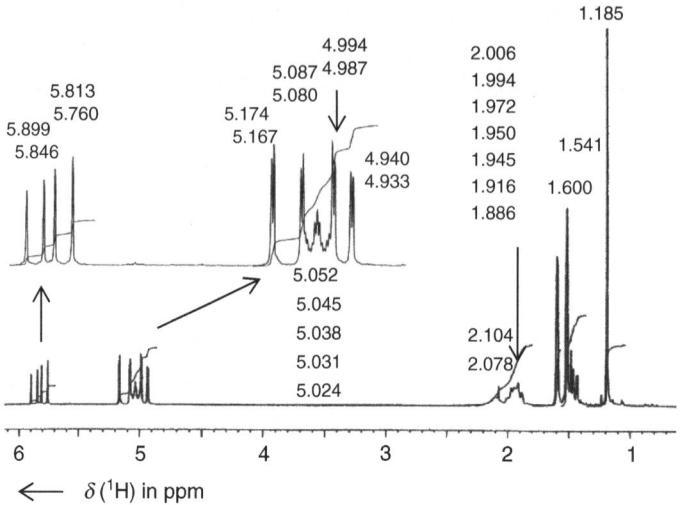

Fig. 5.13C ^{1}H NMR spectrum (v_0 = 200 MHz; solvent: CDCl$_3$)

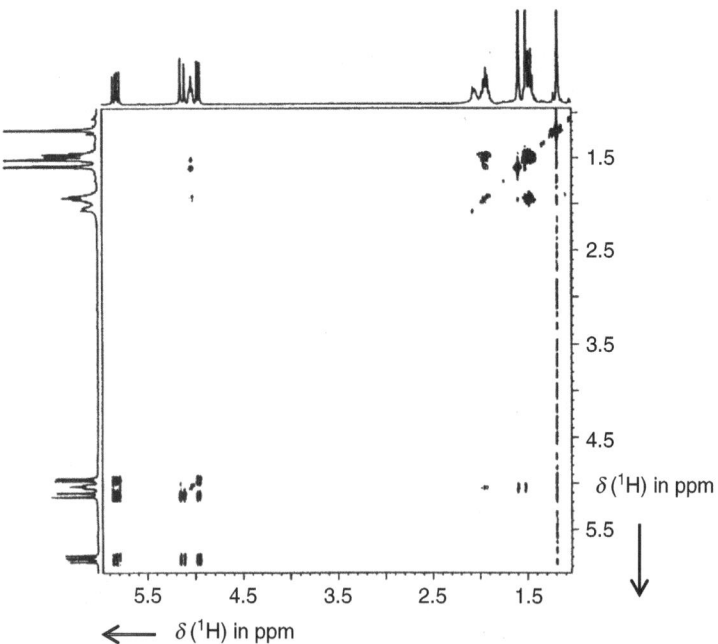

Fig. 5.13D (^{1}H,^{1}H) COSY spectrum

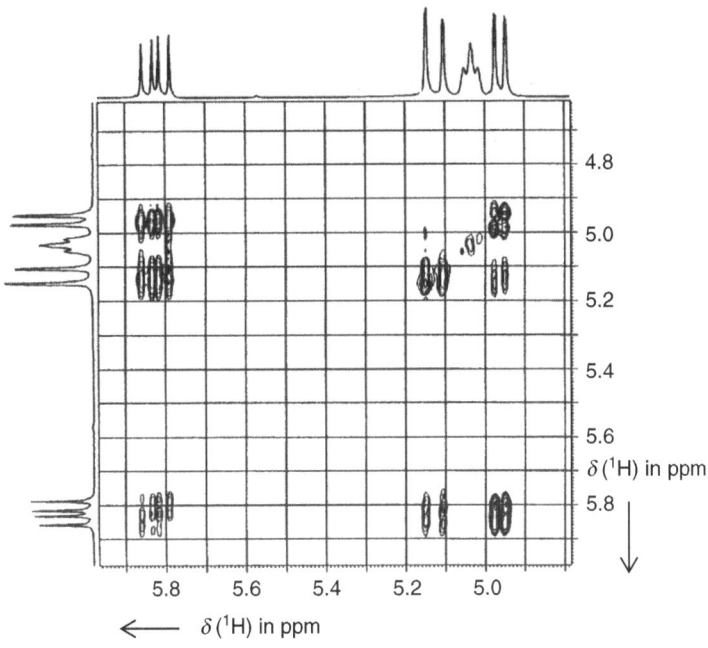

Fig. 5.13E   Section of the (^{1}H,^{1}H) COSY spectrum

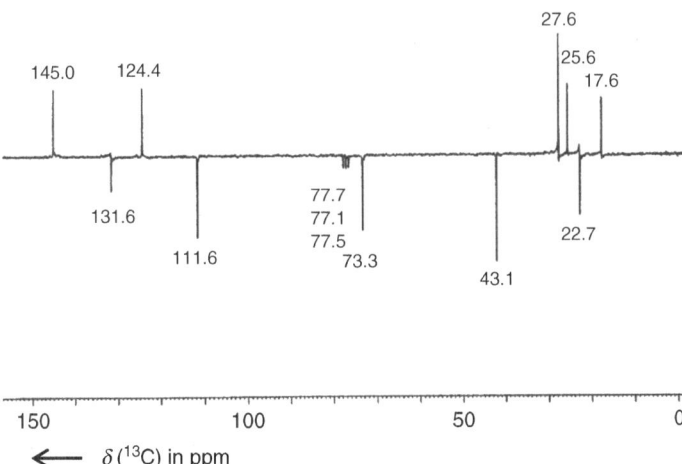

Fig. 5.13F ^{13}C NMR (pendant) spectrum (solvent: CDCl$_3$)

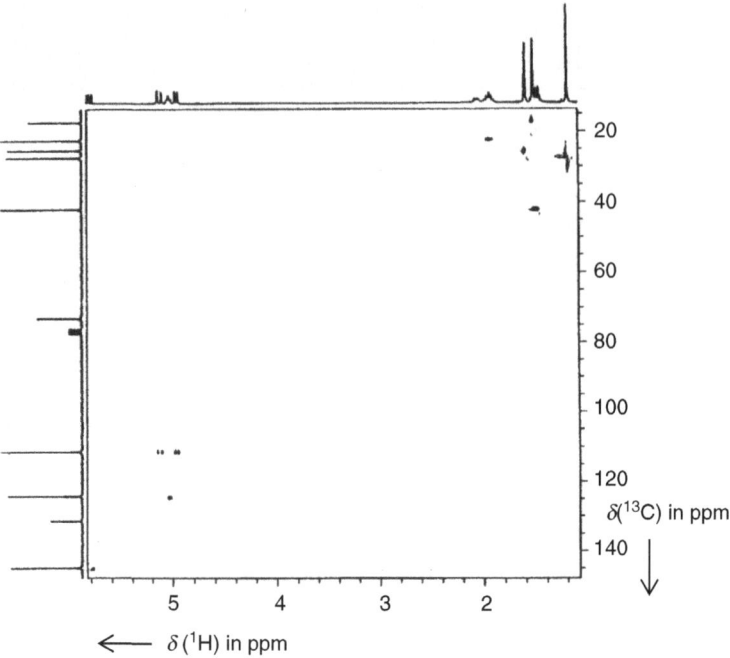

Fig. 5.13G HSQC spectrum

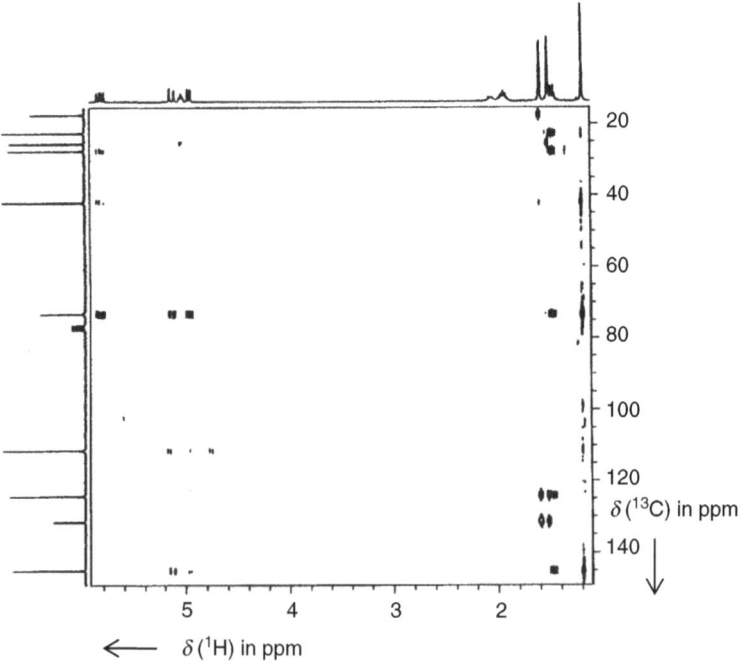

Fig. 5.13H HMBC spectrum

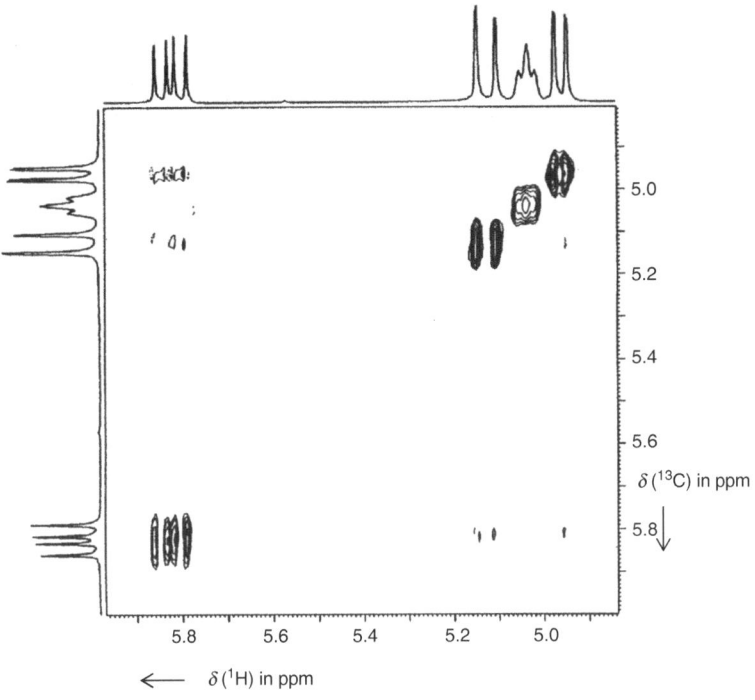

Fig. 5.13I Section of the NOESY spectrum

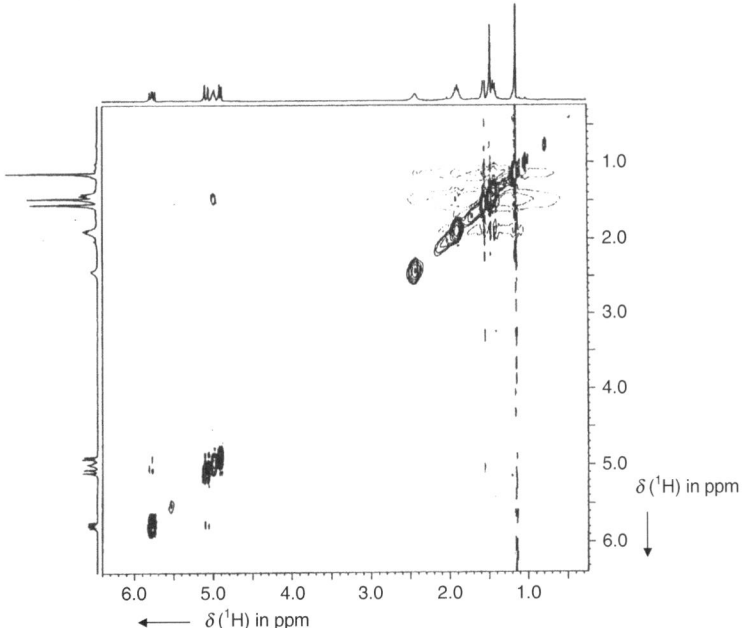

Fig. 5.13J NOESY spectrum

Challenge 5.14
Deduce the structure of the unknown compound by means of the given spectra and assign the spectroscopic data as is presented in Challenge 5.1.
  For solutions to Challenge 5.14, see *extras.springer.com*.

UV spectrum: No absorption in the range $> \lambda = 250$ nm

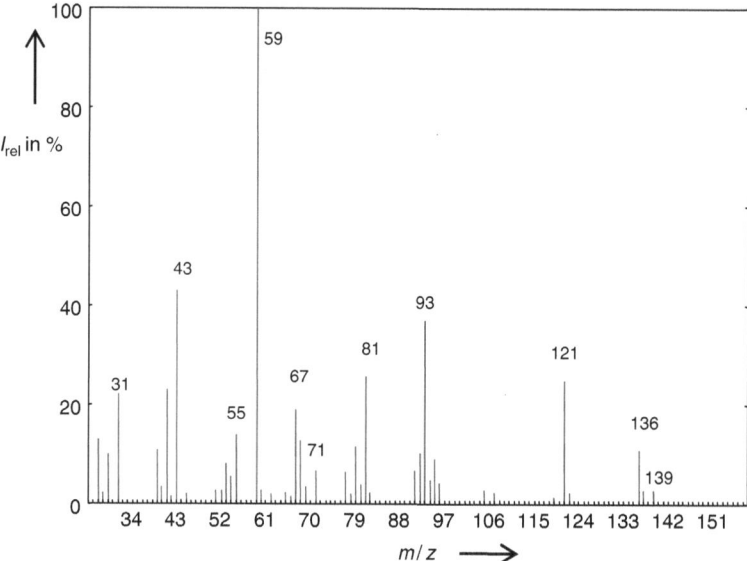

Fig. 5.14A Mass spectrum (EI, 70 eV)

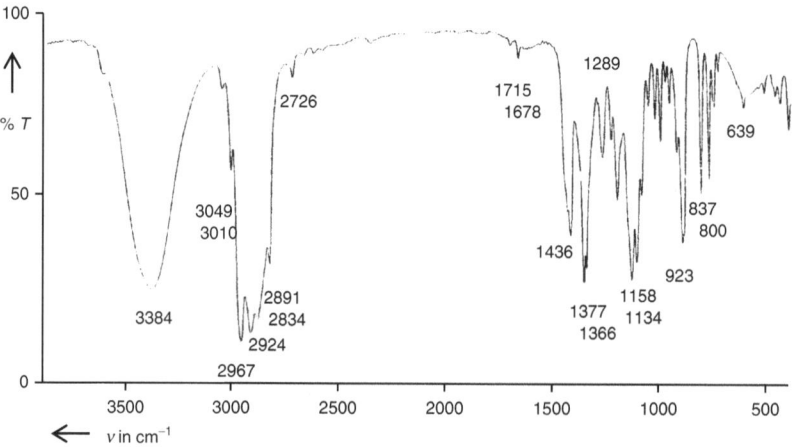

Fig. 5.14B IR spectrum (capillary thin film)

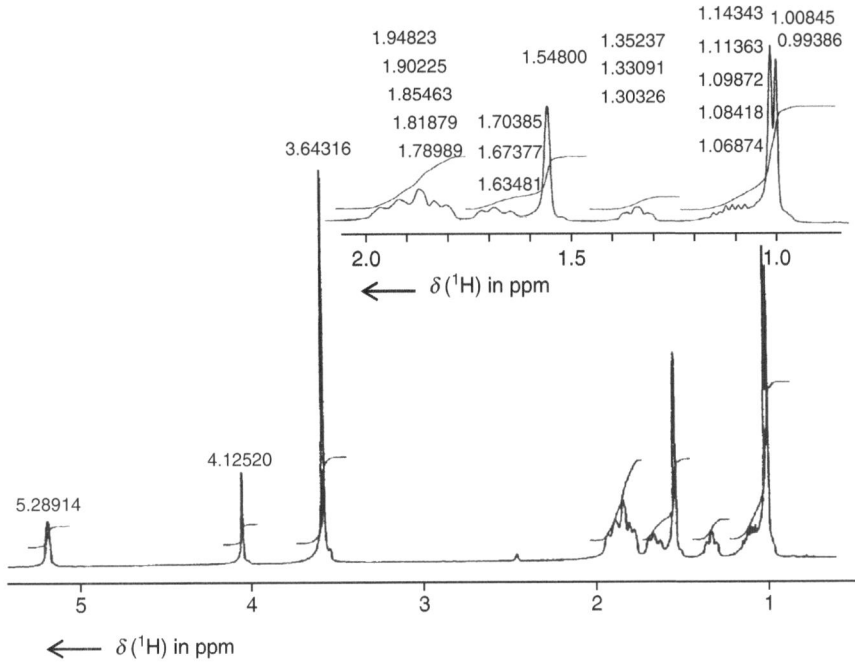

Fig. 5.14C ^{1}H NMR spectrum ($v_0 = 400$ MHz; solvent: DMSO)

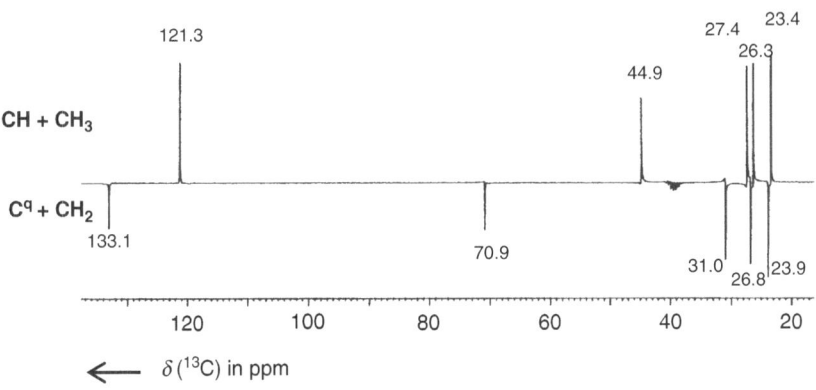

Fig. 5.14D ^{13}C NMR (pendant) spectrum (solvent: DMSO)

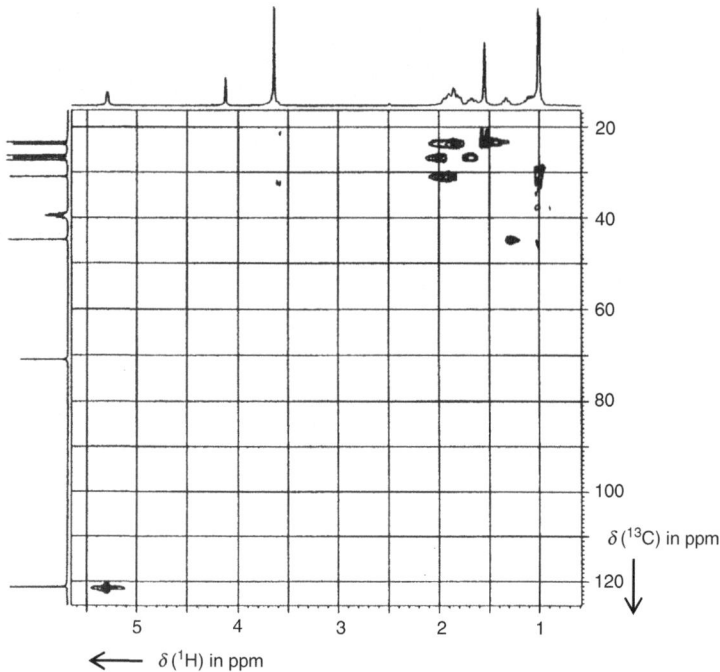

Fig. 5.14E HSQC spectrum

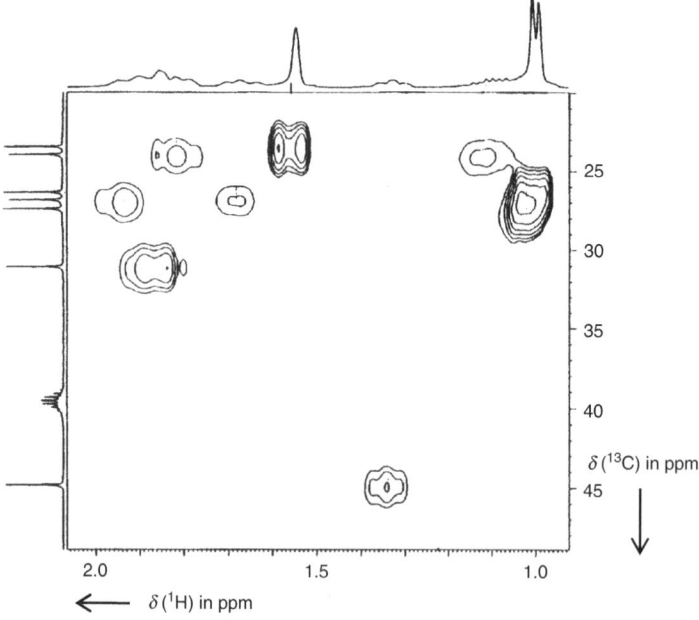

Fig. 5.14F Section of the HSQC spectrum

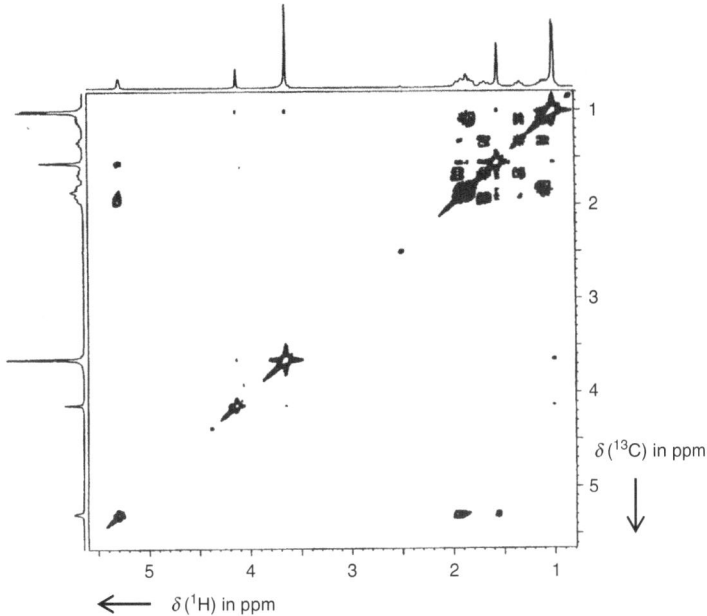

Fig. 5.14G (^{1}H,^{1}H) COSY spectrum

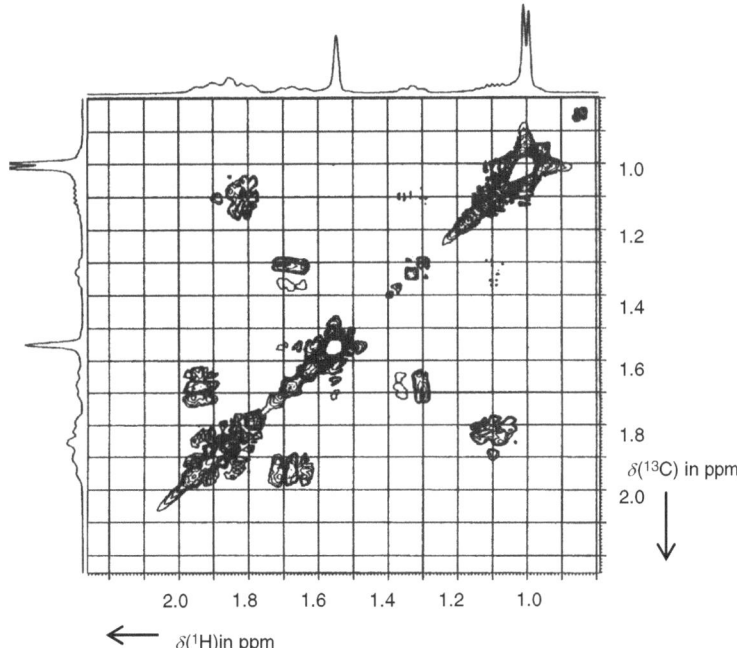

Fig. 5.14H Section of the (^{1}H,^{1}H) COSY spectrum

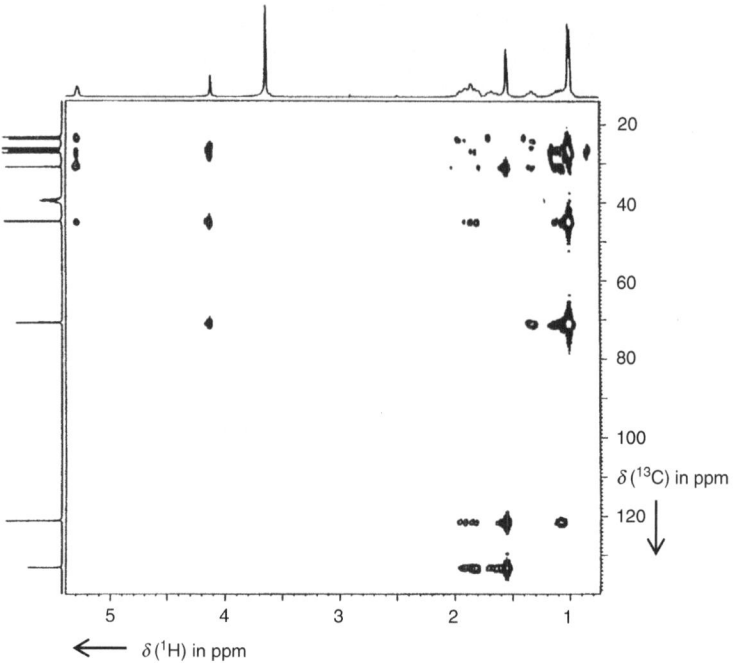

Fig. 5.14I HMBC spectrum

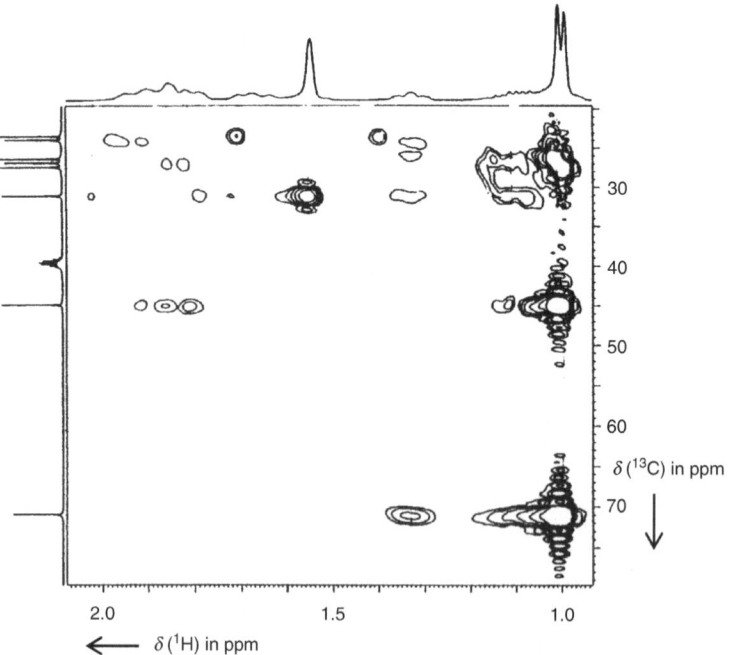

Fig. 5.14J Section of the HMBC spectrum

Challenge 5.15
Deduce the structure of the unknown compound by means of the given spectra and assign the spectroscopic data as is presented in Challenge 5.1.

For solutions to Challenge 5.15, see *extras.springer.com*.

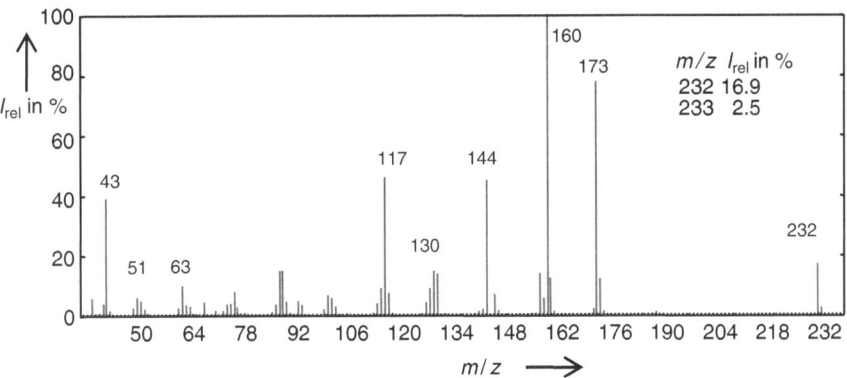

Fig. 5.15A Mass spectrum (EI, 70 eV)

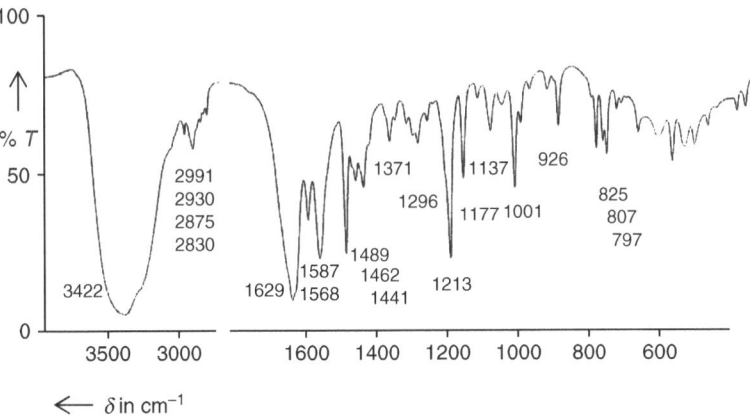

Fig. 5.15B IR spectrum (KBr pellet)

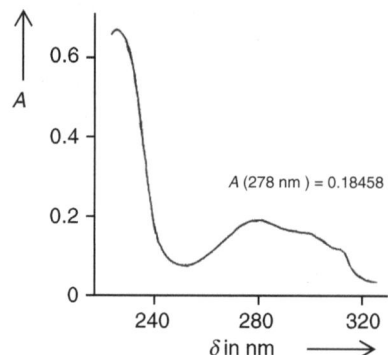

Fig. 5.15C UV spectrum

Stock solution:	0.359 mg in 5 mL
Analytical sample:	1-mL stock solution diluted to 10 mL
Solvent:	Methanol
Cuvette:	l = 1 cm

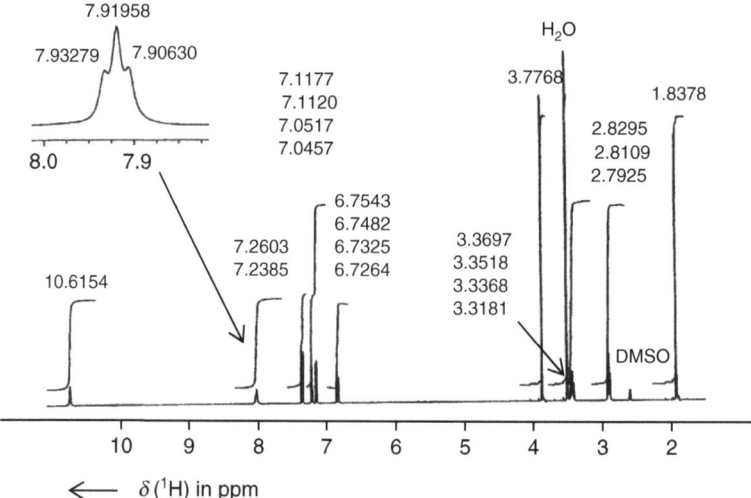

Fig. 5.15D ¹H NMR spectrum (v_0 = 400 MHz; solvent: DMSO)

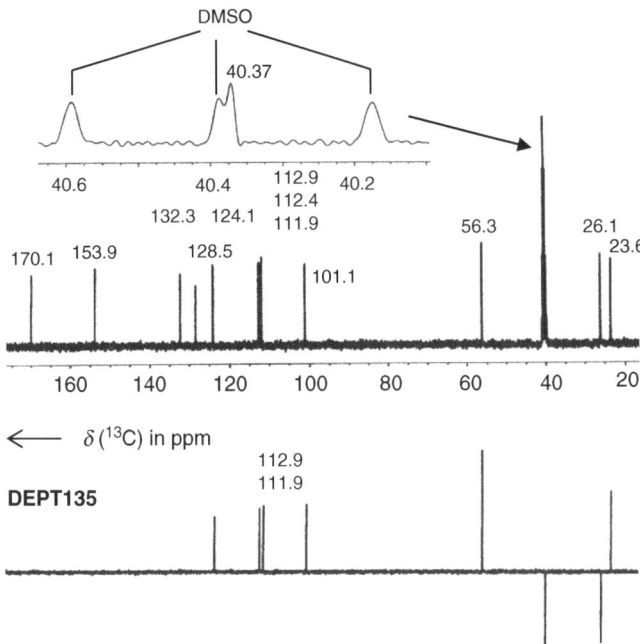

Fig. 5.15E ^{13}C NMR and DEPT spectra (solvent: DMSO)

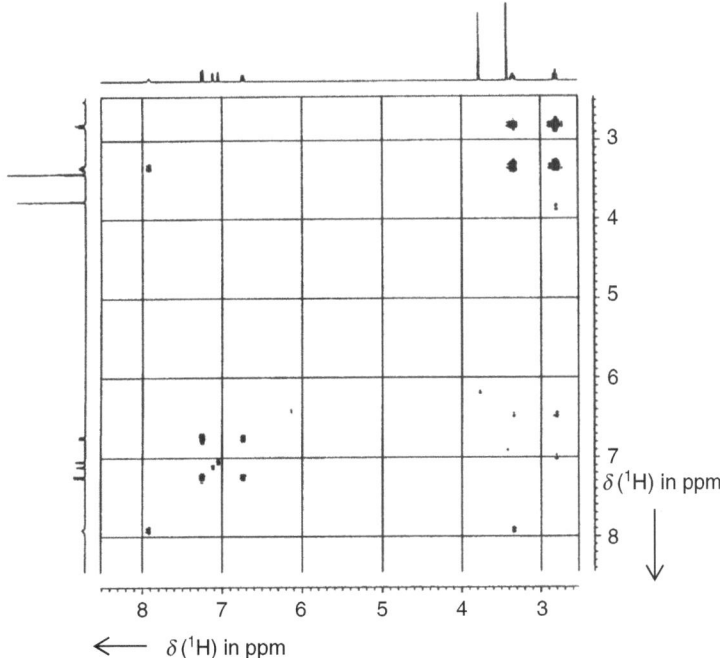

Fig. 5.15F (^{1}H,^{1}H) COSY spectrum (solvent: DMSO)

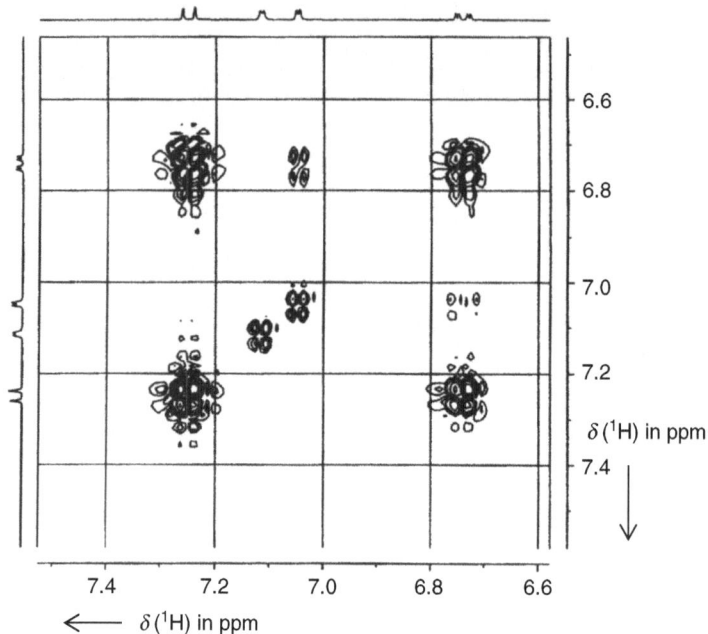

Fig. 5.15G   Section of the (^{1}H,^{1}H) COSY spectrum (solvent: DMSO)

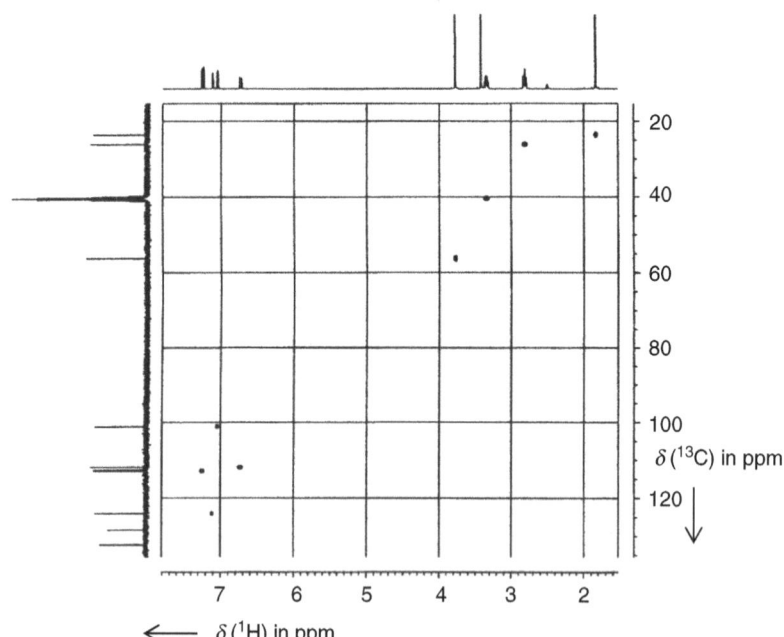

Fig. 5.15H   HMQC spectrum (solvent: DMSO)

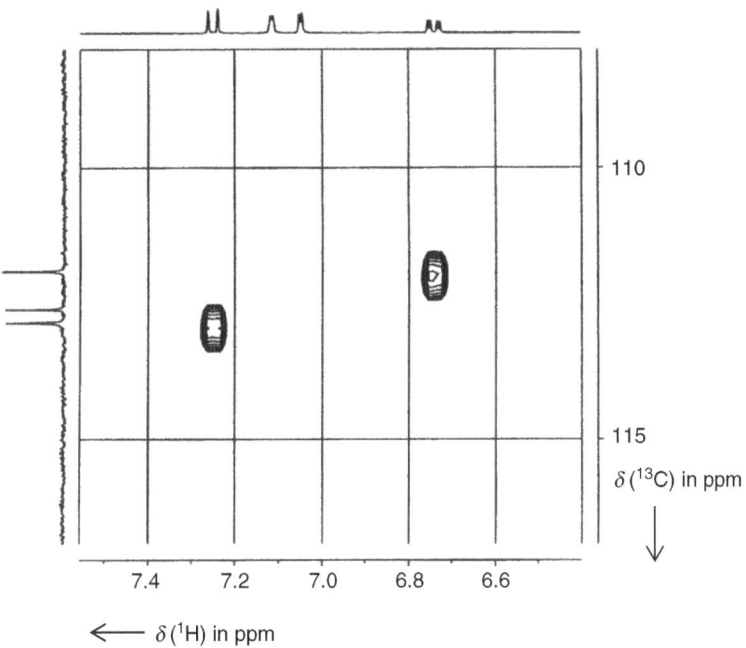

Fig. 5.15I Section of the HMQC spectrum (solvent: DMSO)

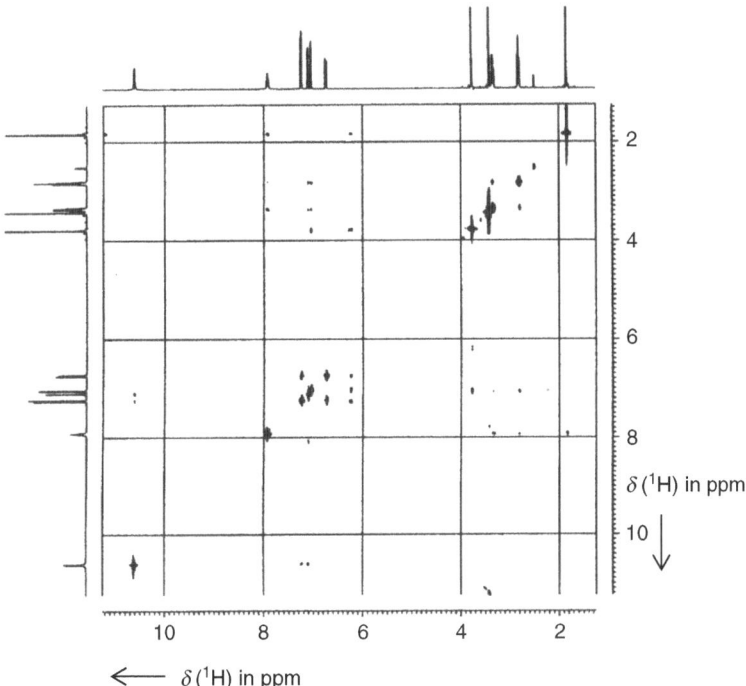

Fig. 5.15J NOESY spectrum (solvent: DMSO)

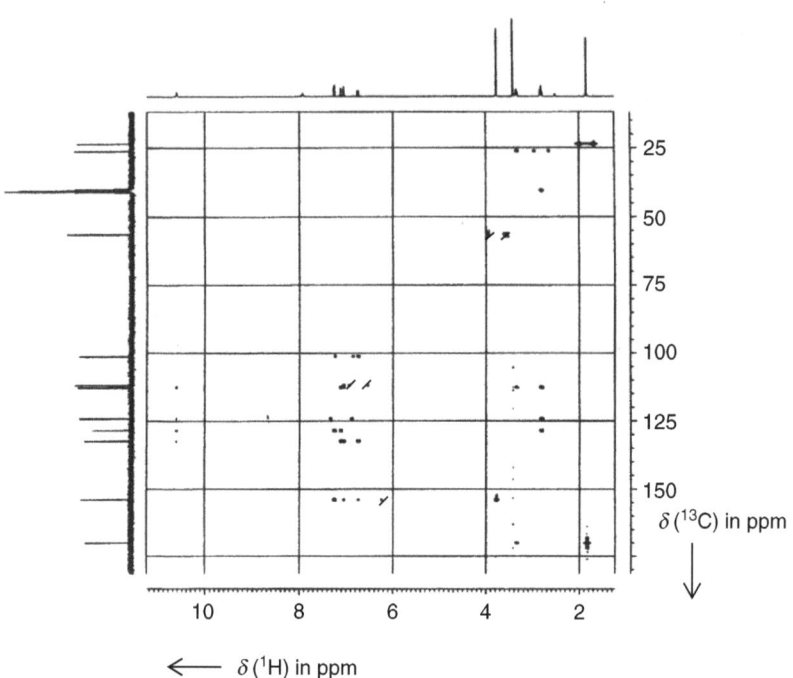

Fig. 5.15K HMBC spectrum (solvent: DMSO)

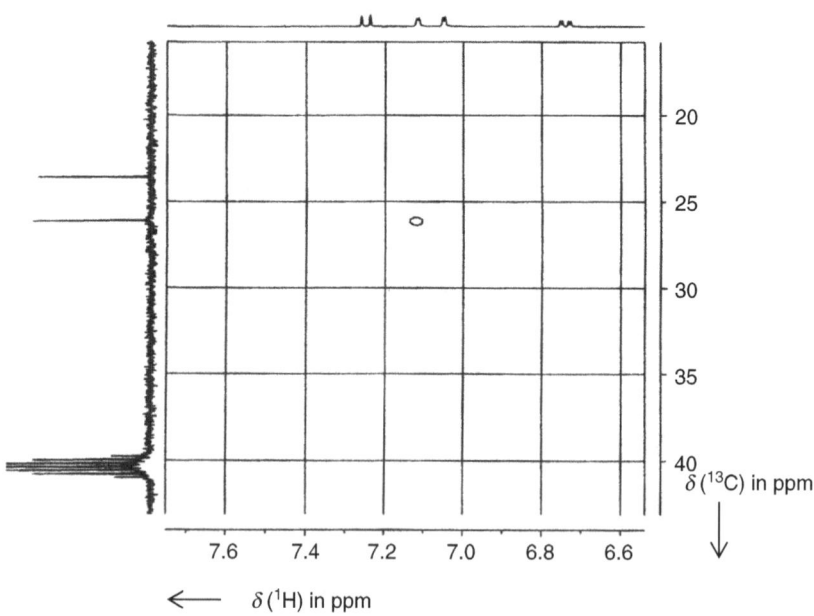

Fig. 5.15L Section of the HMBC spectrum (solvent: DMSO)

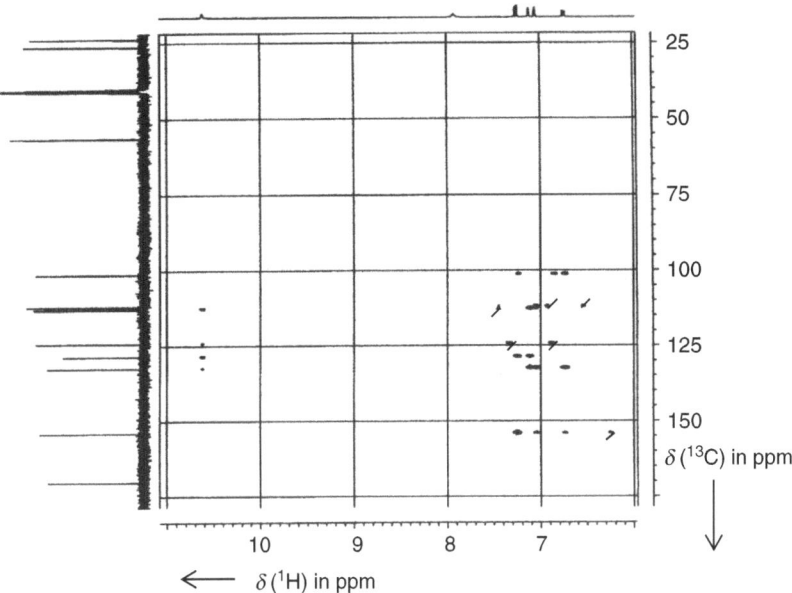

Fig. 5.15M Section of the HMBC spectrum (solvent: DMSO)

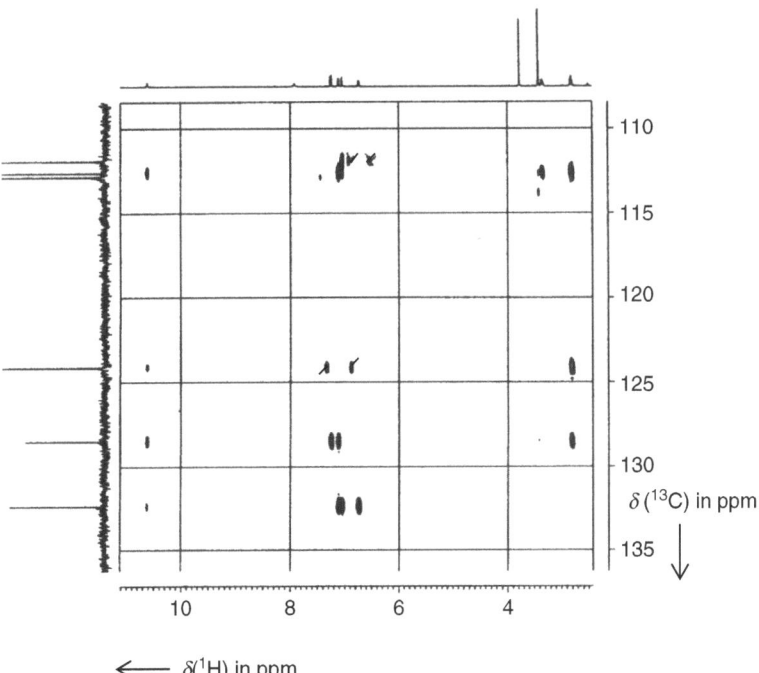

Fig. 5.15N Section of the HMBC spectrum (solvent: DMSO)

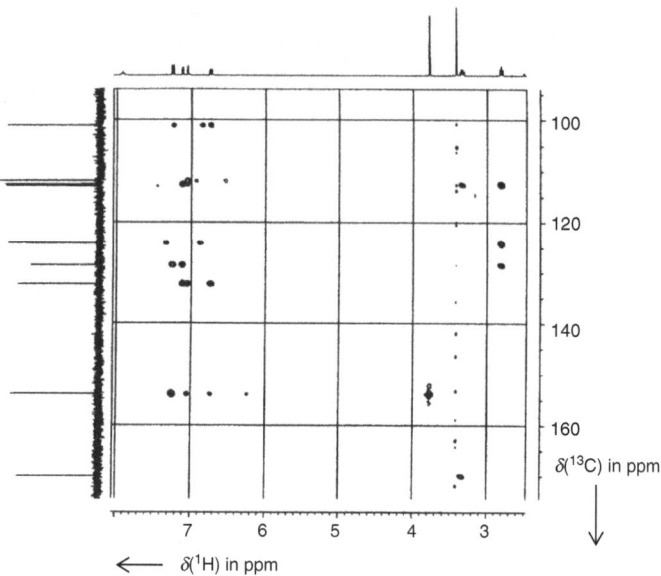

Fig. 5.15O Section of the HMBC spectrum (solvent: DMSO)

Challenge 5.16
Deduce the structure of the unknown compound by means of the given spectra and assign the spectroscopic data as is presented in Challenge 5.1.
   For solutions to Challenge 5.16, see *extras.springer.com*.

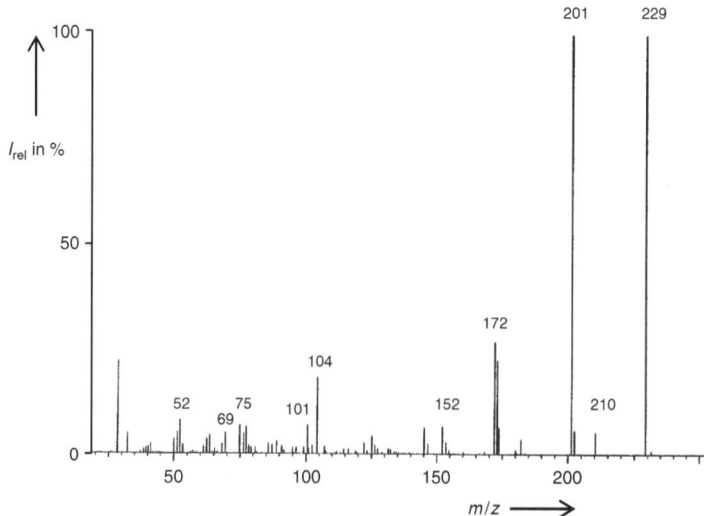

Fig. 5.16A Mass spectrum (EI, 70 eV)

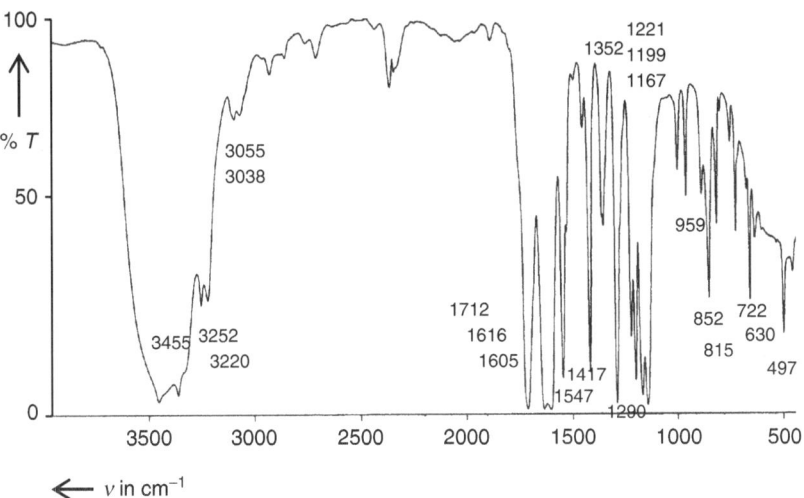

Fig. 5.16B IR spectrum (KBr pellet)

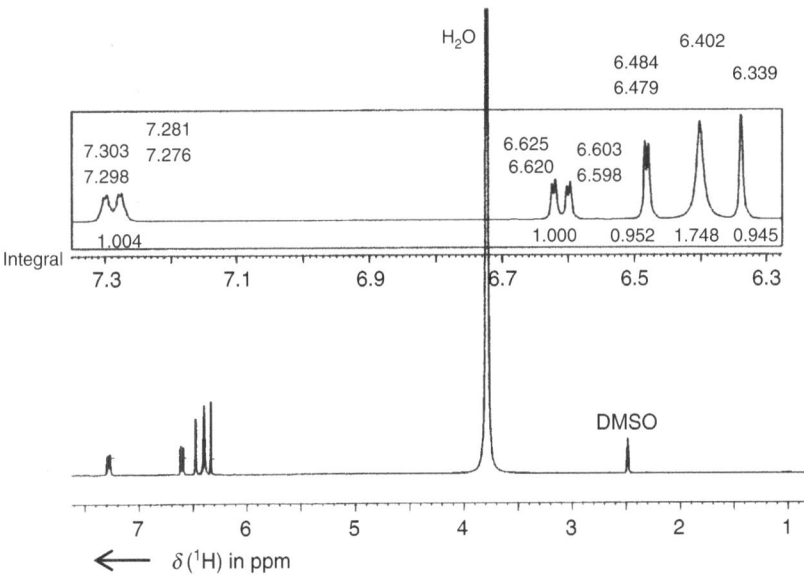

Fig. 5.16C ¹H NMR spectrum ($\nu_0 = 400$ MHz; solvent: DMSO)

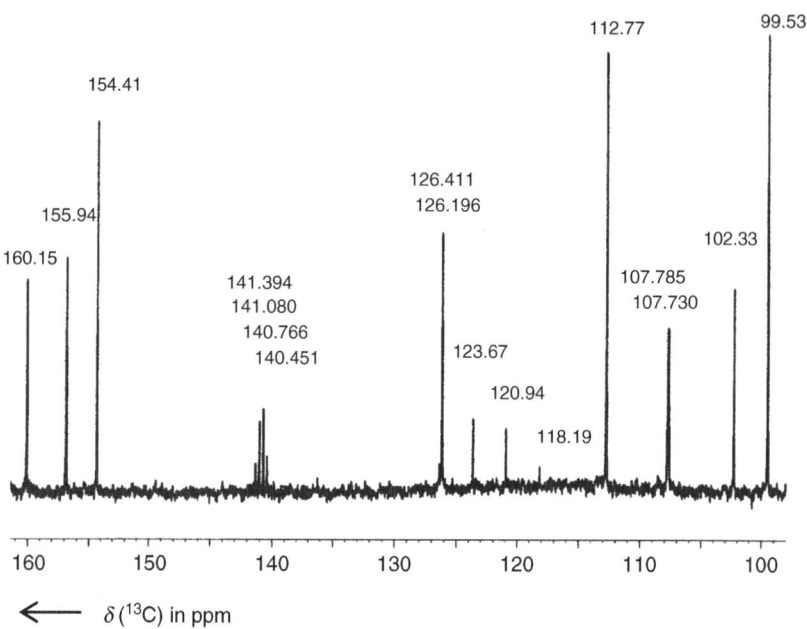

Fig. 5.16D ^{13}C NMR and DEPT spectra ($v_0 = 100$ MHz; solvent: DMSO)

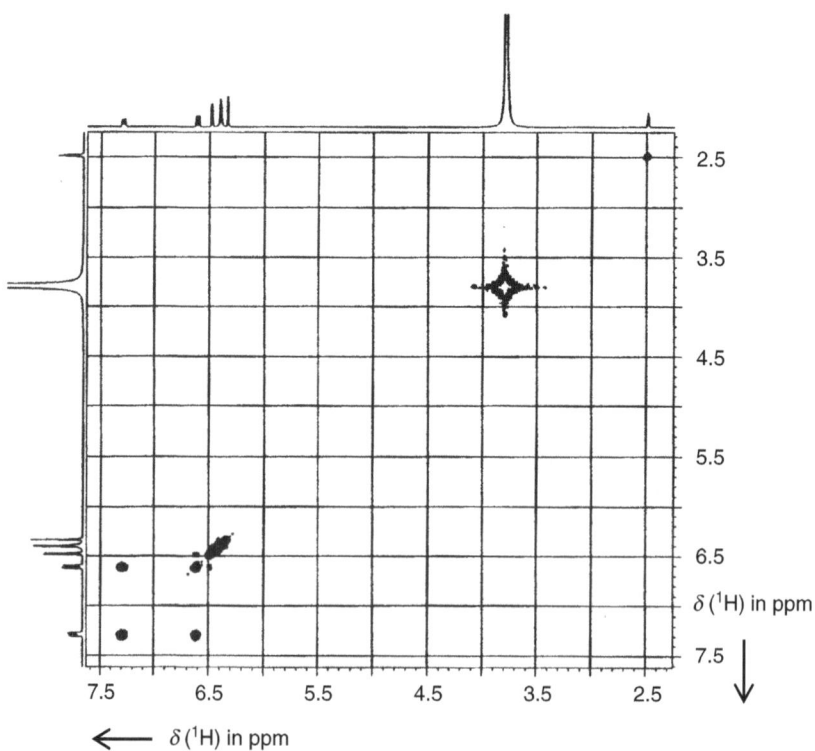

Fig. 5.16E (^{1}H,^{1}H) COSY spectrum

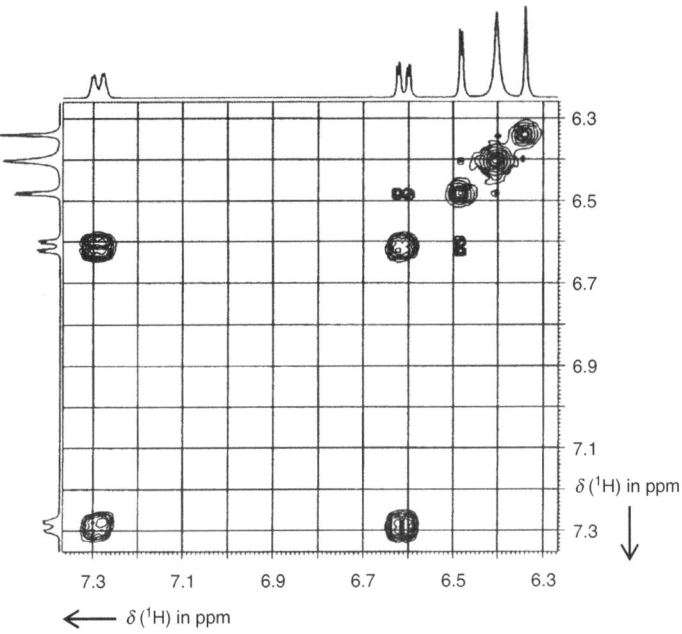

Fig. 5.16F Section of the (^{1}H,^{1}H) COSY spectrum

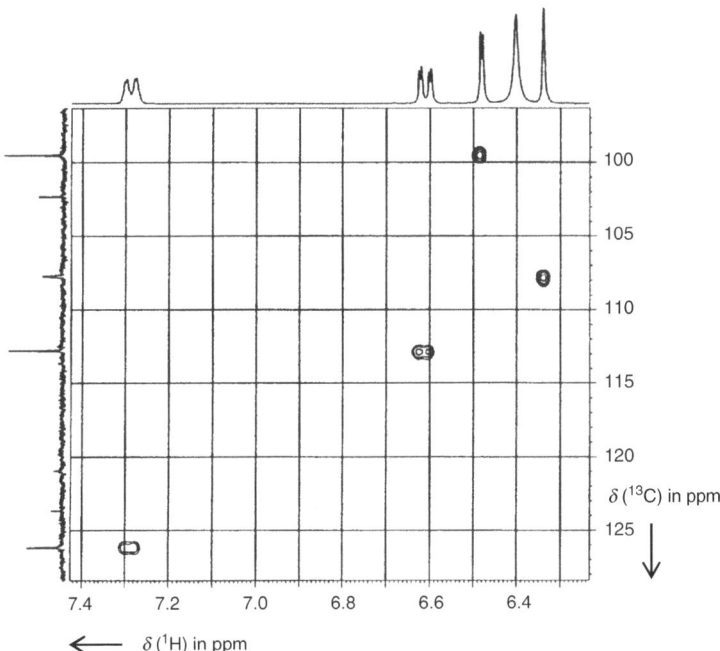

Fig. 5.16G HSQC spectrum

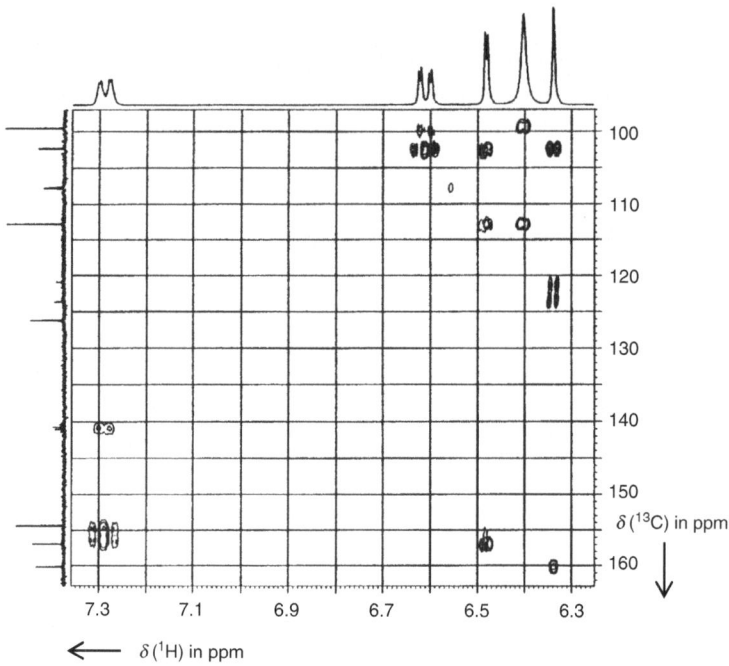

Fig. 5.16H HMBC spectrum

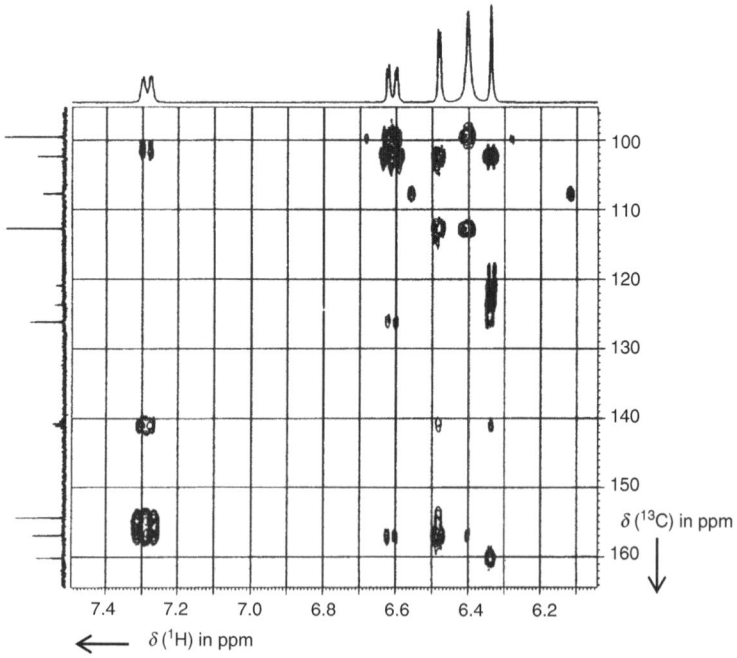

Fig. 5.16I Section of the HMBC spectrum

Challenge 5.17
Deduce the structure of the unknown compound by means of the given
spectra and assign the spectroscopic data as is presented in Challenge 5.1.
  For solutions to Challenge 5.17, see *extras.springer.com.*

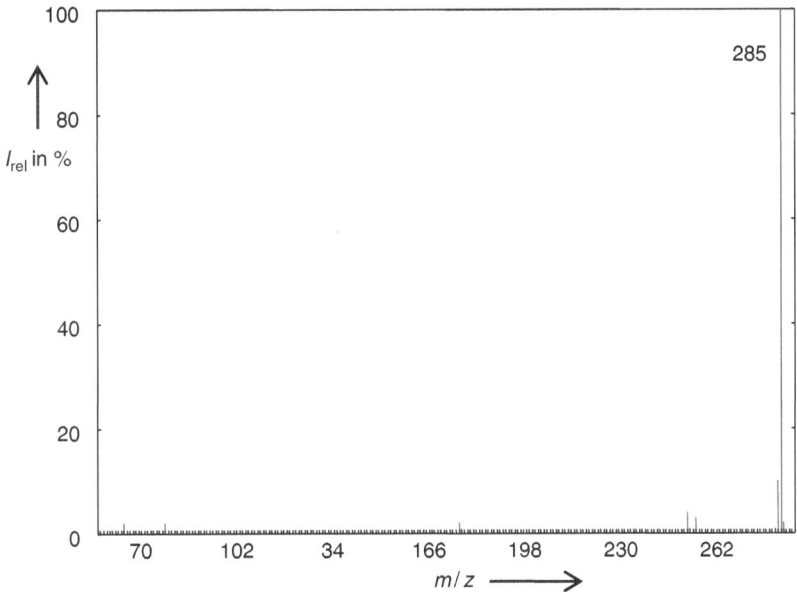

Fig. 5.17A Mass spectrum (CI; methane)

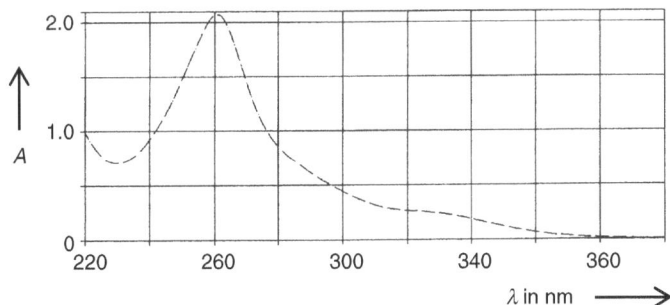

Fig. 5.17B UV spectrum

Analytical sample:	1.09 mg in 10 mL methanol
Cuvette:	$l = 1$ cm

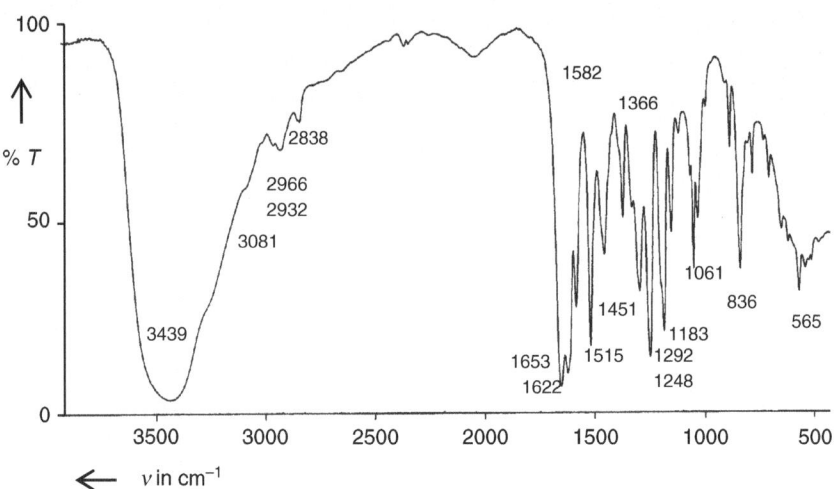

Fig. 5.17C   IR spectrum (KBr pellet)

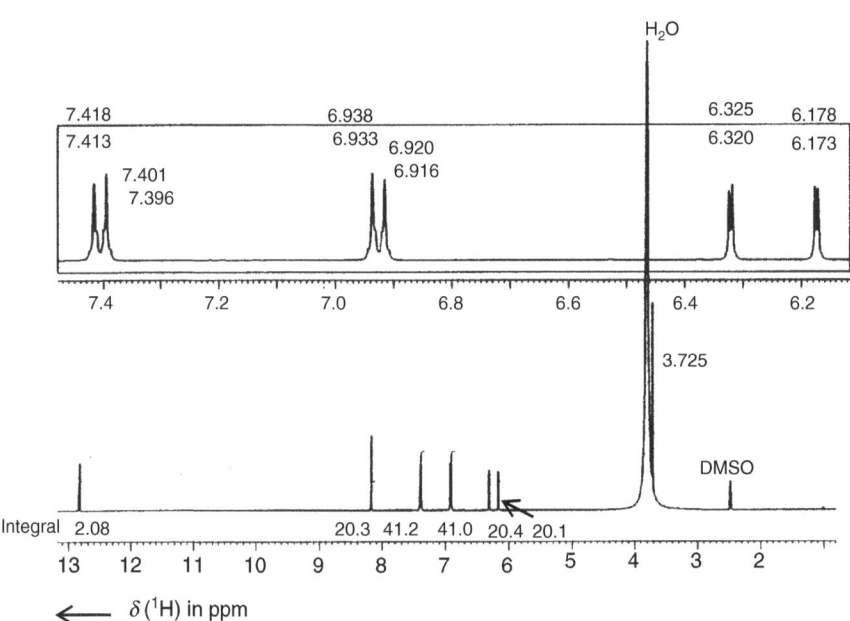

Fig. 5.17D ^{1}H NMR spectrum ($\nu_0 = 400$ MHz; solvent: DMSO)

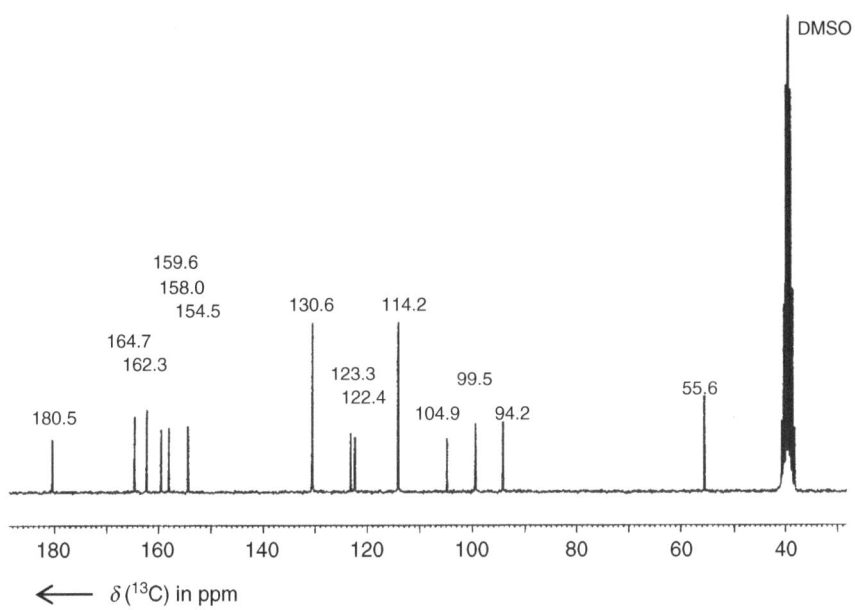

Fig. 5.17E ^{13}C NMR spectrum ($\nu_0 = 100$ MHz; solvent: DMSO)

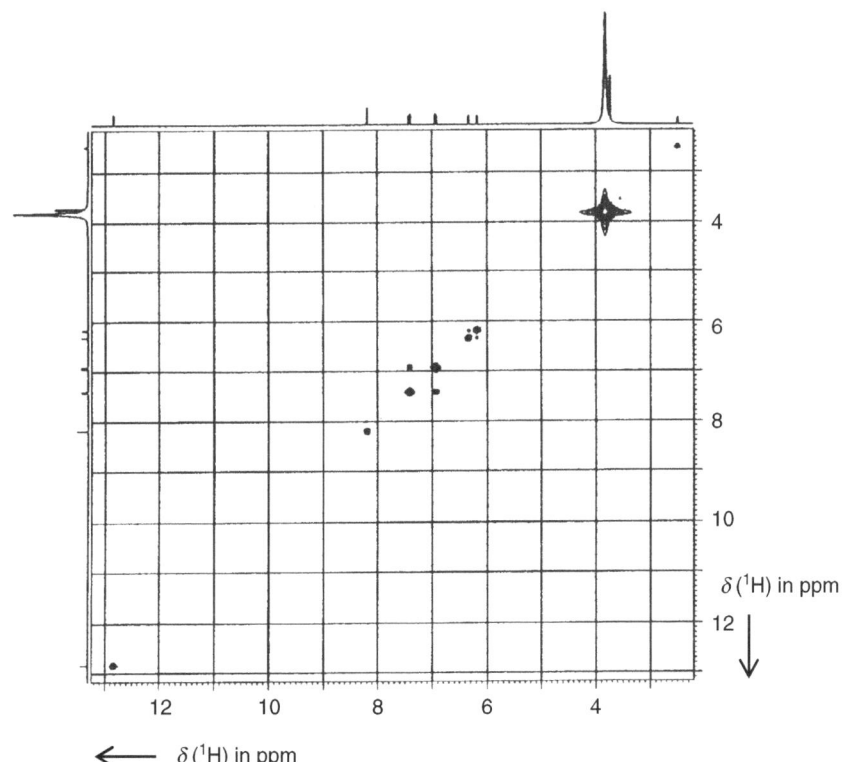

Fig. 5.17F (^{1}H,^{1}H) COSY spectrum (solvent: DMSO)

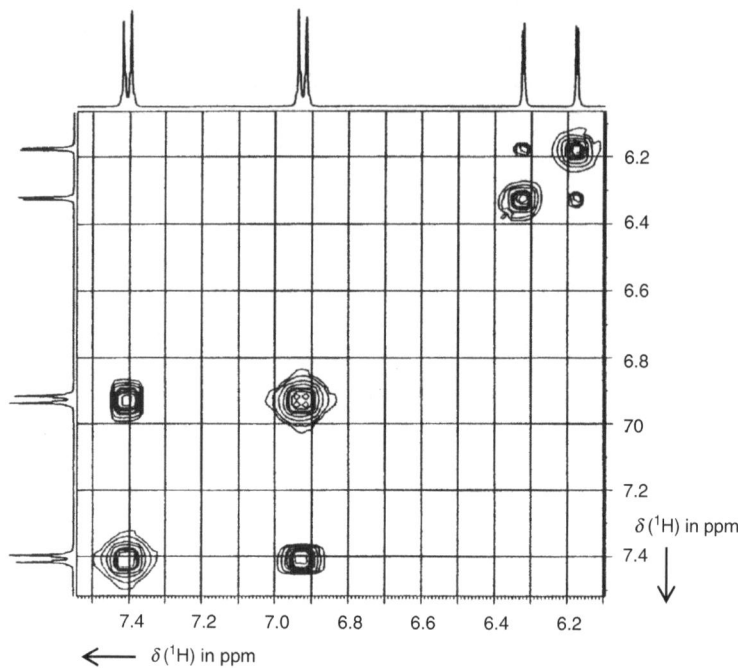

Fig. 5.17G   Section of the (^{1}H,^{1}H) COSY spectrum

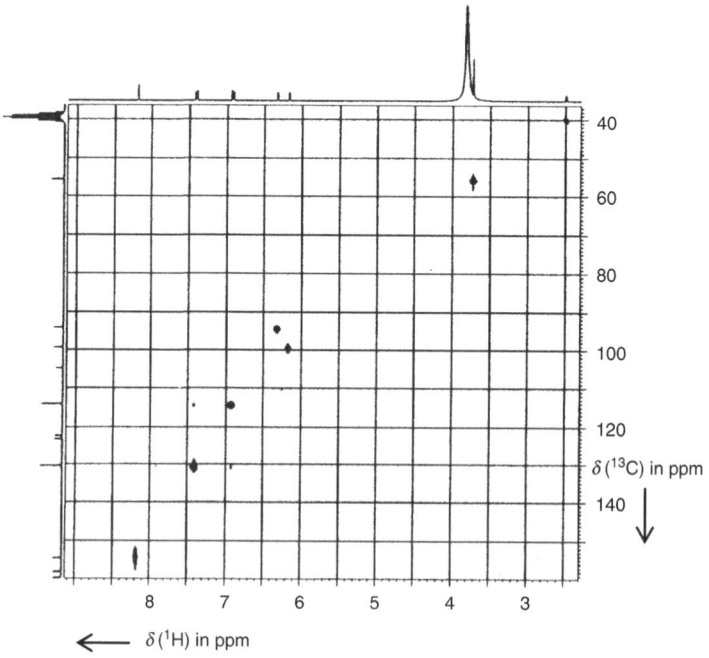

Fig. 5.17H   HSQC spectrum (solvent: DMSO)

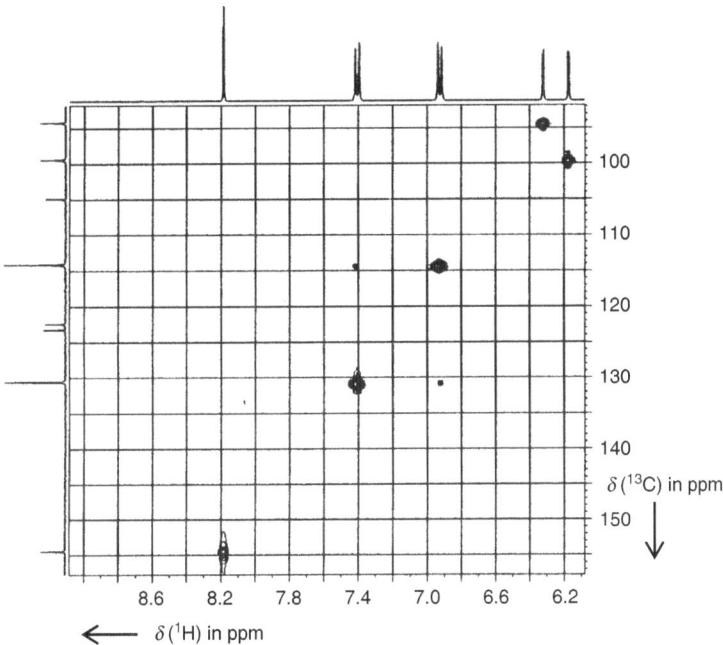

Fig. 5.17I Section of the HSQC spectrum

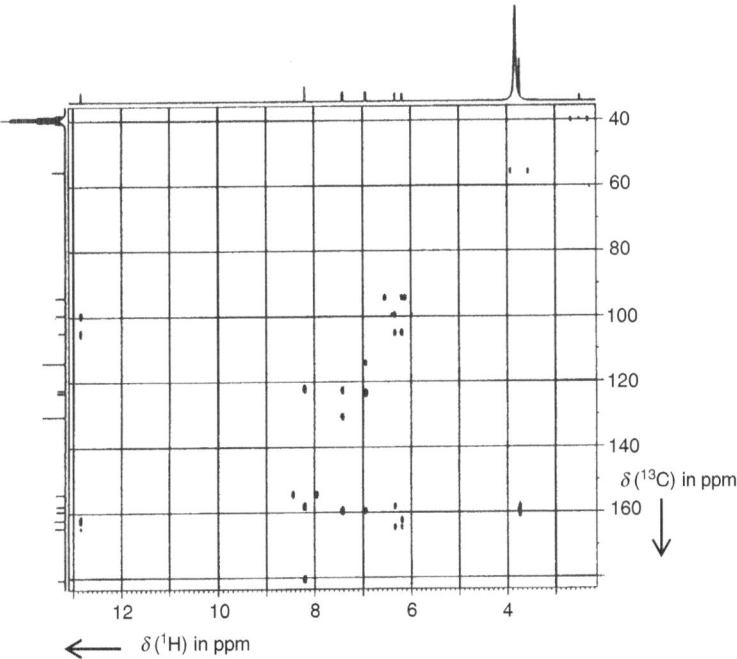

Fig. 5.17J HMBC spectrum (solvent: DMSO)

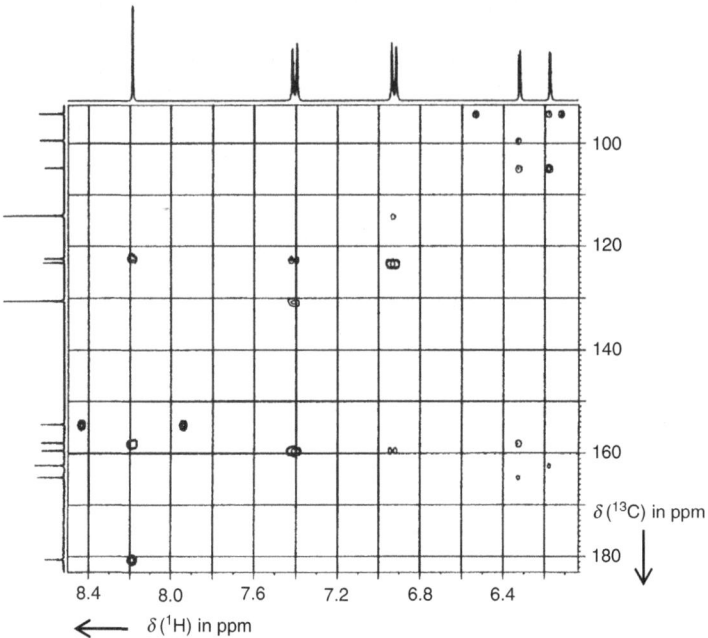

Fig. 5.17K Section of the HMBC spectrum

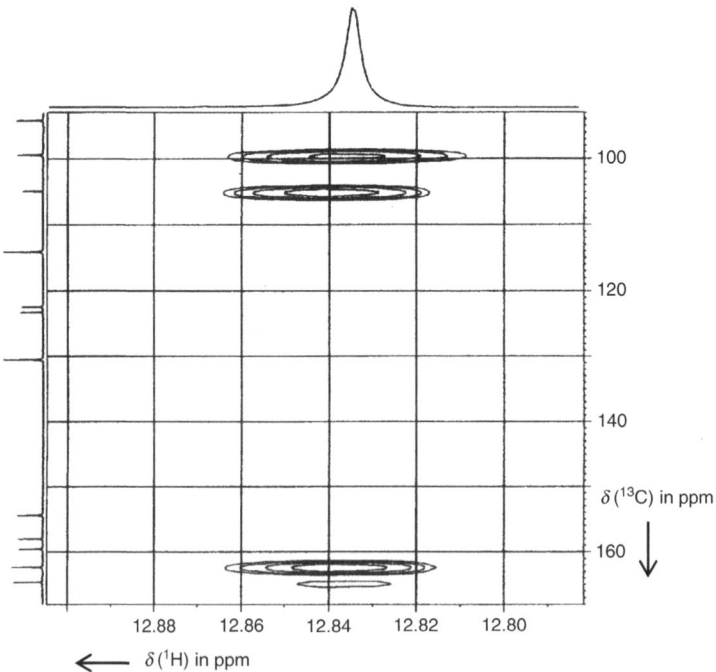

Fig. 5.17L Section of the HMBC spectrum

Challenge 5.18

Deduce the structure of the unknown compound by means of the given spectra and assign the spectroscopic data as is presented in Challenge 5.1.

For solutions to Challenge 5.18, see *extras.springer*.com.

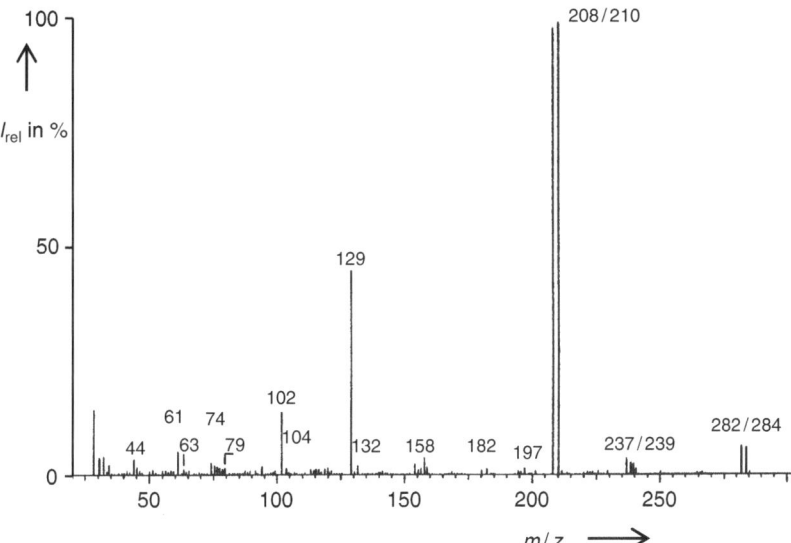

Fig. 5.18A Mass spectrum (EI, 70 eV)

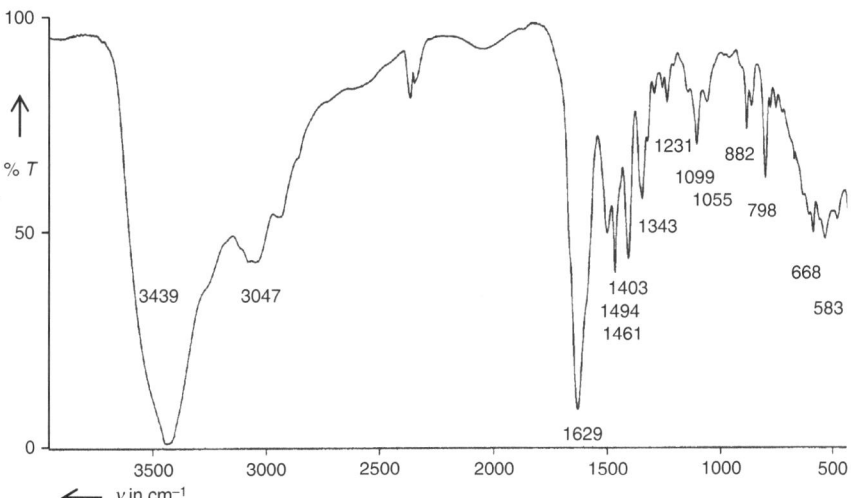

Fig. 5.18B IR spectrum (KBr pellet)

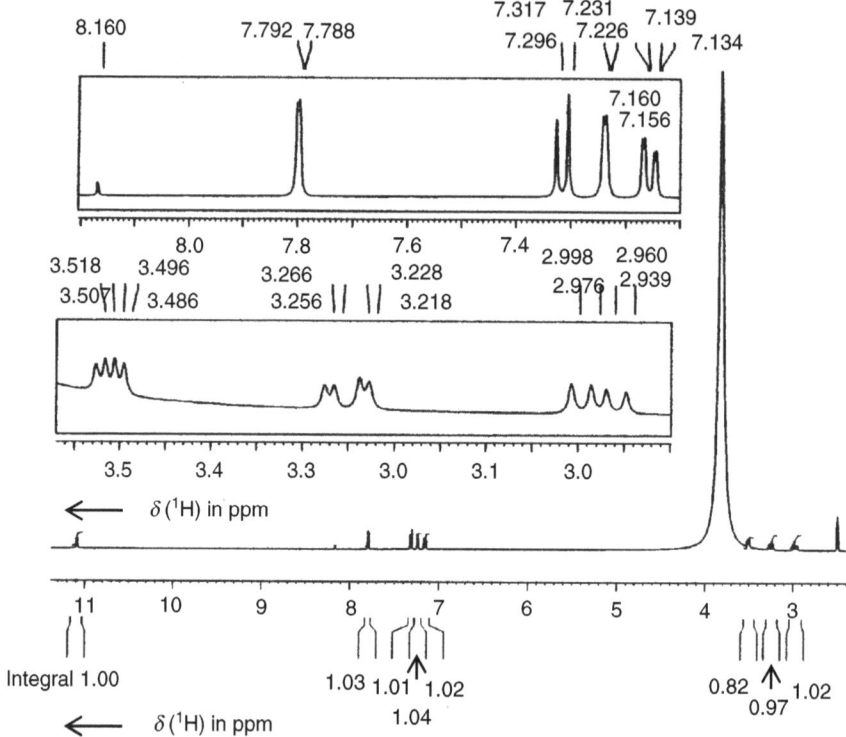

Fig. 5.18C ¹H NMR spectrum ($\nu_0 = 400$ MHz; solvent: DMSO)

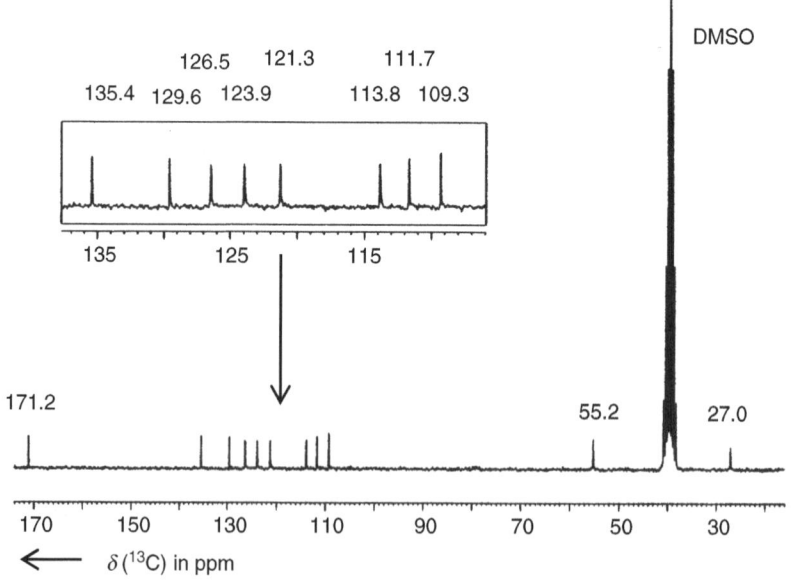

Fig. 5.18D ¹³C NMR spectrum ($\nu_0 = 100$ MHz; solvent: DMSO)

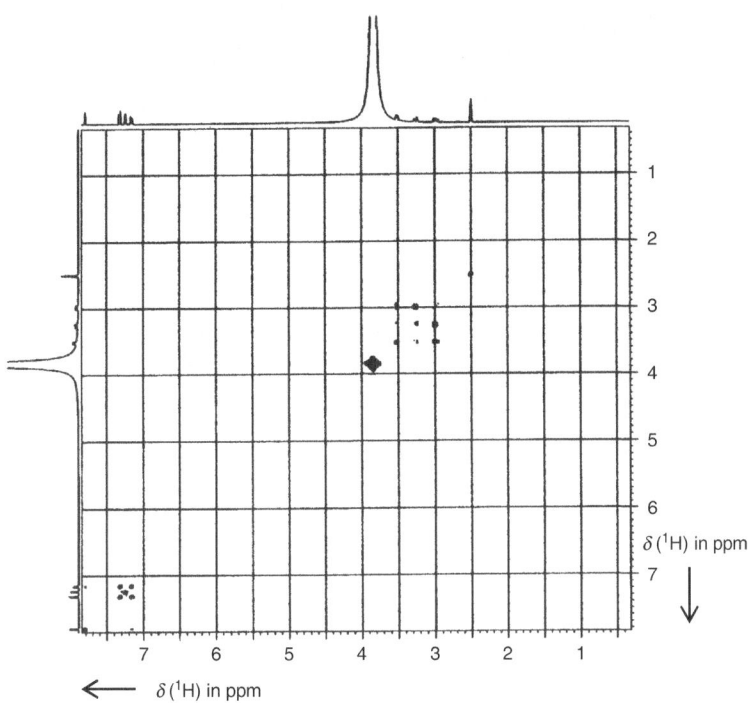

Fig. 5.18E $(^1H, ^1H)$ COSY spectrum (solvent: DMSO)

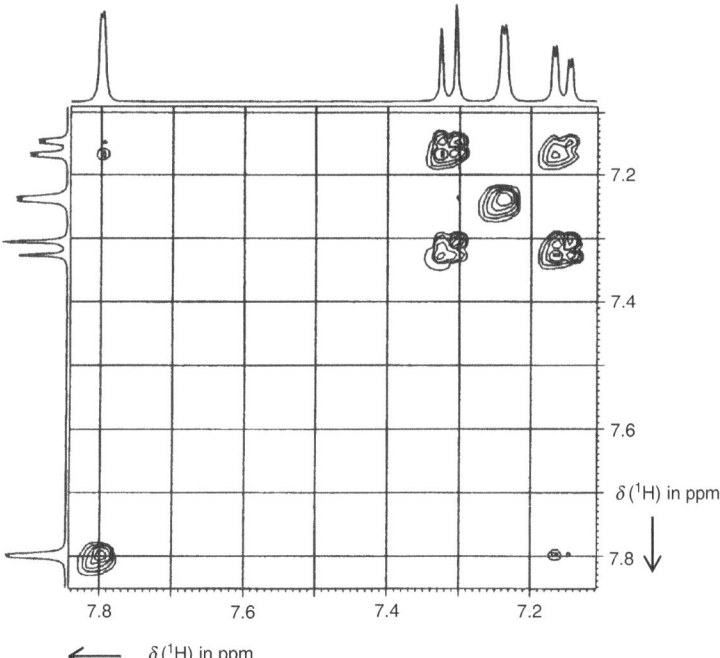

Fig. 5.18F Section of the $(^1H, ^1H)$ COSY spectrum

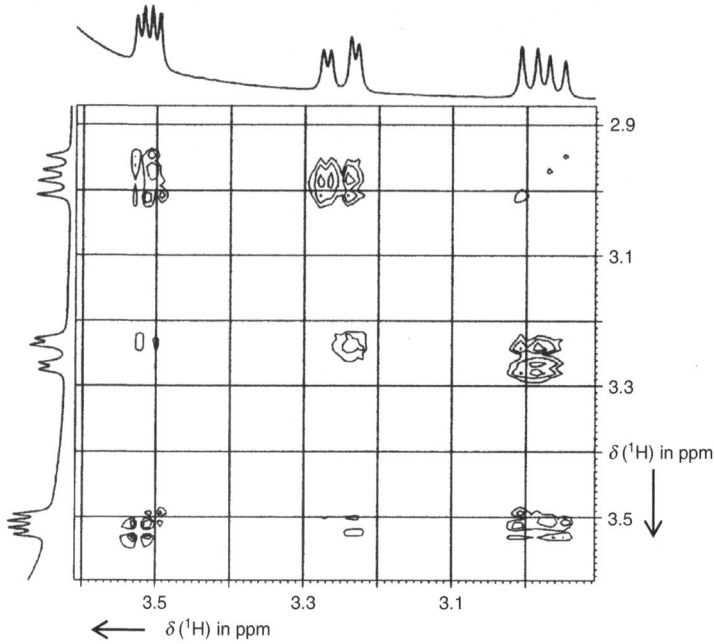

Fig. 5.18G Section of the (^{1}H,^{1}H) COSY spectrum

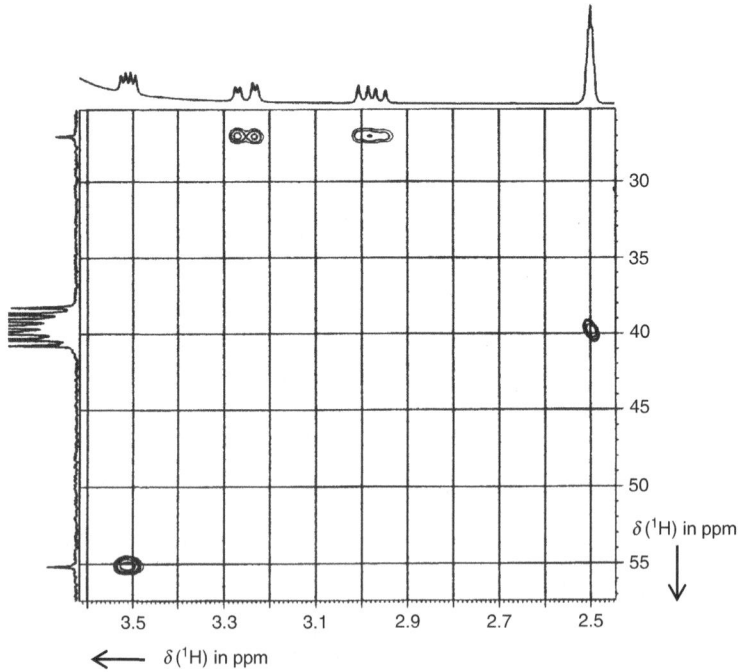

Fig. 5.18H HSQC spectrum (solvent: DMSO)

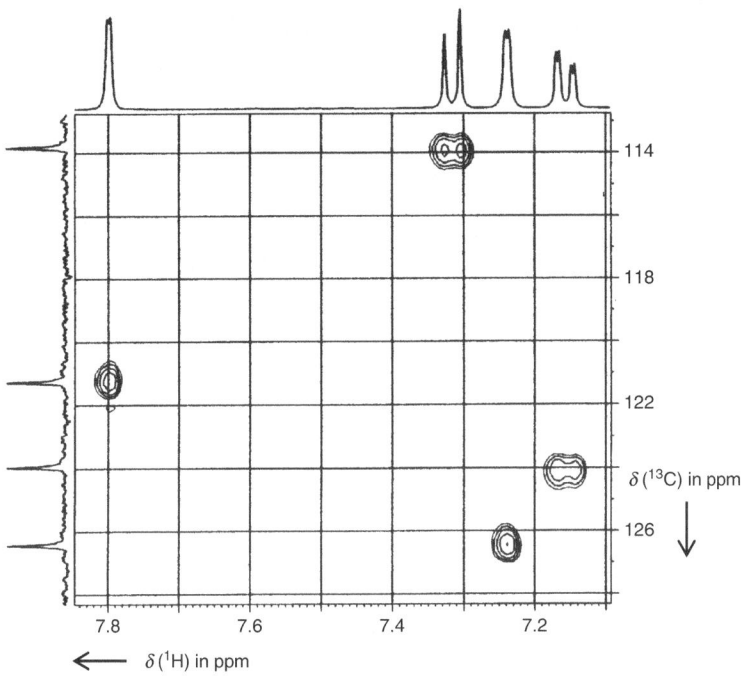

Fig. 5.18I   Section of the HSQC spectrum

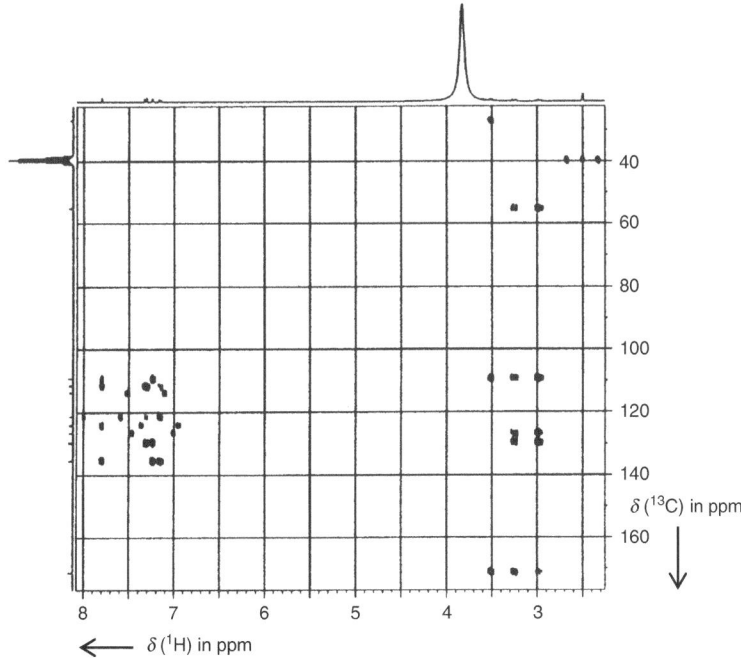

Fig. 5.18J   Section of the HMBC spectrum

Fig. 5.18K Section of the HMBC spectrum (solvent: DMSO)

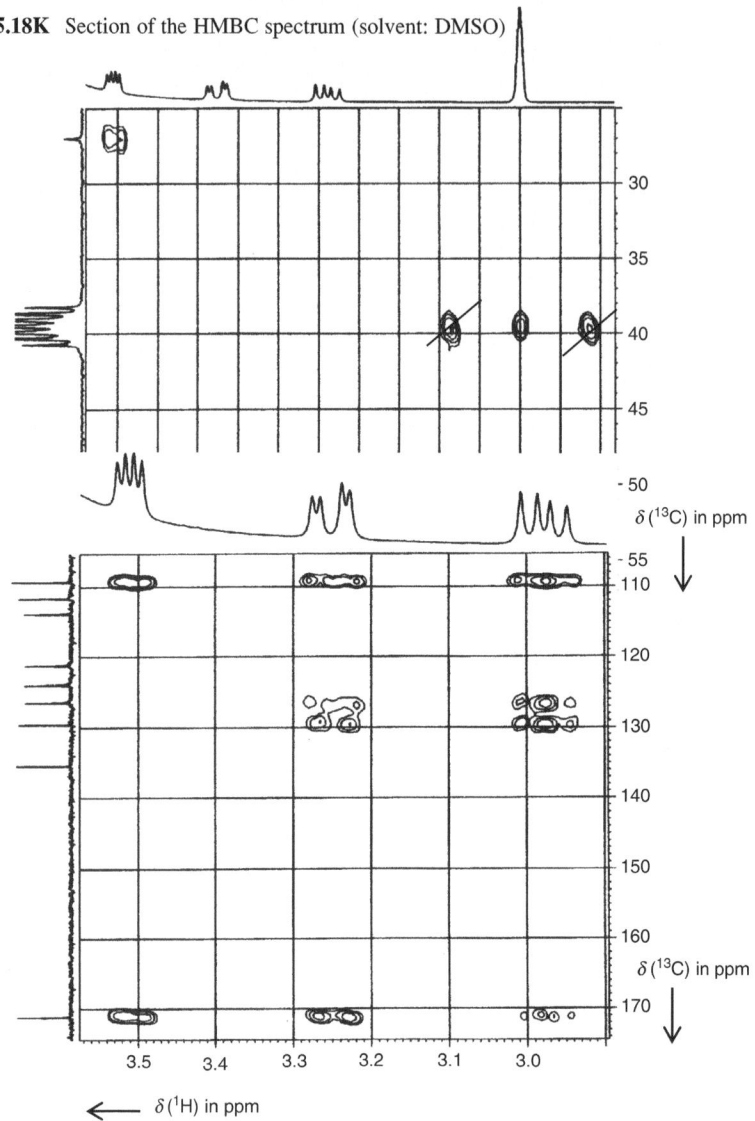

δ (^{13}C) in ppm

δ (^{13}C) in ppm

δ (^{1}H) in ppm

Fig. 5.18L Section of the HMBC spectrum

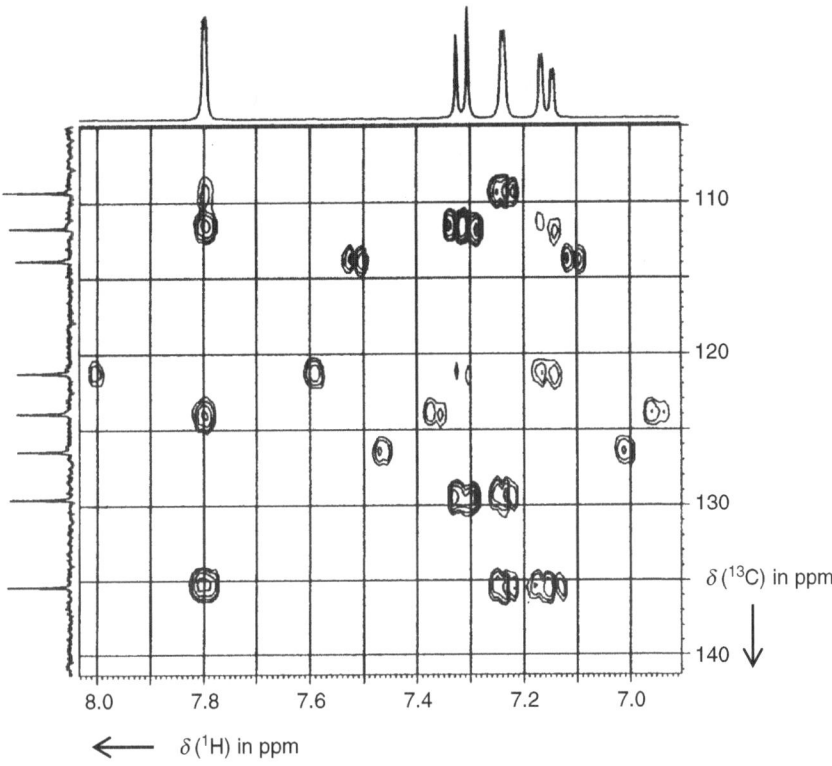

Fig. 5.18M   Section of the HMBC spectrum

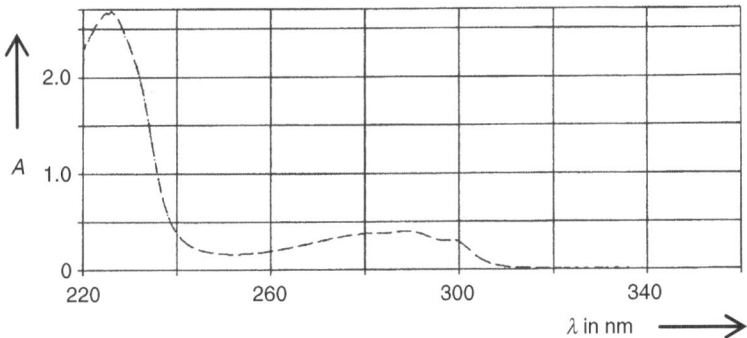

Fig. 5.18N   UV spectrum

Analytical sample:	1.09 mg in 10 mL methanol
Cuvette:	l = 1 cm

Challenge 5.19
Deduce the structure of the unknown compound by means of the given spectra and assign the spectroscopic data as is presented in Challenge 5.1.
    For solutions to Challenge 5.19, see *extras.springer.com*.

Elementary analysis: 74.4% C, 8.20% H, 7.65% N

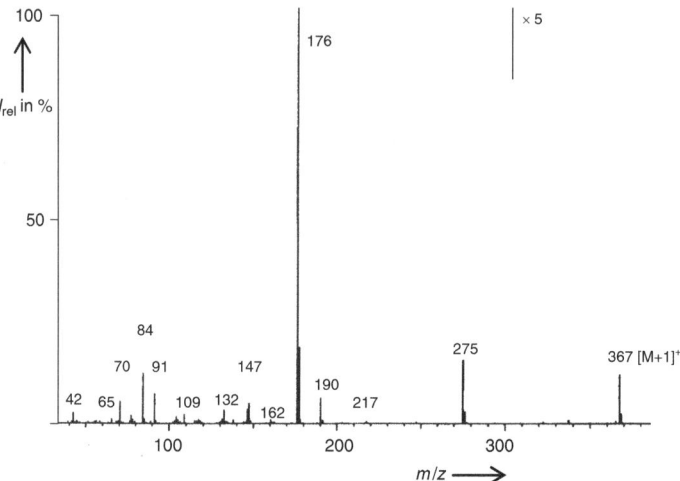

Fig. 5.19A Mass spectrum (EI, 70 eV)

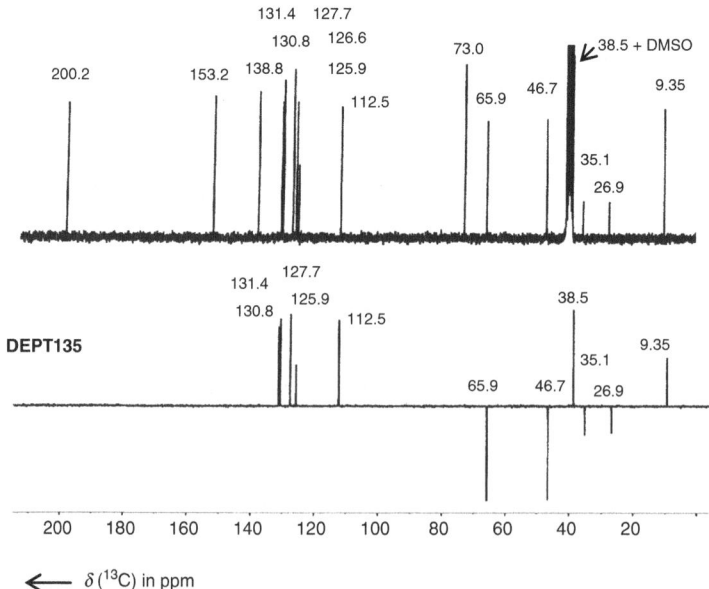

Fig. 5.19B ^{13}C NMR and DEPT135 spectra (solvent: DMSO)

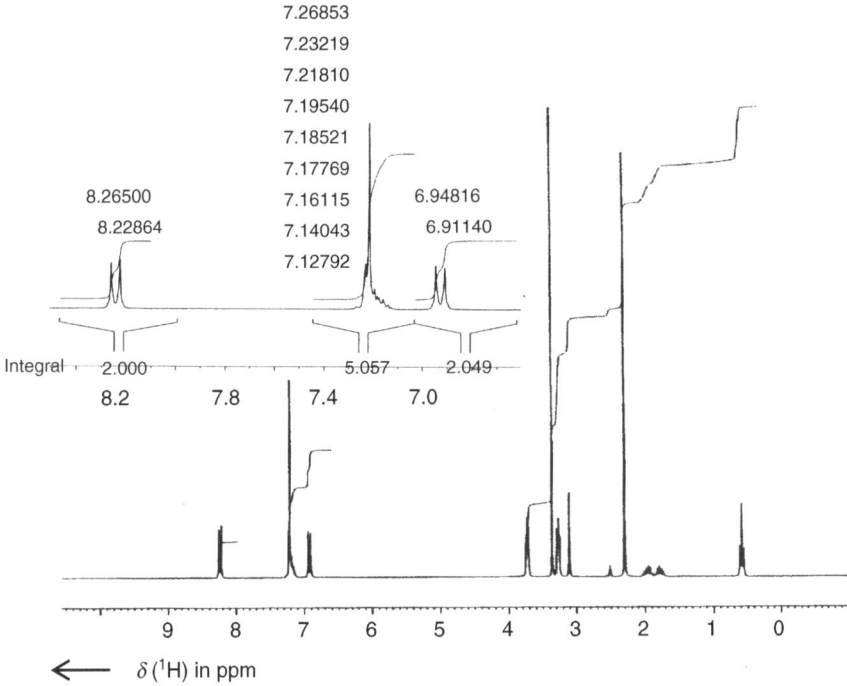

Fig. 5.19C ^{1}H NMR spectrum ($v_0 = 250$ MHz; solvent: DMSO)

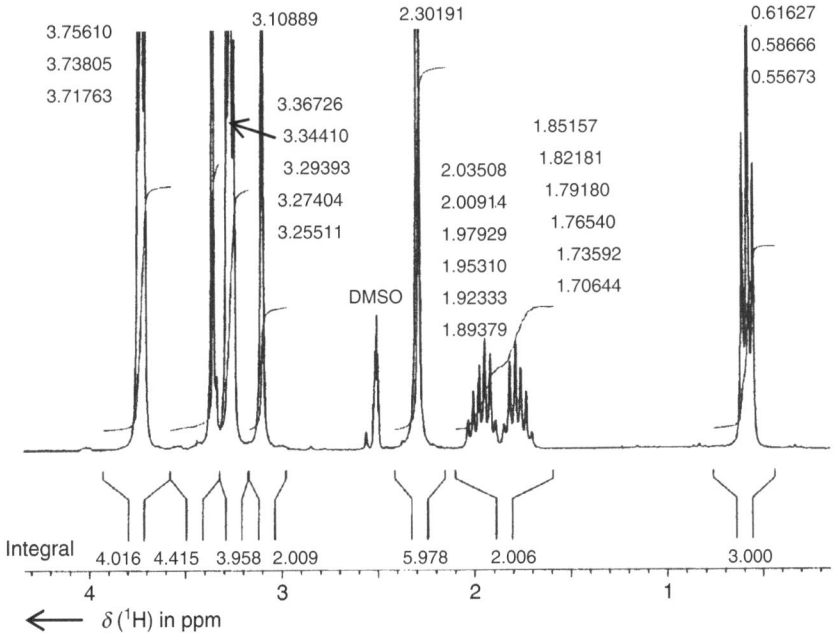

Fig. 5.19D Section of the ^{1}H NMR spectrum

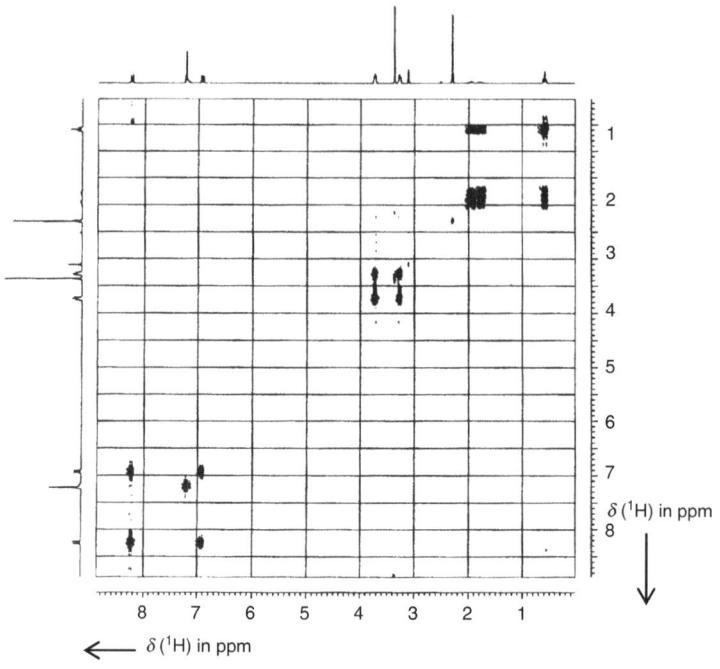

Fig. 5.19E (^{1}H,^{1}H) COSY spectrum (solvent: DMSO)

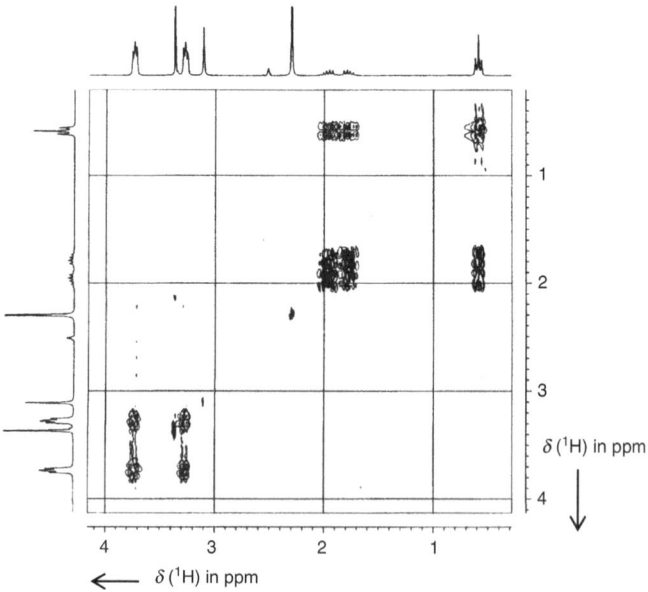

Fig. 5.19F Section of the (^{1}H,^{1}H) COSY spectrum

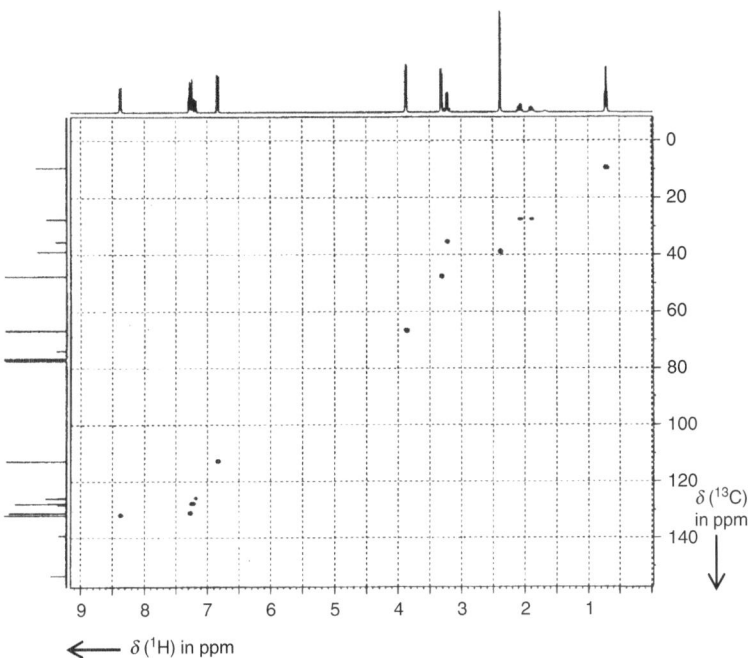

Fig. 5.19G HSQC spectrum (solvent: DMSO)

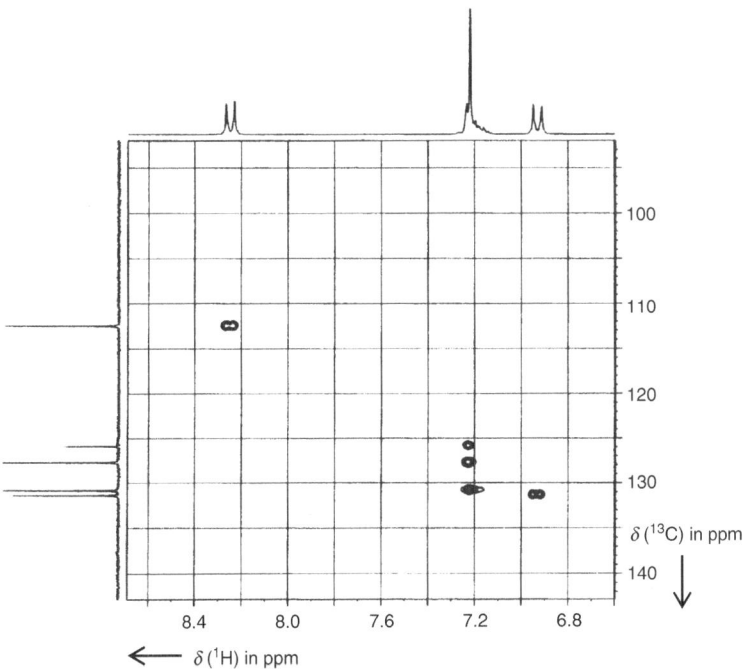

Fig. 5.19H Section of the HSQC spectrum

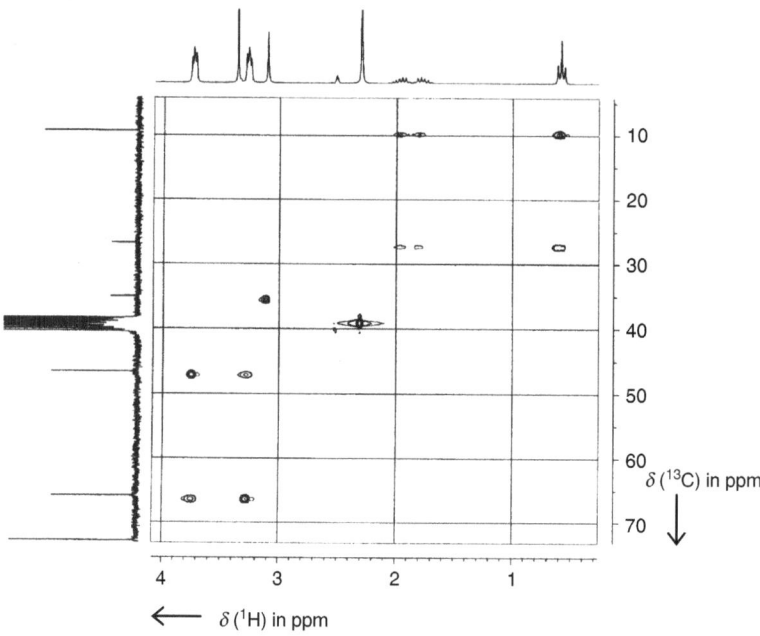

Fig. 5.19I Section of the HSQC-TOCSY spectrum

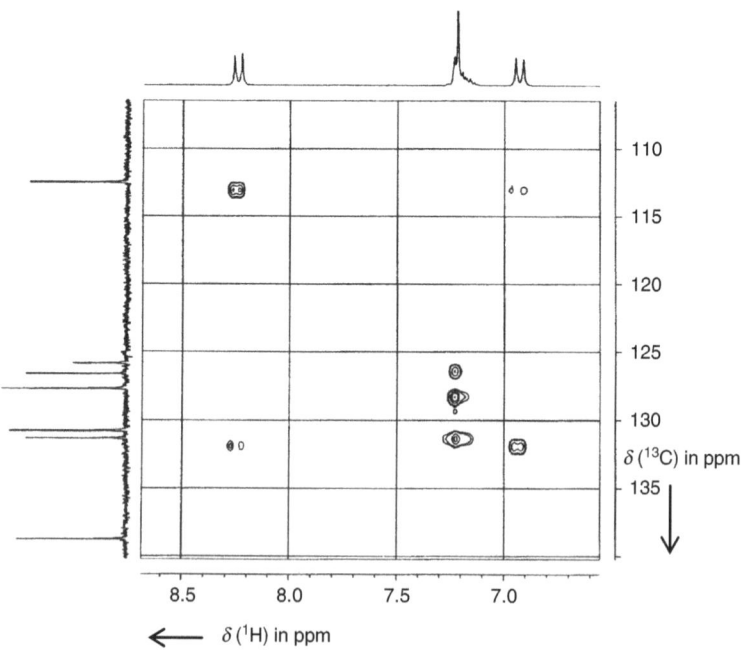

Fig. 5.19J Section of the HSQC-TOCSY spectrum

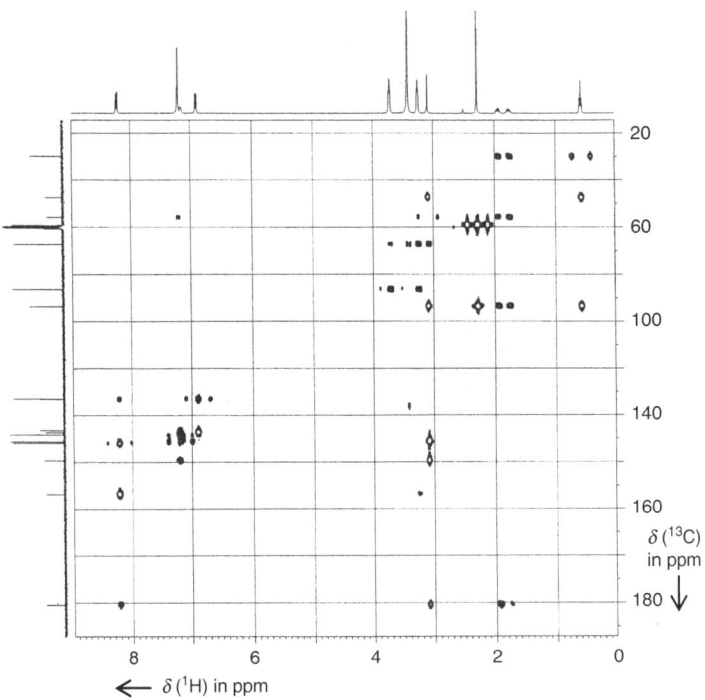

Fig. 5.19K HMBC spectrum

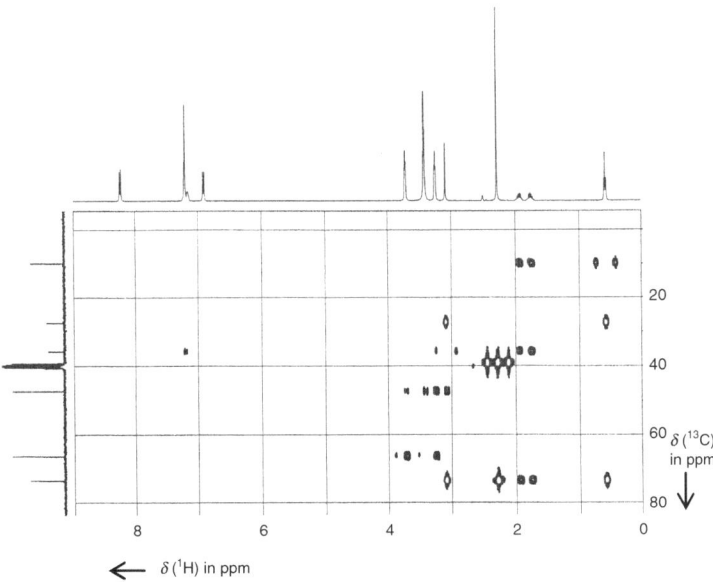

Fig. 5.19L Section of the HMBC spectrum (solvent: DMSO)

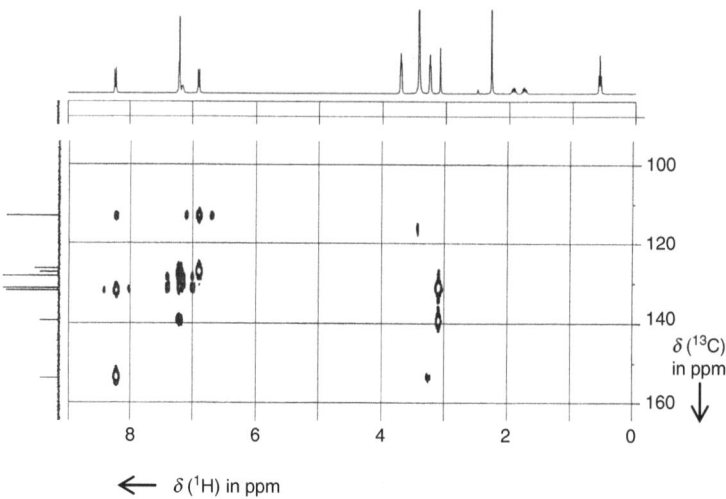

Fig. 5.19M Section of the HMBC spectrum

Challenge 5.20
Deduce the structure of the unknown compound by means of the given
spectra and assign the spectroscopic data as is presented in Challenge 5.1.
    For solutions to Challenge 5.20, see *extras.springer.com*.

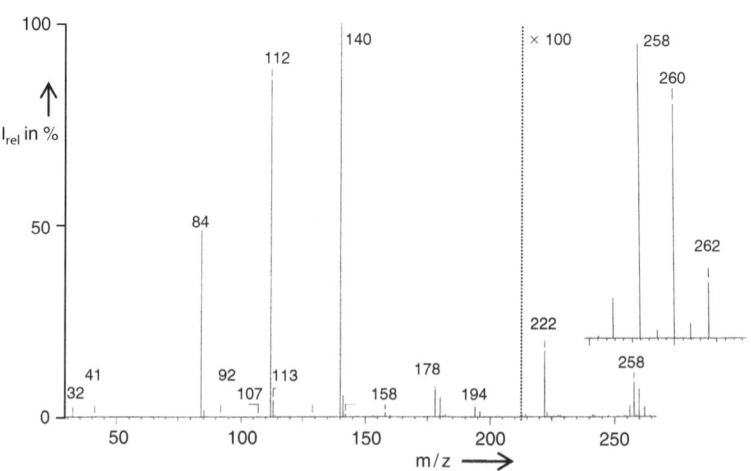

Fig. 5.20A Mass spectrum (EI, 14 eV)

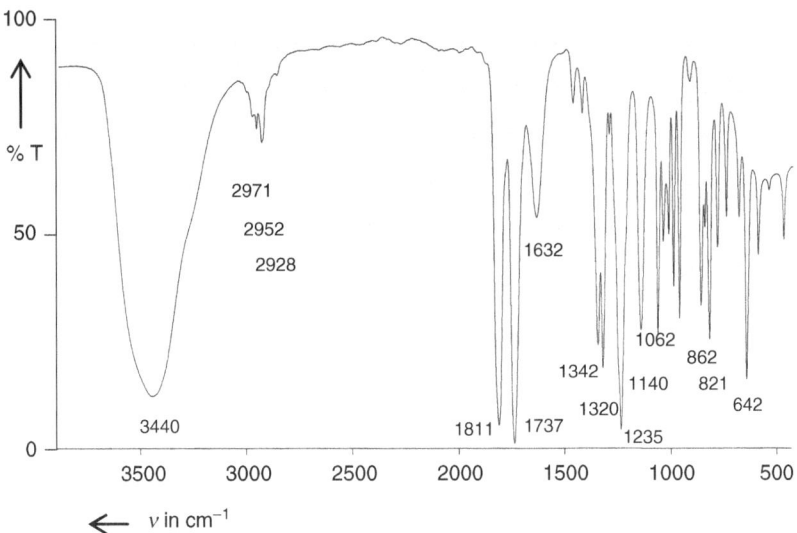

Fig. 5.20B IR spectrum (KBr pellet)

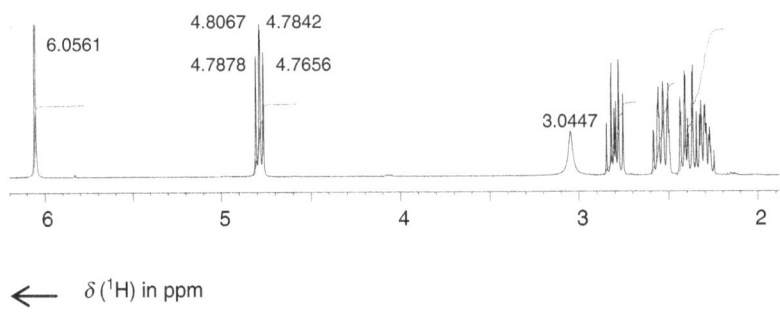

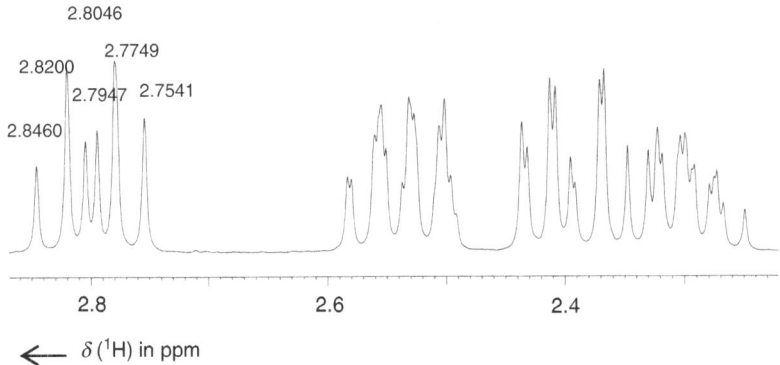

Fig. 5.20C ^{1}H NMR spectrum ($v_0 = 400$ MHz; solvent: DMSO)

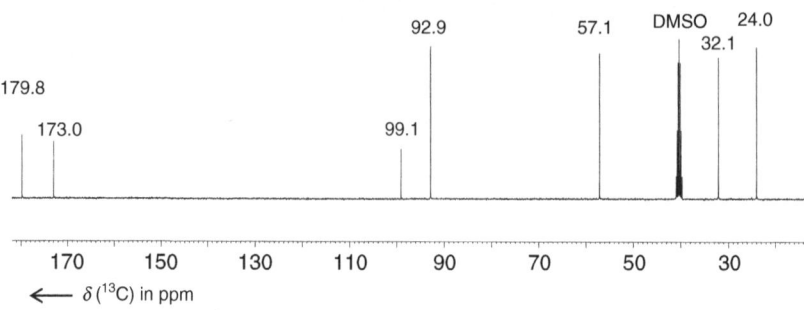

Fig. 5.20D ^{13}C NMR spectrum (solvent: DMSO)

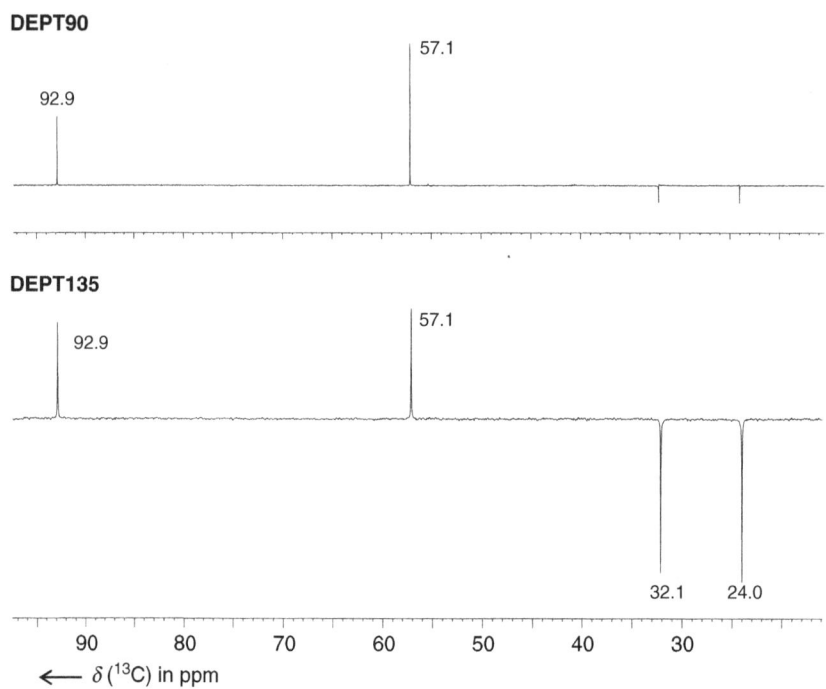

Fig. 5.20E DEPT90 and DEPT135 spectra

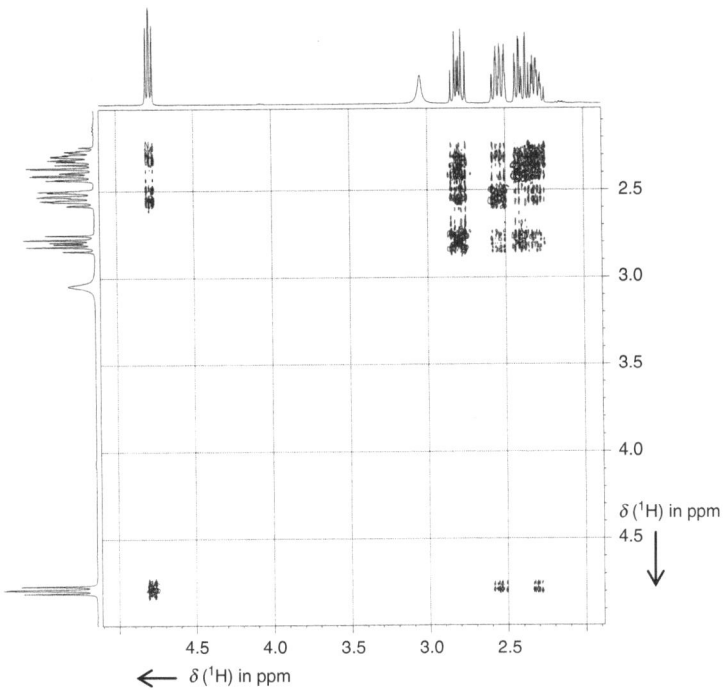

Fig. 5.20F   (^{1}H,^{1}H) COSY spectrum (solvent: DMSO)

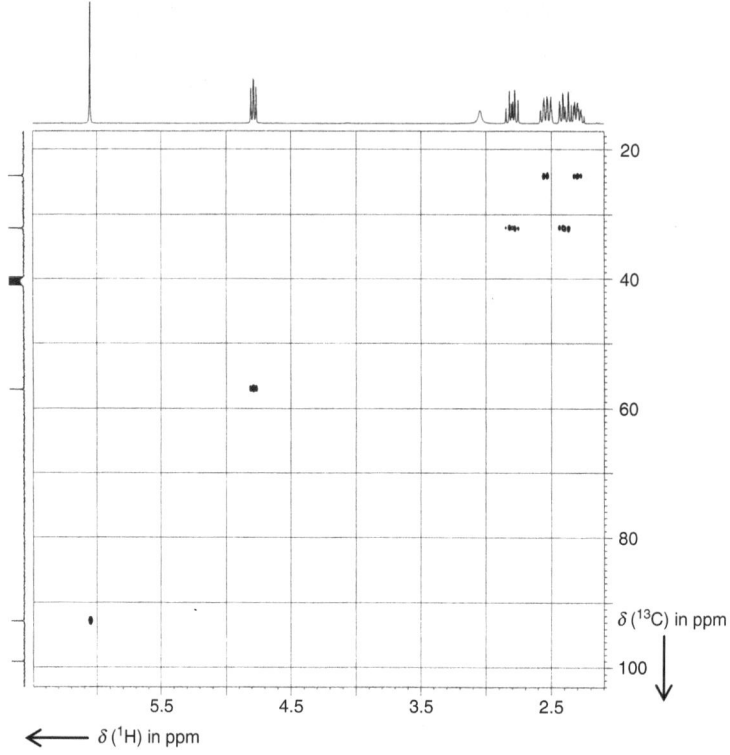

Fig. 5.20G HSQC spectrum (solvent: DMSO)

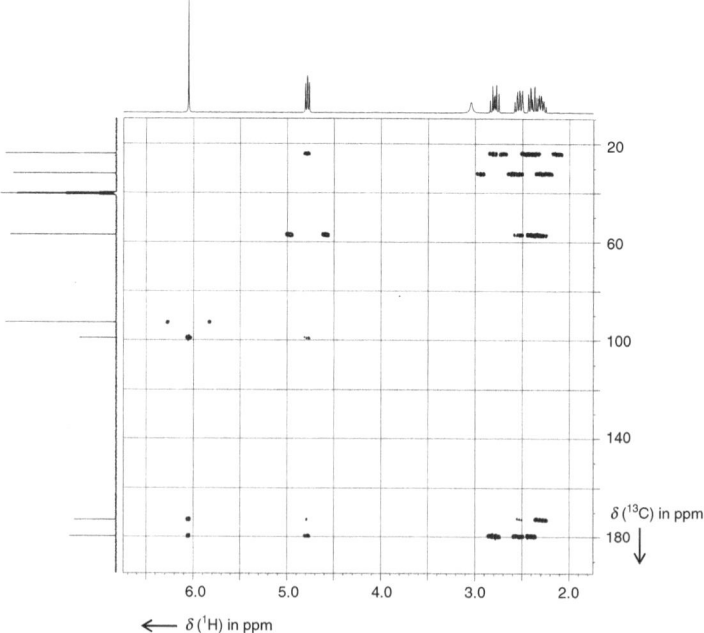

Fig. 5.20H HMBC spectrum (solvent: DMSO)

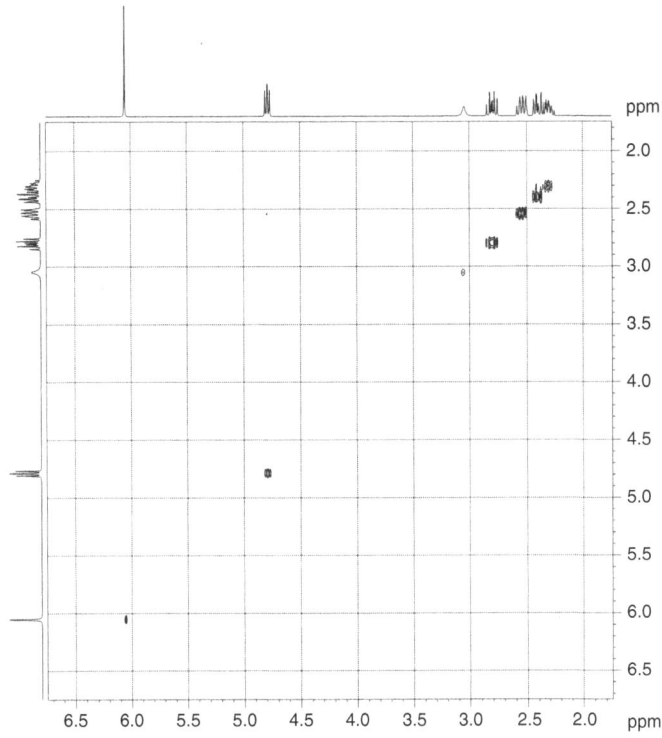

Fig. 5.20I NOESY spectrum

Challenge 5.21

Deduce the structure of the unknown compound by means of the given spectra and assign the spectroscopic data as is presented in Challenge 5.1.

For solutions to Challenge 5.21, see *extras.springer.com*.

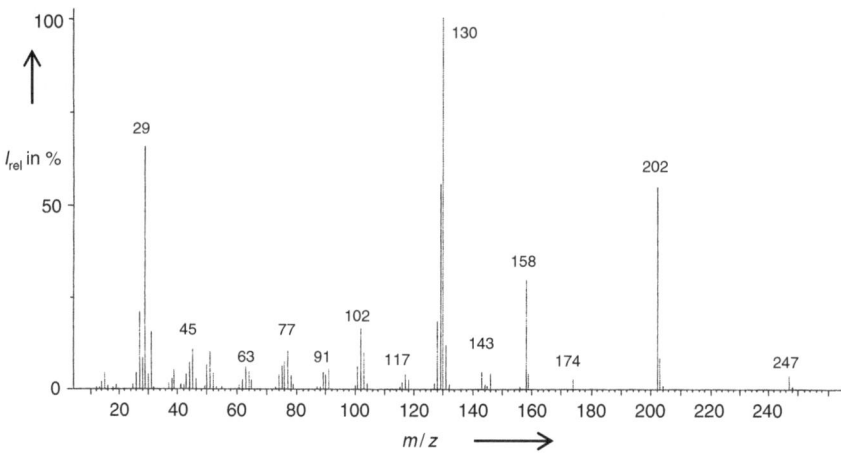

Fig. 5.21A Mass spectrum (EI, 70 eV)

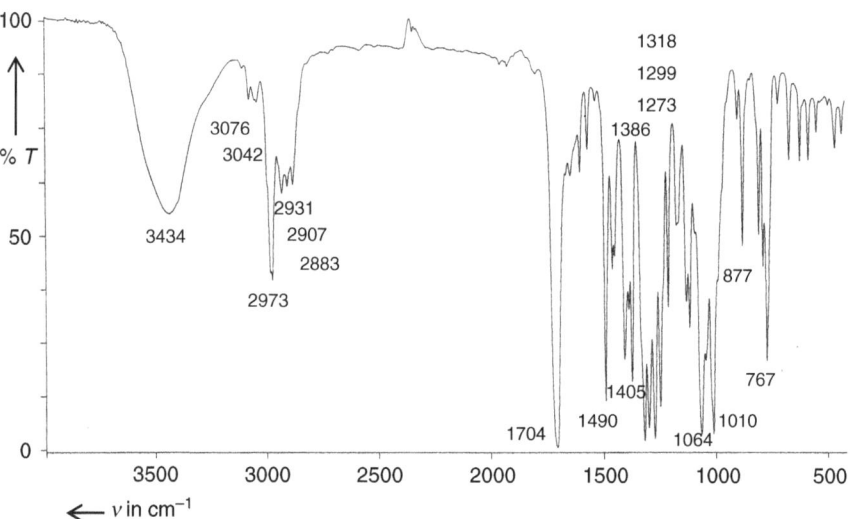

Fig. 5.21B IR spectrum (KBr pellet)

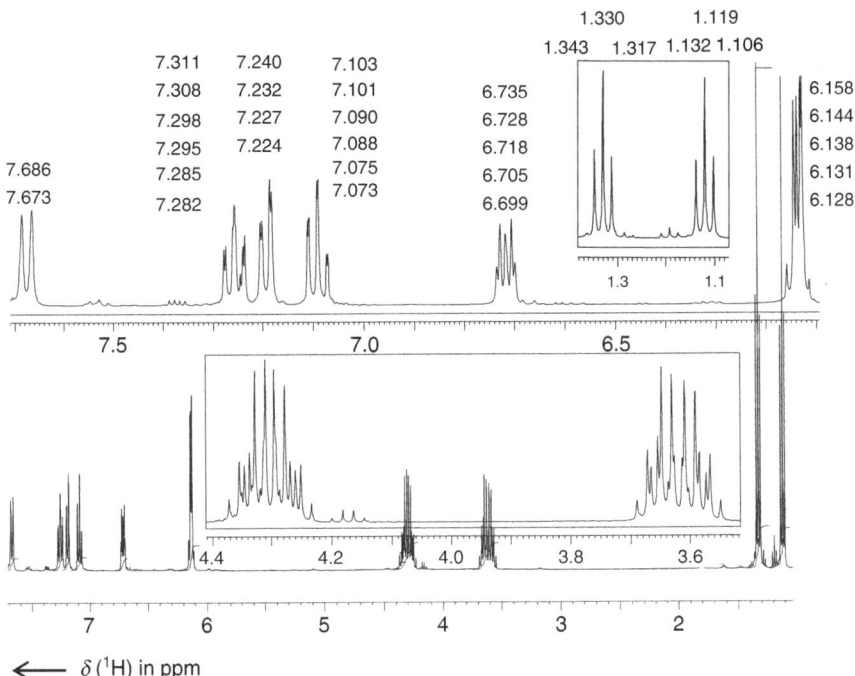

Fig. 5.21C ^{1}H NMR spectrum ($v_0 = 400$ MHz; solvent: CDCl$_3$)

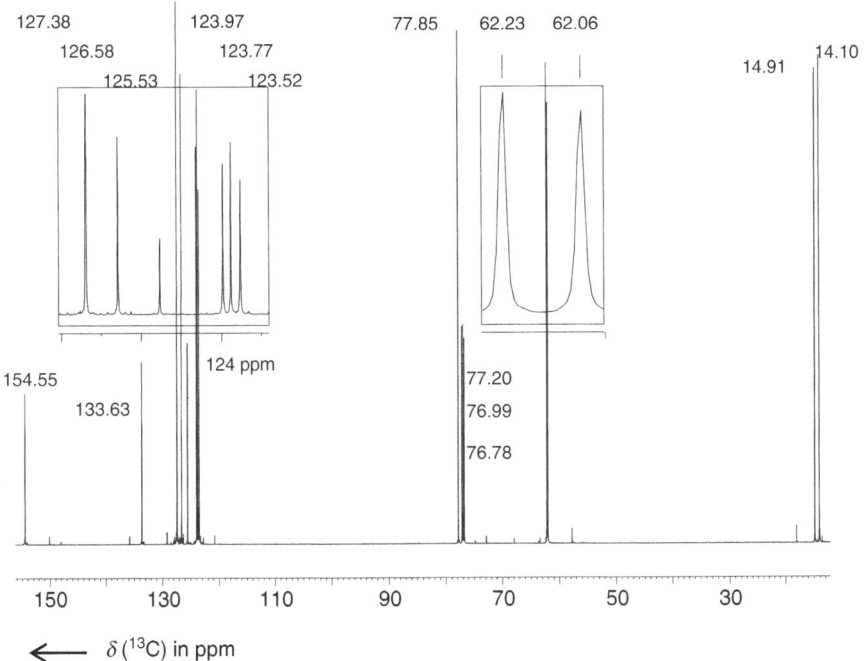

Fig. 5.21D ^{13}C NMR spectrum (solvent: CDCl$_3$)

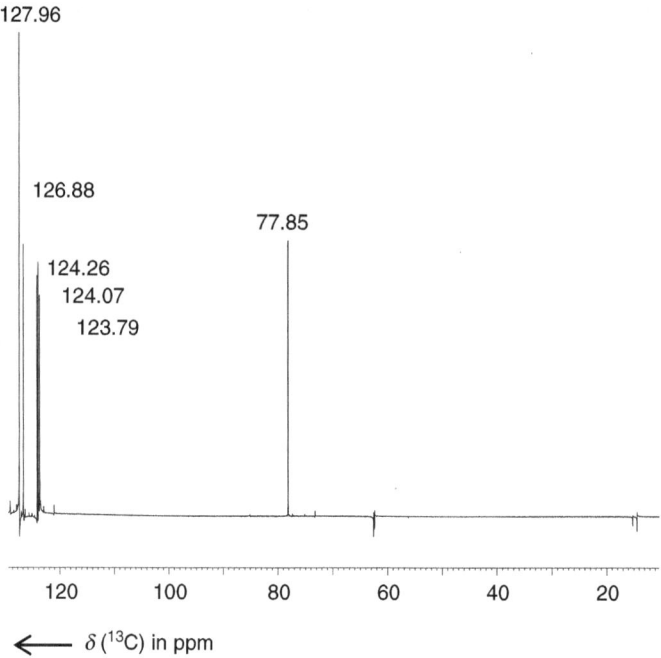

Fig. 5.21E DEPT90 spectrum

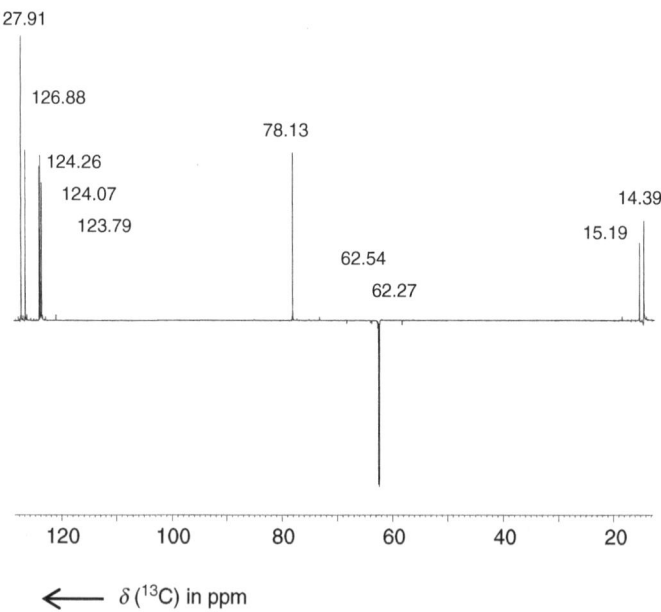

Fig. 5.21F DEPT135 spectrum

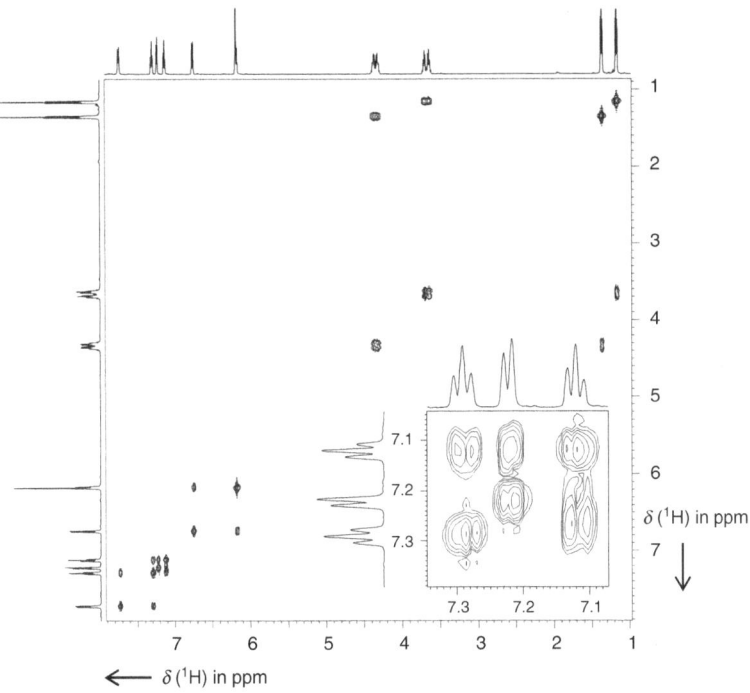

Fig. 5.21G $(^1H,^1H)$ COSY spectrum (solvent: $CDCl_3$)

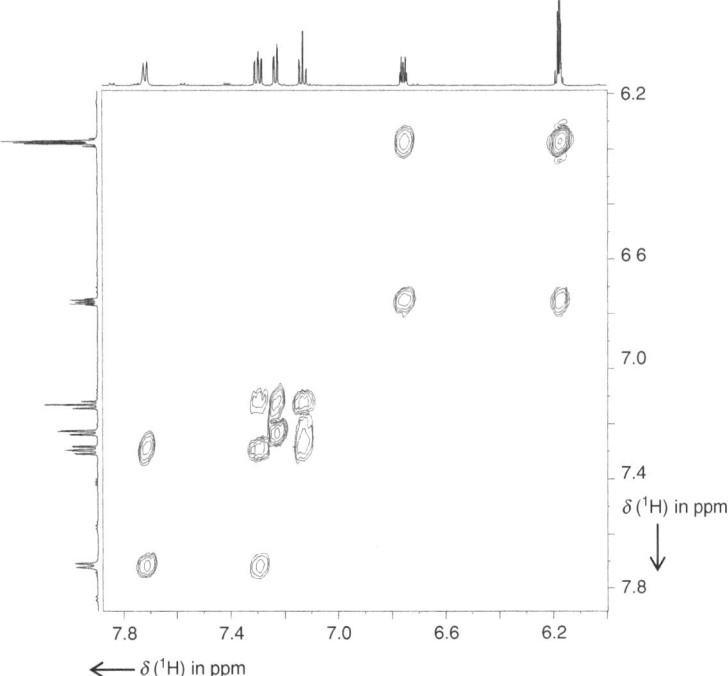

Fig. 5.21H Section of the $(^1H,^1H)$ COSY spectrum

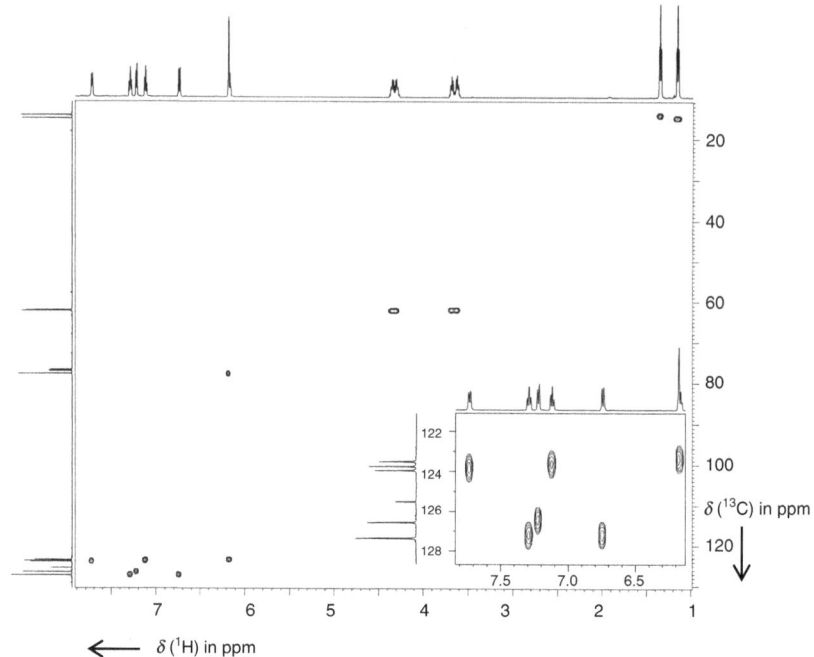

Fig. 5.21I HSQC spectrum (solvent: CDCl₃)

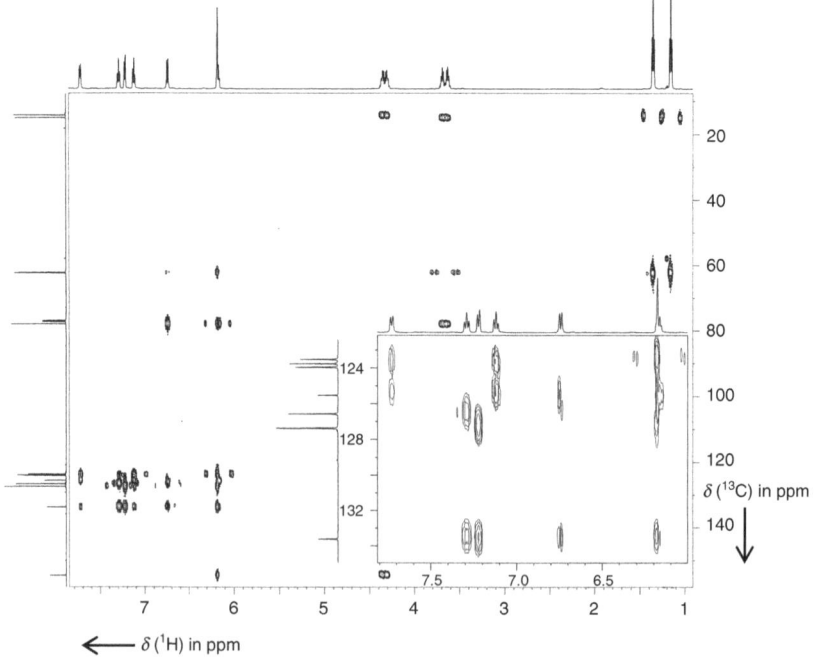

Fig. 5.21J HMBC spectrum (solvent: CDCl₃)

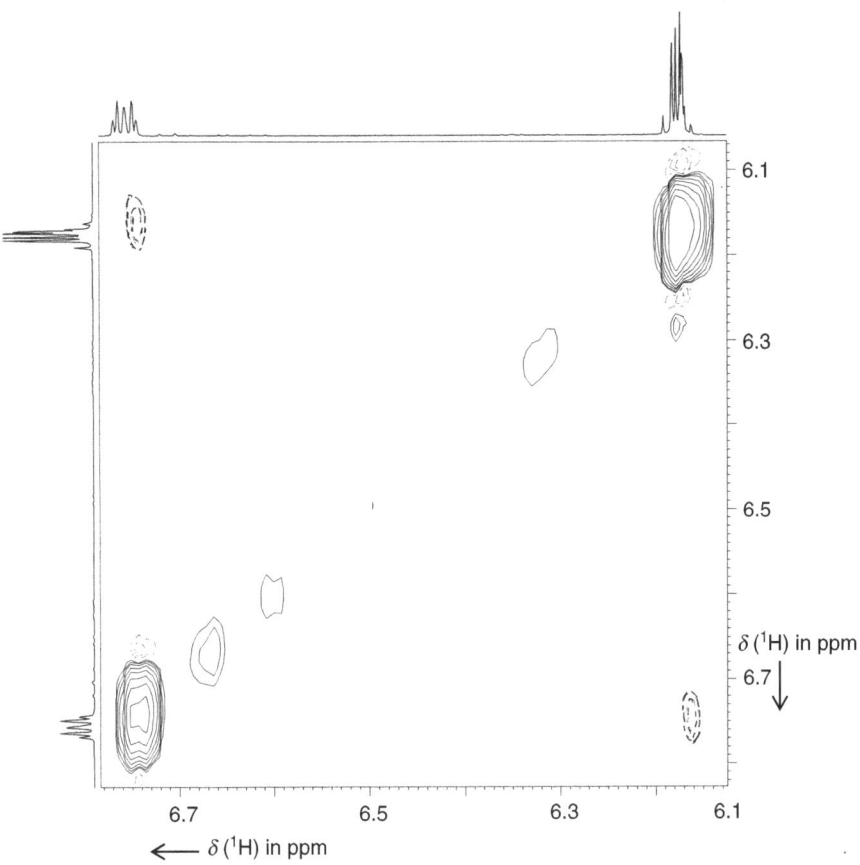

Fig. 5.21K NOESY spectrum (solvent: CDCl₃)

Challenge 5.22

Deduce the structure of the unknown compound by means of the given spectra and assign the spectroscopic data as is presented in Challenge 5.1.

For solutions to Challenge 5.22, see *extras.springer.com.*

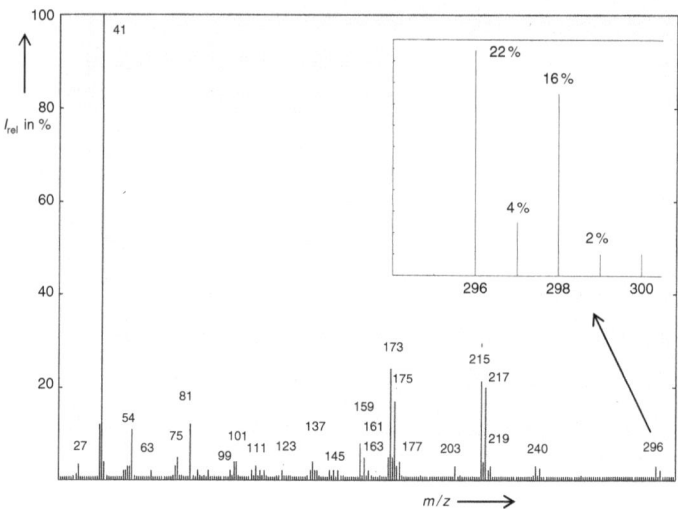

Fig. 5.22A Mass spectrum (EI, 70 eV)

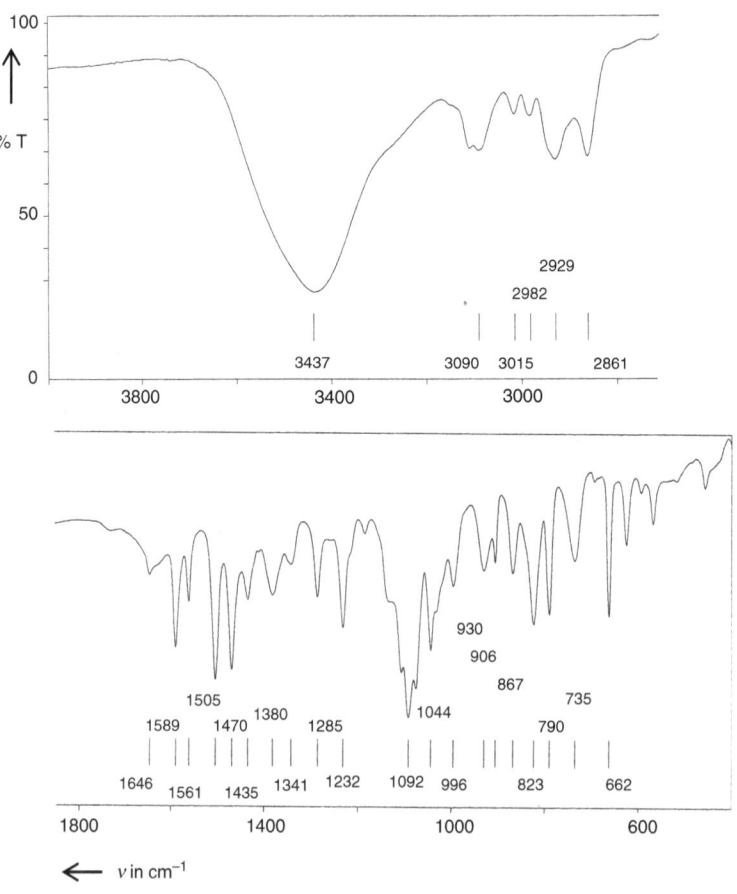

Fig. 5.22B IR spectrum (KBr pellet)

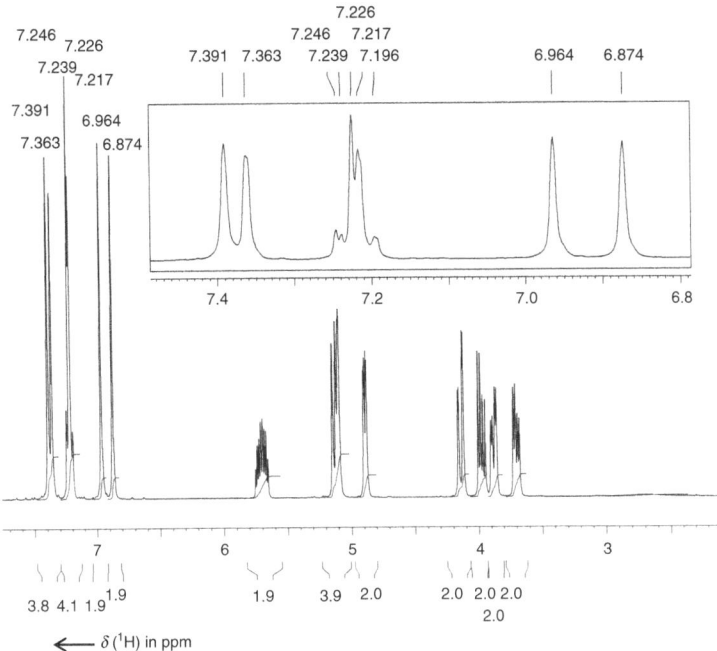

Fig. 5.22C ¹H NMR spectrum ($\nu_0 = 400$ MHz; solvent: CDCl₃)

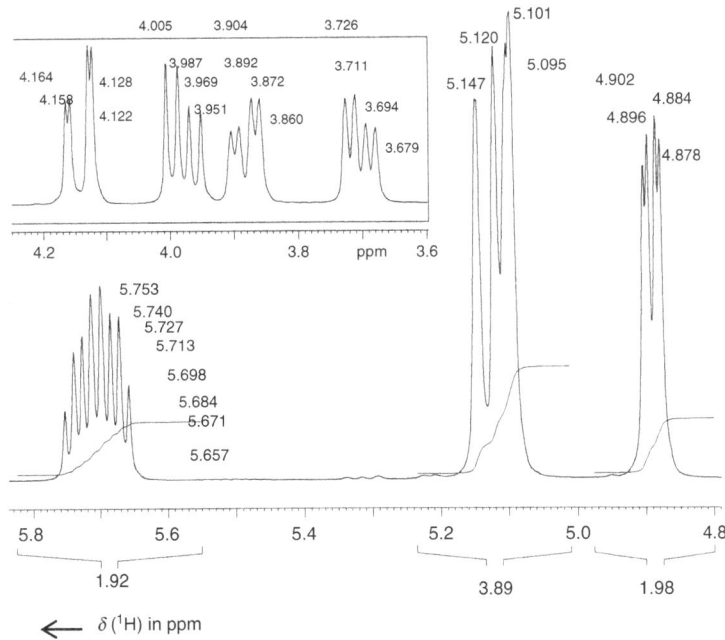

Fig. 5.22D Section of the 400-MHz ¹H NMR spectrum

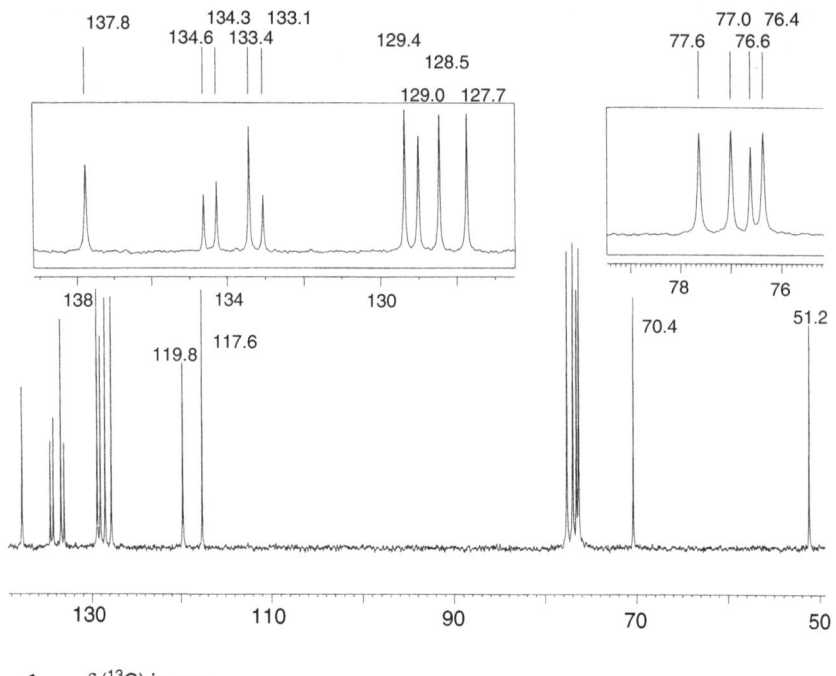

Fig. 5.22E ^{13}C NMR spectrum (solvent: CDCl$_3$)

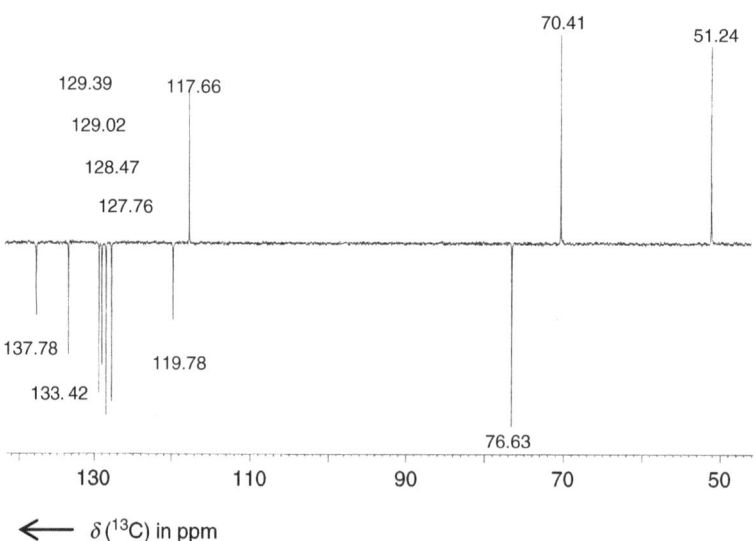

Fig. 5.22F   DEPT135 spectrum

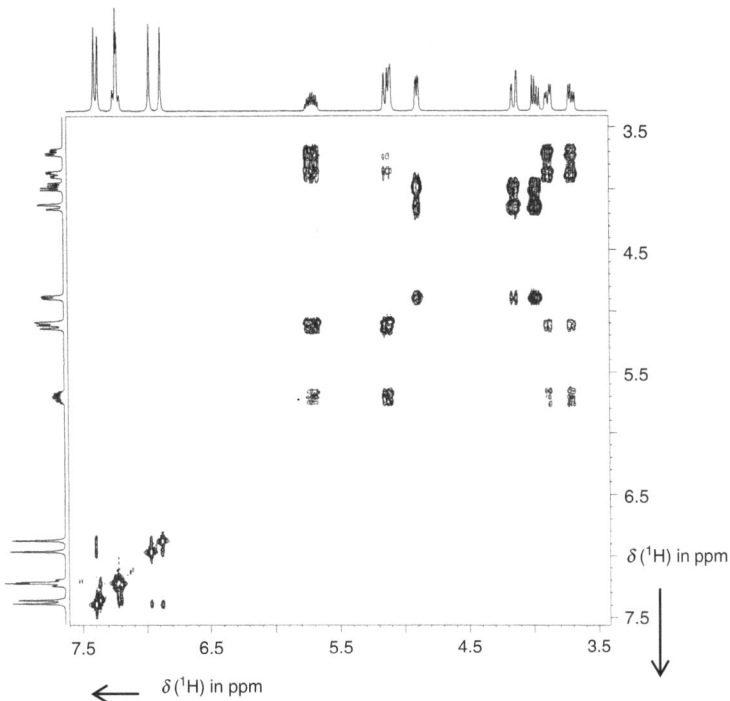

Fig. 5.22G (^{1}H,^{1}H) COSY spectrum

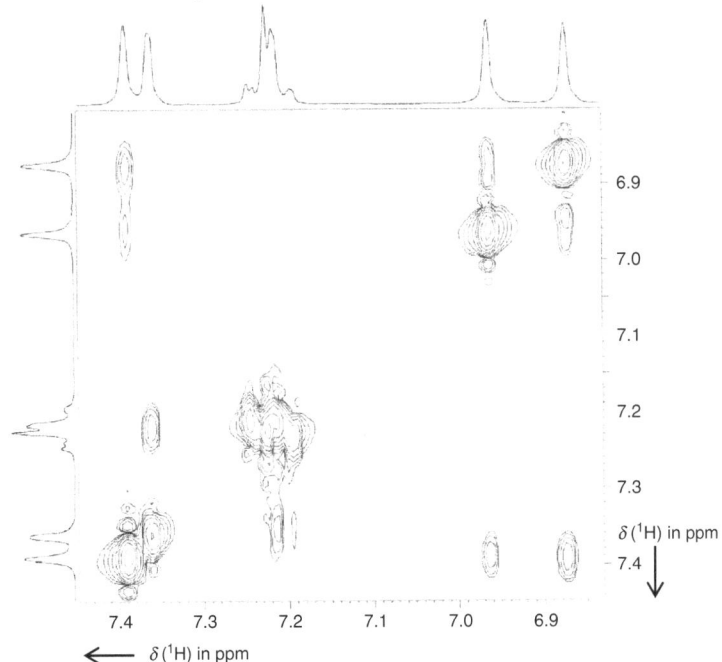

Fig. 5.22H Section of the (^{1}H,^{1}H) COSY spectrum (solvent: CDCl$_3$)

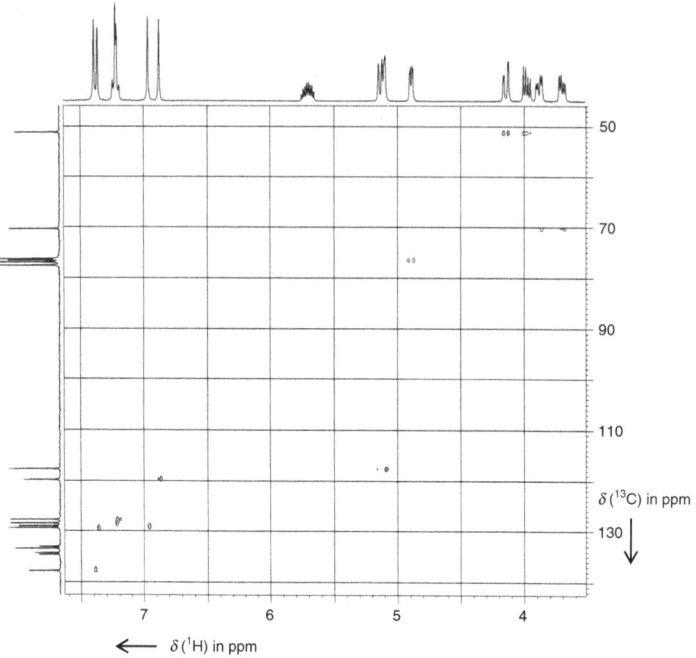

Fig. 5.22I HSQC spectrum (solvent: CDCl$_3$)

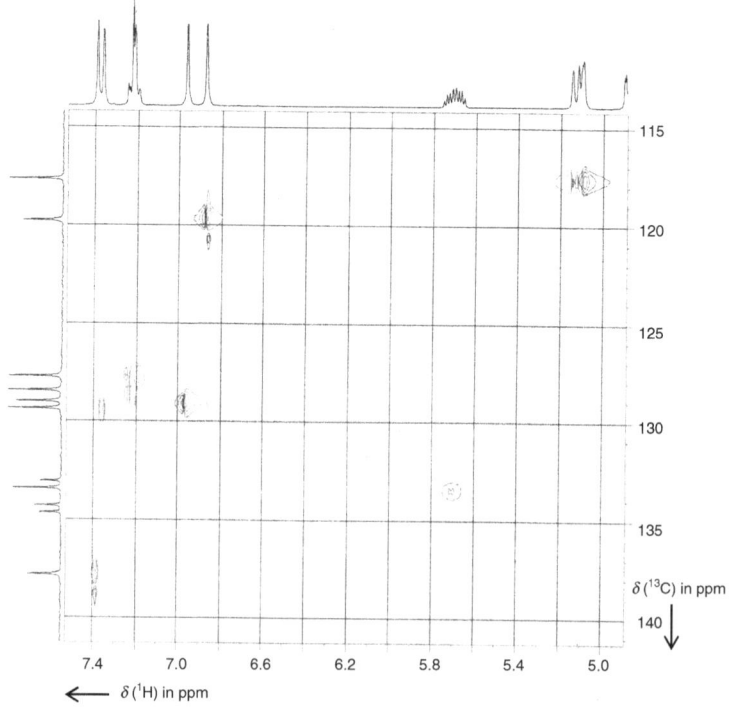

Fig. 5.22J Section of the HSQC spectrum

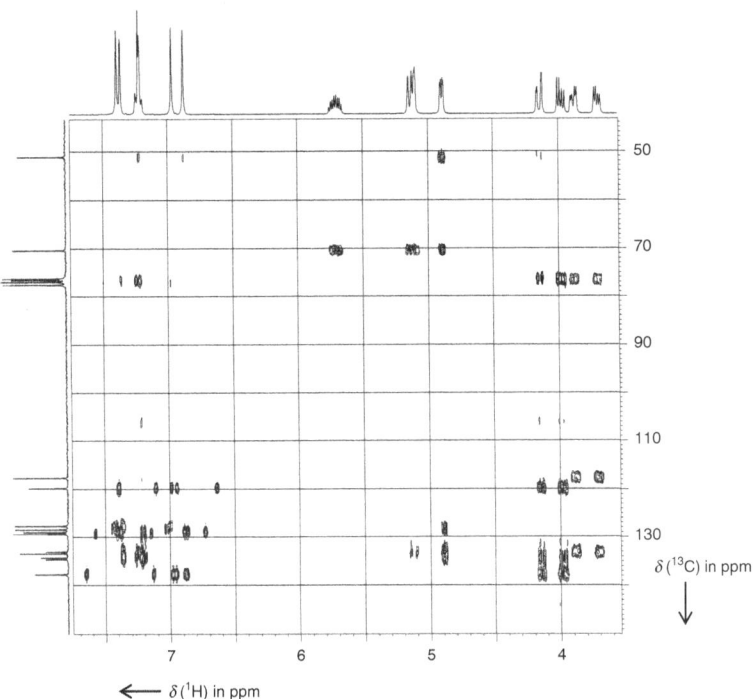

Fig. 5.22K HMBC spectrum

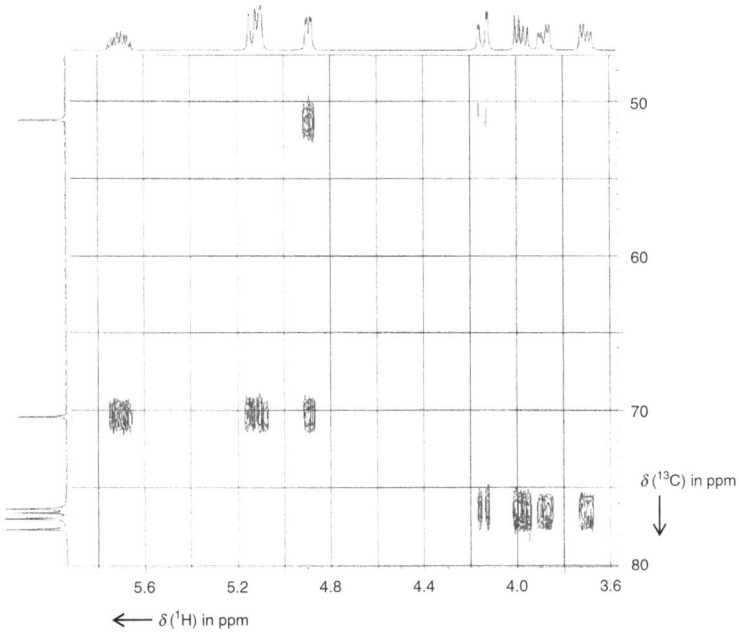

Fig. 5.22L Section of the HMBC spectrum

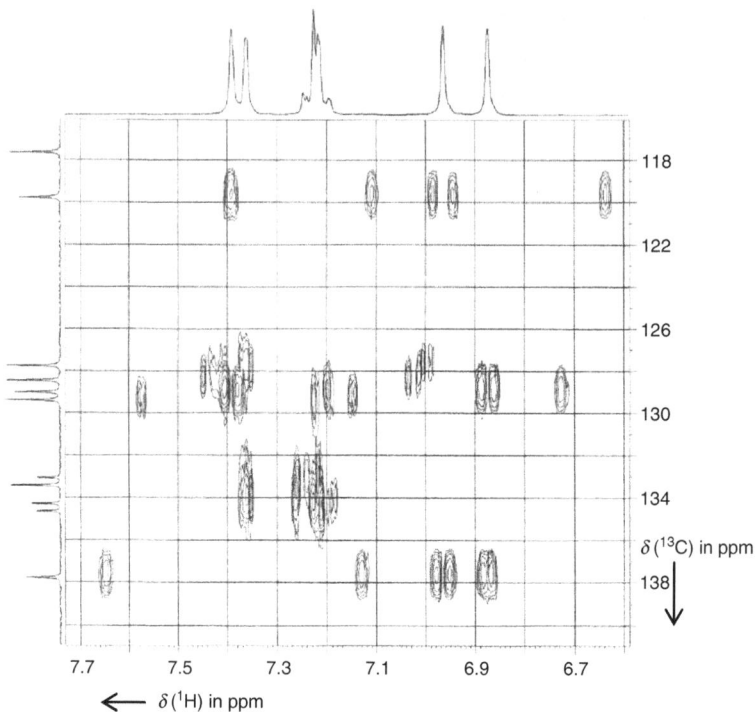

Fig. 5.22M Section of the HMBC spectrum

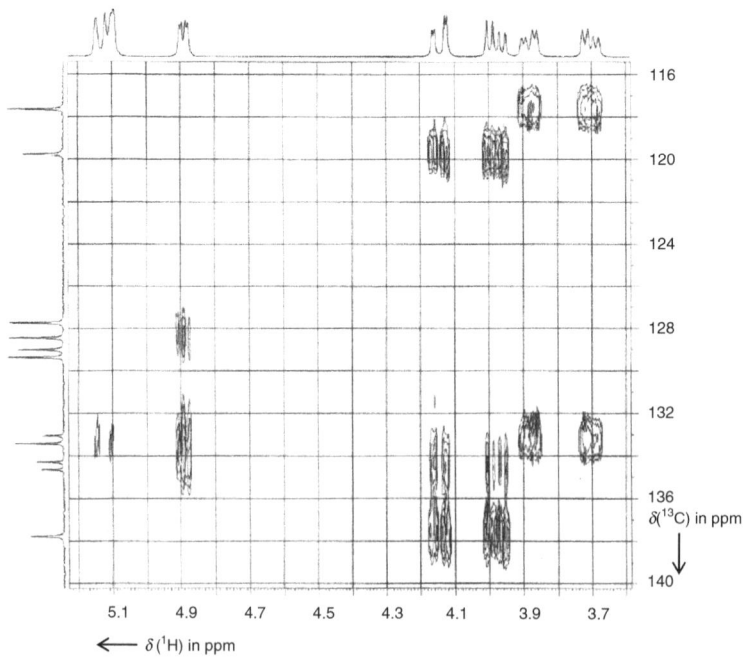

Fig. 5.22N Section of the HMBC spectrum

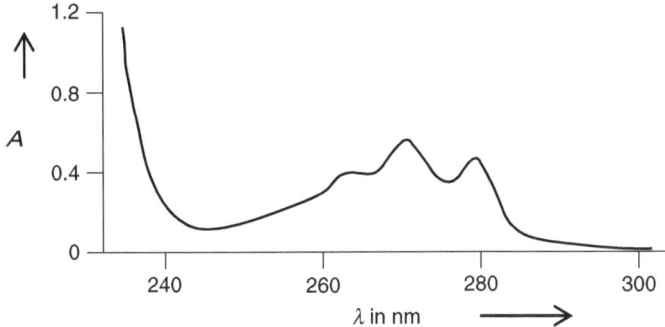

Fig. 5.22O UV spectrum. Sample: 1.028 mg in 5-mL methanol; cuvette: l = 1 cm

Further Reading

1. Pretsch E, Tóth G, Munk ME, Bardertscher M (2002) Computer-aided structure elucidation. Wiley, Weinheim
2. Ningh Y-C (2005) Structural identification of organic compounds with spectroscopic techniques. Wiley, Weinheim
3. Silverstein RM, Webster FX, Kiemle D (2005) Spectrometric identification of organic compounds. Wiley, New York

Chapter 6
Tables

6.1 Mass Spectrometry

Table 6.1 Table of natural isotopes of the important elements of organic compounds

Element	X m	X %	X + 1 m	X + 1 %	X + 2 m	X + 2 %
H	1	100	2	0.015		
C	12	100	13	1.1		
N	14	100	15	0.37		
O	16	100	17	0.04	18	0.20
F	19	100				
Si	28	100	29	5.1	30	3.35
P	31	100				
S	34	100	32	0.79	34	4.44
Cl	35	100			37	32.40
Br	79	100			81	97.94

Table 6.2 Normalized relative abundance of ions with chlorine and bromine atoms

	X	X + 2	X + 4	X + 6	X + 8		X	X + 2	X + 4	X + 6
Cl	100	32				ClBr	77	100	25	
Cl_2	100	64	10			Cl_2Br	61	100	45	7
Cl_3	100	96	31	3		$ClBr_2$	44	1ßß	70	14
Cl_4	78	100	48	10	0.8					
Br	100	98								
Br_2	51	100	49							
Br_3	34	100	98	32						
Br_4	17	68	100	65	16					

M. Reichenbächer and J. Popp, *Challenges in Molecular Structure Determination*, DOI 10.1007/978-3-642-24390-5_6, © Springer-Verlag Berlin Heidelberg 2012

Table 6.3 Common fragments lost; [M-X] peaks

Mass X	Neutral molecules/radicals	Hints to
1	H	Unspecific; intense at Aryl–CHO
15	CH_3	Intense if favorable for fragmentation
16	O	Aryl–NO_2, N-oxide, Sulfoxide
	NH_2	Aryl–SO_2NH_2, R–$CONH_2$
17	OH	O indicator, aryl–COOH
18	H_2O	Alcohols, aldehyde, ketone
19	F	Fluorine compound
20	HF	Fluorine compound
26	C_2H_2	Aromatic hydrocarbons
27	HCN	Nitrile, N-heteroaromatic compounds
28	CO	Quinones, arylketones
	C_2H_4	Arylethylether, -ester-, n-propylketones
29	CHO	Phenoles
	C_2H_5	Ethylketones, n-propylketones
30	CH_2O	Arylmethylethers
	NO	Aryl–NO_2
31	OCH_3	Methylesters
32	$HOCH_3$	o-Substituted arylmethylesters
32	S	S containing compounds
34	H_2S	Thiols
36	HCl	Alkylchlorides
41	C_3H_5	Propylesters
42	CH_2CO	Arylacetates, aryl–$NHCOCH_3$, R–$C(O)CH_3$
	C_3H_6	Butylketones, aryl–O–propyl, aryl–n–butyl
43	C_3H_7	Propylketones, aryl–n–propyl compounds
	CH_3CO	Methylketones
44	CO_2	Lactones, anhydrides, esters
45	COOH	Carboxylic acids
	OC_2H_5	Ethylesters
46	HOC_2H_5	Ethylesters
46	NO_2	Aryl–NO_2
48	SO	Aryl–S=O
55	C_4H_7	Butylesters
56	C_4H_8	Arylpentyl, aryl–O–butyl, pentylketones
57	C_4H_9	Butylketones
	C_2H_5CO	Ethylketone
60	CH_3COOH	Acetates
61	CH_3CH_2S	Thiols, thioethers
64	SO_2	Sulfonic acids and derivatives

Table 6.4 Common fragment ions

m/z	Possible ion	Hint to
19	F^+	Fluorine compounds
29	CHO^+	Aldehydes
	$C_2H_5^+$	Ethyl compounds
30	NO^+	Nitro compounds
	$CH_2 = NH_2^+$	Amines
31	$CH_2 = OH^+$	Alcohols
33	HS^+	Thiols
	CH_2F^+	Fluorine compounds
39	$C_3H_3^+$	Aromatic compounds
43	$C_3H_7^+$	Alkyl groups
	CH_3CO^+	Acetyl compounds
44	CO_2^+	Carboxylic acids
	CH_2CH-OH^+	Aldehydes (McLafferty rearrangement)
	$C_2H_6N^+$	Amines
	$NH_2C = O^+$	Amides
45	CHS^+	Thiols, thioethers
	$COOH^+$	Carboxylic acids
	CH_3CHOH^+	Alcohols
	$CH_3-O = CH_2^+$	Methylethers
46	NO_2^+	Nitro compounds
	CH_2S^+	Thiols, thioethers
47	CH_2SH^+	Thiols, thioether
49/51	CH_2Cl^+	Alkylchloride
51	CHF_2^+	Fluorine compounds
	$C_4H_3^+$	Aromatic compounds
53	$C_4H_5^+$	Aromatic compounds
55	$C_4H_7^+$	Aromatic compounds
	$CH_2 = CH-C \equiv O^+$	Cycloalkanones
57	$C_4H_9^+$	Alkyl groups
	$C_2H_5C = O^+$	Ethylketones, Propionic acid esters
	$CH_2 = CH-CH = OH^+$	Cycloalkanoles
58	$CH_2 = C(OH)CH_3^+$	Alkanones
	$(CH_3)_2 N = CH_2^+,$	
	$C_2H_5NH = CH_2^+$	Amines
59	$CH_2 = C(OH)NH_2^+$	Amides
	CH_3COO^+	Methylesters
	$CH_3C(CH_3) = OH^+$	Alcohols, esters
60	CH_2COOH^+	Carboxylic acid with a γ-H
61	$C_2H_5S^+, CH_3SCH_2^+$	Thiols
65	$C_5H_5^+$	Aromatic compounds (benzylic compounds)
66	$H_2S_2^+$	RS–SR
69	CF_3^+	Trifluormethyl compounds
	$C_5H_9^+$	Alkenes, cycloalkanes
70	$C_5H_{10}^+$	Alkenes
71	$C_5H_{11}^+$	Alkyl groups
	$C_3H_7C = O^+$	Butyric acid esters, propylketones

(continued)

Table 6.4 (continued)

m/z	Possible ion	Hint to
72	$C_3H_7CHNH_2^+$	Amines
	$C_2H_5C(OH) = CH_2^+$	Ethylketones
73	$(CH_3)_3Si^+$	Trimethylsilyl compounds
74	$CH_2-C(OH)OCH_3^+$	Methylesters
77	$C_6H_5^+$	Aromatic compounds
79/81	Br^+	Bromine compounds
80	$C_5H_6N^+$	Pyrrol compounds
80/82	HBr^+	Bromine compounds
81	$C_5H_5O^+$ (furan–CH_2^+)	
83	$C_4H_3S^+$ (thiophene cation)	
85	$C_6H_{13}^+$	Alkyl groups
	$C_4H_9C = O^+$	Butylketones
	$C_5H_9O^+$	Tetrahydropyranes
91	$C_7H_7^+$	Benzyl compounds
91/93	$C_4H_8Cl^+$	Alkylchlorides
92	$C_6H_6N^+$	Alkylpyridines
	(pyridine–CH_2^+)	
93	$C_6H_5O^+$	Phenolethers
93/95	CH_2Br^+	Alkylbromides
94	$C_6H_5O^+ + H$	Phenol derivates
95	$C_5H_3O_2^+$ (furan–$C=O^+$)	
97	$C_5H_5S^+$ (thiophene–CH_2^+)	
99	$C_7H_{15}^+$	Alkyl compounds
	$C_5H_8O_2^+$	Ketales
105	$C_6H_5C = O^+$	Benzoyl compounds
	$C_6H_5-CH_2CH_2^+$	Alkylbenzenes
106	$C_6H_5NHCH_2^+$	Alkylanilines
107	$C_7H_7O^+$	Alkylphenoles
121	$C_8H_9^+$	Alkylphenoles
149	$C_8H_5O_3^+$	Phthalic acid esters
	(phthalic anhydride-OH^+)	

Table 6.5 Formula masses (FM) for various combinations of carbon, hydrogen, nitrogen, and oxygen in the range $m/z = 98$–105 amu calculated by the relative atomic masses of the most frequent ($=$ lightest) isotopes

^{12} C 12.0000		^{1} H 1.0078		^{14} N 14.0031		^{16}O 15.9949	
98	FM	100	FM	102	FM	104	FM
$C_3H_4N_3O$	.0355	$C_2H_4N_4O$	.0386	$C_3H_4NO_3$	.0191	$C_2H_4N_2O_3$	.0222
$C_3H_6N_4$	.0594	$C_3H_4N_2O_2$	.0273	$C_3H_6N_2O_2$	.0429	$C_3H_4O_4$	.0109
$C_4H_4NO_2$	.0242	$C_3H_6N_3O$	.0511	$C_3H_8N_3O$	.0668	$C_3H_6NO_3$	.0348
$C_4H_6N_2O$	.0480	$C_4H_4O_3$	.0160	$C_4H_6O_3$	.0317	$C_3H_8N_2O_2$	.0586
$C_4H_8N_3$	.0719	$C_4H_6NO_2$	.0399	$C_4H_8NO_2$	.0555	$C_3H_{10}N_3O$	.0825
$C_5H_6O_2$	.0368	$C_4H_8N_2O$	.0637	$C_4H_{10}N_2O$	.0794	$C_4H_8O_3$	.0473
C_5H_8NO	.0606	$C_5H_8O_2$	.0524	$C_4H_{12}N_3$	.1032	$C_4H_{10}NO_2$	.0712
$C_5H_{10}N_2$	.0845	$C_5H_{10}NO$	.0763	$C_5H_{10}O_2$	.0681	$C_4H_{12}N_2O$	.0950
$C_6H_{10}O$	.0732	$C_5H_{12}N_2$	.1001	$C_5H_{12}NO$	.0919	$C_5H_{12}O_2$	.0837
$C_6H_{12}N$	.0970	$C_6H_{12}O$	.0888	$C_5H_{14}N_2$	.1158	$C_6H_4N_2$	.0375
C_7H_{14}	.1096	$C_6H_{14}N$	.1127	$C_6H_{14}O$	.1045	C_7H_4O	.0262
		C_7H_{16}	.1253	C_8H_6	.0470	C_8H_8	.0626
99	FM	101	FM	103	FM	105	FM
$C_3H_5N_3O$	.0433	$C_3H_3NO_3$	.0113	$C_3H_3O_4$	.0031	$C_2H_5N_2O_3$	.0300
$C_4H_3O_3$	.0082	$C_3H_5N_2O_2$	.0351	$C_3H_5NO_3$	.0269	$C_2H_7N_3O_2$	.0539
$C_4H_5NO_2$	.0320	$C_4H_5O_3$	.0238	$C_3H_7N_2O_2$	.0508	$C_3H_5O_4$	.0187
$C_4H_7N_2O$	.0559	$C_4H_7NO_2$	.0477	$C_4H_9NO_2$	.0634	$C_3H_7NO_3$	.0426
$C_5H_7O_2$	.0446	$C_4H_9N_2O$	.0715	$C_4H_{11}N_2O$	.0872	$C_3H_9N_2O_2$	.0664
C_5H_9NO	.0685	$C_5H_9O_2$	.0603	$C_5H_{11}O_2$	.0759	$C_4H_9O_3$	.0552
$C_5H_{11}N_2$	.0923	$C_5H_{11}NO$	.0841	$C_5H_{13}NO$	.0998	$C_4H_{11}NO_2$	.0790
$C_6H_{11}O$	.0810	$C_5H_{13}N_2$	.1080	C_7H_5N	.0422	$C_6H_5N_2$	.0453
$C_6H_{13}N$	.1049	$C_6H_{13}O$	.0967	C_8H_7	.0548	C_7H_5O	.0340
C_7H_{15}	.1174	$C_6H_{15}N$	.1205			C_7H_7N	.0579
						C_8H_9	.0705
106	FM	108	FM	110	FM	112	FM
$C_2H_6N_2O_3$	.0379	$C_2H_6NO_4$	.0297	$C_4H_4N_3O$	.0355	$C_4H_4N_2O_2$	.0273
$C_2H_8N_3O_2$	.0617	$C_2H_8N_2O_3$	.0535	$C_4H_6N_4$	.0594	$C_4H_6N_3O$	.0511
$C_3H_6O_4$	.0266	$C_3H_8O_4$	.0422	$C_5H_6N_2O$	.0480	$C_5H_4O_3$	.0160
$C_3H_8NO_3$	.0504	$C_4H_4N_4$	.0437	$C_5H_8N_3$	.0719	$C_5H_6NO_2$	.0399
$C_3H_{10}N_2O_2$	.0743	$C_5H_4N_2O$	.0324	$C_6H_6O_2$	.0368	$C_5H_8N_2O$	.0637
$C_4H_{10}O_3$	.0630	$C_5H_6N_3$	.0563	C_6H_8NO	.0606	$C_6H_8O_2$	.0524
C_6H_4NO	.0293	$C_6H_4O_2$	.0211	$C_6H_{10}N_2$	.0845	$C_6H_{10}NO$	.0763
$C_6H_6N_2$	.0532	C_6H_6NO	.0449	$C_7H_{10}O$	.0732	$C_6H_{12}N_2$	.1001
C_7H_6O	.0419	$C_6H_8N_2$	.0688	$C_7H_{12}N$	.0970	$C_7H_{12}O$	.0888
C_7H_8N	.0657	C_7H_8O	.0575	C_8H_{14}	.1096	$C_7H_{14}N$	.1127
C_8H_{10}	.0783	$C_7H_{10}N$	.0814			C_8H_{16}	.1253
		C_8H_{12}	.0939				
107	FM	109	FM	111	FM	113	FM
$C_2H_5NO_4$	.0218	$C_2H_7NO_4$	.0375	$C_4H_5N_3O$	.0433	$C_4H_5N_2O_2$	.0351
$C_2H_7N_2O_3$	.0457	$C_4H_5N_4$	.0515	$C_4H_7N_4$	.0672	$C_4H_7N_3O$	.0590
$C_2H_9N_3O_2$	.0695	$C_5H_5N_2O$	.0402	$C_5H_5NO_2$	.0320	$C_5H_5O_3$	.0238

(continued)

Table 6.5   (continued)

107	FM	109	FM	111	FM	113	FM
$C_3H_9NO_3$	.0583	$C_5H_7N_3$	.0641	$C_5H_7N_2O$	.0559	$C_5H_7NO_2$	.0477
$C_5H_5N_3$	.0484	$C_6H_5O_2$	.0289	$C_5H_9N_3$	.0789	$C_5H_9N_2O$	.0715
C_6H_5NO	.0371	C_6H_7NO	.0528	$C_6H_7O_2$	.0446	$C_6H_9O_2$	.0603
$C_6H_7N_2$	.0610	$C_6H_9N_2$	.0767	C_6H_9NO	.0684	$C_6H_{11}NO$	.0841
C_7H_7O	.0497	C_7H_9O	.0653	$C_6H_7O_2$	.0446	$C_6H_{13}N_2$	.1080
C_7H_9N	.0736	$C_7H_{11}N$	.0892	$C_7H_{11}O$	.0810	$C_7H_{13}O$	.0967
C_8H_{11}	.0861	C_8H_{13}	.1018	$C_7H_{13}N$	.1049	$C_7H_{15}N$	.1205
				C_8H_{15}	.1174	C_8H_{17}	.1331

114	FM	116	FM	118	FM	120	FM
$C_3H_6N_4O$	.0542	$C_4H_4O_4$	.0109	$C_3H_4NO_4$	.0140	$C_3H_6NO_4$	.0297
$C_4H_4NO_3$	.0191	$C_4H_6NO_3$	.0348	$C_3H_6N_2O_3$	.0379	$C_3H_{10}N_3O_2$	.0774
$C_4H_6N_2O_2$	.0429	$C_4H_8N_2O_2$	.0586	$C_3H_8N_3O_2$	.0617	$C_4H_8O_4$	.0422
$C_5H_6O_3$	.0317	$C_4H_{10}N_3O$	.0825	$C_4H_6O_4$	.0266	$C_4H_{12}N_2O_2$	.0899
$C_5H_8NO_2$	.0555	$C_5H_8O_3$	.0473	$C_4H_{10}N_2O_2$	.0743	$C_5H_4N_4$	.0437
$C_5H_{10}N_2O$	.0794	$C_5H_{10}NO_2$	.0712	$C_5H_{10}O_3$	.0630	$C_5H_{12}O_3$	.0786
$C_6H_{10}O_2$	.0681	$C_5H_{12}N_2O$	.0950	$C_5H_{14}N_2O$	.1107	$C_6H_4N_2O$	.0324
$C_6H_{14}N_2$	.1158	$C_6H_{12}O_2$	.0837	$C_6H_{14}O_2$	.0994	$C_6H_6N_3$	.0563
$C_7H_{14}O$	.1045	$C_6H_{14}NO$	.1076	C_8H_6O	.0419	$C_7H_8N_2$	.0688
C_8H_{18}	.1409	$C_7H_{16}O$	.1202	C_9H_{10}	.0783	C_8H_8O	.0575
		C_8H_6N	.0501			$C_8H_{10}N$	.0814
		C_9H_8	.0626			C_9H_{12}	.0939

115	FM	117	FM	119	FM	121	FM
$C_4H_5NO_3$	.0269	$C_3H_3NO_4$	.0062	$C_3H_5NO_4$	.0218	$C_2H_7N_3O_3$	.0488
$C_4H_7N_2O_2$	.0508	$C_3H_5N_2O_3$	.0300	$C_3H_7N_2O_3$	.0457	$C_2H_9N_4O_2$	.0726
$C_5H_7O_3$	.0395	$C_4H_5O_4$	.0187	$C_3H_9N_3O_2$	.0695	$C_3H_7NO_4$	.0375
$C_5H_9NO_2$	.0634	$C_4H_7NO_3$	.0426	$C_3H_{11}N_4O$	.0934	$C_3H_{11}N_3O_2$	.0852
$C_5H_{11}N_2O$	.0872	$C_4H_9N_2O_2$	.0664	$C_4H_7O_4$	.0344	$C_4H_9O_4$	.0501
$C_5H_{13}N_3$	.1111	$C_4H_{11}N_3O$	.0903	$C_4H_9NO_3$	.0583	$C_4H_{11}NO_3$	.0739
$C_6H_{13}NO$	.0998	$C_5H_9O_3$	.0552	$C_4H_{13}N_3O$	.1060	$C_6H_5N_2O$	.0402
$C_6H_{15}N_2$	.1236	$C_5H_{11}NO_2$	.0790	$C_5H_{13}NO_2$	.0947	$C_6H_7N_3$	.0641
$C_7H_{15}O$	.1123	$C_6H_{13}O_2$	.0916	$C_6H_5N_3$	.0484	C_7H_7NO	.0528
$C_7H_{17}N$	.1362	$C_6H_{15}NO$	.1154	C_7H_5NO	.0371	$C_7H_9N_2$	.0767
C_9H_7	.0548	C_8H_7N	.0579	C_8H_9N	.0736	$C_8H_{11}N$	.0892
						C_9H_{13}	.1018

Table 6.6 Important information from mass spectra of organic classes of compounds

Compound class	Information
Hydrocarbons	
Saturated	$M^{\bullet+}$: mean intensity, but very weak at branched chain
	Ion series: C_nH_{2n+1}; weaker: C_nH_{2n} and C_nH_{2n-1}
	Maximum intensity at C_3/C_4, decreasing in a smooth curve
	Branched chains show "unsteadiness" of the intensity curve
Alkenes	$M^{\bullet+}$: mean intensity
	Ion series: C_nH_{2n-1}; weaker C_nH_{2n} and C_nH_{2n+1}
	No recognition of the position of the double bond because of its migration
	Cyclic alkenes: RDA fragmentations
Aromatic	$M^{\bullet+}$: (very) intense
compounds	Characteristic ions: $m/z = 39, 51, 65, 78, 79, 91$
	Monosubstitution: $m/z = 77$
	Typical ion series: $C_nH_{n\pm1}{}^+$
Acetylides	$M^{\bullet+}$: frequently weak, [M-1] is more intense
Hydroxy compounds	
Alcohols	$M^{\bullet+}$: weak at primary, very weak at secondary, not visible at tertiary alcohols; intense [M-1] peak
	Primary alcohols: intense [M-2] and [M-3] peaks are also visible
	α cleavage: intense $m/z = 31$ peak
	Lost of $H_2O + C_nH_{2n}$ gives rise to ions [M-46], [M-74], etc.
	Mass spectra of long-chain alcohols are similar to that of alkenes
	Secondary, tertiary alcohols: $m/z = 45, 59$, etc.
	α cleavage: $m/z = 45, 59$ etc.; $m/z = 31$ is only weak
	$M^{\bullet+}$: weak, but always present
Alicyclic alcohols	Fragmentations: primary ring cleavage at the OH-bearing carbon atom, loss of $H^{\bullet}$ gives rise to $C_nH_{2n-1}O^+$ ions
	Ion series: $C_nH_{2n-1}{}^+$ and $C_nH_{2n-3}{}^+$
Phenoles	$M^{\bullet+}$: high intensity; [M-1] peak is also visible
	Loss of CO ($m/z = 28$) and CHO ($m/z = 29$) after rearrangement
Ethers	
Aliphatic	$M^{\bullet+}$: weak
	Fragmentations:
	1. α cleavage, followed by a McLafferty rearrangement
	2. Cleavage of the C–O bond followed by charge-induced cleavage to alkylcations $C_nH_{2n+1}{}^+$
Aromatic	$M^{\bullet+}$: intense
	Fragmentations: primary cleavage of the ArO–R bond followed by CO elimination to arylcations
	Loss of alkenes by McLafferty rearrangement (alkyl-C $\geq$ 2)
	Diarylethers: [M-H], [M-CO], [M-CHO]
Aldehydes	
Aliphatic	$M^{\bullet+}$: present
	Fragmentations: [M-1] peak; $m/z = 29$ (CHO^+ or $C_2H_5{}^+$) in long chain ($>C_4$); in long chains $C_nH_{2n+1}{}^+$ peaks dominate
	McLafferty rearrangement for $\geq C_4$ compounds
	Unbranched: [M-18], [M-28], [M-43], [M-44]

(continued)

Table 6.6 (continued)

Compound class	Information
Aromatic	$M^{\cdot+}$: very intense
	α cleavage: [M-1] peak followed by CO elimination and C_2H_2 lost from the arylcations
Ketones	
Aliphatic	$M^{\cdot+}$: mean intensity
	Fragmentations: α cleavage followed by CO loss
	McLafferty rearrangement: elimination of alkenes
	Alkylcations dominate at long-chain ketones
	Cyclic: intense molecular peak
	Fragmentation: α cleavage after ring opening, loss of CO and alkyl radicals to stable ions
Aromatic	$M^{\cdot+}$: high intensity
	Fragmentation: Aryl-C $\equiv$ O$^+$ by α cleavage
Carboxylic acids	
Aliphatic	$M^{\cdot+}$: unbranched mono-carboxylic: weak; also $[M + H]^+$
	Fragmentation: short chain: α cleavage; [M-17], [M-45]
	Long chain: ion series: $C_nH_{2n-1}O_2^+$, $C_nH_{2n\pm1}^+$
	McLafferty rearrangement ($n_C \geq 4$): $m/z = 60$ (base peak!)
Aromatic	$M^{\cdot+}$: intense
	Fragmentation: α cleavage [M-17] (OH), [M-45] (CO_2H)
	Loss of CO_2 ($\Delta m = 44$)
	Ortho-effect: [M-H$_2$O]
Carboxylic acid derivatives	
Aliphatic esters R–C(O) –O–R'	$M^{\cdot+}$: visible
	Fragmentation: Recognition of the alcoholic and α substituent by McLafferty rearrangement
	α cleavage: C(O)–OR'$^+$, R–C(O)$^+$
	Intense R$^+$ in short chain, but weak in long chain
	Long-chain acid part: Ion series like carboxylic acids
	Long-chain alcoholic part: Loss of CH_3COOH and alkenes to R–C(OH)$_2^+$ ions ($m/z = 61, 75$, etc.)
	Benzyl-, phenyl-, heterocyclic acetates eliminate ketene $\Rightarrow$ [M-42] peak
Aromatic esters Aryl–COOR	$M^{\cdot+}$: intense, but the intensity decreases with the chain length rapidly
	Fragmentation: α cleavage to [M-OR], [M-COOR]
	R $\geq$ C$_2$: McLafferty rearrangement with elimination of $CH_2 = CH_2$-R'', $\Rightarrow$ R$^+$ ion series
	Ortho-effect: Loss of ROH
Amides R–CONH$_2$ R–COONHR' R–COONR'R'' Aryl–COONRR'	$M^{\cdot+}$: present
	Fragmentation: McLafferty rearrangement (base peak!)
	Primary: α cleavage: $CONH_2^+$ ($m/z = 44$)
	Secondary, tertiary: McLafferty rearrangement if a γ-H is present
	Aryl–CONRR': α cleavage dominates
N-compounds	
Amines Aliphatic	$M^{\cdot+}$: very weak to invisible; [M-1] mostly present
	Fragmentations: α cleavage dominates
	Secondary, tertiary: Loss of the largest α group
	Primary: Ion series $m/z = 30, 44$, etc., $C_nH_{2n\pm1}$, C_nH_{2n}

(continued)

Table 6.6 (continued)

Compound class	Information
Cycloalkylamines	$M^{\bullet+}$: present
	Alkyl cleavage followed by primary ring opening at the N-containing carbon atom
Aromatic	Ion series: C_nH_{2n}
	Aryl–NH_2: Loss of N as HCN ($m/z = 27$)
Alkyl-C $\equiv$ N	$M^{\bullet+}$: very weak or not present; [M-1] frequently strong
	Fragmentations: McLafferty rearrangement: $m/z = 41$ (base peak!) $\geq C_8$ and higher: $m/z = 97$ (very intense)
	Ion series: $(CH_2)_nC \equiv N^+$ ($m/z = 40, 54$, etc.)
Aryl–NO_2	$M^{\bullet+}$: very intense
	Fragmentations: [M-46] ($NO_2^{\bullet}$), followed by loss of $C_2H_2 \Rightarrow$ [M-72]
	After rearrangement: [M-30] (NO), followed by loss of CO $\Rightarrow$ M-58
	NO^+ ion ($m/z = 30$)
Halogen compounds	Intense $[M + 2n]^+$ peaks: Cl and/or Br (see halogen patterns)
	Unusual mass differences: Δm: 19(F); 20(HF); 127(I); 128(HI)
Aliphatic	$M^{\bullet+}$: Decreasing with increasing molecular mass, branching and number of halogen atoms
	Fragmentations: Loss of halogen radicals
	C_n with n > 6: Very intense $C_4H_8X^+$ peak with X = Cl, Br
Aromatic	$M^{\bullet+}$: Mean intensity
	Fragmentations: Lost of halogen radicals
S-compounds	[M + 2] peak: Recognition of S and determination of the S number
Thiols, thioethers	$M^{\bullet+}$: Mean intensity
	Fragmentations: Loss of SH·($\Delta m = 33$); H_2S ($\Delta m = 34$)
	Fragment ions: CH_2SH^+ ($m/z = 47$), $CS^{\bullet+}$ ($m/z = 44$), CHS^+ ($m/z = 45$)
Sulfonic acids, esters	Loss of SO ($\Delta m = 48$), SO_2 ($\Delta m = 64$)

6.2   Vibrational Spectroscopy

Table 6.7   Algorithm for the determination of symmetry classification (point group)

1. Is the molecule *linear*?
   $\Rightarrow$ yes $\qquad\qquad\qquad\qquad\qquad$ $\Rightarrow$ Question 1.a
   $\Rightarrow$ no $\qquad\qquad\qquad\qquad\qquad$ $\Rightarrow$ Question 2
   1.a Is a horizontal plane σ_h present (symmetric molecule)?
   $\Rightarrow$ yes $\qquad\qquad\qquad\qquad\qquad$ $\Rightarrow$ $D_{\infty h}$
   $\Rightarrow$ no $\qquad\qquad\qquad\qquad\qquad$ $\Rightarrow$ $C_{\infty v}$
2. Does the molecule belong to a *cubic* symmetry?
   $\Rightarrow$ yes $\qquad\qquad\qquad\qquad\qquad$ $\Rightarrow$ Question 2.a
   $\Rightarrow$ no $\qquad\qquad\qquad\qquad\qquad$ $\Rightarrow$ Question 3
   2.a Octahedral molecule $\qquad\qquad$ $\Rightarrow$ O_h
   $\quad$ Tetrahedral molecule $\qquad\quad$ $\Rightarrow$ T_d
3. Does the molecule possess a proper axis C_n (rotation by angle $2\pi/n$ in which n is an integer)
   $\Rightarrow$ yes $\qquad\qquad\qquad\qquad\qquad$ $\Rightarrow$ Question 6
   $\Rightarrow$ no $\qquad\qquad\qquad\qquad\qquad$ $\Rightarrow$ Question 4
4. Is an inversion centre i present?
   $\Rightarrow$ yes $\qquad\qquad\qquad\qquad\qquad$ $\Rightarrow$ C_i
   $\Rightarrow$ no $\qquad\qquad\qquad\qquad\qquad$ $\Rightarrow$ Question 5
5. Is a reflection plane σ present?
   $\Rightarrow$ yes $\qquad\qquad\qquad\qquad\qquad$ $\Rightarrow$ C_σ
   $\Rightarrow$ no $\qquad\qquad\qquad\qquad\qquad$ $\Rightarrow$ C_1
6. Determine the n of the principal rotation axis C_n!
   $\Rightarrow$ Question 7
7. Are there n C_2 proper axes perpendicular to the C_n proper axis with the biggest n?
   $\Rightarrow$ yes $\qquad\qquad\qquad$ $\Rightarrow$ D groups $\qquad\qquad$ $\Rightarrow$ Question 8
   $\Rightarrow$ no $\qquad\qquad\qquad$ $\Rightarrow$ C groups $\qquad\qquad$ $\Rightarrow$ Question 9
8. Is a σ_h (horizontal reflection plane perpendicular to the principal axis C_n) present?
   $\Rightarrow$ yes $\qquad\qquad\qquad\qquad\qquad$ $\Rightarrow$ D_{nh}
   $\Rightarrow$ no $\qquad\qquad\qquad\qquad\qquad$ $\Rightarrow$ Question 8.a
   8. a Are there n σ_d (diagonal reflection plane which contains the principal axis and bisects the angle between two C_2 axes perpendicular to the principal axis)?
   $\Rightarrow$ yes $\qquad\qquad\qquad\qquad\qquad$ $\Rightarrow$ D_{nd}
   $\Rightarrow$ no $\qquad\qquad\qquad\qquad\qquad$ $\Rightarrow$ D_n
9. Is a σ_h present?
   $\Rightarrow$ yes $\qquad\qquad\qquad\qquad\qquad$ $\Rightarrow$ C_{nh}
   $\Rightarrow$ no $\qquad\qquad\qquad\qquad\qquad$ $\Rightarrow$ Question 9.a
9.a Are there n σ_v (vertical reflection plane which contains the principal axis)?
   $\Rightarrow$ yes $\qquad\qquad\qquad\qquad\qquad$ $\Rightarrow$ C_{nv}
   $\Rightarrow$ no $\qquad\qquad\qquad\qquad\qquad$ $\Rightarrow$ C_n

Tables 6.8 Character tables for chemically important point groups

A character table is a two-dimensional table whose rows correspond to irreducible group representations, and whose columns correspond to class of group elements. The entries consist of characters which are the trace of matrices representing group elements of the column's class in the given row's group representation.

The *spectroscopic* information is given in the two last columns of the character table for a certain point group: The *second* to last column contains the symmetry classes of *dipole moment components* (x, y, z) and the *last* column presents the symmetry classes of the *polarizability tensors* (binary products). Note that R in the second to last column denotes the symmetry classes of the rotation, but this is not of interest in this book.

C_s	E	σ			
A'	1	1		x, y, R_z	x^2, y^2, z^2, xy
A''	1	-1		z, R_x, R_y	xz, yz

C_{2h}	E	$C_2(z)$	$\sigma_h(xy)$	i		
A_g	1	1	1	1	R_z	x^2, y^2, z^2, xy
A_u	1	1	-1	-1		
B_g	1	-1	-1	1	z	yz, xz
B_u	1	-1	1	-1	x, y	

C_{2v}	E	$C_2(z)$	$\sigma_v(xz)$	$\sigma_v(yz)$		
A_1	1	1	1	1	z	x^2, y^2, z^2
A_2	1	1	-1	-1	R_z	xy
B_1	1	-1	1	-1	x, R_y	xz
B_2	1	-1	-1	1	y, R_x	yz

C_{3v}	E	$2\,C_3(z)$	$3\,\sigma_v$		
A_1	1	1	1	z	$x^2 + y^2 + z^2$
A_2	1	1	-1	R_z	
E	2	-1	0	(z, y); (R_x, R_y)	

C_{4v}	E	$C_4(z)$	C_2''	$2\,\sigma_v$	$2\,\sigma_d$		
A_1	1	1	1	1	1	z	$x^2 + y^2$
A_2	1	1	1	-1	-1	R_z	
B_1	1	-1	1	1	-1		$x^2 - y^2$
B_2	1	-1	1	-1	1		xy
E	2	0	-2	0	0	(x, y); R_x. R_y	

D_{3h}	E	$2\,C_3(z)$	$3\,C_2'$	σ_h	$2\,S_3$	$3\,\sigma_v$		
A_1'	1	1	1	1	1	1		$x^2 + y^2 + z^2$
A_2'	1	1	-1	1	1	-1	R_z	
E'	2	-1	0	2	-1	0	(x, y)	$(x^2 - y^2, xy)$
A_1''	1	1	1	-1	-1	-1		
A_2''	1	1	-1	-1	-1	1	z	
E''	2	-1	0	-2	1	0	$(R_x. R_y)$	(xz, yz)

D_{4h}	E	$2\,C_4$	C_2	$2\,C_2'$	$2\,C_2''$	i	$2\,S_4$	σ_h	$2\sigma_v$	$2\sigma_d$		
A_{1g}	1	1	1	1	1	1	1	1	1	1		$x^2 + y^2 + z^2$
A_{2g}	1	1	1	-1	-1	1	1	1	-1	-1	R_z	
B_{1g}	1	-1	1	1	-1	1	-1	1	1	-1		$x^2 - y^2$
B_{2g}	1	-1	1	-1	1	1	-1	1	-1	1		xy

(continued)

Tables 6.8 (continued)

D_{4h}	E	$2C_4$	C_2	$2C_2'$	$2C_2''$	i	$2S_4$	σ_h	$2\sigma_v$	$2\sigma_d$		
E_g	2	0	-2	0	0	2	0	-2	0	0	(R_x, R_y)	(xz, yz)
A_{1u}	1	1	1	1	1	-1	-1	-1	-1	1		
A_{2u}	1	1	1	-1	-1	-1	-1	-1	1	1	z	
B_{1u}	1	-1	1	1	-1	-1	1	-1	-1	1		
B_{2u}	1	-1	1	-1	1	-1	1	-1	1	-1		
E_u	2	0	-2	0	0	-2	0	2	0	0	(x, y)	

T_d	E	$8C_3$	$3C_2$	$6S_4$	$6\sigma_d$		
A_1	1	1	1	1	1		$x^2 + y^2 + z^2$
A_2	1	1	1	-1	-1		
E	2	-1	2	0	0		$(2z^2 - x^2 - y^2, x^2 - y^2)$
T_1	3	0	-1	1	-1	(R_x, R_y, R_z)	
T_2	3	0	-1	-1	1	(x, y, z)	(xy, xz, yz)

D_{6h}	E	$2C_6$	$2C_3$	C_2	$3C_2'$	$3C_2''$	i	$2S_3$	$2S6$	σ_h	$3\sigma_d$	$3\sigma_v$	
A_{1g}	1	1	1	1	1	1	1	1	1	1	1	1	x^2 $+y^2$ $+z^2$
A_{2g}	1	1	1	1	-1	-1	1	1	1	1	-1	-1	R_z
B_{1g}	1	-1	1	-1	1	-1	1	-1	1	-1	1	-1	
B_{2g}	1	-1	1	-1	-1	1	1	-1	1	-1	-1	1	
E_{1g}	2	1	-1	-2	0	0	2	1	-1	-2	0	0	$R_x,$ xz, R_y yz
E_{2g}	2	-1	-1	2	0	0	2	-1	-1	2	0	0	$x^2 - y^2,$ xy
A_{1u}	1	1	1	1	1	1	-1	-1	-1	-1	-1	-1	
A_{2u}	1	1	1	1	-1	-1	-1	-1	-1	-1	1	1	z
B_{1u}	1	-1	1	-1	1	-1	-1	1	-1	1	-1	1	
B_{2u}	1	-1	1	-1	-1	1	-1	1	-1	1	1	-1	
E_{1u}	2	1	-1	-2	0	0	-2	-1	1	2	0	0	x,y
E_{2u}	2	-1	-1	2	0	0	-2	1	1	-2	0	0	

O_h	E	$8C_3$	$2C_2$	$6C_4$	$6C_2'$	i	$8S_4$	$3\sigma_h$	$6S_4$	$6\sigma_d$		
A_{1g}	1	1	1	1	1	1	1	1	1	1		$x^2 + y^2 + z^2$
A_{2g}	1	1	1	-1	-1	1	1	1	-1	-1		
E_g	2	-1	2	0	0	2	-1	2	0	0		$(2z^2 - x^2 - y^2, x^2 - y^2)$
T_{1g}	3	0	-1	1	-1	3	0	-1	1	-1	(R_x, R_y, R_z)	
T_{2g}	3	0	-1	-1	1	3	0	-1	-1	1		(xy, xz, yz)
A_{1u}	1	1	1	1	1	-1	-1	-1	-1	-1		
A_{2u}	1	1	1	-1	-1	-1	-1	-1	1	1		
E_u	2	-1	2	0	0	-2	1	-2	0	0		
T_{1u}	3	0	-1	1	-1	-3	0	1	-1	1	(x, y, z)	
T_{2u}	3	0	-1	-1	1	-3	0	1	1	-1		

Table 6.9 Symmetry of fundamental vibrations of important structural types of the respective point group (PG)

Γ_ν and Γ_δ are symbols for the representation of symmetry classes of the stretching and bending vibrations, respectively.

Structural type	PG	Γ_ν	Γ_δ
AX_6	O_h	$A_{1g} + E_g + T_{1u}$	$T_{1u} + T_{2g} + T_{2u}$
AX_5	D_{3h}	$2\,A_1' + A_2'' + E'$	$A_2'' + 2\,E' + E''$
AX_5	C_{4v}	$2\,A_1 + B_1 + E$	$A_1 + B_1 + B_2 + 2\,E$
AX_4	T_d	$A_1 + T_2$	$E + T_2$
AX_4	D_{4h}	$A_{1g} + B_{2g} + E_u$	$A_{2u} + B_{1g} + B_{2g} + E_u$
AX_3Z	C_{3v}	$2\,A_1 + E$	$A_1 + 2\,E$
AX_3Z_2	D_{3h}	$2\,A_1 + A_2'' + E$	$A_2'' + 2\,E' + E''$
AX_2Z_2	C_{2v}	$2\,A_1 + B_1 + B_2$	$2\,A_1 + A_2(\tau) + B_2(\rho) + B_1(\omega)$
AX_3	D_{3h}	$A_1' + E'$	$A_2''\,(\gamma) + E'$
AX_3	C_{3v}	$A_1 + E$	$A_1 + E$
AX_2Z	C_{2v}	$2\,A_1 + B_1$	$A_1 + B_1 + B_2(\gamma)$
AX_2Z	C_s	$2\,A' + A''$	$2\,A' + A''$
AX_2	C_{2v}	$A_1 + B_1$	A_1
AXZ	C_s	$A' + A''$	A'

6.2.1 Tables of Group Wave Numbers of Inorganic Compounds

Table 6.10 Inorganic ions

Ion	Wave numbers region (in cm^{-1})
CO_3^{2-}	1,450–1,410 880–800
ClO_4^-	1,140–1,060
CrO_4^{2-}	950–800
MnO_4^-	920–890 850–840
NH_4^+	3,340–3,030 1,485–1,390
NO_2^-	1,400–1,300 1,250–1,230 840–800
NO_2^+	1,410–1,370
PO_4^{3-}	1,100–950
SO_4^{2-}	1,130–1,080 680–610
HSO_4^-	1,180–1,160 1,080–1,000 880–840
SO_3^{2-}	$\approx 1,100$

Table 6.11 Silicon compounds

Stretching vibration	Structural moiety	Wave numbers region (in cm^{-1})
Si–H	R_2Si–H	2,160–2,120
	$-OSi$–H	2,230–2,120
Si–C	C–Si–CH_3	760–620
	O–Si–CH_3	800–770
Si–O	R_3Si–OH	900–810
	Si–OC	1,110–1,000
	Si–O–Si	1,090–1,030

Table 6.12 Sulfur compounds

Stretching vibration	Structural moiety	Wave numbers region (in cm^{-1})	
S–H	RS–H	2,590–2,530	
S–C	RCH_2–SH	730–570	
S–O	RSO–OH	870–810	
	RO–SO–OR	740–720 710–690	
S = C	R_2C = S	1,075–1,030	
	$(RO)_2$ C = S	1,120–1,075	
S = O	R_2S = O	1,060–1,015	
	$(RO)_2$ S = O	1,225–1,195	
		ν_{as}	ν_s
>SO_2	R_2SO_2	1,370–1,290	1,170–1,110
	RSO_2–OR	1,375–1,350	1,185–1,165
	RSO_2–NR_2	1,365–1,315	1,180–1,150
	RSO_2–Hal	1,385–1,375	1,180–1,170
–SO_3	RSO_2–OH	1,355–1,340	1,165–1,150
	$RSO_3^-$$Me^+$	1,250–1,140	1,070–1,030

Table 6.13 Phosphorus compounds

Stretching vibration	Structural moiety	Wave numbers region (in cm^{-1})
P–H	Phosphine	2,320–2,275
P–C	aliphatic	700–800
	aromatic	1,130–1,090
P–O	P–OC$_{aliphatic}$	1,050–970
	P–OC$_{aromatic}$	1,260–1,160
P = O	O = P(Alkyl)$_3$	$\approx$1,150
	O = P(Aryl)$_3$	$\approx$1,190
P = S	P = S	700–625

6.2.2   *Tables of Group Wave Numbers of Organic Compounds*

Table 6.14 Characteristic wave numbers of several functional groups (group frequencies)

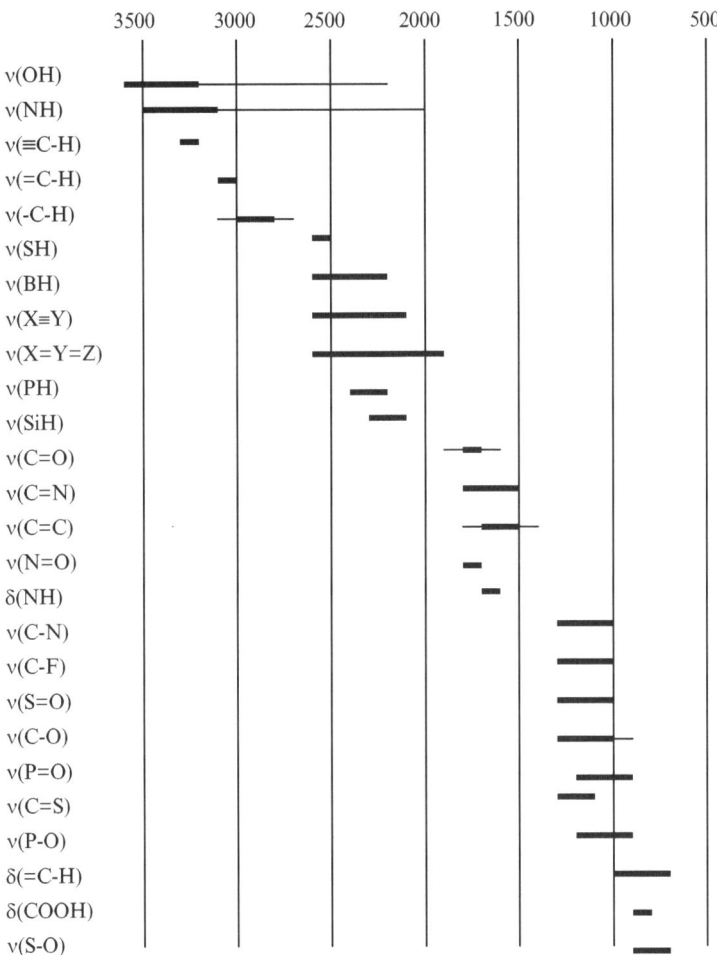

Table 6.15 Wave number regions of classes of organic compounds (in cm^{-1})

I. Hydrocarbons

1. Alkyl groups

1.1 *CH stretching vibrations*, $v(sp^3 C–H)$ **3,000–2,840**

Common region

1.1.1 CH$_3$ v_{as} 2,965 ± 10 m v_s 2,875 ± 10 m Intensity(v_{as}) > Intensity(v_s)

1.1.2 CH$_2$ v_{as} 2,925 ± 10 m v_s 2,855 ± 10 m Intensity(v_{as}) > Intensity(v_s)

1.1.3 CH   2,890–2,880 w

Deviations from the common region can be caused by electronic and ring stress effects.

1.1.4	Aryl–CH$_3$	v_s + Fermi-resonance (OT δ_{as} + v_s) 2,925 ± 5 + 2,865 ± 5
1.1.5	CH$_3$O v_{as}	v_s 2,830 ± 10 (sharp, m) 2,945 ± 25
1.1.6	CH$_2$O v_{as}	v_s 2,855 ± 20 2,938 ± 17
1.1.7	CH$_3$N–Alkyl 2,795 ± 10 CH$_3$N–Aryl 2,815 ± 5 (CH$_3$)$_2$ N–Alkyl ≈ 2,820 + ≈ 2,770	
1.1.8	O–CH$_2$–O	2,880–2,835 + 2,780–2,750
1.1.9	3-membered ring	3,055–3,000 m

1.2 *CH bending vibrations*   **1,480–1,370**

1.2.1   CH$_3$ δ_{as} 1,470–1,430 m δ_s 1,390–1,365 m (umbrella)

1.2.2   CH$_2$ δ 1,465–1,445 m (scissoring) ρ ≈ 720 m-w

Splitting of δ_s(CH$_3$) at branched chains:

−C(CH$_3$)$_3$,−CH$_2$CH(CH$_3$)$_2$:   1,395–1,365 2 bands with asymmetrical intensity

−CH(CH$_3$)$_2$,−C(CH$_3$)$_2$−:   1,385–1,365 2 bands with equal intensity

2. Alkenyl groups

2.1 *CH stretching vibrations*, $v(sp^2 C–H)$ **3,150–3,000**

2.2 *Out of plane* vibrations, γ_{CH} **1,005–675**

2.2.1 *trans*-CH = CH (conjugated) 965 ± 5 (≈975) s

2.2.2 *cis*-CH = CH (conjugated) 690 ± 25 (≈820) m

2.2.3 − CH = CH   990 ± 5 s + 910 ± 5 s + 1,850–1,800 w (OT)

2.2.4 > C = CH$_2$ 890 ± 5 s + 1,850–1,800 (OT)

2.2.5 C = CH 840–800 m

2.3 *C = C stretching vibrations*, $v_{C=C}$ **1,1,690–1,635**
(Variable intensity up to inactive, i. a.)

2.3.1 *trans*-CH = CH   1,670 ± 5 w to i. a.

2.3.2 *cis*-CH = CH   1,645 ± 15 w-m

2.3.3 -CH = CH$_2$, >C = CH$_2$ 1,645 ± 5 m-w

2.3.4 C = CH, C = C   1,672 ± 7 vw–i. a.

3. Alkynyl groups

3.1 *CH stretching vibration*, $v(spC–H)$ **3,340–3,250** m-s (sharp!)

(continued)

Table 6.15 (continued)

3.2 C ≡ C- *stretching vibration*, $v_{C≡C}$	**2,260–2,100** m-w (like $v_{C≡N}$!)
3.3 *Bending*, $\delta(C ≡ C–H)$	**700–600** s, b + 1,370–1,220 w, b (OT)
4. Aromatic compounds	
4.1 *CH stretching vibrations*, $v(sp^2C–H)$	**3,100–3,000** m (mostly several bands in the same region like alkenes and small alicyclic rings)
4.2 *C = C stretching vibrations*, $v(C = C)$	**1,625–1,575** m; w if *i* is present,doublet at conjugation **1,525–1,475** m and **1,470–1,400** m; neither band is ever visible
4.3 CH *bending vibrations in plane*, $\delta_{in\text{-}plane}$	1,250–950 (several bands of variable intensity but meaningless for structural determination)
4.4 *CH and CC bending vibrations*, $\gamma(CH)$ + oop CC skeleton	**900–650** m-s Region like $\gamma(CH)$ alkenes and $v(CCl)$!
4.5 *Overtones* and *combination vibrations*	**2,000–1,650** w-m

Overtones (OT) and combination vibrations (CV) for various benzene derivatives are presented in Fig. 6.1. Note there are deviations from the "ideal" patterns at derivatives with strongly electronegative substituents, like NO_2 and C ≡ N. Note that the patterns are obtained from alkyl-substituted derivatives.

II. Alcohols, Phenols	
1. O–H *stretching vibrations*, $v(OH)$	**3,650–3,200** (2,500)
$v(OH)_{free}$	3,650–3,590 sharp!
$v(OH)_{associated}$	3,550–3,450 very broad chelates, enols: 2,300–2,500
2. O–H *bending vibrations in plane*, $\delta_{in\text{-}plane}$	1,450–1,200 m, b (without meaning)
3. O–H *bending vibrations out of plane*, δ_{oop}	< 700 m, b (without meaning)
4. *Asymmetrical C–C–O stretching vibrations*, $v_{as}(CCO)$	**1,280–980** s (frequently split)
Primary alcohols, CH_2–OH	1,075–1,000
Secondary alcohols, CH–OH	1,150–1,075
Tertiary alcohols, C–OH	1,210–1,100
Phenols	1,275–1,150
5. *Asymmetrical C–C–O stretching vibrations*, $v_{as}(CCO)$	1,000–800
Primary and secondary alcohols	900–800
Tertiary alcohols	≈1,000
III. Ether	
1. *Asymmetrical C–O–C stretching vibrations*, $v_{as}(C–O–C)$	
Saturated, unbranched	1,150–1,070 (one band)
Saturated, branched	1,210–1,070 (two or several bands)
Alkyl/aryl mixed	1,300–1,200 + 1,050–1,000
di-aryl	1,300–1,200

(continued)

2. *Symmetrical C–O–C stretching vibrations, v_s(C–O–C)*

Saturated, unbranched	890–820
Saturated, branched	890–820

IV. Amine

1. NH *stretching vibrations, v(NH)*   **3,500–3,280** (up to 3,200 at NH bridges v(NH)$_{as}$ is smaller than v(OH)!)

Primary amines NH$_2$	v_{as}(NH) alkyl: 3,380–3,350	aryl: 3,500–3,450
	v_s(NH) alkyl: 3,310–3,280	aryl: 3,420–3,350
Secondary amines NH	v(NH) alkyl: 3,320–3,280	aryl: 3,400
NH$_3^+$, NH$_2^+$	3,000–2,000 m, b (strongly structured)	

2. NH *bending vibrations, $\delta_{scissoring}$ (NH$_2$)*   **1,650–1,580** s

3. NH *bending vibrations, δ_{oop}(NH$_2$)*   850–700 m

4. CN *stretching vibrations, v(CN)*   1,350–1,020 w-m

NH$_2$	Alkyl: 1,250–1,020
	Aryl: 1,350–1,250
NH	Alkyl: 1,180–1,130
	Aryl: 1,350–1,250

V. Aldehydes

1. C = O stretching vibrations, v(C = O)   **1,750–1,650** vs

Alkyl–CHO	1,740–1,720
Aryl–CHO	1,715–1,685 (up to 1,645 at intramolecular bridges)
α,β unsaturated	1,700–1,660

2. Fermi resonance   **2,900–2,800** + 2,700–2,680 (frequently two Bands)

3. Bending vibrations, δ_{HCO} $\approx$1,390

VI. Ketones

1. C = O *stretching vibrations, v(C = O)*   **1,725–1,640** vs

Dialkyl	1,725–1,705
Aryl/alkyl mixed	1,700–1,680
Diaryl	1,670–1,660
Enone	1,765–1,725
$\geq$ 6-membered ring, saturated	1,725–1,710
5-membered ring, saturated	1,750–1,740
4-membered ring, saturated	1,780–1,770
α chlorine	1,740–1,720
Quinones	1,690–1,615

2. C–C–C *stretching vibrations, v_{C-C-C}*

Dialkyl:	1,230–1,100
Aryl/alkyl mixed, diaryl	1,300–1,230

(continued)

VII. Carboxylic acids

1. OH *stretching vibration*, $v(OH)$	**3,000–2,500** vs, very broad!
2. C = O *stretching vibration*, $v(C = O)$	**1,800–1,650** s
Saturated	1,730–1,700
α,β unsaturated	1,715–1,690
α halogen	1,740–1,720
Aryl	1,710–1,680
3. OH *bending vibrations*	
$\delta_{\text{in-plane}}$	1,440–1,395
δ_{oop}	960–900 (but also $\delta_{=CH}$, δ_{NH}, γ_{Aromat})
4. Carboxylate anion, COO	$v_{as}(CO_2)$ 1,650–1,540
	$v_{s}(CO_2)$ 1,450–1,360

VIII. Carboxylic acid derivatives

1. *Acid chlorides*

1.1 C = O *stretching vibrations*, $v(C = O)$	**1,820–1,750** vs
Saturated	1,820–1,790
Aryl	1,790–1,750 (split band because of Fermi resonance)

2. *Anhydrides*

2.1 C = O *stretching vibrations*, $v(C = O)$	**1,870–1,700** vs	
	acyclic	cyclic
$v_{s}(C = O)$, saturated	1,825–1,815 (stronger)	1,870–1,845 (weaker)
$v_{as}(C = O)$, saturated	1,755–1,745 (weaker)	1,800–1,775 (stronger)
$v_{s}(C = O)$, unsaturated	1,780–1,770 (stronger)	1,860–1,840 (weaker)
$v_{as}(C = O)$, unsaturated	1,725–1,715 (weaker)	1,780–1,760 (stronger)

2.2 CO *stretching vibrations*, $v_{as}(C–O) + v_{s}(C–O)$	
Acyclic	$\approx$1,040
Cyclic	960–880

3. *Esters*

Three band rule: $\approx$**1,700** $v(C = O)$, $\approx$ **1,200**, $v_{as}(C–C(=O)–O$, $\approx$**1,100** $v_{as}(O–C–C(=O)$

3.1 C = O *stretching vibrations*, $v(C = O)$	1,750–1,700 vs
Saturated	1,750–1,735
Aryl	1,730–1,710
3.2 C–C(=O)–O *stretching vibrations*, $v_{as}(C–C(=O)–O)$	
Saturated	1,210–1,160
Aryl	1,330–1,250
3.3 O–C–C(=O) *stretching vibrations*, $v_{as}(O–C–C(=O))$	
Saturated	1,100–1,030
Aryl	1,130–1,000

(continued)

4. *Lactones*

4.1 6-membered rings	1,760–1,730 s
4.2 5-membered rings	1,775–1,730
α,β unsaturated	1,770–1,740
β,γ unsaturated	$\approx 1,800$
4.3 4-membered rings	$\approx 1,840$

5. *Amides, Lactams*

5.1 NH *stretching vibrations*, v(NH)		**3,500–3,100 s**
CONH$_2$	v_{as}(NH)	$\approx 3,350$
	v_s(NH)	$\approx 3,180$
CONH	v(NH)	3,400–3,100
		(two bands if associated)

5.2 AMIDE bands

AMIDE I	OCNH$_2$	free: $\approx 1,690$
		associated: $\approx 1,615$
	OCNH	free: $\approx 1,695$
		associated: $\approx 1,660$
	Lactam	
	6-membered ring:	1,675–1,650
	5-membered ring:	1,725–1,715
	4-membered ring:	$\approx 1,750$
AMIDE II	OCNH$_2$	free: $\approx 1,610$
		associated: $\approx 1,630$
	OCNH	free: $\approx 1,530$
		associated: $\approx 1,540$
	Lactam	IR inactiv
AMIDE III (coupling of AMIDE II with δ(NH)):		$\approx 1,250$

5.3 NH *bending vibration*, δ_{oop}	**800–600**
OCNH$_2$	750–600
OCNH	≈ 700
Lactam	≈ 800

IX. Nitriles, isonitriles, diazonium salts (containing an $X \equiv Y$ functional group)

1. $C \equiv N$ stretching vibration, $v(C \equiv N)$	**2,260–2,200** m (frequently weak!)
2. $N \equiv C$ stretching vibration, $v(N \equiv C)$	**2,150–2,130** m
3. $N \equiv N$ stretching vibration, $v(N \equiv N)$	**2,330–2,130** m

X. Azomethines, azines (containing an $X = N$ functional group)

1. $C = N$ stretching vibration, $v(C = N)$	**1,700–1,520** w-vw
2. CH $= $ N–N $= $ CH stretching vibration	1,670–1,600

(continued)

XI. Nitro- and nitroso compounds

1. NO_2 stretching vibration,	$v_{as}(NO_2)$	**1,660–1,500** vs
	$v_s(NO_2)$	**1,400–1,250** s
Alk–NO_2	$v_{as}(NO_2)$	1,570–1,540
	$v_s(NO_2)$	1,390–1,340
Aryl–NO_2	$v_{as}(NO_2)$	1,560–1,500
	$v_s(NO_2)$	1,360–1,300 (frequently two bands)
2. $O = N$ stretching vibration,	$v(N = O)$	**1,680–1,450** vs

XII. Organic halogen compounds

1. C–halogen stretching vibration, v(C–hal)

Halogen	alkyl–Halogen	aryl–Halogen
F	1,365–1,120 s	1,270–1,100 s
Cl	830–560 s	1,100–1,030 s
Br	680–515 s	1,075–1,030 s
I	≈500 s	≈1,060 s

XIII. Organic sulfur compounds

(Mercaptanes, thioethers, thionyl and sulfuryl compounds)

1. S–H stretching vibration, v(S–H)		**2,600–2,550** m-w
2. S–S stretching vibration, v(S–S)		≈500 w (meaningless in IR) Intense line in Raman!
3. C–S *stretching vibration*, v(C–S)		710–570 (meaningless)
4. $S = C$ *stretching vibration*, $v(C = S)$		**1,250–1,050** s
Thioketones		1,075–1,030
Thioesters		1,210–1,080
5. Thionyl compounds		
$O = S$ *stretching vibration*, $v(S = O)$		**1,200–1,000** s
6. Sulfuryl compounds		**1,400–1,000** vs
Alkylalkyl′SO_2	$v_{as}(SO_2)$	1,370–1,290 vs
	$v_s(SO_2)$	1,170–1,110 s
$-SO_2-N<$	$v_{as}(SO_2)$	1,370–1,290 vs
	$v_s(SO_2)$	1,180–1,150 vs
$-SO_2-O-$	$v_{as}(SO_2)$	1,420–1,330 s
	$v_s(SO_2)$	1,200–1,140 s

XIV. Organic phosphorus compounds

1. P–H stretching vibration, v(P–H)	2,450–2,250 m-w
2. P–O stretching vibration, v(P–O)	1,260–855
3. $O = P$ stretching vibration, $v(P = O)$	1,300–950 s
$R_3P = O$	1,190–1,150
$R_2(R'O)P = O$	1,265–1,200
$R(R'O)_2P = O$	1,280–1,240
$(RO)_3P = O$	1,300–1,260

Substitution type	*Mono*	*1,2-di*	*1,3-di*
γ(CH) + δoop CC skeleton	770 – 730 710 – 690	770 – 735	900 – 860 (865 – 810) 810 – 750 725 – 680

OT + CV

Substitution type	*1,4-di*	*1,2,3-tri*
γ(CH) + δoop CC skeleton	860 – 780	800 – 770 720 – 685 (780 – 760)

OT + CV

Substitution type	*1,2,4-tri*	*1,3,5-tri*
γ(CH) + δoop CC skeleton	900 – 860 860 – 800 730 – 690	900 – 840 850 – 800 730 – 680

OT + CV

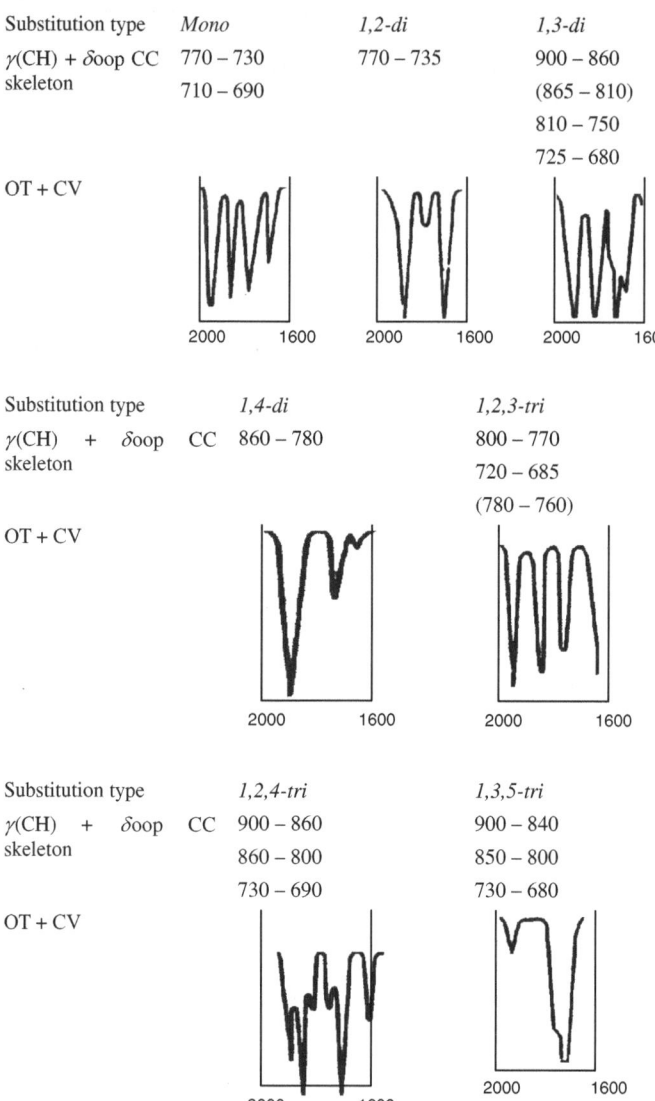

Fig. 6.1 Overtones (OT) and combination vibrations (CV) for benzene derivatives

6.3   Electronic Absorption Spectroscopy

Table 6.16 Woodward diene and enone rules

Dienes	
Base structure	λ_{max} in nm
Acyclic 	217
Heteroannular 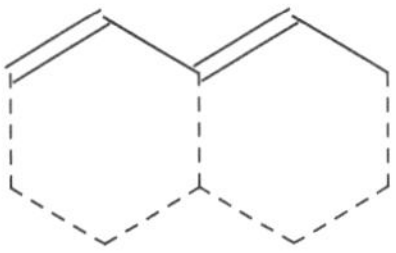	214
Homoannular 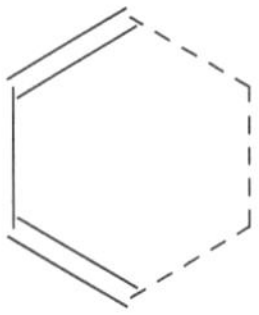	253

Increments	in nm
Each further conjugated double bond	+ 30
Each exocyclic double bond	+ 5
C substituent	+ 5
O–acetyl	0
Cl or Br	+ 5
O–alkyl	+ 6
S–alkyl	+ 30
N(alkyl)$_2$	+ 60

Enones

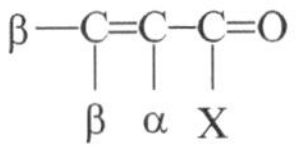

$$\beta - C = C - C = O$$
$$\quad \beta \quad \alpha \quad X$$

Base structure	λ_{max} in nm
X = H	207
X = alkyl	217

(continued)

Table 6.16 (continued)

Dienes	
Base structure	λ_{max} in nm
H = OH, O–alkyl	193
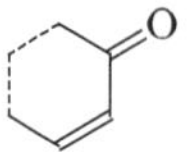	215

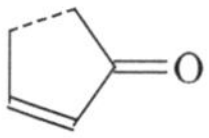	202

Increments	in nm
Each further conjugated double bond	+ 30
Each exocyclic double bond	+ 5
Homoannular pattern of double bonds	+ 39

Increment values for substituents at the π system (in nm) which must be added to the base values.

Substituent	α	β	γ	δ	Further
C	10	12	18	18	18
OH	35	30		50	50
O–alkyl	35	30	17	31	31
S–alkyl		85			
N(alkyl)$_2$		95			
Cl	15	12			
Br	25	30			

Solvent correction (in nm)

Water	+8	Alcohol	0	CHCl$_3$	+1
Dioxane	−5	Ether	−7	Hexane	−11

Table 6.17 Scott rules for the estimation of λ_{max} values in aryl–α-carbonyl compounds (structure 6.1)

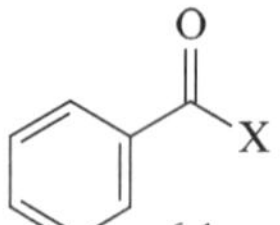

Base values λ in nm

X = H	250	X = alkyl	246	X = OH (OR)	230

Increment values for substituents at the aromatic ring (in nm) which must be added to the base values.

Substituent	ortho	meta	para
Alkyl	3	3	10
OH, O–alkyl	7	7	25
O^-	11	20	10
Cl	0	0	10
Br	2	2	15
NH_2	13	13	58
$NH(COCH_3)$	20	20	45
$N(CH_3)_2$	20	20	85

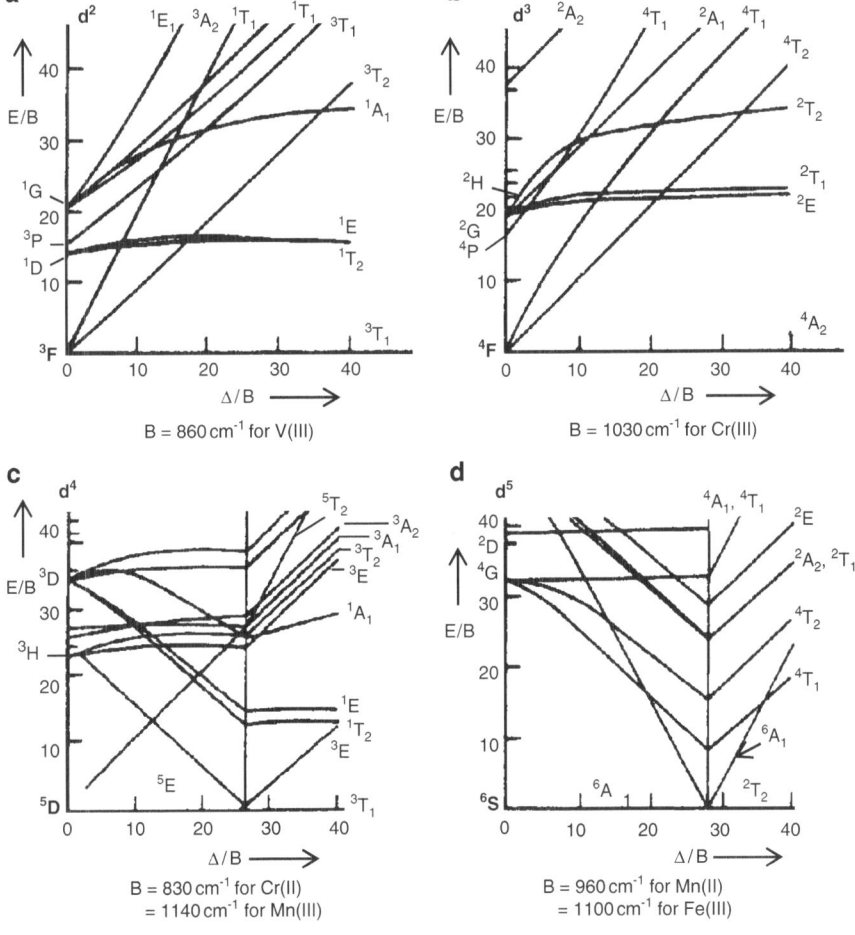

Fig. 6.2 (continued)

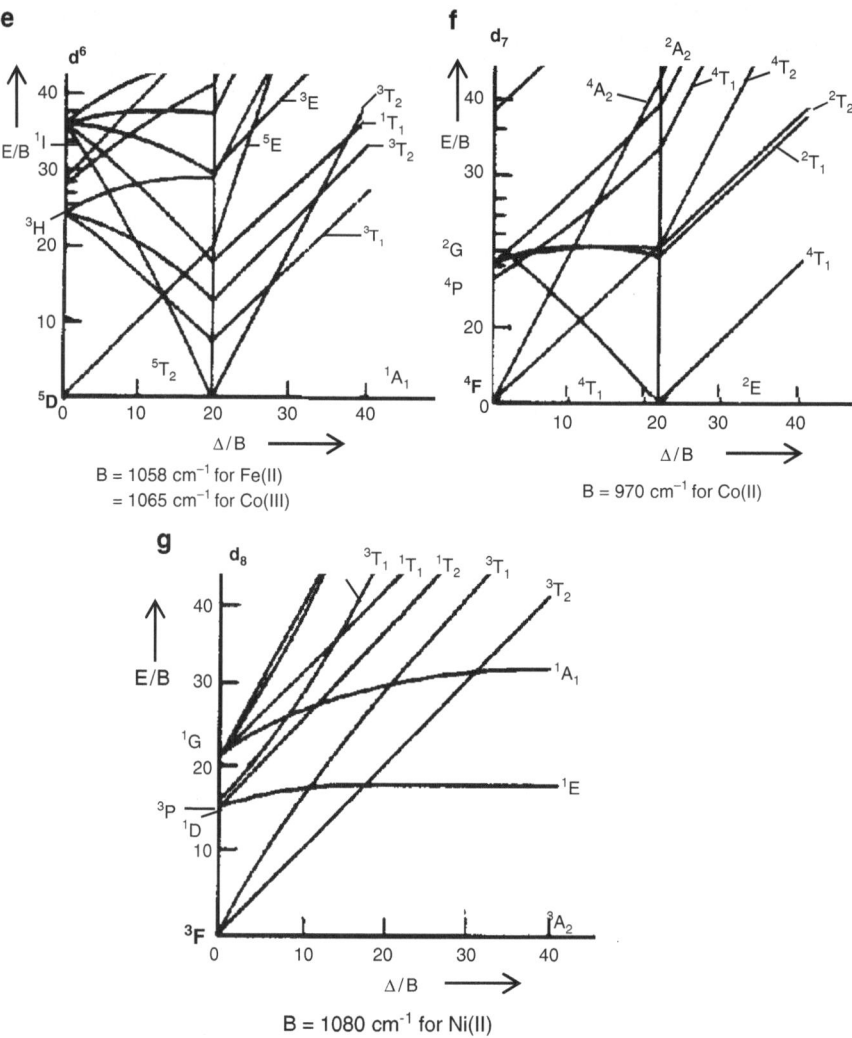

Fig. 6.2 Section of the Tanabe Sugano term diagrams for the electron configurations d^3 to d^8 of 3 d elements with octahedral structure

6.4   Nuclear Magnetic Resonance Spectroscopy

6.4.1 ^1H NMR Spectroscopy

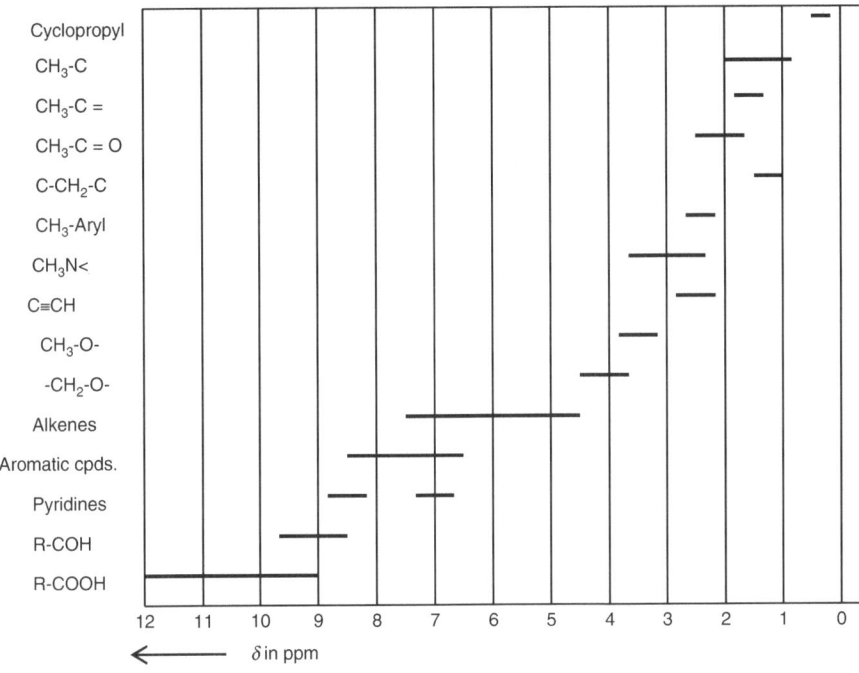

Fig. 6.3 Chart of the chemical shifts of protons

Table 6.18 Chemical shifts of XH protons with X = O, N, and S

Note that the chemical shifts are dependent on concentration, temperature, and water content. The protons can be exchanged with D_2O.

OH	δ in ppm	NH	δ in ppm	SH	δ in ppm
Alcohols	1–6	Amines	3–5	Aliphatic	1–2.5
Phenols	4–10	Amides	5–8.5	Aromatic	3–4
Enols	10–17				

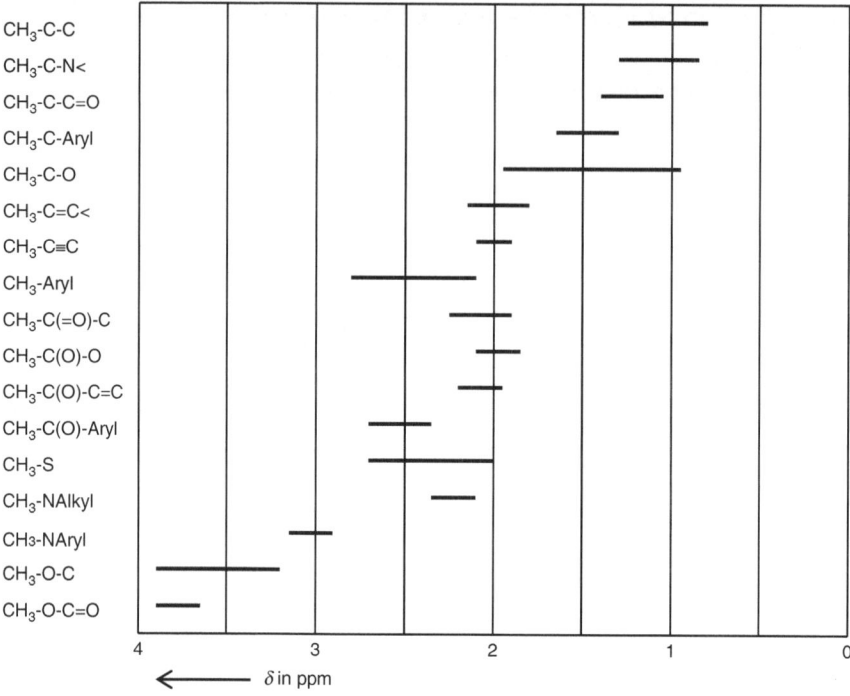

Fig. 6.4 Chart of the chemical shifts of *methyl* protons

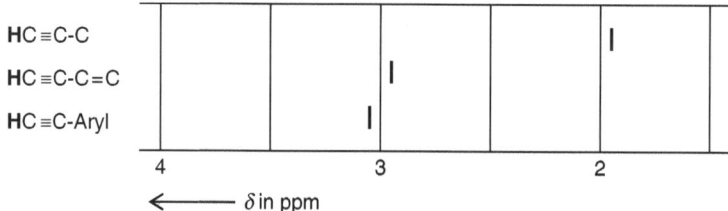

Fig. 6.5 Chart of the chemical shifts of *acetylide* protons

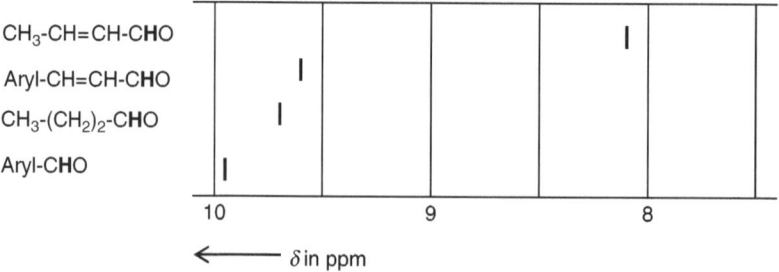

Fig. 6.6 Chart of the chemical shifts of *aldehyd* protons

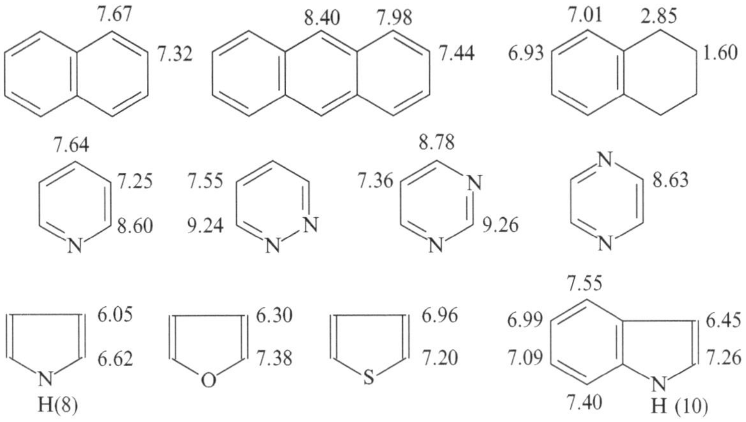

7.67 7.32   8.40 7.98 7.44   7.01 2.85 6.93 1.60

7.64 7.25 8.60   7.55 9.24   7.36 8.78 9.26   8.63

6.05 6.62 H(8)   6.30 7.38   6.96 7.20   7.55 6.99 6.45 7.09 7.26 7.40 H (10)

Fig. 6.7 Chemical shifts of the protons in *aromatic* and *heteoaromatic* compounds

Table 6.19 Incremental shifts S_i of *methyl* and *methylene* protons (in ppm)

$S_1S_2CH_2$ $S_1S_2 \, S_3$ **CH**

$$\delta(\mathbf{CH_2}) = 1.25 + S_1 + S_2 \tag{6.1}$$

$$\delta(\mathbf{CH}) = 1.50 + S_1 + S_2 + S_3 \tag{6.2}$$

Substituent	S	Substituent	S
Alkyl	0	$-C = C$	0.8
Phenyl	1.3	$-C \equiv C$	0.9
$-OH$	1.7	$-C \equiv N$	1.2
$-O$–alkyl	1.5	$-COH, -CO$–alkyl	1.2
$-O$–phenyl	2.3	$-COOH$	0.8
$-O$–CO–alkyl	2.7	$-CO$–O–alkyl	0.7
$-O$–CO–phenyl	2.9	$-NO_2$	3.0
$-SH, -S$–alkyl	1.0	$-Br$	1.9
$-NR_2, -NH_2$	1.0	$-Cl$	2.0

Table 6.20 Incremental shifts (S_{gem}, S_{cis}, S_{trans}) of the *alkene* protons (in ppm)

$$\delta(\mathbf{C=CH}) = 5.25 + S_{gem} + S_{cis} + S_{trans} \qquad (6.3)$$

Substituent	S_{gem}	S_{cis}	S_{trans}
Alkyl	0.45	−0.22	−0.28
−CH$_2$–aryl	1.05	−0.29	−0.32
−CH$_2$–O	0.64	−0.01	−0.02
−CH$_2$–halogen	0.70	0.11	−0.04
−CH$_2$–NR$_2$	0.58	−0.10	−0.08
−CH = CH– (conj.)	1.24	0.02	−0.05
−CHO	1.02	0.95	1.17
−COOH	0.80	0.98	0.74
−CONH$_2$	1.37	0.98	0.46
−C $\equiv$ N	0.27	0.75	0.55
−O–CO–R	2.11	−0.35	−0.64
−O–alkyl	1.22	−1.07	−1.21
−NR$_2$	0.80	−1.26	−1.21
−F	1.54	−0.40	−1.02
−Cl	1.08	0.18	0.13

Table 6.21 Incremental shifts (S_{ortho}, S_{meta}, S_{para}) of the aromatic protons of *mono*substituted *benzenes* (in ppm)

$$\delta(\mathbf{H}) = 7.25 + S_{ortho} + S_{meta} + S_{para} \qquad (6.4)$$

Substituent	S_{ortho}	S_{meta}	S_{para}
−CH$_3$	−0.20	−0.12	−0.22
−CH$_2$–CH$_3$	−0.14	−0.05	−0.18
−CH(CH$_3$)$_2$	−0.13	−0.08	−0.18
−CH$_2$–OH	−0.07	−0.07	−0.07
−CH = CH$_2$	0.04	−0.05	−0.12
−C $\equiv$ C–	0.47	0.38	0.12
−C $\equiv$ N	0.35	0.18	0.30
−CHO	0.60	0.25	0.35
−COCH$_3$	0.62	0.14	0.21
−COOH	0.87	0.21	0.34
−COOCH$_3$	0.74	0.07	0.21
−OH	−0.53	−0.17	−0.44
−O–CH$_3$	−0.49	−0.10	−0.44
−O–CO–CH$_3$	−0.25	0.03	−0.13
−NO$_2$	0.95	0.26	0.38

(continued)

Table 6.21   (continued)

Substituent	S_{ortho}	S_{meta}	S_{para}
$-NH_2$	−0.80	−0.25	−0.64
$-N(CH_3)_2$	−0.66	−0.18	−0.67
$-SH$	−0.08	0.16	−0.22
$-F$	−0.26	0.02	−0.21
$-Cl$	0.03	−0.06	−0.10

Table 6.22   Incremental shifts of the protons of *pyridines*

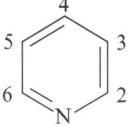

$$\delta(\mathbf{H-2}) = 8.59 + S_{i,2} \tag{6.5}$$

$$\delta(\mathbf{H-3}) = 7.38 + S_{i,3} \tag{6.6}$$

$$\delta(\mathbf{H-4}) = 7.75 + S_{i,4} \tag{6.7}$$

$$\delta(\mathbf{H-5}) = 7.38 + S_{i,5} \tag{6.8}$$

$$\delta(\mathbf{H-6}) = 8.59 + S_{i,6} \tag{6.9}$$

$i = 2$ and 6, respectively	$S_{23} = S_{65}$	$S_{24} = S_{64}$	$S_{25} = S_{63}$	$S_{26} = S_{62}$
$-CH_3$	−0.11	−0.01	−0.16	0.08
$-CHO$	0.93	0.42	0.50	0.44
$-COO$–alkyl	0.86	0.39	0.35	0.34
$-CO$–NH_2	1.05	0.47	0.43	0.30
$-CN$	0.88	0.38	0.55	0.39
$-NH_2$	−0.68	−0.31	−0.78	−0.48
$-Cl$	0.32	0.29	0.29	0.20
$i = 3$ and 5, respectively	$S_{32} = S_{56}$	$S_{34} = S_{54}$	$S_{35} = S_{53}$	$S_{36} = S_{52}$
$-CH_3$	−0.02	−0.06	−0.09	−0.02
$-CHO$	0.45	0.42	0.12	0.20
$-COO$–alkyl	0.62	0.60	0.23	0.34
$-CO$–NH_2	0.58	0.57	0.25	0.25
$-CN$	0.63	0.72	0.43	0.50
$-NH_2$	−0.06	−0.49	0.02	−0.36
$-Cl$	0.20	0.24	0.19	0.09

(continued)

$i = 4$	$S_{42} = S_{46}$	$S_{43} = S_{45}$
$-CH_3$	0.01	-0.01
$-CHO$	0.47	0.58
$-COO-alkyl$	0.34	0.54
$-CN$	0.46	0.62
$-NH_2$	-0.15	-0.74
$-Cl$	0	0.05

Absolute values of proton–proton coupling constants

(Typical values are given in brackets.)

Table 6.23 Geminal coupling constants $^2J(^1H/^1H)$ in Hz

$C = CH_2$	0–3	$=C-CH_2-C$	16–18
$=C-CH_2-C=$	0–20	$X-CH_2- (X = O, N)$	≈ 10

Table 6.24 Vicinal coupling constants $^3J(^1H/^1H)$ in Hz

Alkanes	7–8		
Cyclohexanes	ax/ax	6–14 (8–10)	ax/eq = eq/eq 0–5 (2–3)
$=CH-CH=$	9–13 (10)		
$-C = CH-CH = O$	5–8 (6)		
$-CH = CH-$	*trans*	12–18	*cis* 6–12
$>CH-CHO$	2–3		

Table 6.25 Coupling constants for *aromatic* and heteroaromatic compounds (in Hz)

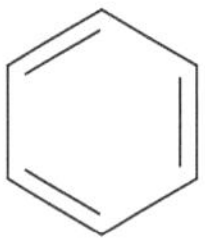

$^3J_{ortho}$ 6–10 (8–9)
$^4J_{meta}$ 1–3 (2)
$^5J_{para}$ 0–1

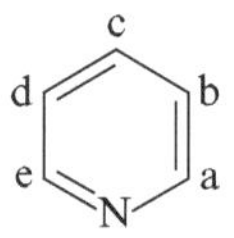

$^3J_{ab}$ 4–6 (5)
$^4J_{ac}$ 0–2.5 (1.5)
$^5J_{ad}$ 0–2.5 (1.5) $^4J_{ac}$ 0–0.5 ($\approx$0)
$^3J_{bc}$ 7–9 (8)
$^4J_{bd}$ 0.5–2 (1.5)

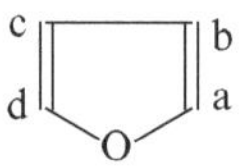

$^3J_{ab}$ 1.5–2.0 (1.8)
$^4J_{ac}$ 0–1 ($\approx$0)
$^4J_{ad}$ 1–2 (1.5)
$^3J_{bc}$ 3–3.8 (3.5)

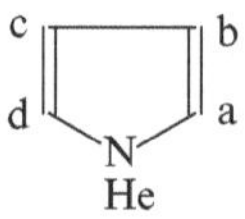

$^3J_{ab}$ 2–3
$^4J_{ac}$ 1–2
$^4J_{ad}$ 1.5–3
$^4J_{ae}$ 2–3
$^3J_{bc}$ 3–9
$^4J_{be}$ 2–3

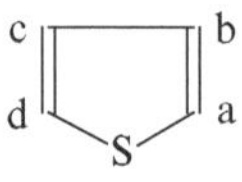

$^3J_{ab}$ 5–6 (5.4)
$^4J_{ac}$ 1.2–1.7 (1.5)
$^4J_{ad}$ 3.2–3.8 (3.5)
$^3J_{bc}$ 3.5–5 (4.0)

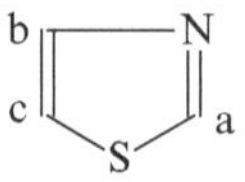

$^4J_{ab} \approx 0$
$^4J_{ac}$ 1–2
$^3J_{bc}$ 3–4

Table 6.26 Coupling constants $J(^1H/^{19}F)$ (in Hz)

>CHF	$^2J(^1H,^{19}F)$ 45–80	>CH–CF<	$^3J(^1H,^{19}F)$ 3–25
>CH–C–CF<	$^4J(^1H,^{19}F)$ 0–4	–HC = CF–	$^3J(^1H,^{19}F)_{trans}$ 12–40
			$^3J(^1H,^{19}F)_{cis}$ 1–8
–CH$_2$–C(=O)–CF < HHH	$^4J(^1H,^{19}F)$ 4–5		
Aromatic fluorine compounds			$^3J(^1H,^{19}F)$ 6–10
			$^4J(^1H,^{19}F)$ 5–6
			$^5J(^1H,^{19}F)$ 2

Table 6.27 Coupling constants $J(^{19}F/^{19}F)$ (in Hz)

>CF$_2$	$^2J(^{19}F,^{19}F)$ 150–160
	In 6-membered rings 240–250
>CF–CF<	$^3J(^{19}F,^{19}F)$ 0–20
–CF = CF–	$^3J(^{19}F,^{19}F)_{trans}$ 115–300
	$^3J(^{19}F,^{19}F)_{cis}$ 10–60

Table 6.28 $J(^{19}F/^{13}C)$ coupling constants (in Hz)

$^1J(^{19}C\text{-}1,^{19}F)$ 245
$^2J(^{19}C\text{-}2,^{19}F)$ 21
$^3J(^{19}C\text{-}3,^{19}F)$ 7.8
$^4J(^{19}C\text{-}4,^{19}F)$ 3.2

$^1J(^{19}C\text{-}7,^{19}F)$ 272
$^2J(^{19}C\text{-}1,^{19}F)$ 32
$^3J(^{19}C\text{-}2,^{19}F)$ 4
$^4J(^{19}C\text{-}3,^{19}F)$ 1
$^5J(^{19}C\text{-}4,^{19}F)$ ≈ 0

Table 6.29 $J(^{19}F/^{31}P)$ Coupling Constants (in Hz)

=CFP	≈80

6.4.2 ^{13}C NMR Spectroscopy

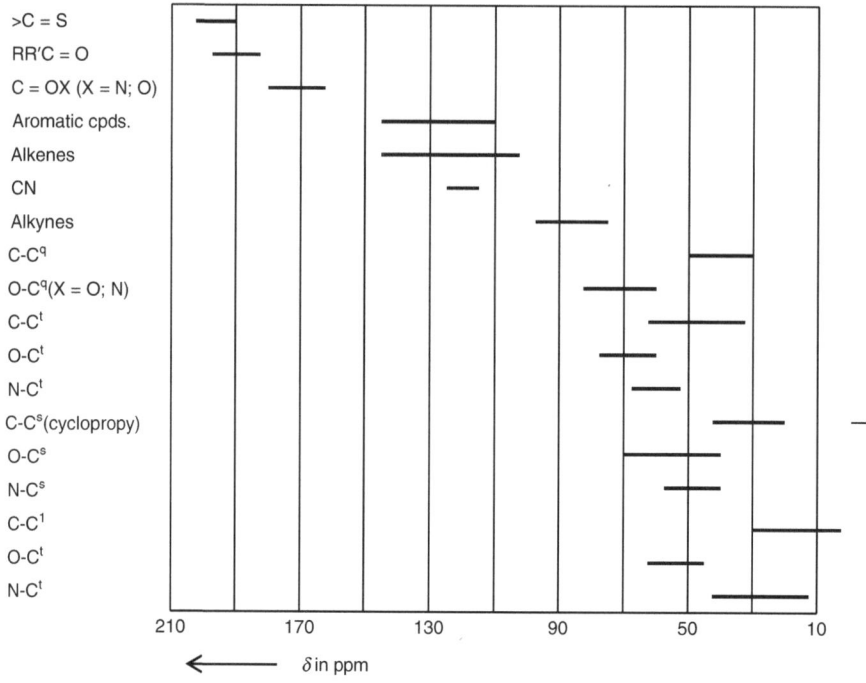

Fig. 6.8 Chart of the ^{13}C chemical shifts
C^s, C^t, C^q mean secondary, tertiary, quaternary carbon atoms.

Table 6.30 Incremental shifts (S_α, S_β, S_γ, S_δ) of the carbon atoms of *alkanes* and *substituted alkanes* (in ppm)

$$\delta_i(^{13}C) = -2.3 + 9.1n_a + 9.4n_\beta - 2.5n_\gamma + 0.3n_\delta + \sum S_{i,j} \qquad (6.10)$$

n_α, n_β, n_γ, n_δ are the number of carbon atoms in α, β, γ and δ position related to the considered carbon atom C_1.
$S_{i,j}$ are the steric factors.
Steric factors $S_{i,j}$
Note that only sp^3 hybridized carbon atoms are considered.
i is the respective carbon atom and j is the neighboring carbon atom.

j	Primary	Secondary	Tertiary	Quaternary
i				
Primary	0	0	−1.1	−3.4
Secondary	0	0	−2.5	−6.0
Tertiary	0	−3.7	−8.5	−10.0
Quaternary	−1.5	−8.0	−10.0	−12.5

Substituent	S_α	S_β	S_γ	S_δ
$-CH_3$	9.1	9.4	−2.5	0.3
$-CH = CH_2$	22.3	6.9	−2.2	0.2
$-C \equiv C-$	4.4	5.6	−3.4	−0.6
$-C \equiv N$	3.1	2.4	−3.3	−0.5
Phenyl	22.3	8.6	−2.3	0.2
$-CHO$	30.9	−0.5	−2.7	0
$-C = O$	22.5	3.0	−3.0	0
$-COOH$	20.8	2.7	−2.3	1.0
$-OCOCH_3$	51.1	7.1	−4.8	1.1
$-OH$ (primary)	48.5	10.2	−5.8	0.3
$-OH$ (secondary)	44.5	9.7	−3.3	0.2
$-OH$ (tertiary)	39.7	7.3	−1.8	0.3
$-OR$	58.0	8.1	−4.8	1.5
$-NH_2$	28.3	11.3	−5.1	0
$-NO_2$	61.6	3.1	−4.6	−1.0
$-F$	70.1	7.8	−6.8	0
$-Cl$	31.2	10.1	−4.6	0
$-Br$	18.9	11.0	−3.8	−0.7
$-S-$	10.6	11.4	−3.6	−0.4
$-SH$	11.1	11.8	−4.8	1.1

Table 6.31 Incremental shifts (S_1, S_2) of the carbon atoms of *mono*substituted *ethylenes* (in ppm)

$$\delta_i\left(^{13}C\right) = 123.3 + S_1 + S_2 \tag{6.11}$$

Substituent	S_1	S_2
$-CH_3$	10.6	−7.9
$-CH_2-CH_3$	15.5	−9.7
Phenyl	12.5	−11.0
$-CH = CH_2$	13.6	−7.0
$-COOH$	4.2	8.9
$-COOCH_3$	13.8	4.7
$-CHO$	13.1	12.7
$-O-CH_3$	29.4	−38.9
$-OCO-CH_3$	18.4	−26.7
$-NO_2$	22.3	−0.9
$-F$	24.9	−34.5
$-Cl$	2.6	−6.1
$-Br$	−7.9	−1.4
$-S-CH_2-phenyl$	18.1	−16.4

Table 6.32 Incremental shifts (S_{ypso}, S_{ortho}, S_{meta}, S_{para}) of the carbon atoms of *mono*substituted
benzenes

$$\delta_i(^{13}C) = 128.5 + S_{ypso} + S_{ortho} + S_{meta} + S_{para} \qquad (6.12)$$

Substituent	S_{ypso}	S_{ortho}	S_{meta}	S_{para}
$-CH_3$	9.2	0.7	−0.1	−3.0
$-CH_2-CH_3$	15.7	−0.6	−0.1	−2.8
$-CH(CH_3)_2$	20.2	−2.2	−0.3	−2.8
$-CH_2-Cl$	9.3	0.3	0.2	0
$-CH_2-Br$	9.5	0.7	0.3	0.2
$-CH_2-NH_2$	14.9	−1.4	−0.2	−2.0
$-CH_2-OH$	12.6	−1.2	0.2	−1.1
$-CH = CH_2$	8.9	−2.3	−0.1	−0.8
$-C \equiv CH$	−6.2	3.6	−0.4	−0.3
$-CHO$	8.2	1.2	0.5	5.8
$-CO-CH_3$	8.9	0.1	−0.1	4.4
$-CO-phenyl$	9.3	1.6	−0.3	3.7
$-CO-OH$	2.1	1.6	−0.3	5.2
$-CO-OCH_3$	2.0	1.2	−0.1	4.3
$-CO-Cl$	4.7	2.7	0.3	6.6
$-CO-NH_2$	5.0	−1.2	0.1	4.3
$-OH$	26.9	−12.6	1.4	−7.4
$-O-CH_3$	31.4	−14.4	1.0	−7.7
$-O-phenyl$	27.6	−11.2	−0.3	−6.9
$-OCH = CH_2$	28.2	−11.5	0.7	−5.8
$-OCO-CH_3$	22.4	−7.1	0.4	−3.2
$-CF_3$	2.5	−3.2	0.3	3.3
$-NH_2$	18.2	−13.4	0.8	−10.0
$-N(CH_3)_2$	22.5	−15.4	0.9	6.1
$-NHCO-CH_3$	9.7	−8.1	0.2	−4.4
$-NO_2$	19.9	−4.9	0.9	6.1
$-SH$	2.1	0.7	0.3	−0.3
$-S-CH_3$	10.0	−1.9	0.2	−3.6
$-F$	34.8	−13.0	1.6	−4.4
$-Cl$	6.3	0.4	1.4	−1.9

Table 6.33 Incremental shifts ($S_{i,j}$) of the carbon atoms of *mono*substituted *pyridines*

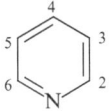

$$\delta(\mathbf{C-2}) = 149.8 + S_{i,2} \tag{6.13}$$

$$\delta(\mathbf{C-3}) = 123.7 + S_{i,3} \tag{6.14}$$

$$\delta(\mathbf{C-4}) = 135.9 + S_{i,4} \tag{6.15}$$

$$\delta(\mathbf{C-5}) = 123.7 + S_{i,5} \tag{6.16}$$

$$\delta(\mathbf{C-6}) = 149.8 + S_{i,6} \tag{6.17}$$

$i = 2$ and 6, respectively	$S_{22} = S_{66}$	$S_{23} = S_{65}$	$S_{24} = S_{64}$	$S_{25} = S_{63}$	$S_{26} = S_{62}$
$-CH_3$	8.8	-0.6	0.2	-3.0	-0.4
$-CHO$	3.5	-2.6	1.2	4.2	0.7
$-CO-alkyl$	-1.7	1.5	1.1	3.3	0
$-CO-NH_2$	-0.1	-1.2	1.5	2.8	-1.5
$-CN$	-15.9	4.8	1.1	3.2	1.4
$-NH_2$	11.3	-14.7	2.3	-10.8	-0.9
$-F$	13.9	-14.0	5.4	-2.5	-2.0
$i = 3$ and 5, respectively	$S_{32} = S_{56}$	$S_{33} = S_{55}$	$S_{34} = S_{54}$	$S_{35} = S_{53}$	$S_{36} = S_{52}$
$-CH_3$	1.3	8.9	0.2	-0.8	-2.3
$-CHO$	2.4	7.8	-0.1	0.6	5.4
$-CO-alkyl$	-0.6	1.0	-0.3	-1.8	1.8
$-CO-NH_2$	2.7	6.0	1.3	1.3	-1.5
$-CN$	3.6	-13.8	4.2	0.6	4.2
$-NH_2$	-11.9	21.4	-14.4	0.8	-10.8
$-F$	-11.5	36.1	-13.1	0.8	-3.9
$i = 4$	$S_{42} = S_{46}$	$S_{43} = S_{45}$	S_{44}		
$-CH_3$	0.5	0.8	10.6		
$-CHO$	1.7	-0.7	5.5		
$-CO-alkyl$	-1.0	-0.7	1.6		
$-CO-NH_2$	0.4	-0.8	6.4		
$-CN$	2.1	2.2	-15.7		
$-NH_2$	0.9	-13.8	19.6		
$-F$	2.7	-11.9	32.8		

Further Reading

1. Pretsch E, Bühlmann P, Affolter C, Badertscher M (2009) Structure determination of organic compounds. Springer, Berlin/New York
2. Silverstein RM, Webster FX, Kiemle D (2005) Spectrometric identification of organic compounds. Wiley, Hoboken

Index

M. Reichenbächer and J. Popp, *Challenges in Molecular Structure Determination*,   477
DOI 10.1007/978-3-642-24390-5, © Springer-Verlag Berlin Heidelberg 2012